The Changing Landscape in Eastern Europe

A Personal Perspective on Philanthropy and Technology Transfer

Richard E. Quandt

OXFORD
UNIVERSITY PRESS

2002

OXFORD

Oxford New York
Auckland Bangkok Buenos Aires Cape Town Chennai
Dar es Salaam Delhi Hong Kong Istanbul Karachi Kolkata
Kuala Lumpur Madrid Melbourne Mexico City Mumbai Nairobi
São Paulo Shanghai Singapore Taipei Tokyo Toronto

and an associated company in
Berlin

Published by Oxford University Press, Inc.
198 Madison Avenue, New York, New York 10016

www.oup.com

Library of Congress Cataloging-in-Publication Data
Quandt, Richard E.
The changing landscape in Eastern Europe : a personal perspective on philanthropy and technology transfer / by Richard E. Quandt.
p. cm.
Includes bibliographical references and index.
ISBN 0-19-514669-7
1. Technology transfers—Europe, Eastern. 2. Economic assistance—Europe, Eastern. 3. Management—Europe, Eastern—Endowments. 4. Economics—Europe, Eastern—Endowments. 5. Library science—Europe, Eastern—Endowments. I. Title.
HC244.Z9 T4724 2001
338.947—dc21 2001034649

9 8 7 6 5 4 3 2 1

Printed in the United States of America
on acid-free paper

Contents

Foreword

In recent years, the Mellon Foundation has followed the practice of identifying a limited number of "senior agents" (more properly called "senior advisors"). These are highly accomplished individuals who would not be interested in full-time appointments at the Foundation, but who can be interested in providing leadership in targeted program areas where they have special expertise. Richard Quandt is an outstanding case in point.[1] A widely respected econometrician in the economics department at Princeton University who is of Hungarian descent, Professor Quandt was an obvious choice to direct a new initiative in Eastern Europe that the Foundation launched in 1989. Added qualifications were Professor Quandt's familiarity with information technology, his knowledge of higher education, and his sense of how organizations work. This book on philanthropy and technology transfer in Eastern Europe illustrates the effectiveness of this model—when the right leader is available.

Substantively, the projects described here fall into three broad areas: (1) higher education, with a special emphasis on advanced training in economics and business management; (2) the applications of information technology in research libraries; and (3) the stimulation of entrepreneurship. In the aftermath of the Communist control of Eastern Europe, there was no lack of opportunity to meet pressing needs in each of these areas. The legacy of a deadly mix of ideology and bureaucratic controls is everywhere in evidence in this narrative, which reminds us once again that it is dangerous to take for granted the freedoms that we enjoy in so much of the world today. The importance of incentives is another theme, with the inability to reward leading academics properly a major factor impeding progress in university education. Against this backdrop, the successes described in this book are quite remarkable, and they testify to another kind of legacy: the history of culture and learning in these countries that of course antedates the period of Communist rule and could not be eradicated. In short, there was much to build on in attempting to restore academic standards, to reinvigorate libraries, and to encourage risk taking in a wide range of business pursuits.

It is the individual stories that are, to me at least, the most compelling. These include the extraordinary progress made in developing CERGE, which is today the leading program in Ph.D. training in economics in all of Eastern

Europe. The receptivity of the Czechs and other Eastern Europeans to Western concepts (including the role played today by mathematical modeling and econometrics in the teaching of economics) was matched to the willingness of some very able economists from U.S. universities to devote considerable time to planning programs and participating in teaching students and guiding research projects. Efforts are now underway to secure the base of endowment funding that is needed to establish a secure foundation for CERGE. What must be seen as modest investments of foundation funds (given the scale of the enterprise) have yielded very large dividends.

The tensions between governmental mandates and academic freedom are illustrated extremely well by the struggles of Academia Istropolitana Nova (AIN) in Slovakia. It was far from clear at many junctures that AIN would survive, but survive it did, while transforming itself into a private institution.

The extended discussion of automation and research libraries will be of interest to academics and librarians in many countries, since it demonstrates the potential of new technologies and the key role that can be played by collaborative arrangements. The holdings of a particular library are less and less important; what matters much more is access to the holdings of many libraries and, for that matter, to electronic resources such as JSTOR. In working through the myriad problems encountered in trying to put sophisticated systems in place, Professor Quandt had the invaluable help of Professor Andrew Lass of Mount Holyoke College. Professor Lass' patient persistence is a story all its own. Today, Czech and Slovak libraries are collaborating through CASLIN, efforts are still underway to develop a Polish Union Catalog, Hungarian libraries are automated in a way that would have been hard to imagine a decade ago, and substantial progress has been made in Estonia and Latvia, as well. In the long run, it is hard to imagine a more important contribution than this dramatic strengthening of library resources.

The work of the Foundation in the sphere of practical business initiatives has also been rewarding, though less extensive. The development of banking and finance, combined with a broader commitment to legal protections and a stable business environment, have been critical ingredients in allowing small business, in particular, to establish itself. One of my favorite stories is the saga of rabbit farming in Poland, which illustrates the pitfalls that must be anticipated in establishing new models and the results that can be achieved if one doesn't give up too soon.

The final chapter of the book will be the most valuable for readers interested not just in this region of the world but in the much broader questions of how private philanthropy can have the best chance of being useful. In describing "Lessons Learned," Professor Quandt provides sage advice not only for foundation trustees and others working in unfamiliar settings, but also for programs right at home. One central point is the need for having sustainability in mind at every stage of one's work. It is often tempting to move from one appealing idea to the next, assuming that somehow the hard work of sustaining what has been established will be handled by someone else. As Professor Quandt argues, it is essential that recipients of grant funds demonstrate a willingness to invest their own resources, even in limited amounts, in projects. It is equally clear that donors should insist on having in place a plan that will allow successful innovations to build on themselves rather than atrophy.

An even broader lesson is the importance of betting on those institutions and projects that have committed local leadership that truly believes in the value of what is being attempted. When efforts are being made to transfer technology, there is inevitably a technical aspect to the project, but failures almost never occur because technical problems can't be solved. In Professor Quandt's words, "when projects really fail, they almost always fail because of people, and when they succeed, they do so because of the competence, imagination, enthusiasm, and mutual courtesy of people." A sad reality of philanthropic activity is that the lure of a grant can cause a potential grantee to work overtime finding ways of saying what he or she thinks the donor wants to hear, pretending to embrace a concept that is not really of interest. There is nothing to be said for forcing projects on people who don't really want them, and the arrogance involved in such "forcing" is dangerous in and of itself. A central adage is: "No reluctant dragons!"

A final point that is made more by inference than by direct assertion, and that it is easier for me to make explicitly, is that there is also no substitute for a program officer who is in tune with the local culture and who is willing to take some personal risks in getting involved in what can be complicated situations. Just "leaving the dollars on the stump" may work in some situations, but in others much more involvement is required. The accounts of the library automation projects detailed in this volume demonstrate how important it can be for donors to work with high-level governmental bodies and with vendors, as well as with the direct grantees. Care has to be taken in not overstepping one's bounds, but care also has to be taken not to be too passive. Sometimes a third party, from outside an area of local conflict or confusion, may make a real difference in getting an idea to work.

It is, then, with a real appreciation for the leadership provided by Professor Quandt, and an appreciation too for the commitment of so many in Eastern Europe to overcome obstacles of all kinds, that I commend this book to those interested in technology transfer, in Eastern Europe, and in how a professor-turned-philanthropist can make a difference.

William G. Bowen
March 10, 2001

Introduction

It was my very great fortune to have been asked to formulate and to direct in late 1989 The Andrew W. Mellon Foundation's East European Program. Initially, the program was restricted geographically to Hungary, Czechoslovakia (later Czech Republic and Slovakia) and Poland and functionally to economics and management education, and higher education, with research libraries singled out for special consideration. Later, the program was expanded geographically to include Estonia and Latvia and functionally to encompass agriculture and the humanities. During the eleven years from 1990 through 2000, I made about 30 trips to the region and had ample opportunity to observe economic and political developments, market reforms, and the revival of higher education. In 1996, I also became associated with the Foundation's South African program, efficiently and imaginatively directed by Thomas Nygren, where higher education was again the prime target of the Foundation's grantmaking, and where once again, research libraries were accorded special attention.

While the present volume deals only with Eastern Europe, it is noteworthy that both Eastern Europe and South Africa were, and in many ways still are, transitional societies. And while the details of their circumstances are drastically different, they share a number of characteristics. Both regions had autocratic political systems that deprived all or large segments of citizens of civil rights; the former Communism, the latter Apartheid. Both had higher educational systems in which significant numbers of educational institutions were denied the resources needed to develop high-quality education. Both were subject to various export embargoes; Eastern Europe as a result of the U.S. Department of Commerce's COCOM regulations and South Africa because of the Apartheid embargo. Both had significant indigenous groups that strongly opposed the regimes in power. Finally, both regions emerged from their darkest periods in recent history and undertook massive political reforms.

In approaching the task of placing in context the work of Western foundations, and in particular, of The Andrew W. Mellon Foundation, I thought it desirable to weave together three distinct stories: first, the story of what happened, broadly speaking, in the region; second, the story of how foundations attempted to reach their philanthropic objectives and what happened to the initiatives they underwrote; and third, the story of my own wanderings through the region and an account of my observations. There will be no great effort expended to keep these stories separate; rather, I consider the boundaries between them to be highly permeable and will make no apology for straying from one approach or point of view to another.

Indeed, even in the functional areas in which the Foundation was active, it would be a disservice to keep to certain boundaries too rigidly. Nowhere is this more evident than in the general area that may be called "economics, management training, and economic restructuring." Economics is an academic subject, culminating in the United States in doctoral degrees, with a strong theoretical component, but also strong applied and policy-oriented fields. Economists teach at universities, advise government agencies, and hold important government and industry positions.[1] Management training in the United States is usually taken to refer to business school training, leading to the MBA (master of business administration) degree, and holders of these are to be found everywhere in the private sector. But management training in Eastern Europe is also understood to mean the training of entrepreneurs in small, and medium-sized businesses, in formats different from the usual, highly structured university format. Such courses of study may be of very much shorter duration than the standard MBA course. The restructuring of the economy may involve concrete programs intended to create new institutions; that is, the creation of new markets or new financial institutions. But ultimately, many of these activities overlap; one cannot create new markets without training at least some people, and the shorter term training that has characterized much of the educational efforts in Eastern Europe is just one particular spot in the spectrum of training activities from the very short to the very long and detailed. I shall try, whenever possible, to draw useful distinctions when it seems to make sense to do so; but I shall not feel compelled to be a purist in categorizing every activity as belonging under one or the other label.

Much western philanthropy and assistance has worked well, but as one would expect, some projects were more successful than others. We will understand more about both the region and the assistance efforts by learning what has or has not worked and why. But the fact that a project "worked," that is, accomplished its objectives, does not necessarily mean that it was executed in a cost-effective fashion. As long as resources, even the resources of western donors, remain limited, this must always be a paramount consideration.

Much of this volume is about technology transfer, which I interpret to be a broader concept than may be commonly understood. Ordinarily, the term may refer to the adoption of new hardware or new processes in industrial contexts, and its propagation is often by diffusion. The spread of the Internet in Eastern Europe may be a fairly classical example of technology transfer in this sense. But in the context of what foundations and government agencies attempted to do in Eastern Europe, I consider the term "technology transfer" to denote a wider variety of actions, all designed to carry out innovations, ranging from the provision of equipment (for examples, computers) to all kinds of knowledge or skill transfers, including training in economics or management techniques, library modernization, creation of new market-oriented institutions, establishment of civil society organizations to foster collaboration among people, self-help, ecological awareness, respect for democracy and the rule of law, and a myriad of similar things. All of these activities are aimed at providing the knowledge of how to do something that was deemed desirable or important in the context of a democratic market economy. Ex post, of course, we all have grappled with the problem of assessing how effective or successful our attempts to transfer some particular piece of technology (in this broad sense) turned out to be. And that, unfortunately, is a nontrivial problem in comparison with assessing the effectiveness or productivity of technology transfer in the narrower sense in which output per man-hour or average cost per item produced may provide very satisfactory measures. In monitoring the effectiveness of the spread of the Internet, we may reasonably examine the increase in the number of end users, the volume of incoming and outgoing messages, the number of sites connected, subjective satisfaction, response time, and system availability. But even in something as easily described as making the Internet more broadly available, one would ultimately want to know how this has affected, say, scholarly productivity.[2] The measure of effectiveness in modernizing a library is not primarily how many computers have been placed in the library—although that is surely a relevant datum. Slightly closer to the mark might be a measure of cost savings in running a library (for instance, cataloging books and providing circulation, and reference services for patrons). But ultimately, what we would like to know is how library modernization has affected teaching and research in a university: since the outputs of a university are research and the teaching of students, we would like to know how technology transfer has affected them.[3] It may be even harder to measure the effectiveness or productivity of a short seminar or workshop designed to teach the techniques of conflict resolution or of collaboration. These difficulties, which are practical as well as conceptual, will not deter me from trying to make assessments of effectiveness, productivity, and success. The reader will have to exercise his or her judgment whether my evaluations, more informal than I would like, are on the mark.

It is also not clear from mere observation that a project attained its manifest objective that people's outlook on the world has been affected, that their Weltanschauung has been modified. It was, of course, quite unavoidable that worldviews in Eastern Europe change to better reflect the prevailing views among the western industrialized nations: the planned economy systems of the countries in the orbit of the former Soviet Union were not sustainable. The only practical alternative was a largely privatized economy with—it is fervently hoped—an adequate provision for human welfare and social justice. While I am emphatically not suggesting that donors ought to try to make others look at the world through their own glasses and slavishly adopt everything the donor himself believes, there are two areas in which a change of views was desirable and in which East Europeans may have been slow to learn. First, to realize that consumers express preferences by voting with dollars, which gives signals to providers of goods and services that these goods and services are needed. And second, to come to grips with the fact that when useful institutions are created, their long-term sustainability is of prime importance. It is not surprising that these insights had to be learned. Neither of these factors was of importance in the planned economies. Consumer sovereignty did not exist, voting with dollars was largely irrelevant, and sustainability was determined and ordained by the state—again without input from private citizens.

The following episode illustrates both of these elements. The Mellon Foundation had made several grants to TechnoServe (see chapter 11),[4] an excellent not-for-profit organization specializing in agricultural technology transfer, for starting a small business advisory organization in Poland that would assist farmers and small agrobusinesses in solving various problems, such as preparing business plans and securing bank loans. TechnoServe did establish such an organization in the town of Sandomierz, in southeastern Poland, which was doing extremely well and fulfilling all expectations one could reasonably have. TechnoServe, realizing the importance of eventually letting the Poles do all this on their own, started to loosen the reins just around the time I visited Sandomierz. I talked to local businessmen and women and was much pleased by what I saw and heard. Then I was invited to a lunch, at which various local dignitaries were present, including the president of the local bank, the deputy mayor, and several others. They all praised the activities of this small business advisory group. The banker assured me that when he received a loan application, accompanied by a business plan that had the imprimatur of the TechnoServe advisory service, he was very confident that the loan would be repaid and did not hesitate to extend the loan. The deputy mayor argued that the welfare of the town had increased significantly as a result of the activities of the service, because there were more and more viable businesses in town. They all joined in asking that the Foundation continue to support this organization for the sake of the town's welfare.

I replied that we were prepared to think about this, but now, after several years of support, and in the light of the increased profitability of the bank due to this advisory organization, it seemed to me that the bank should make some contribution to its survival, perhaps modest at first, but increasing later. The banker sadly informed me that the bank had no money. I then asked the deputy mayor whether, in the light of the great benefit that this business service organization was bestowing on the town, the town might lend a hand and provide it with some modest support. Alas, the deputy mayor said equally sadly that the town had no money either. I then tried to explain that if they did not contribute even a penny to the cost of some great and beneficial enterprise, I would not be able to judge how much they really wanted it. For example, if they welcomed the construction of the largest opera house in the world in Sandomierz but were unwilling to bear any part of the cost, I would have to think twice before I invested in something that they did not really seem to want or need. The obvious point I was trying to get across was that we can judge needs better or more reliably by what people are willing to do to fill the need than just by what they say. Moreover, unless they were willing to contribute to the effort, it would not be sustainable in the long run.

They listened politely, but I was not certain they got the point. Fortunately, the principals in many other projects did get the point: a good example is the Rectors' Conference in the city of Kraków, Poland, which had decided at a certain point in the life cycle of the Kraków library consortium to shoulder its ongoing costs. In any event, I firmly believe that at some point in the life of a project the donees must begin to pay for at least a portion of the costs, both to avoid moral hazard and to raise the probability that the activity will be sustainable.

This episode goes to the heart of the matter, which is that "culture matters."[4] Projects succeed—and in this respect there is really no difference between projects undertaken for philanthropic reasons or for the sake of profit—when the cultural environment is right; that is, when the "attitudes, values and beliefs" of the participants are appropriate for progress and growth. Among the most important of these are the beliefs that productivity is a key factor, that competition fosters productivity, that time is a precious resource, and that transparency and accountability help to keep unenlightened self-interest at bay. The subsequent chapters report numerous instances in which these beliefs played a major role and a few in which their absence placed projects in great jeopardy.

The number of people who have helped me during the years that the program was in high gear and who have subsequently helped me in writing this book is enormous. First and foremost, I am indebted to William G. Bowen, president of The Andrew W. Mellon Foundation, who entrusted the direction of the program to me and gave me many years of strong support. Profound thanks are due to John C. Whitehead, chairman of the board of trustees of the Mellon Foundation in 1990, whose vision was responsible for the creation of an East European program at the Foundation. An important person at the Foundation who helped me over the years and taught me a great deal about the Mellon Foundation and how foundations operate in general is Kellum Smith, former vice president and then senior advisor of the Foundation. Richard Ekman, secretary of the Mellon Foundation for a number of years and currently president of the Council for Independent Colleges, was always ready to help with sage counsel and was a much valued colleague. Virginia Simone, who rules over the Foundation's archives, has helped me immeasurably when my own files proved inadequate. Andrew Lass, professor of anthropology at Mount Holyoke College, received from me the thankless task of supervising all Czech and Slovak library projects, which he accomplished brilliantly and, in the process, gave me much excellent advice; he also read the manuscript of this volume and provided many important contributions. Valters Nollendorfs, emeritus professor of German at the University of Wisconsin and, until recently, resident director of the Riga office of the Association for the Advancement of Baltic Studies, was instrumental in introducing to me the academic world of the Baltic countries and helped me in many important ways, including the reading of this manuscript; he also supervises the Latvian Library Information Network Consortium (LINC) funded by the Mellon Foundation. Gundar King, Janīs Gaigulis, Barbara Heldt, Gerry Smith, Kevin Quigley, Thomas Nygren, Christine Borgman, Nadia Caidi, Gene Spiro, Malcolm Getz, Joseph Brada, Leslie E. Grayson, Thomas Cochran, and Randall Filer read all or parts of the manuscript and made many valuable comments. Others who have made important contributions to my understanding and knowledge include Irina Rybacek, formerly of the American Czech-and-Slovak Education Fund, William Moody of the Rockefeller Brothers Fund; Kevin Quigley, formerly of the Pew Charitable Trusts; and Paul Balaran and Joseph Schull, formerly of the Ford Foundation. Finally, a number of grantee representatives have made supererogatory efforts to keep me informed about the post-grant period evolution of their projects; I particularly want to thank Linda Gasser of Cornell University, Joanna Regulska of Rutgers University, Robert Hisrich of Case Western Reserve University, Zbigniew Przasnyski of Loyola Marymount University, Loren Tauer of Cornell University, Tadeusz Borkowski of Jagiellonian University, Grzegorz Nowak of Marie Curie Skłodowska University, Maria Śliwińska of Nicholas Copernicus University, Stanisław Chwirot, chairman of the Polish University Accreditation Committee, Adam Manikowski of the University of Białystok, John Hogan of Georgia State University, Anikó Soltész of the SEED Foundation, László Muraközy and Irén Lévay of Debrecen University, Béla Mader of József Attila University, and Brack Brown of George Mason University. To all of them I owe profound thanks for their help, patience, and understanding.

All views expressed in this volume and any errors are my own and not those of The Andrew W. Mellon Foundation.

Princeton, N.J. Richard E. Quandt
May 21, 2001

The Changing Landscape in Eastern Europe

1

The Winds of Change

The Start

In the autumn of 1989, I was doing exactly what I had been doing every other autumn: teaching my classes in economics at Princeton University, where I was a professor, writing articles for learned journals, and participating in the life of the university in various ways. But while it was business as usual for me, the times were unusually exciting and the atmosphere heady, because significant political changes were occurring in Central and Eastern Europe.

In early November, I received a telephone call from William G. Bowen, the president of The Andrew W. Mellon Foundation. Bowen had been the president of Princeton University until January, 1988, at which time he assumed the presidency of the Mellon Foundation. Obviously, I had known him for many years—he is also an economist and, in fact, I was one of his examiners on his final Ph.D. oral examination many years earlier. He wanted to know whether I would be willing to go to Budapest on behalf of the Mellon Foundation in order to evaluate two projects that had been called to the Foundation's attention by Mark Palmer, the U.S. Ambassador to Hungary. The intermediary between Palmer and the Foundation was John C. Whitehead, former deputy secretary of state (July 1985–January 1989), who was a member of the Mellon Foundation's board of trustees. It is noteworthy that the Foundation had no programs at the time in Hungary or any other Central or East European country.[1] I replied that I would be happy to do so, since I was scheduled to go to Budapest anyway in a few of weeks in order to attend a scholarly meeting of economists. Stephanie Bell-Rose, who had joined the Foundation recently as its general counsel, came down to Princeton to explain in more detail what the Foundation wanted me to find out about the two projects and to assure me that she would arrange appointments for me with Ambassador Palmer and with a number of Hungarian experts who were reported to be knowledgeable about the projects.

The first of these consisted of creating an organization initially referred to as either the Business Development Foundation (BDF) or the Foundation for Enterprise Development; eventually it became the Small Enterprise Economic Development (SEED) Foundation (see chapter 4). Its charter members comprised an illustrious list of organizations,[2] and its charter stated its objectives as follows:

> It is the objective of the Foundation to promote, within the broad framework of Hungarian and international cooperation, the creation and effective functioning of enterprises and, in the interests of this objective, to participate in the formation of the infrastructure that will aid and support enterprises. . . .

I managed to interview a number of people about this project. At the U.S. Embassy, I talked with Ambassador Palmer, with Donald B. Kursch, the deputy chief of mission, and with Sandy Dempsey, the head of the economic section. Palmer defined a three-pronged objective for BDF: enterprise creation, enterprise development, and venture capital formation. During the next two days I succeeded in seeing János Palotás, the president of the National Association of Entrepreneurs (VOSZ), Dr. Csaba Csaki, the rector[3] of Karl Marx University; Zsuzsanna Ránki, the director of the International Management Center; and several others. There was substantial agreement that large enterprises needed to be privatized and would eventually be privatized, but at that moment nobody had any idea how this was going to be accomplished. In the meantime, it was essential to encourage the growth of small enterprises by providing technical know-how of a fairly elementary sort for small firms, trouble-shooting assistance, and small-scale funds for start-up or even continuing expenses. The idea of BDF appears to have been Palmer's brain-child. When he first suggested it to some of the founding members, they considered the concept to be fairly radical, but by now they had all accepted it as a desirable thing to do. Some early discussions about BDF had taken place even before the first President Bush's visit to Budapest (July 11–13, 1989) during which he revealed the creation of a $25 million aid package for Hungary, which was ultimately to become the Hungarian Enterprise Fund. But in fact, BDF did not yet exist. While it did have an executive director (Ágnes Tibor), it had no offices and was not a legal entity as yet. In spite of assurances that all the legal paperwork was going to be done in two to three weeks, I had doubts whether BDF could start significant activities in the near term. I also thought that the hopes entertained by several people that BDF would be selected as a conduit for dispensing some of the $25 million promised by President Bush were unrealistic. But in spite of my doubts, I was quite convinced that there was a genuine and important role to be played by organizations such as BDF in providing technical advice to small enterprises and providing them with "incubator" services. Ambassador Palmer was hoping for Mellon Foundation support in the amount of $500,000 per year for three years; I thought that the range of $60–80,000 suggested by Ránki was much more reasonable. I recommended in my report to the Foundation that it keep an eye on the development of BDF. If BDF did, indeed, succeed in getting established as a legal an operating entity, an initial grant of $60,000 would be reasonable, with continuation in future years contingent on establishing a reasonable track record.

The second project was a proposal to establish a new and private university in Hungary. The pro tempore head of this effort was Iván Horváth, a scholar of literature at the Hungarian Academy of Sciences, and he was assisted in this by György Geréby, an assistant professor of philosophy at the University of Pécs,[4] and Katalin Molnár, about whom I could find out nothing more than that she taught English at the United States Embassy.

Private universities, bye and large, did not exist in Eastern Europe.[5] The arguments that the people behind this project advanced for it were that (1) the number of university-trained people in Hungary was too low for what was required for the efficient functioning of a society; (2) the existing universities used antiquated methods of teaching that did not permit students to fulfil their potential; (3) the quality of teaching staff at the existing universities was extremely low—a condition permitted by the history of the past 40 years, in which politically correct behavior provided job security; (4) the existing universities were poorly equipped for training a scholarly elite; and (5) the state universities could benefit from the competition that would be provided by a new private university.

All this struck me as undeveloped and inchoate. While the principals had mentioned that they would soon have a meeting with the president of the Hungarian Academy of Sciences, Iván Berend, and with certain government officials, whose support for such a venture was very important, they had no money, no offices, no staff, no academic program, and no idea how much organization and money it would take to start a university, even if it avoided teaching the "equipment-intensive" disciplines in the sciences, engineering, and medicine.[6] In short, they had no plan, and even if they could succeed in building a high-quality institution, I was not convinced that this was a good idea. It would just draw the best teachers and scholars away from the existing universities, and thus make those institutions even worse than they already were. I parted from them with the request that they keep me posted concerning important developments on their project. The idea for this project, eventually named Corvin University,[7] languished for some years, but neither it, nor another private university initiative undertaken by the Philosophers' Society in Miskolc ever amounted to much, in spite of the fact that by December I had received a letter from Berend, to the effect that the Academy was definitely supporting the idea of Corvin University.[8]

Upon my return to the United States, I submitted a written report to Bowen, who then asked me to attend the next quarterly meeting of the trustees of the Mellon Foundation and to deliver an oral report to them. I did so, and they graciously thanked me for my effort. A week or so later, Bowen telephoned me and told me that the trustees had decided to undertake a program of support in Eastern Europe and that he wanted me to be the director of this program. He said that the trustees were agreed that small and medium enterprise development, privatization, and general economic development had a high priority in Eastern Europe. I also agreed with this assessment, but suggested that, particularly in the light of the Mellon Foundation's historic commitment to higher education, it would also be desirable to support higher education and particularly higher educational infrastructure, and he readily assented to the inclusion of these objectives. He also said that the Foundation wished to place a precise geographical restriction on its East European activities: it would be prepared to support projects in Hungary, Czechoslovakia, and Poland, but not in the other countries, on the triple grounds that (1) one had to concentrate one's resources in order to be truly effective; (2) these were the most developed countries in the region and thus best able to take advantage of the type of support that the Foundation would be interested in giving; and (3) they were the countries where the political risks were acceptable.[9] We also agreed that I would send him a written report about my doings and comings and goings every six weeks or so, and that I would operate as a consultant (later named senior advisor) to the Foundation. All this seemed to promise to become a very exciting adventure, and while wisdom may have suggested that I think about it, I accepted the appointment on the spot.

Antecedents

What were my qualifications for the tasks ahead of me? I clearly had no foundation experience except as the recipient of the largesse of various foundations that had supported my own research in the previous decades. But I had a more than passing familiarity with the region. I was born in Hungary in 1930 and spoke fluent Hungarian, in no small measure due to the fact that I had continued to read Hungarian literature throughout my adult life. There was a long hiatus between my emigration from Hungary in 1947 and my first post-emigration return in 1973, but after that I started to visit more frequently. Most importantly, I started to become acquainted with important economists in the region.

Between 1968 and 1971, I had had the bad judgment to accept the position of chairman of the Princeton economics department. During this period, we hired a young assistant professor, Richard Portes, who was a specialist in "Soviet-type economies" and who was to have a considerable impact on the field, both as a scholar and as the founder of the Centre for Economic Policy Research in London. At his house, I had the good fortune of meeting János Kornai, arguably the most influential economist in the entire region. Kornai, a member of the Institute of Economics of the Hungarian Academy of Sciences, provided the most complete characterization of socialist economic systems. He introduced the concept of the "soft budget constraint," a condition applicable to enterprises in socialist systems that is equivalent to removing the capitalist system's market discipline from the enterprise by providing frequent bail-outs rather than permit the enterprise to fail. He analyzed the consequences of the socialist economic system, such as endemic shortages, in a number of profound works.[10] While I was not tempted in those early years to devote my energies to the topics that occupied Kornai, much later, in the late 1980s and in the 1990s, I threw myself enthusiastically into providing mathematical formalizations and extensions for some of Kornai's insights. My late colleague, Stephen M. Goldfeld, and I even named a particular phenomenon the "Kornai Effect."[11]

A most significant event for me occurred in 1981. I had sabbatical leave from Princeton and had decided to spend six months in London. Richard Portes, who by now had moved to Birkbeck College, University of London, where he held a chair in economics, graciously invited me to spend some time at Birkbeck and offered me a comfortable office. By that time, I was deeply immersed in the study of the econometric problems of estimating disequilibrium models. The standard economic models of markets are "equilibrium models" in the following sense: if at any one time the price were such that the demand for a good were greater (smaller) than the supply, forces would be set in motion to raise (lower) the price, until it was just at the right level to equate demand and supply. Thus, in the case of an excess demand, potential purchasers would bid more vigorously for the scarce good, and drive its price up. In the case of excess supply, sellers would compete against one another, cutting prices so as to get rid of their excess inventories. In any event, a state is then achieved in which equilibrium reigns, that is there are no further incentives for the market participants to raise or lower prices. Circumstances in which prices are not flexible are represented by disequilibrium models. In these models the market is assumed not to clear; in other words, they are markets in which there may be a chronic excess demand (shortage) or excess supply (glut). It seemed reasonable that the new statistical techniques that we were discovering for dealing with models of this kind might be particularly suitable for calibrating models dealing with the socialist countries, as indeed Portes had tried to do in an earlier article.[12] Over many weeks, Portes, David Winter of the University of Bristol, Stephen Yeo (later with the Centre for Economic Policy Research which Portes directed), and I had discussed just how we could get a suitable dataset that would allow a fairly ambitious model to be estimated and what the nature of this model would be. To make a long story short, by the end of my stay the dataset was almost completed, and by the end of the summer we were well advanced in computer work, which eventually, after a great deal of additional effort, led to a number of papers.[13] But it was also important that during my stay I met one day in Portes' office Professor Władisław Welfe, of the University of Łódż, Poland, who invited us in 1983 to present our preliminary findings at a conference that he had organized in Łódż.

The year before I came to London, I had received a paper from an economist at the University of Gdańsk, Wojciech Charemza. I believed that I could improve on his draft, and wrote a separate paper on his topic and sent it to him for his comments. It did not take long for us to decide that our two papers together were better than either one alone, and we decided to write a joint paper. I was also working on that during the London stay, and our joint paper was published in due course.[14] Emboldened by our success, we were seeking other topics for possible collaboration, and the next time it was practical for us to meet was at Welfe's conference in Łódż in December 1983. This was only a few months after Martial Law had been lifted and Poland was unspeakably grim. I had the good fortune of being driven by car from Warsaw to Łódż; Portes took the train, and his passport was promptly stolen on the journey. In my hotel in Łódż, the hot water faucet in my bathtub did not work; so to take a bath I would quarter fill it with cold water, and then transfer hot water from the sink by using a cup—a time-consuming enterprise, to say the least. The shops were empty, here and there a drunk would be lying on the street in a stupor, and the entire visit was most depressing (except for a banquet given by Professor Welfe, at which the only thing to drink was vodka, but in at least half a dozen varieties). The quality of the conference was also depressing and the papers by East European authors were largely (but by no means uniformly) dreadful. I could barely wait for the discussion of one particular paper to begin so I could shred it verbally, but my old friend David Hendry from Oxford got there before me and all that was left for me was to administer the coup de grâce. (What was even more depressing was that the author delivered the same paper, with essentially no changes, two years later at a similar conference, having learnt nothing from our critical comments.) It is worth noting that the Anglo-American style of discussing and criticizing papers at a scientific conference is quite different from the Continental European style. We tend to rush in with six-shooters blazing while East and West Europeans tend to be more understated. I recall a 1976 conference in Urbino, Italy, chaired by the distinguished Sir Austin Robinson, who encouraged us in his opening remarks to engage in vigorous debate: he wanted to see the "thrust and parry of intellectual swords." There were three of us at the conference from the United States and we obliged, but we were soon asked to tone down the vigor of our criticism because some of the Europeans had complained that it was too harsh.

But I made some important contacts at the Łódż conference. I met Vladimír Dlouhý, who subsequently became deputy prime minister and then the minister of trade and industry of Czechoslovakia and after that, of the Czech Republic; Karel Dyba, who became minister of the economy; and Václav Klaus, who after 1989 became minister of finance and then prime minister of Czechoslovakia and then of the Czech Republic. All three of them had been economists working at the Institute of Forecasting of the Czechoslovak Academy of Sciences, and I had important further contacts with all three them. At that time, but not five years later, Klaus affected a grossly deferential mien, and I said to myself, "so this is what a Czech Uriah Heep looks like," not suspecting the transformation that would take place a few years later. In any event, I also met a friend of Charemza, Mirek Gronicki, who would become a collaborator of ours on our next paper, and the three of us did some useful work outlining its key elements.[15] I recall that one afternoon

I had scribbled some notes that all three of us should have had a copy of and I asked whether we could find a Xerox machine. I was told that that was not so simple. Copying was a privilege, strictly controlled by a university officer, lest the copying machine be used for nefarious purposes. Alas, this officer happened to be on vacation, and hence no copying of papers could be done at the university.

This visit led to others that also proved to be very useful. In 1985, Welfe organized another conference, this time in Szczrk, in the mountains in the south, and in 1987 I was invited by Vladimír Dlouhý to visit him at the Institute of Forecasting in Prague. The general level of economic sophistication at the Institute is illustrated by Václav Klaus' request that I give an impromptu lecture to a study group that was studying microeconomic theory on its own. When I asked what text they were using, I was told that it was the elementary textbook by Samuelson and Nordhaus—a fine book, but strictly on the introductory undergraduate level. I also met Dušan Tříška, a well-trained economist specializing in mathematical programming, and Josef Zieleniec, who later became the foreign minister of the Czech Republic. They asked me whether I would be willing to be coorganizer with Tříška of a conference, to be held the next year and to be coeditor of a volume containing selected papers from the conference.

I agreed, and the conference took place as scheduled in the spring of 1988. It had participants from the United States, Western Europe, and Eastern Europe. The quality of papers was sufficiently high that we ultimately selected 25 for inclusion in a volume that was published in 1990.[16] The conference took place in a lovely baroque palace in Liblice, near Prague. The most memorable part of the conference was the closing remarks by Václav Klaus, in which he spoke of Czechoslovakia's past and the people's current aspirations, and in which he referred to the Russian intervention in the Prague Spring of 1968 as "an invasion." I remember turning to the person sitting next to me in the audience and muttering something like "Václav will be in jail before the day is over." But the times were changing and, in spite of the fact that many were still in jail, Klaus escaped that fate.

When Bowen had asked me to lead the Foundation's efforts in Eastern Europe, I felt moderately confident that my general familiarity with the region and my many contacts would stand me in good stead in developing a useful portfolio of projects, even though I was somewhat apprehensive because of my lack of foundation experience. It was also a highly serendipitous moment for me, because much of my research during the previous decade dealt with disequilibrium phenomena in economics. The existence of disequilibria in western capitalist economies was strongly doubted by most western economists, but even the most doctrinaire of these was willing to admit the relevance of the concept for planned economies. Now, at one stroke, the planned economies disappeared and my 1988 book[17] became an epitaph not only for the planned economies, but for my own research in that area. But I could now start a new kind of engagement with the region.

Whether I was fit for that or not was unclear, but I thought I would know soon enough if I was a fish out of water.

The Context

The Background. It had been understood for a long time that the Soviet-type economic system was grossly inefficient. Ignoring for the moment the differences among the various countries and the relaxations of economic planning that were introduced here and there, in its crassest form the system allocated resources by administrative decisions and did not permit the price system to reveal where shortages might exist; it thus prevented resources from flowing to where they might be most needed. In a system in which prices are flexible, an increase in the demand for some commodity usually leads to an increase in its price; hence it becomes more profitable to produce that commodity and resources are shifted towards its production. This mechanism cannot work if prices are controlled—as they typically were in planned economies—and an unavoidable consequence is that chronic shortages (and, in principle, also chronic surpluses) develop; hence long queues at stores, empty shelves, and rude sales personnel that never has to ingratiate itself to sell an item.[18]

Target fulfillment by enterprises was also an important feature of planned economies, whether this was economically desirable or not, and whether this could be accomplished profitably or not. As a result, a complex system of industrial subsidies came into being, lest enterprises suffer bankruptcies, in which case workers might have to be dismissed. The result was that government budget deficits were substantial, and in Hungary the accumulated government debt amounted to 56.8% of GDP at the end of 1990, and consumer price inflation was 25% in 1989,[19] while the inflation rate had reached 60% in Poland.[20] Inefficiencies also resulted from the very large scale of industrial enterprises, which were generally much more highly concentrated than in, say, Western Europe. Blazyca reports, for example, that between 1974 and 1977, the percentage of enterprises employing more than 500 persons (that is, "large" enterprises) amounted to 85% in Hungary and 81% in Poland, with the corresponding figures for Denmark, France, and Netherlands being 38, 57, and 44%, respectively.[21] High industrial concentration was encouraged on the basis of the mistaken and uncritically accepted belief that size implied efficiency, but in reality, it was often encouraged for the sake of administrative convenience. A large proportion of international trade was intra-CMEA,[22] and trade had to be bilaterally balanced and was conducted in nonconvertible rubles. As a result, many of the advantages derived from free international trade had to be foregone by the East European countries.[23]

It does need to be said that there were distinct differences among the economic systems in place in the various countries. Substantial reforms were introduced in Hungary in 1969 and are known as the New Economic Mechanism.[24] Through these reforms, the physical allocation of commodities by government agencies had been discontinued, consumer prices became flexible, and, perhaps most importantly, profits were permitted to provide incentives for enterprises by tying managerial compensation to profits.[25] Some early economic reforms in Poland (1956–59) were not particularly successful, and it was not until 1982 that some trade liberalization, modest price decontrol, and somewhat enhanced independence for enterprises were instituted (Crane, p. 319). While these reforms may have blunted some of the worst aspects of a command economy in Hungary and Poland, Czechoslovakia remained largely in the original mold.

A small ray of sunshine was also provided by the fact that some private sector activity grew up in both Hungary and Poland. Kornai has estimated that in the early 1980s as much as 30% of Hungarian GNP might have originated in the private sector.[26] Dynamic growth in the private sector was also occurring in Poland in the 1980s. It is particularly noteworthy that about 70% of agricultural land was privately owned throughout the Communist period.

In spite of these improvements in the largely doctrinaire and uncompromising economic systems, it is not surprising that economic performance on the whole was miserable. Growth rates that were as high as 10% in Czechoslovakia during the first Five Year Plan, above 5% in the 1960s in Romania and Bulgaria, and 3% in Hungary, had fallen to below 2% by the 1980s in all the countries of the region.[27] GNP per capita in 1989 was only $4,565 in Poland, $6,108 in Hungary, $7,878 in Czechoslovakia, and $3,445 in Romania.[28] By the same time, Poland had accumulated a foreign debt of about $38 billion (with a total GDP of $72.9) and Hungary a foreign debt of $20 billion (with a total GDP of $29 billion). Economic growth was small, real wages declined, inflation increased, and shortages continued to exist throughout the 1980s.[29] The decline may have started in Poland with the wildly optimistic economic plans conceived under the Gierek regime in the second half of the 1970s, which led to a host of miscalculations and increasing reliance on foreign imports, to be followed by a foreign exchange crisis in 1979 and by the economic catastrophe inflicted by Martial Law, during and after which GNP fell by some 25%.[30] The story of Hungary may have begun with the softening of its western export markets and consequent pressure on its balance of trade in the 1970s and by the increasing difficulty of securing western loans, caused at least in part by increasing doubts about the credit-worthiness of the eastern bloc. The periodic attempts to achieve balance in Hungary's foreign trade led to restrictive macroeconomic policies, which succeeded in reducing investment while allowing consumption to grow. The military preparations and expenditures demanded by the pressures of the Cold War and the oil shocks may well have exacerbated the problems of the socialist economies, but serious problems would have existed in any event. As Kornai points out, the socialist countries lost the productivity race.[31] It is against the backdrop of economies misfiring on all cylinders that the dramatic political changes of 1989 began to occur. One of the very first steps was the elections in the Soviet Union on March 26, 1989, in which multiple candidates could run for seats.

Political Changes. The recognition that the system was not sustainable began to dawn even in official government circles. Imre Pozsgay, a Hungarian with a vision of reform, is quoted as saying that the socialist model "was the wrong path in its entirety."[32] In fact, the Hungarian Communist Party committed itself to accepting the existence of more than one party, and on October 18, 1989, the Hungarian Parliament approved constitutional changes that would restore multiparty democracy.[33]

In Poland, Lech Wałęsa began Round Table discussions in February 1989 with the minister of the interior, General Czesław Kiszczak. It is noteworthy that both parties had agreed to "some basic rules of self-restraint in a period of political transition,"[34] and on April 5, the government and Solidarity signed an agreement that legitimized the outlawed Solidarity, as a way of enlisting the membership of Solidarity in restoring the health of the economy. This ended the political and economic monopoly of the Communist Party. Such a move was not without risks for the leaders of Solidarity: they might have been perceived by the rank and file as selling out and might also have had to shoulder some of the blame if the economy were to perform even worse.[35] The Central Committee of the Communist Party had no choice but to accept the notion of a multiparty system and elections were called for June. Two-thirds of the seats in the lower house (Sejm) were reserved for the Communist Party, with the rest freely contestable by Solidarity; all the seats in the upper house (Senate) were freely contestable. The results were a rout: Solidarity won all 161 seats that it was allowed to contest in the Sejm and won 99 out of the 100 seats in the Senate.[36] General Wojciech Jaruzelski became president of Poland, and after a brief interlude during which the former Communist minister of the interior, Czesław Kiszczak was prime minister, Tadeusz Mazowiecki, a Catholic and emphatically not a Communist, assumed that post on August 19.

Events in Hungary were equally dramatic. The demonstration on March 15, 1989, with the enormous symbolic meaning of being the anniversary of the beginning of the Hungarian uprising in 1848,[37] was followed on June 16, 1989 by the reburial of Imre Nagy, the prime minister of Hungary during the 1956 revolution who had subsequently been executed, without police interference, in Budapest in a ceremony that was attended by some 250,000 people.[38] In short order, some seven different political groups started to play a significant role and the official Hungarian Communist leadership was in retreat (Ash). As early as May, Hungary started to dismantle the border fence between Austria and Hungary. Other East Europeans, particularly East Germans, started to take "vacations" in Hungary, with a view towards moving on through the newly opened border to Austria. After negotiations between the Hungarian and East German governments, Hungary unilaterally decided to permit East Germans to pass through Hungary, and some 55,000 East

Germans did so in September and 21,000 in October.[39] Hungary held genuinely free elections during the summer in which the Communists were convincingly defeated. The difference between Poland and Hungary is described by Timothy Garton Ash as follows: "In the race for freedom of speech and freedom of enterprise, the Hungarians are currently in the lead, although Poland is coming up fast. In the political stakes, Hungary leads in words, but Poland in deeds."

Czechoslovakia, economically probably the most advanced but politically the most repressive of the three countries, was ambling rather than rushing towards democracy. As Kornai points out, "The strongest contrast to Hungary is offered by Czechoslovakia. Until the change of the political system, it had one of the most ossified regimes in the socialist region, resolutely resistant to change and repressing all opposing opinions," (Kornai, p. 16). As recently before the dramatic changes as February 21, 1989, Václav Havel was sentenced to a nine-month term of imprisonment.[40] A demonstration in Wenceslas square on November 17, in commemoration of Jan Opletal, a Czech student murdered by the Nazis, led to hundreds of arrests, and "became the spark that set Czechoslovakia alight."[41] The Velvet Revolution was brewing. Opposition groups, led by Charter 77, held meetings and the Civic Forum organization was set up in November 1989; 200,000 people demonstrated on Wenceslas Square on November 20 and again on the 21st. And while the police still beat hundreds of demonstrators in that month, the number of demonstrators grew and they were joined by workers in a two-hour strike on November 27.[42] The government had no choice but to start talks with the opposition. After another mass protest in Wenceslas Square on December 4 against Prime Minister Ladislav Adamec's proposals for a government with 75% Communists, Marian Calfa replaced Adamec as Prime Minister on December 10 and a new cabinet was sworn in which non-Communists had a majority (Whipple, pp. 18–20). In late December, Václav Havel was elected president and soon Václav Klaus, a man with a distinctly Milton Friedman-type outlook on the economy, became minister of finance. Followers of Friedman characteristically believe that the market does a good job in allocating resources under all circumstances—an assertion doubted by most economists.[43] Klaus tended to have strong disagreements with people who advocated any kind of intervention: George Soros, because his philanthropic activities "interfered" with what the market would have done, and Joseph Stiglitz,[44] because he was not sufficiently libertarian.[45]

The events of 1989 were surprising not only because they happened at all, but because they happened rapidly, their scope was enormous, they took place nearly synchronously in many countries, and they occurred in countries with substantially different economies, ethnic compositions, and political traditions.[46] Hungary, as I indicated before, had been experimenting with looser forms of economic planning since 1968, and the freedom of writers was arguably greater there than in some other countries. But Hungary did not have a broad-based movement such as the Polish Solidarity, with which the Polish government had to compromise from time to time, nor a more narrowly based one such as Charter 77, which the Czechoslovak government resolutely persecuted throughout the 1980s. Would the events of 1989 have happened without glasnost and perestroika, if there had been good reason to fear another Soviet intervention? Probably not, and even if they had happened, the outcome would have been drastically different. But Gorbachev was no Breznev, and changes were occurring even in the Soviet Union. And everywhere, in the West as well as in Eastern Europe, the scale and the abruptness of the changes were surprising and even shocking. It is difficult to fathom the full significance of the change in mindsets through which a Manichaean worldview of "good" and "evil" was replaced by a vastly more confusing and chaotic world of apparent contradictions.[47]

Changing the Political and Economic Environment. By the end of 1989, it was clear that the movement towards political democracy and pluralism was irreversible and that fundamental changes in the economic systems had to be brought about. In the political arena, Robert Dahl's six-part test of democracy was quickly passed in large measure in Hungary, Czechoslovakia, and Poland: the highest officials of the land were subject to elections, elections were free and fair, freedom of expression was safeguarded, the right to organize politically was protected, citizens could access nongovernmental sources of information, and participation was universal.[48] In the economic sphere, macroeconomic stability, privatization of state enterprises, and the termination of endless subsidies to enterprises were among the principal tasks. And while macroeconomic stability is an area on which a great deal had been written by economists, privatization in the East European context was a relatively new and fertile area for scholarship.[49] Many distinguished economists wrote about the key components of an economic agenda. David Kemme, for example, suggested that four things were required for the economic transformation: (1) appropriate monetary and fiscal policies had to be introduced to stabilize the economies and to induce growth; (2) appropriate legislation had to be passed in order to convey economic power from the state to individual economic agents; (3) new types of individual behavior needed to be fostered so that individuals would respond to economic incentives; and (4) western countries had to participate in the process.[50] Much learned discussion was devoted to the proper sequence by which the necessary changes were to be brought about: would one try for macroeconomic stability first and worry about privatization later, or would it be better first to privatize and then try to achieve stability? There was clearly some disagreement among western economists, with Kemme recommending a priority action list beginning with price liberalization and followed by the creation of more sophisticated market institutions, creation of an appropriate safety net, privatization of state-owned assets, and expanded foreign investment. Brada and King, on the other hand, argued that "on the domestic front a liberalization of the capital market and privatization must be among the early objectives," and "an early objective of the reforms should be the opening of the country to direct foreign investments."[51]

One of the biggest questions debated among economists was how best to achieve macroeconomic stability; should one try for "shock treatment," as did Poland, or would a more gradual approach be equally satisfactory? The advocates of slow price liberalization sometimes argued for relative prices to be adjusted sector by sector, eventually to be followed by economy-wide adjustments of price levels to reflect the disappearance of state subsidies and the decontrol of prices, which would be politically more tolerable, particularly when price changes affected food products and other essential consumer items. Rapid price liberalization of the Polish variety can work sudden changes in the income distribution and can cause (and has caused) massive changes in peoples' real incomes. From the economic point of view, I have no doubt that the former model is the correct one. Under the slow-change model, prices will be adjusted over longer periods of time and there is no reason to believe that prices will be "right" during the course of the long adjustment period; hence, we just postpone the time when prices can begin to exert their proper allocative functions. Also, there is a strong possibility that inflation will continue to creep along during an extended adjustment period, with clearly unfavorable effects on the inflation expectations of the public, if for no other reason that shortages are unlikely to have disappeared during the long adjustment period and that hoarding would thus fail to become irrelevant. These observations can be illustrated by comparing GNP growth rates and inflation rates for Poland, the country that employed the most radical shock treatment, and Russia, which was the most gradualist in its approach to restructuring. The comparison overwhelmingly favors Poland.[52] But at the same time, the hardships inflicted and the short-term welfare loss suffered during a rapid adjustment cannot be ignored, which makes this question ultimately difficult to resolve.

But in terms of sheer economic effectiveness, there is much to be learned from the case of Poland. Finance Minister Leszek Balcerowicz, with the advice of Jeffrey Sachs and David Lipton, initiated his reforms in September, 1989 with significant budget cuts, tightened bank credit, and severe tax penalties for producers who allowed wage increases to exceed 30% of the increases in the consumer price index. The złoty–dollar exchange rate was fixed at a level that implied a 40% depreciation, and prices were liberalized.[53]

All countries faced a myriad of ancillary problems: tax reform and the imposition of fiscal discipline on enterprises, the question of balance of payments equilibrium, exacerbated by the virtual disappearance of COMECON trade, the high degree of industrial concentration, the immense need for new investment to rebuild the antiquated and run-down industries, and the intransigeant problems of the environment. These and other questions about economic priorities were occupying the attention of economists both in Eastern Europe and in the West. A formal occasion devoted to a detailed discussion of a research agenda took place at CEPR (Centre for Economic Policy Research) in London on February 17, 1990, and was attended, in addition to a large number of West European experts on the region, by noted East

European economists such as Dariusz Rosati, Grzegorz Kołodko, István Székely, and others. Major topics of discussion were the integration of the East European economies into the world economy, domestic financial markets, competition policy and privatization, fiscal reform, social security, housing markets, energy and the environment, and the sequencing of reforms.[54]

A much more extensive effort, but restricted to Hungary, was that of the Blue Ribbon Commission, chaired by Márton Tardos and Sylvia Ostry, which worked under the auspices of the Hudson Institute at Indiana University, as well as of the Hungarian Academy of Sciences, the Hungarian Chamber of Commerce, the East-West Forum, and the Nomura Research Institute. The Commission consisted of 20 Hungarian and western experts who parceled out the various topics to teams of economists. The Commission released its report, entitled *Action Program for Hungary's Transformation to Freedom and Prosperity*, which dealt with an agenda of privatization, foreign investments, entrepreneurship, taxation, currency convertibility, and political changes.[55] However, it is not clear to what extent the report was effective in the long run.[56] The Blue Ribbon Commission was disbanded after its work was complete, but was then reconstituted a year later to study longer run problems of the Hungarian economy.

Economic reform, in whatever sector and whatever form, was not thinkable without individuals trained in economics or management or business. While an army of United States and West European consultants streamed into Eastern Europe, it was clear that their role was to advise and possibly to teach, but they would not be able to bear the brunt of the effort needed to change the system. (They soon became known in Poland as the "Marriott crowd," because they came to Warsaw for a few days and stayed at the Marriott.) There were a few centers of learning where respectable mathematical economics or econometrics was practiced,[57] but hardly any economist in the region could be said to know macroeconomics or modern finance, accounting, and so on. In designing the course of economic reforms, Poland, with the assistance and advice of Jeffrey Sachs of Harvard University, chose the shock-treatment method, while Hungary and Czechoslovakia were more gradualist.[58] The shock treatment in Poland, by allowing prices to be freely determined and the value of the złoty to fall to the black market level, was quite successful in bringing about macroeconomic stability. But the cost was high, and it is perhaps ironic that the much heralded new age in Eastern Europe began with drastic declines in the standard of living, even if, as can be reasonably argued, the reestablishment of equilibrium in commodity markets drastically reduced the disutility that the system of planning was inflicting on the population through the endless queueing that it created. The fall in output was probably unavoidable because of a combination of circumstances; because many of the countries were following restrictive monetary policies, because intra–East European trade and trade with the Soviet Union were collapsing[59] and other reasons, but none of this mitigates the pain that was suffered.

The story of privatization presents a different and less successful picture.

Initially, the view was that at least as far as Poland and Hungary were concerned, privatization was less successful than in Czechoslovakia.[60] That country (and later the Czech Republic) adopted a variety of privatization methods, including restitution to original owners, public auctions and tenders, and direct sales, but the bulk of property value transferred employed a relatively direct method based on turning enterprises into joint-stock companies and offering the shares of these companies in exchange for vouchers distributed to the population, with which they could bid for shares in enterprises.[61] Voucher privatization proceeded in two waves, with about 6 million Czech participating in each of them.[62] In fact, the benefits of the speed with which assets could be transferred by the voucher method was counteracted by the fact that investment funds succeeded (often by questionable methods such as offering a ten-fold return) in purchasing large quantities of vouchers and they, many owned by banks, became the principal holders of Czech companies.[63] The poor economic performance, indeed the serious collapse of the Czech stock market, can be traced to the absence of stock market regulations (which was in the best traditions of the no-holds-barred capitalism that Václav Klaus advocated) and it has been suggested that if effective stock market regulations had been in place, the Czech economy would have grown 7% in 1997, rather than 1.8%.[64] But the question of what the appropriate method for privatization was had no obvious answer, and was even a sensitive issue in some ways. Even privatization by restitution to original owners could be said to be controversial: thus, for example, in Hungary, Viktor Orbán, then the leader of the political party FIDESZ (Federation of Young Democrats) and later prime minister, vigorously opposed restitution in a parliamentary speech on February 4, 1990.[65] Questions of who should be the primary beneficiaries of privatization, whether foreigners should be permitted to own sizeable portions of former state assets, and many similar questions had to be answered.[66] It is not surprising, therefore, that Poland and Hungary resorted to privatization in more halting and uncertain ways, often taking the form of spontaneous privatization in which enterprise management tended to end up the prime beneficiary. Both of those countries adopted a variety of methods for privatizing large state enterprises, including restitution to original owners, spontaneous privatization by enterprise managers, sales of firms to outside investors, management and employee buy-outs, and asset privatization by liquidation.[67] Abuses were frequent, as in cases in which managers set up new enterprises in which they were the owners, and then sold the assets of the state enterprises to the new companies at low prices. Moreover, the entire process was slowed down by administrative inefficiencies, by uncertainties as to how the privatization processes were going to play themselves out, and by the difficulties in obtaining credit. Some segments of the populations feared privatization because they (correctly) foresaw that it would lead to layoffs. Consequently, while by 1997, 65–75% of GNP was generated in the private sector, the process of privatization was slower than had been anticipated and almost certainly slower than would have been desirable.

Subsidies to enterprises began to be cut, as were social and welfare ben-

Table 1-1 Inflation rate (consumer price index).[a]

	1990	1991	1992	1993	1994	1995	1996	1997
Hungary	30%	34%	23%	23%	21%	28.3%	20%	18%
Poland	250%	60%	44%	37%	30%	21.6%	18.8%	15%
Czechoslovakia	9%	52%						
Czech Rep			12.5%	19%	10.2%	9.1%	8.7%	10%
Slovakia			8.7%	23%	12%	7.5%	5.5%	6%
Romania	50%	215%	200%		62%	25%	56.9%	151%

General Note:

a. See *The World Factbook*, 1991–1997 for all tables in this section. These figures differ somewhat from those reported by Quigley, Kevin F. F., *For Democracy's Sake: Foundations and Democracy Assistance in Central Europe*, Washington, D.C.: The Woodrow Wilson Press, 1997, pp. 14, 30, 44, 60, which are based on *PlanEcon*, and the figures for Hungary also differ from those reported by Kornai, János, *Struggle and Hope: Essays on Stabilization and Reform in a Post-socialist Economy*, Cheltenham, U.K.: Edward Elgar, 1997, p. 107.

efits. But even the drastic reduction of subsidies was not enough for macroeconomic control, if enterprises continued to extend to each other inter-enterprise credits—a serious problem in Russia, for example, even today.[68] In any event, many political groups were concerned about the erosion of living standards. A poll taken in 1991 in Hungary indicated that 63% of the respondents felt that "the country was going in the wrong direction," (Rondinelli and Fellenz, p. 82).

What was the impact of all of these changes? We provide some summary figures in tables 1-1, 1-2, and 1-3 in which we also include, for comparative purposes, Romania.[69]

The gradual termination of state subsidies to unprofitable industries had an immediate impact on the unemployment rate, which grew from essentially nothing (0.4% in Hungary and 0.9% in Czechoslovakia in 1989) to the very substantial figures in table 1-2. The decontrol of prices that occurred in all countries led to an immediate price inflation (the 1989 inflation rate was 25% in Hungary and 1.5% in Czechoslovakia, although it was 640%

Table 1-2 Unemployment rate.

	1990	1991	1992	1993	1994	1995	1996	1997
Hungary	1.7%	8.0%	12.3%	13.0%	10.4%	10.4%	11.0%	9.0%
Poland	6.1%	11.4%	13.6%	15.7%	16.1%	14.9%	13.3%	12%
Czechoslovakia	0.8%	6.7%						
Czech Rep		4.1%	3.1%	3.3%	3.2%	2.0%	3.3%	5.0%
Slovakia		12.0%	11.3%	14.4%	14.6%	13.0%	12.0%	12.8%
Romania		40%	9.0%	11.0%	10.9%	8.9%	6.1%	8.8%

Table 1-3 Private sector share of GDP.[a]

	1990	1991	1992	1993	1994	1995	1996	1997
Hungary		33%	44%	52.4%	58%	60%	70%	75%
Poland	31.4%	45.3%	48.2%	53.5%	56%	60%	60%	65%
Czech Rep	12.3%	17.3%	27.7%	45.1%	56.3%	70%	75%	75%
Slovakia		25%	32.4%	39%	58.2%	60%	70%	75%
Romania	16.4%	23.6%	26.4%	32%	35%	40%	60%	60%

General Note:

a. SEED *Act Implementation Report Fiscal Year 1998*, Washington, D.C.: Department of State, March 1999.

in Poland), which the various countries combatted with differential success. Per capita GNP started to decline immediately and industrial production fell in each country and recovered to the 1989 level only in 1998 in Hungary and in 1997 in Poland. The short-term welfare impact of the economic changes was clearly unfavorable,[70] although the private sector share of GDP exhibited reasonable growth (table 1-3).

Not unexpectedly, a concomitant of the restructuring of the economies and privatization, the inequality of the income distribution increased after 1990. While before 1990, the Gini coefficients in Hungary, Poland, and Czechoslovakia were lower than in any West European country (except for Poland, where the Gini coefficient marginally exceeded that of Finland and Sweden), by the end of the 1990s these had increased to near West European levels.[71]

It seemed in 1990 and 1991 that the political battles had been won, but the economic foundations on which the future democracies would have to rest were still shaky. Real GDP declined in the region for several years; in the Visegrad countries the period of decline was shortest in Poland (two years) and longest in Hungary and Slovakia (four years), although in other countries the period of decline was even longer (six years in Bulgaria and five years in Estonia and Lithuania). But by 1995, most countries in the region had growing real GDPs. By 1997, inflation rates had substantially abated in most countries, although no great gains were made in reducing unemployment rates. One is tempted, by the end of the decade, to regard most countries' economic developments with some optimism; however, this optimism needs to be tempered in the light of the number of years it is expected to take for the countries in the region to catch up with the lowest per capita GDP members of the European Union. If GDP growth rates were to exceed those of Spain, Portugal, and Greece by 7%, the number of years required for GDP convergence is shown in table 1-4. Thus, for example, the Czech Republic, which was the most favorably situated country among the East European countries shown in the table, is expected to take 11 years to catch up with Portugal, which is the

Table 1-4 Estimated years for catch-up.[a]

	Spain	Portugal	Greece
Hungary	17	13	14
Czech Republic	15	11	12
Poland	19	15	16

General Note:

a. Morita, Tsuneo, "The Hidden Growth Potential of EU Candidates," *Transition*, 10, No. 5, Oct. 1999, 5–7.

least favorably placed country in the EU, on the rather sanguine expectation of a 7% excess growth.

Considering that the source for the table was written ten years after the transition began, the best-case scenario involves an over 20-year wait to approach minimal EU levels, which is entirely consistent with the casual remarks many economists were making in 1989 and 1990 that it would take 30 or more years for reasonable economic recovery.

What should have been the aspirations of the East European countries? Ralf Dahrendorf argues most convincingly that the search for a middle ground between democratic capitalism and socialism is utopian, and that this search is itself a utopian system which, like all systems, is an enemy of liberty and open societies.[72] The fundamental first step is the constitutional step of establishing private property and markets and creating room for appropriate economic incentives. The sentimentalists will want to preserve the social welfare achievements of socialism under the new systems; but in fact these achievements were largely illusory. Nevertheless, substantial misery did ensue upon the reduction or cessation of state subsidies to large, inefficient enterprises, the liberalization of prices, and deflationary policies by governments. With hindsight, everything is foreseeable, and so was this; but in any event it was unavoidable in the short run and certainly played a part in socialists being returned to power after some years, to wit Aleksander Kwaśniewski in Poland, Gyula Horn in Hungary, and Miloš Zeman in the Czech Republic (Vladimír Meciar, in Slovakia, is entirely *sui generis* and cannot be explained in these terms as easily). Restoration or creation of welfare services is a task left to what Dahrendorf calls "normal politics," and one would hope that in the not-too-distant future democratic leaders will assign high priority to these matters. But the matter is clearly not to be taken lightly. As Dahrendorf puts it,

> The formal process of constitutional reform takes at least six months; a general sense that things are moving up as a result of economic reform is unlikely to spread before six years have passed; the third condition of the road to freedom is to provide the social foundations which transform the constitution and the economy from fair-weather into all-weather institu-

> tions which can withstand the storm generated within and without, and sixty years are barely enough to lay these foundations. (Dahrendorf, pp. 92–93)

Has democracy taken permanent roots in the region? If Dahl is right, that democracies older than 20 years are usually stable, at the time of this writing we are at the half-way point, and we can be confident but not certain. One's confidence should certainly be bolstered by such exhilarating expressions of political dissent as witnessed by the mass demonstration in Wenceslas Square in early December 1999, demanding the resignation of the prime minister, Miloš Zeman, and the speaker of the parliament, Václav Klaus, and calling for a renewal of civic values and democracy.[73] It is perhaps less confidence inspiring, but not unexpected, that Zeman is reported to have characterized the demonstration as "adolescent" and Klaus is reported to have said that he was not going to react to "children's hysterical outburst." The economies of the region are performing much better than in the early 1990s and are in the neighborhood of the Dahrendorf target. But it will surely take more time to ensure that the East European institutions are "all-weather institutions which can withstand the storms."

2

A Survey of Assistance Programs for Eastern Europe

The political changes that took place in 1989 mobilized extraordinary efforts by governments and private philanthropic organizations to provide assistance to the East European countries, and a little later to the Newly Independent States of the former Soviet Union, so that the worst and most dangerous legacies of the previous regime could be eliminated. If the countries of the region were to become stable democracies, which could take their place in the community of other European countries as constructive partners, their political systems, economies, infrastructure, educational systems, provision of health care, and other sectors all had to be drastically reformed and a sound civil society created. These reforms had to be carried out in a reasonably expeditious manner, lest lingering thoughts of the possible advantages of an *ancien regime* and a recalcitrant nomenclatura contribute to undoing what was accomplished in 1989. Hence, the U.S. government, the European Union, and many American and European foundations addressed themselves to providing assistance for the perceived needs.

Much has been written about the antiquated nature of the systems and processes employed in the socialist economies, whether they were in industrial enterprises, in research libraries, or elsewhere. While there were many reasons for that state of affairs, two related ones need to be mentioned here. First, outdated methods can be swept away only if there is a sufficiently vigorous flow of innovations. But the educational system in the socialist countries tended to produce narrow specialists, and that is not conducive to the kind of innovation that imaginatively combines elements from different fields. This is precisely the reason why American business schools were advised to concentrate on educating generalists rather than narrow specialists,[1] and why some of the more innovative research in American universities in recent decades has emerged from interdisciplinary approaches. Second, innovations tend to sweep away the old in a Schumpeterian "creative destruction," which was not compatible with five-year plans and required too much risk taking to be comfortable in a hierarchical system.

Much of this volume is about technology transfer. As commonly discussed in the literature and generally understood, technology transfer has three key characteristics: (1) it is about processes that take place at the level of business firms; (2) it deals with information and the communication of information; and (3) it is undertaken with a view toward more effective competition and is an essential part of the process of innovation.[2] For our purposes, technology needs to be defined fairly broadly so as to include embodiments in hardware, software, processes, methods, and methodologies;[3] In other words, one may equate technology transfer with knowledge transfer, where the knowledge becomes embodied in some activity and may be represented by some tangible artifact to a greater or lesser extent.

The assistance that western agencies provided in Eastern Europe were essentially of two kinds. On the one hand, direct economic assistance was delivered when a western agency (such as one of the Enterprise Funds or the European Bank for Reconstruction and Development) took an equity position in an East European firm, or when a western corporation purchased an East European enterprise, or when direct humanitarian aid was provided through one of several agencies specializing in refugee relief. On the other hand, numerous western agencies delivered technical and educational assistance. In many instances, this type of assistance was directed toward business enterprises, not so much toward improving the efficiency of the production line, but rather toward improving the operations of the business as a business. But overall, the technical and educational assistance differed from classical technology transfer in at least three ways. First, the targets were frequently not business firms but educational institutions or not-for-profit organizations. Second, the ultimate source of the transfer was often a philanthropic organization and not a business firm, educational institution, or government,[4] although these often played a crucial role as the intermediaries in actually performing the transfer. Third, the objective was not competition in the sense in which profit-making enterprises compete (although the objective was competition to the extent of trying to make the East European societies and economies more competitive, that is, allowing them to catch up with the West). As a result, the transferers of technology did not worry about inadvertent spillovers to parties other than the immediate target,[5] which in a business context might create unwanted competition. On the contrary, they were hoping to maximize spillovers as much as possible.

It has been noted that it is desirable for technology transfer to be managed and that management strategies may well be different, depending on the type of transfer that is occurring, that is, whether it corresponds to the point-to-point model or the diffusion model (Williams and Gibson, p. 44). It is fair to say that there was no central conception of managing the technology transfer in Eastern Europe at all; it just happened, largely as the result of the uncoordinated actions of many players of highly disparate sizes. What is noteworthy is that diffusion often developed in almost classical ways, and nowhere is this seen more clearly than in library automation. Initially, many potential users were wary because of the high perceived costs of automation, both in terms of money and effort, and the uncertain benefits. But the library

benefits libraries; acquisitions of materials, library automation, and training are the most important activities under this heading. Among the geographic headings, the only one that requires an explanation is the category "other"; this refers to either grants that have benefited more than one country, or to the relatively few grants that were for the benefit of Estonia, Latvia, Romania, or Russia.

Grantmaking geared up rapidly in 1990 and 1991, achieved steady and high-level expenditures in 1992, 1993, and 1994, and gradually tapered off after that. The largest category of grantmaking was in libraries, accounting for 32.5%, with management education and other academic grants second and third, with 16.9 and 16.2%, respectively.

In table 2-2, we show the number of grants made, as distinct from the dollar amounts of grants. The number of grants was high through 1993, then dropped to a lower level for the next three years (even though the dollar volume of grantmaking was still high in 1994), and finally tapered off after 1996 (only five grants were made in 2000). A total of 790 grant requests were declined from 1992 to 1999 (grants that were declined were not recorded in 1990 and 1991). Hungary, Poland, Slovakia, and the Czech Republic account for only 315 of these, Russia and Ukraine for 190, and all other countries for the rest. The temporal variation reflects the fact that when the Foundation let it be known in 1995 that it was beginning gradually to diminish its grantmaking in the region, potential grantees quickly understood that they were better served by applying to other sources.

Table 2-2 Number of grants made.

	1990	1991	1992	1993	1994	1995	1996	1997	1998	1999	2000	Total
Bulgaria						1						1
Czech Republic				7	2	5	3	1			1	19
Czech and Slovak	11	12	16	6	1	1		1	1	1		50
Estonia							1	1	1	1		4
Hungary	18	10	11	11	12	9	9		2	1		83
Latvia							1	2	2	2	2	9
Poland	6	13	12	15	10	12	12	3	4	1		88
Romania		1										1
Slovakia				2	6	5	7	1		1	1	23
NIS		1	1					1				3
Multicountry	6	5	13	8	5	4	5	3		1	1	51
Total	41	42	53	49	36	37	38	13	10	8	5	332

Other Foundations

It may be instructive to look briefly at the scale and the type of activities that other donors to the region engaged in. We shall not endeavor to be exhaustive, but merely to provide the flavor of the assistance that was provided. In particular, we note that comparable data, collected and recorded on an identical or even roughly similar basis, is not obtainable for all foundations. It is difficult to analyze the content and size of individual grant categories, because we would have to classify particular grants under more general headings, which is often not easy from the brief descriptions given. Second, foundations do not classify their grants under the same headings; thus, what may be "democracy development" for one foundation, may well be "political reform" for another, which sound similar; yet foundation A may classify something as "democracy development" which foundation B would not classify as "political reform." Third, the periods for which data are available are not complete, and often data are reported for periods that are fractions of a year or periods of more than 12 months, overlapping more than one calendar year. Even the imputation of a grant to a particular country may be inaccurate when the grant deals with more than one country: in classifying the grant it may be "split" between the countries or, occasionally attributed to both countries, which is likely to cause double-counting. Finally, in the present section we are not concentrating on any one period. Thus, the focus is neither on the early period nor the later years; rather we report what we could find. It is therefore important that the data reported in what follows be regarded as suggestive, even illuminating, but not definitive. The overall scale of foundation activity is suggested by the fact that from 1989 through 1994, foundations gave over $450 million for Central European projects.[11]

The Soros Foundation(s). Most foundation activity in Eastern Europe prior to 1989 was on a smaller scale than after the turning point and some was episodic, with the exception of the activities of George Soros, who viewed his primary mission as opening up closed societies.[12] He began in 1980 by supporting East European dissidents with scholarships and also assisted Solidarity in Poland and Charta 77 in Czechoslovakia. It was not until 1984 that he formally established the Soros Foundation in Hungary, and because the Communist government was still very much in control, in the early years expenditures also had to be approved by the Hungarian side. While the budgeted expenditures were not all spent in the first few years, the level of the early annual budgets for the Hungarian foundation was an impressive $3 million and the operations of the foundations were extremely successful: "The foundation enabled people who were not dissidents to act, in effect, like dissidents. Teachers, university professors, and researchers were able to indulge in their nongovernmental activities while keeping their jobs. So it was really a very successful operation and a wonderful spirit prevailed.

Nothing that we have done since quite compares with it," (Soros, pp. 119–21).

The activities of the Soros Foundations are administered by a complex set of organizations. First, after 1989, "country foundations" were set up in many countries, which have names such as "Open Society Institute–Azerbeijan" or "Open Estonia Foundation" or "Soros Foundation–Hungary," and a very substantial volume of activity is centered on these. In addition, regional programs are administered by the Open Society Institute in Budapest and U.S.-based activities are administered by the Open Society Institute in New York. In 1998, there were 30 country programs, foundations, or initiatives in existence, as well as the various Open Society Institute programs, other International Initiatives, and the Central European University (CEU).

CEU, a remarkable initiative, was established in 1990 in Prague, opened additional sites in Budapest and in Warsaw, and eventually moved the bulk of its operations to Budapest, with some remaining in Warsaw. It now has seven departments and four interdisciplinary programs in fields such as economics, environmental sciences, history, legal studies, medieval studies, political science and sociology. Since 1996, the Board of Regents of the State of New York has recognized the university as a doctorate-granting institution.[13] In the 1998–99 academic year, CEU enrolled 710 students from almost 40 countries, with the majority in master's programs and some in doctoral programs.[14] The scale of the Soros Foundations' activities can be seen in Table 2-3. We show the totals for the country foundations, followed by the totals for the Open Society Institute–New York. From the total for OSI-Budapest, we break out on separate lines and set in italics the Higher Education Support Program (HESP) and the Library Programs (included in the total of OSI–Budapest), since these programs are of greatest relevance for

Table 2-3 Soros Foundations activity in thousands of dollars.[a]

	1994	1995	1996	1997	1998	Total
Total Country Found	139,010	146,561	173,188	238,281	280,161	977,201
OSI–NY	36,593	81,098	87,791	57,249	118,455	381,186
HESP	*8,691*	*12,043*	*14,614*	*16,080*	*21,588*	*73,016*
Library Programs	*3,238*	*2,084*	*2,444*	*2,742*	*3,349*	*13,857*
OSI–Budapest	22,734	33,632	80,693	91,502	137,798	336,359
CEU	39,877	36,629	18,805	15,290	17,700	128,301

General Note:

a. *Building Open Societies Soros Foundations 1994 Report*, New York: Open Society Institute, 1995; *Building Open Societies: Soros Foundations 1995 Report*, New York: Open Society Institute, 1996; *Building Open Societies: Soros Foundations Network 1996 Report*, New York: Open Society Institute, 1997; *Building Open Societies: Soros Foundations Network 1997 Report*, New York: Open Society Institute, 1998; *Building Open Societies: Soros Foundations Network 1998 Report*, New York: Open Society Institute, 1999.

parallel Mellon activities. The last row shows the expenditures on the Central European University. Total Soros expenditures in the 1994–1998 period amounted to $1,823,047,000.

Typical program categories for the country foundations are education, children and youth, media and communications, civil society, human rights and humanitarian aid, science and medicine, arts and culture, economic restructuring, and legal reform and public administration. Over a five-year period, these activities have consumed nearly $1 billion, with OSI and CEU activities accounting for another $845 million. No other foundation comes near this level of expenditures.

The Ford Foundation. The Ford Foundation has made grants in Eastern Europe since 1957, although substantial activity began only in the late 1960s. The development of its grantmaking exclusive of the former Soviet Union is shown in table 2.4.

Poland received one half of all the grant moneys and grantmaking reached its peak in 1996. The Ford Foundation had a largely consistent set of objectives in its grantmaking; while the early years saw some grants made to alleviate rural poverty, the objectives that were regularly funded were international affairs, human rights, civil and political liberties, peace and social justice, governance and public policy, and in the last two years also a category labeled "education, media, arts, and culture." An illustrative list of grants is the following:

- *European Roma Rights Center*, Hungary, $600,000.
- *Slovak Foreign Policy Association*, Slovakia, $120,000.

Table 2-4 Ford Foundation grants in thousands of dollars[a] (excluding Soviet Union).

	Hungary	Poland	Czech Republic	Slovakia	Other	Total
1990	100	371			75	546
1991		500				500
1992	301	494		63	2,857	3,714
1993	444	498	100			1,042
1994		1,402	243	125		1,770
1995	400	1,409	696			2,505
1996	119	3,618	875	293	2,074	6,979
1997	595	1,506	340	150	615	3,207
1998	810	2,249	165	587		3,811
Total	2,769	12,047	2,419	1,218	5,621	24,074

General Note:

a. *Approved Grants and Projects*, Ford Foundation. Row totals do not necessarily add up due to rounding.

- *Jagiellonian University*, Kraków, $54,500.
- *Research Center for Marketization and Property Reform*, Poland, $300,000.
- *Invisible College Foundation*, Budapest, $120,000.
- *Stefan Batory Foundation*, Warsaw, $186,000.
- *CERGE Foundation*, Prague, $750,000.

The Charles H. Mott Foundation. For the Mott Foundation we report data for the period 1988–99. The aggregate expenditures for Central and Eastern Europe and for Russia (for 1988–90) and for the Former Soviet Union (1993 through June 30, 1999) are shown in table 2-5.

The primary headings under which the Mott Foundation has made grants are the nonprofit sector, citizen rights and responsibilities, race and ethnic relations, political and economic structures and special projects. The fundamental objective behind these activities was to strengthen civil society through technical assistance and institution building.[15] Since the Mott Foundation has made many multicountry grants, the classification of projects is fairly complicated.

Among its grants is one for $1.8 million to Partners for Democratic Change for conflict management centers in CEE and Russia, and grants of $1,942,500 to the Eurasia Foundation for NGO development in Russia.

The nonprofit sector received the lion's share of grants in the Visegrad countries; the largest of these was to the German Marshall Fund of the United States ($1,564,500) for a program called Environmental Partnership for Central Europe, which has the objective of improving environmental policies in the region by assisting local NGOs. To illustrate the flavor of grants in each of the several categories, we list the following:[16]

- *Central Europe Institute*, Košice, Slovakia, $75,000 to assist five communities in the Czech and Slovak Republics create chambers of commerce for enabling business enterprises to advance their goals and community well-being.
- *Foundation in Support of Local Democracy*, Warsaw, $109,545, to provide technical assistance to several Polish communities interested in community foundations, including networking with other Polish initiatives involved in promoting community philanthropy. (See also chapter 12)
- *Polish Ecological Club—National Board*, Kraków, $140,000, "to support a network of environmental NGOs in seven CEE nations established to influence environmental policies and actions undertaken by multilateral development banks.
- *Econnect*, Prague, $79,500, to strengthen an innovative internet services organization that provides services to NGOs, toward a goal of helping smaller NGOs and creating a model of NGO internet use that is applicable to other countries.
- *Reflex Environmental Protection Society*, Győr, Hungary, $25,000, to work with local government, citizens, and technical experts to assess the impacts and generate solutions for problems caused by the diversion of the Danube River in the region.

Table 2-5. Aggregate spending by Mott Foundation.[a]

1988	$35,000	1994	$4,598,000
1989	$478,000	1995	$6,840,000
1990	$1,000,000	1996	$7,765,000
1991	$1,668,000	1997	$9,457,000
1992	$1,972,000	1998	$10,231,000
1993	$3,870,000	1999	$10,800,000[b]

General Note:

a. *Mott Foundation Civil Society Grants in Central Eastern Europe/Russia and Central Eastern Europe/FSU*, Flint, MI: *The Charles H. Mott Foundation*, 1999.

b. Projected figure.

The German Marshall Fund of the United States. The Fund has concentrated its activities on civil society, political reform and development, economic reform, the media, environmental reform, fellowships, and policy and opinion leaders. Political development and environmental reform were clearly important pillars of the Fund. A 1995 document[17] describes the political development program as (1) providing skills in cooperative planning, conflict resolution and community organizing; (2) encouraging public participation in local and national decisionmaking; and (3) promoting cooperation between governmental organizations and NGOs. The document describes the environmental reform program as supporting institutional capacity building by assisting NGOs in addressing environmental problems and improving their ability to assess needs. Grantmaking in the Visegrad countries from 1994 to 1998 totaled $7.7 million. A sample of the grants made by the Fund gives the flavor of the projects that it supported (see also table 2-6):

Table 2-6 Visegrad country projects in thousands of dollars 1994–98.[a]

Year	Poland	Hungary	Czech Republic	Slovakia	Multi-country	Total
1994	464	121	78	151	68	1,787
1995	341	161	50			1,572
1996	548	165	61	70		1,468
1997	724	20	45	45	174	1,665
1998	236	18		181	409	1,210
Total	2,133	485	234	447	651	7,702

General Note:

a. *German Marshall Fund of the United States Annual Report* for 1994, 1995, 1996, 1997, 1998. The column "Total" indicates the total for all regions, not only the Visegrad countries.

- *American Journalism Center*, Budapest, $27,500, To fund the second of a three-month internship in the United States for five Hungarian journalism students in the ELTE Media Studies Center program."[18]
- *Foundation in Support of Local Democracy*, Białystok, Poland, $168,570, To support the expansion of DIALOG in five cities in Poland. DIALOG/ Białystok will provide overall coordination and direct technical assistance to the other five cities.[19]
- *Hungarian American Coalition*, Washington, D.C., $20,000, "For a NATO Education project in Hungary to involve and inform citizens and political leaders through conferences, exhibits, polls, publications and the media."[20]
- *BORIS*, Warsaw, $22,000, To support the Minnesota Council on Nonprofits and BORIS in establishing the Center for Social Mobilizations to aid cooperation between NGOs and local governments.[21]

DIALOG and BORIS, mentioned in two of the above grants, are among the important projects supported by the Fund. Both of these projects attempt to broaden the reach of civil society and to create links between the citizenry and local governments that are more personal. In reviewing the DIALOG project, Peter Szanton wrote,

> I had little doubt of the importance of DIALOG's goals. They were two: to encourage citizens to respond actively to problems that concerned them; and, through such responses, to build—or rebuild—civil society. A society in which ordinary citizens trust each other, organize voluntarily to achieve common ends, expect local government to respond to their needs and participate generally in the public life of the community.[22]

BORIS, in turn, is a Polish NGO created in 1992 for the purpose of improving the ability of community-based NGOs dealing with the provision of social services to have satisfactory links with local governments. BORIS is intended as a mechanism that will help overcome the lack of skills in NGO management, the potential mistrust that may exist among different NGOS, and will help create links among them.[23]

The Pew Charitable Trusts. Pew made relatively few grants prior to 1990. Like the Carnegie Corporation, it makes grants only through U.S. intermediaries and its grantmaking has stressed economic reform, civic education, higher education, and the development of democratic institutions. It has provided training, technical assistance and workshops, and funds for research (Quigley, p. 46) Eastern Europe fell into the Public Policy Program whose its formal objectives were as follows: (1) to provide current and future leaders with critical skills and knowledge that will facilitate the process of democratic, market-oriented reform; (2) to promote the establishment or further development of nongovernmental public policy centers that provide independent sources of policy advice; (3) to advance policy research on issues central to the reform process; and (4) to promote greater understanding of the values essential to democracy.[24] The activities of Pew after 1989 are

Table 2-7 Pew projects from 1990 to 1997 in thousands of dollars.

	1990	1991	1992	1993	1994	1995	1996	1997	Total
Poland	115	450	200	325	2,065	250			3,405
Hungary	65	245	60	735	50			200	1,355
Czech Republic	425	100		500	85				1,110
Slovakia				50	150	650			850
Czech & Slovak			350	200	450				1,000
CEE	1,145	1,255	2,150	2,000	5,320	400	3,015	1,500	16,785
FSU	200	100	375	475		200	250	500	2,100
CEE & FSU	200		300	275		500			1,275
Baltic		570	200	1,050	370	1,000	300	200	3,690
Other	405	3,425	2,400	4,680	650	1,415	525	4,200	17,700
Total	2,555	6,145	6,035	10,290	9,140	4,415	4,090	6,600	49,270

summarized in table 2-7.[25] The category "Other" represents grants in which a country of countries could not be identified unambiguously. In this category, $6.3 million was for the Pew Economic Freedom Fellows program, which provided intensive training for midcareer fellows of high promise. Pew's support leaned in the direction of NGO and environmental support to a lesser extent than that of the Mott Foundation or the German Marshall Fund, and more in the direction of technical assistance, policy analysis, and related matters. A brief and illustrative list of grants is below.

- *California Institute of Technology*, $200,000, To develop policy proposals for further currency convertibility in Eastern Europe and the Soviet Union.
- *Columbia University*, $200,000, To upgrade the Economics department at Warsaw University through an integrated program of faculty workshops, joint research, and fellowships. (See also chapter 6)
- *American Council of Learned Societies*, $50,000, In support of the Fourth Bratislava Symposium on Constitutionalism and Politics.
- *Resources Development Foundation*, $125,000, In support of a technical assistance project to help create a Municipal Infrastructure Fund in Poland. (See also chapter 12)
- *University of Pittsburgh*, $500,000, In support of faculty and student exchanges as a part of a strategy of building the institutional capacity of the Center for Economic Research and Graduate Education in Prague. (See also chapter 6)

The Rockefeller Brothers Fund. RBF has been active in Eastern Europe since the early 1980s. Much of its early interests focused on questions of East–West security issues and it contributed substantially to the creation of the Institute for East–West Security Studies. It invested heavily in supporting projects in the area of sustainable resource use and more recently, in fostering the development of the NGO sector. Finally, RBF was committed to improving communications and relations among Hungary, Poland, the Czech Republic, Slovakia, and their eastern and southern neighbors. The summary totals of grants from 1989 through 1999 are given in table 2-8. RBF grant activity is characterized by considerable steadiness in the amounts allocated to individual program categories as well as to the total grant making in the region. Some illustrative grants over the years are as follows:

- *Institute for Sustainable Development*, Warsaw, $225,000, start-up support for the institute. Its mission will be to promote economic and social development in Poland in the context of sound management of natural resources.[26]
- *The Czechoslovak Management Center Foundation*, Prague, $300,000, to establish the Czechoslovak Management Center, the first independent management training center in Czechoslovakia. The center, formed in response to the urgent need for managers with market economy skills, will offer a Master of Business Administration program, advanced management seminars, business English, internships in Western Europe and North America, and special courses for government administrators.[27]
- *Institute for Environmental Policy*, Prague, $150,000, to launch the Institute for Environmental Policy, the first independent policy research institute established by Czechoslovaks to provide policy analysis regarding issues of sustainable development.[28]
- *International Management Center Foundation*, Budapest, $150,000, continued general budgetary support for the center, established in 1988 as the first Western-style business management training center in East Central Europe.[29]
- *Environmental Training Project Slovakia Foundation*, Bratislava, $26,000, for an assessment of the viability of community philanthropy in Slovakia, in an effort to enlist indigenous sources of support for the local nonprofit sector.[30]

Table 2-8 RBF grant activity, 1989–90, in thousands of dollars.

	Security East/West	Resource Use	Nonprofit Sector	Total
1989	560	575	35	1,170
1990	760	800	85	1,645
1991	860	900		1,760
1992	860	1,075		1,935
1993	860	1,075	242	2,177
1994	860	1,050	200	2,110
1995	535	1,030	596	2,161
1996	190	1,000	595	1,785
1997		1,075	600	1,675
1998		1,300	597	1,897
1999		1,400	600	2,000
Total	5,485	11,280	3,551	20,316

Other U.S. Foundations. It is impossible to list all the American foundations that have had some involvement in Eastern Europe; but for the sake of comprehensiveness we mention a few others. First, the John D. and Catherine T. MacArthur Foundation has concentrated its efforts on the countries of the former Soviet Union.[31] It has supported human rights initiatives, environmental programs, higher education in economics, market reform, energy efficiency, and related projects. Its 1997, expenditures in this area totaled $3,054,868. Two of its most significant grants were to the New Economic School ($750,000) and to the Moscow School of Social and Economic Sciences ($420,000). It has also supported IREX, human rights programs such as Helsinki Watch, and the Helsinki Citizens Assembly and the European Studies Center in Prague (Quigley, p. 40)

The Carnegie Corporation has funded, through U.S. intermediaries, mostly training programs in Hungary and Slovakia, has recently initiated programs in Russia and other post-Soviet states, and

> will make a modest contribution toward strengthening Russia's ability to prepare a new generation of leaders in sectors of the community that are critical in building a new Russia. With respect to policymakers, leaders in banking, business and finance, and upper-level military personnel, the Corporation will support discrete projects that foster exchanges with U.S. counterparts. . . . [another] initiative will address the deteriorating conditions of scholarship in the humanities and social sciences in the former Soviet Union. There is an urgent need to nurture a new generation of scholars and scientists in the post-Soviet states . . . the Corporation will assess the needs of the social sciences and humanities and seek ways in which . . . it can strengthen institutions of higher learning and research . . . [32]

A program of fellowships for humanist is in place and is administered by the American Council of Learned Societies.

The Lynde and Harry Bradley Foundation has supported the Institut für die Wissenschaften vom Menschen in Vienna, an organization that does research on East European issues, the Central Europe Institute (also supported by Mellon) and some others (Quigley, p. 26.)

Concluding Comments. While the figures in the preceding tables are not directly comparable because they do not always cover the same countries, they are adequate for establishing that foundation activity in Eastern Europe was very substantial between 1990 and 1998. The foundations' grantmaking covered an extraordinarily wide range of initiatives, and it is particularly interesting to note that many foundations discovered—sometimes by accident and sometimes by design—that their views concerning the value of certain donee organizations were shared by other foundations. The table 2-9 shows the cumulative grant levels by the various foundations to a particular sample of donee organizations where, for the sake of comparison, we also

cesses of democratic governance and improving public administration, and for the U.S. Information Agency (USIA). USIA itself conducted a number of programs such as establishing and maintaining American libraries in East European countries and funding the National Endowment for Democracy, which funded projects on pluralism and democratic governance, education, culture, research, and international cooperation.[38] Support in this category also includes a variety of educational initiatives, such as educational reforms funded through USIA. USAID was the largest funder of the American University in Blagoevgrad an outstanding liberal arts college on the American model. (Bulgaria) with $8.067 million. In the education area, management and economics training, English teaching, book donation programs, and training for media managers, editors, and reporters figured importantly in its program.[39]

The U.S. government–funded programs for strengthening democracy are reviewed articulately by Paula R. Newburg and Thomas Carothers.[40] They find that most of the programs were run through American intermediaries, which had the advantage of bringing many knowledgeable and committed Americans to the region, but had the disadvantage that they often imported structures and approaches peculiarly American in flavor. The American programs were welcome in Romania, where they were deemed grossly inadequate in volume, and were regarded as unnecessary and perhaps patronizing in the Czech Republic. On balance, they find that these programs did not have a dramatic effect.

Economic Restructuring. Support under this heading consists of macroeconomic assistance, privatization, improving the business climate, technical assistance to enterprises, administering the Enterprise Funds which provided direct capital for enterprises in the region,[41] and improving human resources, specifically management and economics training. Support was also given to improving telecommunications and energy efficiency, to creating an understanding of antitrust as well as bankruptcy issues, and to reforming financial intermediaries. This latter objective was pursued by USAID through support for the Financial Services Volunteer Corps which specializes in providing high-level advice for central and commercial bankers, insurance programs, and pension funds, as well as for the Institute for EastWest Studies, which created and maintained the Banking Finance Assistance Center in Budapest for related purposes (see also chapter 12).

Improving the Quality of Life. Assistance in this category focused on a number of areas such as child health (Romania), war victims (Bosnia and Croatia), various NGO-oriented humanitarian grants (numerous countries), environmental protection (numerous countries), air and water improvement in the Kraków region,[42] the manifold activities in the region of the World Wildlife Fund, "hot spots" in Estonia, forestry in Albania, and so on.

Tables 2-10 and 2-11 provide a statistical overview of the program from for the fiscal years 1990 to 1996 for the Visegrad countries.[43] Slovakia re-

Table 2-10 Fiscal year obligations in thousands of dollars.

	Czech Rep	Hungary	Poland	Slovakia	Total
1990	746	7,874	259,583	584	268,787
1991	28,270	51,027	108,630	20,105	208,032
1992	35,712	46,821	159,793	28,335	270,661
1993	17,852	30,892	66,933	14,137	129,814
1994	38,460	42,951	129,432	44,370	255,213
1995	19,935	27,342	83,514	29,654	160,445
1996	6,027	20,785	48,943	17,346	93,101
Total	147,002	227,692	847,828	154,531	1,386,053

ceived the largest amount of assistance per capita of population, the Czech Republic the smallest, with Hungary and Poland in the middle receiving approximately the same level of per capita assistance. This is probably quite compatible with the general perception that the Czech Republic was economically the strongest of the four, and while the form of Communism there was of a very rigid and repressive variety, it was the only country that had lived under a parliamentary democracy in the years between the two world wars. Economic restructuring was the most favored functional category, whereas the strengthening of political institutions received the smallest amount of assistance.

Funding by European Countries and the European Union

Some of the most important European sources of funds for Eastern Europe were the European Bank for Reconstruction and Development, the PHARE Program, the TEMPUS Program, and European foundations.

The European Bank for Reconstruction and Development (EBRD). While much of our concern here is with assistance programs that aim at social and political reform and higher education, it is important to mention EBRD, which has provided massive assistance to the economies of the region. The level of assistance is shown in table 2-12.[44]

Assistance peaked in 1993, and the form of the assistance was either loans or equity positions. A sampling of projects supported is as follows:

- Čokoládovni, a food manufacturing company in Czechoslovakia; equity position in the amount of 24.1 million ECU in 1991;
- Westel Radiotelefon Kft, expansion of the national mobile cellular telecommunications system in Hungary; loan of 8.25 million ECU in 1992;

Table 2-11 Fiscal year 1990–96 obligations in thousands of dollars by category.

Main Category	Subcategory	Czech Republic	Hungary	Poland	Slovakia	Total
Democratic Institutions	Democratic Pluralism	3,237	6,609	8,806	7,995	26,647
	Political Process	3,675	5,278	20,529	1,950	31,432
	Subtotal	*6,192*	*11,887*	*29,335*	*9,946*	*57,360*
Economic restructuring	Macro support	9,999	10,000	199,227	5,001	224,227
	Privatization	34,596	40,558	191,115	34,508	300,777
	Business Climate	3,945	8,046	21,100	6,110	39,201
	Investment & Trade	29,765	75,018	263,592	40,513	408,888
	Human Resources	17,809	20,827	38,794	12,795	90,225
	Agriculture	5,248	9,627	32,880	5,117	52,872
	Energy	7,674	7,743	31,264	5,731	52,412
	Subtotal	*109,038*	*171,822*	*687,974*	*109,777*	*1,078,611*
Quality of Life	Emergencies	2,718	3,415	10,917	2,825	19,875
	Social Safety	224	4,696	18,954	1,042	24,916
	Housing	7,885	10,726	36,620	8,626	63,857
	Health	8,428	7,279	9,101	8,618	33,426
	Environment	10,326	12,101	33,476	12,350	68,253
	Subtotal	*29,582*	*38,219*	*109,071*	*33,463*	*210,335*
Miscellaneous		1,472	5,765	21,451	1,346	30,034

Table 2-12 EBRD funds committed in millions of ECU.

	1991	1992	1993	1994	1995	1996	1997	1998
Czechoslovakia	31.6	147.1						
Czech Republic			216.3	202.1	94.0	50.3	405.4	67.0
Hungary	98.0	150.3	418.4	166.8	406.3	133.3	99.7	40.9
Poland	88.9	403.1	233.6	350.3	106.5	234.5	294.9	272.3
Slovakia			237.9	89.3	41.0	93.6	19.9	31.5
Total	218.5	700.5	1106.3	808.5	647.8	511.7	819.9	412.7

- Huta Sandomierz–Pilkinton, a joint venture to privatize a sheet glass manufacturer in Poland; loan of 27.4 million ECU in 1992;
- International Road Corridor Project, to improve the condition of the road network in Slovakia; loan of 15 million ECU in 1993;
- Rema 1000, to establish a grocery franchise chain in the Czech Republic; loan of 11.8 million ECU in 1996;
- EBRD took a significant equity position in the Hungarian Foreign Trade Bank in the summer of 1994, which marked the onset of bank privatization in Hungary.

Essentially all projects funded by EBRD are of this variety, which sharply differentiates its assistance from that of the PHARE Program.

The PHARE Program. One of the largest assistance programs for Eastern Europe is the PHARE program, established by European Council Regulation 3906/89 in 1989. (The corresponding program for the Newly Independent States and Mongolia is Tacis, which has committed 2,804 million Ecu from 1991 to 1996, but will not be covered in more detail here.[45]) PHARE was originally designed to promote social and economic reform in Hungary and Poland. It is currently active in two groups of countries: those that have applied for membership in the European Union,[46] and those that have not.[47] In the former, the principal objective is to prepare the countries' accession to EU, while in the latter, PHARE continues promote the transition to democracy and a market economy.[48] Accession criteria for the Central and East European countries agreed to at the Copenhagen European Council in 1993 were as follows:

- Stability of institutions guaranteeing democracy, the rule of law, human rights, and respect for and protection of minorities;
- The existence of a functioning market economy as well as the capacity to cope with competitive pressure and market forces within the Union;
- The ability to take on obligations of membership including adherence to the aims of political, economic and monetary union.[49]

The accession program for the first group of countries has two main parts. The first is institution building, that is,

Table 2-15 Participation in JEPs, 1990–1996.

Column 1: Western Partner	Column 2: JEPS	Column 3: % of Column 2	Column 4: % of EE country JEPS
Austria[a]	69	Hungary: 30%	Hungary: 7.2% Slovakia: >25%
Belgium[b]	373	Poland: 30%	Uzbekistan: 50% Czech Rep., Hungary, Romania: 33%
Denmark[c]	69	Poland: 37% Hungary: 21%	Lithuania: 55%
Finland[d]	84	Poland: 22%	Estonia: 42% Latvia: 29% Lithuania: 20%
France[e]	586	Poland: 33% Hungary: 23%	Kyrgyzstan: 100% Romania: 72%
Germany[f]	674	Poland: 32% Hungary: 29%	Hungary: >50% Slovakia: >50% Lithuania: >50%
Greece[g]	219	Poland, Hungary, Bulgaria, Romania 16–27%	Albania, Moldova: 46–50%
Ireland[h]	186	Poland: 38%	Poland: 14% Slovakia: 23%
Italy[i]	390	Poland: 27%	Albania: 67%
Netherlands[j]	430	Hungary: 33% Poland: 29%	Lithuania: >38% Slovakia: >38%
Portugal[k]	158	Poland: 36% Hungary: 23%	Czech Republic: 30%
Spain[l]	313	Poland: 30% Hungary: 28%	Czech Rep.: >25% Hungary: >25% Romania: >25%
Sweden[m]	91	Poland: 33% Hungary: 20%	Estonia: >30% Latvia: >30% Lithuania: >30%
United Kingdom[n]	869	Poland: 34%	Czech Republic: 80% Belarus: 67%

Notes:

a. http://europa.eu.int/en/comm/dg22/tempus/eumember/austria/austria.htm
b. http://europa.eu.int/en/comm/dg22/tempus/eumember/belgium/belgio.htm
c. http://europa.eu.int/en/comm/dg22/tempus/eumember/denmark/denmark.htm
d. http://europa.eu.int/en/comm/dg22/tempus/eumember/finland/finland.htm
e. http://europa.eu.int/en/comm/dg22/tempus/eumember/france/france.htm
f. http://europa.eu.int/en/comm/dg22/tempus/eumember/germany/germany.htm
g. http://europa.eu.int/en/comm/dg22/tempus/eumember/greece/greece.htm
h. http://europa.eu.int/en/comm/dg22/tempus/eumember/ireland/ireland.htm
i. http://europa.eu.int/en/comm/dg22/tempus/eumember/italy/italy.htm
j. http://europa.eu.int/en/comm/dg22/tempus/eumember/netherl/nether.htm
k. http://europa.eu.int/en/comm/dg22/tempus/eumember/portugal/portugal.htm
l. http://europa.eu.int/en/comm/dg22/tempus/eumember/spain/spain.htm
m. http://europa.eu.int/en/comm/dg22/tempus/eumember/sweden/sweden.htm
n. http://europa.eu.int/en/comm/dg22/tempus/eumember/uk/uk.htm

partner has substantial involvement in Poland and/or Hungary (column 3), and that East European countries receive very substantial assistance from those western countries that have historical relations with or geographic proximity to them (column 4). Thus, the Baltic countries' JEPs place heavy reliance on the Nordic countries, Hungary and Slovakia have some dependence on Austria and Germany, Albania is involved in projects with Italy, and so on. In order to give the flavor of the JEPs, we list a few below.

- The University of Salzburg, Austria, is implementing a project which aims to develop a new course in environmental sciences at Jagiellonian University in Kraków.
- The Karl-Franzens University in Graz, Austria, participates in a project at Janus Pannonius University in Pécs, Hungary, to reform legal studies at the Ph.D. level and to develop a new master's degree in law.
- Strategic development of the Academia Istropolitana in Bratislava, Slovakia, includes performing quality assessments of academic programs, obtaining a legally recognized status, and developing its administrative structure.
- A network of education in human rights in Poland and other countries is being developed, in order to strengthen the rule of law and to foster economic and social development.
- The University of Saarland (Germany) is undertaking a project of assistance to introduce courses in business information systems at Warsaw University and the Academy of Economics in Poznań, Poland.
- Education in romance languages and cultures at Charles University in Prague is being modernized.
- The University College Dublin is undertaking a project of assistance to establish a course in rural development at Pannon Agricultural University in Keszthely, Hungary.

The British Council. Funded primarily by the foreign and commonwealth office of the United Kingdom, The British Council is, very roughly speaking, the British equivalent to USIS. Its objectives include the provision of educational opportunities for people in other countries and to make the U.K. a partner of other countries in carrying out reforms and promoting sustainable development.[56] It has maintained over 200 information centers and libraries, supported scholarly exchange visits, and has, in particular, supported the development of Hungarian library services. It also manages the British Know How Fund, which is the U.K. program for technical assistance in the countries of Eastern Europe and the former Soviet Union. The first of these funds was established in Poland in June 1989 with a start-up capital of £25 million, which was increased in November of that year to £50 million. Similar funds were established for Hungary and Czechoslovakia in 1990.[57] The Fund has managed many projects in the countries of the region. A brief, illustrative list of projects follows.[58]

European integration: assistance to Polish government ministries to adapt and implement EU Internal market rules, drafting customs regulations, implementing EU health and safety regulations.

Management training: a project to train management trainers in the region of Kraków, concentrating on individuals rather than institutions. The project emphasizes regional development and encourages partnerships among the academic, public and private sectors.

JICAP: the Joint Industrial and Commercial Attachments Program in the Czech Republic brings middle and senior managers to the U.K. for short-term management training. It has similar objectives in Estonia, where the program consists of a 7–10 day introductory course on management and market economy, followed by a three-and-a-half week secondment to British companies for practical experience. In Slovakia, JICAP provides five weeks of management training for Slovak managers.

European Foundations. In the present subsection we give a brief sampling of some of the more important European foundations' activities in Eastern Europe.

- The Volkswagen Stiftung in Germany has supported in Central and Eastern Europe a significant number of research groups in the sciences and mathematics and is now refocusing its attention on support for young scientists, so that they may intensify their research activities. It expects to fund up to 15 new projects annually, with a maximum amount of 300,000 DM per project.[59]
- The Robert Bosch Stiftung in Germany has supported partnerships between East European and German organizations in the area of health care, has supported lectureships in German as a foreign language and in the humanities and social sciences, and has particularly stressed the teaching of German in Poland, the Czech Republic, Slovakia, Estonia, Latvia, and Lithuania.
- The Friedrich Naumann Stiftung, the Friedrich Ebert Stiftung and the Konrad Adenauer Stiftung are German foundations associated with particular political parties and they have, severally, supported building democratic institutions, political education, international understanding, and liberal media.
- The Humboldt Stiftung provided substantial moneys for education and exchanges between 1989 and 1994, particularly in Hungary, Poland, and Czechoslovakia; these amounted to $33.56 million.[60]
- The Fritz Thyssen Stiftung has contributed to the establishment of Collegium Budapest, an institute for advanced research along the lines of the Wisssenschaftskolleg zu Berlin, and has given for this purpose $2.6 million.[61]
- The Deutscher Akademischer Austauschdienst (DAAD) has participated in numerous joint projects with universities.

Concluding Comments

The scale of financial assistance by American and European organizations has been massive. Private philanthropy is unlikely in such cases to be able to match governmental assistance in scale and in reach, but what foundations can do and have done is to set themselves fairly precise guidelines and thus have an effect that may be disproportionately large given their size. They may be particularly effective, as Quigley has argued, by being less bureaucratic than government agencies and by being able to act more quickly,[62] an observation also made by Czermiński[63] and by Cochran.[64] But as is clear from the many examples cited in this chapter, there has been an enormous overlap among the program objectives of the many foundations and government agencies. Nearly all of these projects have come about in a completely uncoordinated way. Because of the many agencies involved, the number of potential recipients in each country, and the number of similar objectives, it would have been inconceivable to coordinate even a modest fraction of all these activities. This raises at least three difficult-to-answer questions. (1) To what extent have activities been pursued in such a way *in the aggregate* that the marginal return to the last dollar invested was the same across all activities (which is a necessary condition for the activities having optimal amounts invested in them)? The answer has to be that the relative levels of expenditures in the various activities were almost certainly far from optimal. (2) To what extent have the same organizations and individuals appeared and reappeared as grant recipients while other deserving institutions or individuals have been left out altogether? The answer very likely is that although some worthy potential recipients may well have been forgotten, so many worthwhile organizations have received grants, and often several grants, that it would probably not have been cost-effective to expend the effort to make sure that all worthwhile organizations could participate in western grants. And finally, (3) to what extent have organizations and individuals engaged in "double-dipping," that is, obtained funds from different donors for exactly the same activity? My sense is that there was very little of that, but it is difficult to prove one way or another. But unfavorable answers to any of these questions would point to some efficiency losses or inequities or both in the aggregate support activities of western donors. In spite of these possibilities, the overall effect has clearly been extremely productive and beneficial, although some observers such as Quigley have pointed to serious shortcomings, at least in the programs that aimed at political reforms and enhancements of civil society. As we shall see in subsequent chapters, the Mellon Foundation, in particular, has had a significant effect in promoting economics and management training, in improving the computing environment, and in modernizing and developing research libraries.

3

The Status Quo and Formulation of an Agenda

Competition among Objectives

As I pondered what I had gotten myself into, I began to experience genuine feelings of panic. Would I be able to find the right people at the right institutions in Eastern Europe for developing projects that fell into the Mellon Foundations areas of interest and that would be of substantial benefit to the countries of the region? How many initiatives did I have to encourage and what fraction of those would turn out to be viable? How ambitious could I be about the size of the grants that I would recommend to the Foundation? And what was the total budget for grantmaking in Eastern Europe? In January 1990, I had no answers to these questions, and I had not received many guidelines from the Foundation. But in retrospect, I think this was all right, because it would have been premature to propose firm answers to these questions at such an early stage.

I got some preliminary insights into the "big picture" by reading a report about a December 15–16, 1989 meeting at the Carnegie Corporation[1] and by attending a meeting on January 4, 1990, at the Ford Foundation devoted to a discussion of Czechoslovakia and another meeting on January 11, 1991, dealing with all of Eastern Europe. The first meeting was especially useful for me, a newcomer on the foundation scene, because it allowed me to meet the staff of other foundations, such as Enid Shoettle and Paul Balaran from the Ford Foundation, Colin Campbell and William Moody from the Rockefeller Brothers Fund, Kevin Quigley from the Pew Charitable Trusts, Allan Kassoff from IREX, William Luers, former U.S. Ambassador to Prague, George Soros, and many others. Countless foundations as well as distinguished West and East Europeans had been invited to the third meeting on January 11, 1991. Lord Dahrendorf, at that time warden of St. Anthony's College at Oxford, was a featured speaker. Wendy Luers, president of the Charter 77 Foundation in New York, also played a prominent role at that meeting and asked me to address the group on the Mellon Foundation's approach to providing assistance to the region. I did so, with trepidation;

the final section of my remarks concerning the cost-effective things that foundations can do to support economics and management training was based on the following points:

(a) Training those that will train others is more cost-effective in the long run than training those who will not. Empowering East Europeans to do things for themselves is important.
(b) Short-term training efforts have a value that is directly proportional to the degree that they are specific and have a precise focus, but they may be getting less important overall for the upper and middle management levels.
(c) University-based management and business programs are particularly effective ways of delivering training.
(d) Establishment of regular, university-based business programs benefit from having a U.S. or Western European institutional partner for a limited number of years.
(e) Programs that train central and local government officials are important.
(f) Supporting experts from the west who are seconded to East European government agencies at their request is an effective way of helping with the transformation.
(g) Providing training to East Europeans in North America is generally not as cost-effective as providing it in Eastern Europe.
(h) There is no reason why western providers (that is, teachers) of skills should or need be from the most prestigious American institutions.
(i) By the same token, honoraria for American personnel teaching in Eastern Europe should be moderate.
(j) Conferences are often not cost-effective because of the speakers' predilection to recycle papers from the previous conference; they tend to be more effective if they are addressing very specific issues and have a sharp focus.
(k) The cost-effectiveness of research institutes and think-tanks is directly proportional to the degree that their mandate is specific and focused.
(l) Providing teaching materials (books, PCs, business software) is very useful if done as part of a coherent plan.
(m) We need to be sensitive to what needs to be taught in management training—courses that start out by teaching leveraged buyouts and option pricing formulas are nonsense.
(n) Communication problems in grantmaking are difficult; supporting communications hardware and software, including EARN/BITNET is useful.[2]

By that time, I also believed that whatever American foundations did, it was important that they not concentrate solely on the capital cities, but undertake useful projects in the provinces.[3] Everybody firmly believed that economics and management training were extremely important, partly because the economies of Eastern Europe were going to transform themselves into capitalist economies with privately owned businesses that desperately needed qualified managers, but partly also because experts would be needed to carry out the necessary institutional reforms. They would need to phase out subsidies; privatize the state sector; create a small enterprise sector; eliminate controlled prices; create a social safety net appropriate for the new circumstances; achieve macro stability; introduce currency convertibility; reform tax systems; create a banking sector; and create securities markets. Many other useful things were said at that meeting: that it was important to support independent policy institutes, that quality control was essential, that the southern tier of countries (Romania, Bulgaria, Yugoslavia) was being neglected, that new institutions had to be created, that parliamentary and media training had to be provided, and so on.

I had no doubts that economics education and management training were essential. I had had enough of experience in the previous decade to realize that East European economists did not have the depth of training in microtheory, macrotheory, and econometrics that is not only customary in western graduate schools, but is also essential for doing high-quality research and policy analysis. I recall a dreadful argument in the late 1980s with a Hungarian economist whom I tried to convince that modern statistics permits us not only to determine the influence of one observable variable on another observable variable, but may also permit us to estimate the influence on unobservable variables. For example, if the demand for a good is greater than its supply, then the amount traded in the marketplace, which is an observable variable, will be the supply, and the demand is unobserved. In other words, if all we know is that at the prevailing price consumers would have liked to buy more than they did, we do not in fact observe their demand. But that does not mean that we cannot statistically determine the demand curve. Just as we cannot see a black hole through a telescope but are able to verify its existence and influence on other heavenly bodies by the gravitational force it exerts, so we can make inferences about the unobserved demand without actually being able to measure it. This economist had never heard of latent or qualitative dependent variable analysis—well known among western economists for some time—and he must have thought that I was pulling his leg and making this all up out of whole cloth.[4] Ten years later, it may be difficult to recall the massive unpreparedness of the populations at large to deal with even the elementary concepts of economics and business. Todd Jagerson, who later managed some Mellon-funded training projects under the auspices of the State University of New York in Albany, reported that

> many Hungarians are unaware of even the most basic mechanics of how a market-driven economy works, let alone how to succeed within one. The little they know has often been distorted by the past regime. (An example: In our short course presentation in Budapest last week a great deal of audience interest focused on the question "What is profit?" More surprisingly, this was followed by an intense argument over the question "*Why* profit?" This from a group of aspiring entrepreneurs.) From the

> Hungarian executive who is unable to compute his firm's costs, to the Hungarian worker who sees no connection between hard work and personal self interest, the consequences of this lack of basic knowledge and skills are devastating to the restructuring process.[5]

And Hungarians were among the region's most sophisticated in business matters! Another example is recounted by Tom G. Palmer:

> As an example of the lack of experience with markets, I can recount a conversation I had with a very bright and strongly committed student demonstrator in Prague in early December of 1989. His field of study at the Prague School of Economics was "foreign trade," i.e., the management of trade between the state monopoly firms and foreign firms; I asked him, however, what he wished to do now that so many things were possible that were unthinkable just a few weeks ago, such as choosing what you wished to study or to do after school. He told me that he wanted to become a private businessman and set up a travel office to assist both Czechs to travel abroad and poorer foreigners to find accommodations in Czechoslovakia (where, at the time, lodging was handled by the state travel monopoly, which charged huge sums to foreigners). I asked whether he had considered opening his own hotel or restaurant and he replied that the problem with that was that "there is always a shortage of food." When I suggested that there might be a reason for shortages, and that perhaps a privatized system of food production, distribution, and sales would generate a different outcome, he mused, "A private food store . . . I don't know. How would that work?" He was not opposed to the idea, but he could not imagine how the managers would know what food people wanted to eat, where to get it, how much to charge, and so forth.[6]

In any event, I was grateful that my marching orders were confined to economics and management training and to higher education, more broadly speaking. The number of potential functional areas in Eastern Europe in which useful work could be done was just too large to be undertaken by a one-person program or even by a single foundation. It should be said that we all believed as an axiom that we would be effective, that we would succeed in improving technology and infrastructure and changing people's minds and behavior. At least I was glad that I was charged with functional areas in which practical results might be easier to attain than in more nebulous realms in which the test of success would be some broad changes in Weltanschauung. In any event, I shall briefly discuss some of these potential areas below.

Strengthening Democracy. It was clear that democracy had to be strengthened in the region. The fact that these countries had formally adopted new forms of government, which provided explicitly for free elections, was no ironclad guarantee that former Communists or other malcontents would not want to try to turn the clock back. One need only think of the recent history of Belarus or what remains of Yugoslavia to understand the potential threats to democracy. The concept of civil rights was rudimentary in the region and the abuse of minorities was becoming worse as nationalism was supplanting Communism: Hungarian minorities were at a disadvantage in Romania and Slovakia and the Roma (gypsies) were discriminated against everywhere. Natives of these countries have had a tendency to downplay the importance of racial or ethnic discrimination, but there is no doubt that the problems had been serious for some time. Czech Roma were emigrating in increasing numbers and seeking asylum in the West. Czech newspapers bemoaned the international embarrassment that "Roma hysteria" was causing.[7] In May 1998, in Ústí nad Labem (Czech Republic), a major dispute about noise and disorderliness broke out on Matiční Street between residents of family dwellings and the Roma inhabitants of apartment houses. After protracted negotiations, which reached up to the central government, the mayor announced plans to build a 12-foot-high wall or fence to separate the Roma housing from non-Roma family homes. A wall was actually built, in effect, creating a ghetto, and was torn down only in November 1999, after it was suggested that entry into the European Union might have to be postponed under the circumstances.[8] In the Czech Republic, only 6% of Roma youth are enrolled in secondary schools and 70% are unemployed; in Hungary, only half of all Roma children complete primary education.[9] In Slovakia, the unemployment rate among Roma is close to 100% in some regions.[10] To cite another example, a few years earlier, some conversations I had with a Hungarian convinced me that this was a serious problem in Hungary, as well. One year, while visiting Budapest, I had asked him whether there was any antisemitism in Hungary. He said that there was not; but clearly my question must have made him think about this issue, because a year later, when I visited again, he turned to me and said, "last year you asked me whether there was antisemitism in Hungary. I can now prove to you that I was right saying that there is none. The proof is that Hungarians hate the gypsies even more than they hate the Jews." The building of democratic institutions and of civil society had to be placed high on the agenda and it was generally perceived to have high priority by both private and government organizations preparing to aid the transformation of Eastern Europe.

But a decision to strengthen democracy does not point unambiguously to a single set of activities and does not provide an immediate blueprint. Brack Brown noted that some conservative economists equate democracy with the presence of a market economy, while the political view equates democracy with regular elections, a representative parliament, and a constitution that protects freedoms. According to a civil society interpretation, democracy consists of the freedom of association in economic, cultural and political spheres, with attendant rights to cooperate and criticize the government.[11] Kevin Quigley argued that a broad array of programs could be considered to represent democracy assistance, including assistance to parliaments or judiciary arms of governments, political parties, trade unions, independent media, civil society, civic education, community development, and many others (Quigley, pp. 8–9). The important thing to note is that different philanthropic organizations may have defined for themselves the appropriate target differently, which in part explains the very substantial richness of programs encountered in the region.

A final, cautionary observation made by Schöpflin is that democracy has to be established on two levels: on the level of institutions and as a discourse. One can, more or less, easily create institutions, but that does not guarantee the existence of a discourse, and he notes that "the uneven functioning of institutions, the low level of grassroots organization of political parties, and the seemingly endless political debates tended to be about disenchantment and cynicism."[12]

Reforming the Legal Profession. Not only did legislatures in the region have to enact new commercial codes and other laws, the legal profession itself needed to be drastically reformed. While there were clear differences among the various countries, the past 50 years have stunted the legal profession. There were very few, if any, legal scholars who understood what legal systems, legal thinking, and rule of law are in a western sense, because lawyers were predominantly an arm of the Party and of the State and did not, on the whole, understand such elementary concepts as the western idea of the client–lawyer relationship. In the commercial sphere, it is important that rules of conduct be clear and respected by government, private individuals, and enterprises alike. Transactions also need to be governed by a rule of law, that is, "a system in which the laws are public knowledge, are clear in meaning, apply equally to everyone, and uphold the political and civil liberties often referred to as human rights,"[13] conditions that were far from being satisfied under the Communist system. Lawyers were undoubtedly needed in large numbers to draft new laws for every imaginable sphere of society, and reforming the legal profession is not far from the related problem of enhancing trust in the legal system, which has been significantly lower in post-Communist Eastern Europe than in Western Europe because violation of laws had become routine.[14]

Environmental Awareness and Protection. Communist ideology stressed industrial production and growth, and environmental concerns were perceived to be an obstacle to these objectives. Concern about industrial or automobile emissions was discouraged and as a result, for example, the Kraków region in Poland—especially in Nova Huta because of metal refining—became one of the most polluted in Europe. A black cloud on a street meant that an East German Trabant had passed. In the 1950s, the Hungarian and Czechoslovak governments began to devise a hydroelectric generation project that would divert the Danube at Gabčikovo in Slovakia and at Nagymaros in Hungary.[15] Agreements were signed in 1977 and Austria entered the project as a financial partner in 1985. Environmentalists in Hungary began to express concerns, because the plan would have resulted in lowering the water table over a large area; it would have threatened the habitat of many species of birds and destroyed the beauty of the land in the vicinity of the project. Finally, in 1989, the Hungarian government halted construction at Nagymaros, an act that *The Economist* called "unthinkable" a mere one year earlier. But the Gabčikovo part was completed by the Slovak government and bio-diverse and precious lands abutting the Danube were flooded, to the indignant outcry of environmental groups as well as the Hungarian government. There was a palpable demand coming from the newly democratized countries for environmental change: for raising public consciousness about the importance of the environment and for the introduction of environmental courses at universities. An early concept paper prepared by the German Marshall Fund of the United States argued that an environmental assistance program in the region needed to stress five objectives: (1) the building of institutional capacity by assisting nongovernmental organizations to deal with environmental problems; (2) the transfer of knowledge and technology to individuals; (3) the provision of assistance for collecting relevant information; (4) the building of a consensus between NGOs, industry and government; and (5) the strengthening of the environmental media.[16] And while a substantial consensus about the environment has developed during the first ten years of the post-socialist world, negligence and cynicism continue to take a terrible toll. For example, on January 31, 2000, a gold mine operated by the Australian company Esmeralda Exploration, Ltd. in Baia Mare, Romania, released through negligence vast quantities of cyanide into the Tisza River in Hungary, essentially killing all fishlife in the river and substantial portions of avian life in the area.[17] To add insult to injury, on March 10, another disaster occurred when a dam burst in northern Romania and 20,000 tons of mud contaminated with heavy metals reached Hungary, leading to a ban on the use of water from the Tisza.[18]

Strengthening the Nongovernmental Sector. Progress on the environmental front or in civil rights in the United States owes a great deal to nongovernmental organizations (NGOs), those spontaneous agglomerations of citizens that are essential to the functioning of a civil society. It is also clear that the strengthening of democracy and of civil society is hardly thinkable without a vigorous nongovernmental sector. We might have tried to spend major resources on fostering the growth of this civil society sector that literally did not exist under Communism, because spontaneous groupings of citizens with an agenda were deeply suspect and viewed as a potential threat to the state. Fortunately, some foundations regarded the difficult task of fostering the growth of NGOs as a prime objective, and the Rockefeller Brothers Fund clearly became a leader in this activity and in bolstering environmental concerns.

Strengthening Independent Media. One cannot imagine a democracy functioning without independent media, and the countries in the region had zero experience in this (except for some samizdat literature, which was thriving in at least Poland).[19] Private newspapers, radio and television stations, and responsible journalism as we know it, had been nonexistent. One could have devoted major resources to encourage the growth of independent media.

While all of these objectives were interesting and important, I was glad that the Mellon Foundation had decided to limit is functional interest to economics and management training and the higher education sector more generally. It is not possible to do everything equally well, and I was grateful that my mandate provided me with the opportunity to develop some expertise in a limited number of fields.

Mellon was relatively unusual among foundations in having a strong interest in higher education. At the second Ford Foundation meeting, Allen Kassoff, then the executive director of IREX, spoke briefly and suggested that the foundations' agenda include higher education. While I completely agreed with his remarks, they evoked no sympathetic echo in the audience. In any event, while the former—economics and management training—is reasonably well defined (and I felt that I knew how to encourage and organize such training), the latter, "higher education more generally," was still a very large area. Early in the game I started to receive, for example, proposals for the study of Slavic linguistics, or of interstellar dust, or of nuclear engineering; mostly subjects about which I knew nothing and some about which I knew next to nothing.[20] I realized that I would have to develop a whole army of referees to tell me whether a particular proposal made sense or not, and that did not seem to be a sensible use of my time. Then again, we could have helped higher education with the construction of buildings; fortunately the Mellon Foundation had a policy of never spending money on "bricks and mortar." Out of all these possibilities, the focus was ultimately narrowed to assisting higher education, mostly in two infrastructure areas: computing capabilities and libraries. The concentration on economics and management training, computing capabilities, and libraries was to be observed flexibly and creatively rather than rigidly, and so it was.

Improvements in Education, Culture, and Research. The meeting at the Carnegie Corporation in December 1989 covered a very broad ground and dealt with the areas discussed above; but it also dealt with potential programs in the area of education, culture, and research. Proposed actions by western and East European government and private sector agencies were ably summarized by Robert Beschel.[21] What is remarkable in the report's section on education, culture, and scientific activities is the remarkable paucity of ideas beyond the mundane, albeit useful. Polish and Hungarian proposals recommended financing visits by foreign lecturers in Poland; training domestic teaching staff to implement training programs; organizing apprenticeships for trainees in banks, offices, and brokerage firms; purchasing modern teaching equipment; and expanding the number of scholarships available for students. The European Commission recommended technical training of various sorts. The SEED Act authorized science and technology exchanges ($5.5 million for Poland and $2.5 million for Hungary) and the U.S. Executive Branch established the Alexander Hamilton Fellowships in Management. We must be clear about the fact that all of these activities and proposed programs were useful, but they were piecemeal and none of them came to grips with the fundamental problems of the educational sectors in the region.

But perhaps there was nothing better to be done in the short run than to foster exchanges and provide training. It is clear that foundations, including the Mellon Foundation, found it useful to engage in these otherwise humdrum activities.

Scrambling for Contacts

It somehow became clear that we would not have to make great efforts to generate proposals for economics and management training: most of these would come to us smoothly from American universities and from academics who had spontaneously developed contacts with East European counterparts as early as 1989 or even earlier, and had given serious thought to providing training in the region. The buzzword was "training the trainers," and the many teaching programs in Eastern Europe would permit vast numbers of academics to travel to more or less exotic places.

Such an inflow of proposals was unlikely in the computing or library areas, and librarians were probably less experienced in writing proposals to foundations than were professors. So it was obvious to me that I had to develop contacts at East European universities. I started to write to friends such as Imre Csekő in Hungary (who had spent almost all of 1987 in Princeton on a Fulbright grant) and Vladimír Benaček in Czechoslovakia (whom I had met at the Institute of Forecasting) and asked them to provide me with the names of the rectors of universities, of librarians, and other relevant officials. In the first month or two I wrote an inordinate number of letters to rectors and librarians to introduce myself and the Mellon Foundation to them. With the exception of Dušan Tříška, with whom I had coedited a volume of essays earlier, and František Janouch, the president of the Czech Charter 77 Foundation whom I had met at the Ford Foundation meeting and who lived in Stockholm, I concentrated on Hungary at this early stage, because I knew it best.[22] A typical letter, addressed to the rector of József Attila University in Szeged, was as follows:

> January 17, 1990
>
> Dear Professor Csákány:
>
> The Andrew W. Mellon Foundation, a very large foundation with a long history of supporting educational institutions, is interested in extending its charitable activities into Central and Eastern Europe. I have recently become the Foundation's Advisor for Eastern Europe and I am using this opportunity to inform you that the Foundation would be interested in considering proposals from Hungarian universities for improving those

aspects of university education and university facilities that you find are in greatest need of assistance.

I do not want to seem to have preconceived ideas as to what might be useful to your University. I have recently spoken with the Rector of Karl Marx University, Dr. Csáki Csaba, whom I have known for some time, and he suggested that university libraries in Hungary are often in need of professional journals published in the West, the acquisition of which necessitates foreign exchange that is hard to obtain. He himself will prepare a proposal for us that we fund the acquisition of library materials. But I could well imagine that there are numerous other types of materials and equipment that might be useful; computer hardware and software also come to mind. In any event, I do not want you to be limited by my possibly narrow outlook and would welcome your communicating to me the manner in which the Mellon Foundation might be of assistance.

If we begin a dialogue about these matters and if it should turn out that the Mellon Foundation would want to proceed further, we would want to work out a detailed budget with you. At the same time, we need to obtain certification from you, which should not be difficult in the case of a university, that your institution qualifies under U.S. Internal Revenue Code as a not-for-profit institution, since the targets of the Mellon Foundation's giving have to be restricted to that category. As far as budgets are concerned, I need to tell you that for requests up to $50,000, approval can be fairly swift. Proposals that are larger than that have to be approved by the full Board of Directors of the Foundation, which makes approval in these cases a somewhat more time-consuming project.

I look forward to hearing from you whether we may be of assistance.

Sincerely yours,

The certification of the not-for-profit status of a potential grantee turned out to be a continuing nightmare. Each grantee had to complete a five-page questionnaire, called the Foreign Organization Questionnaire. In addition to inoccuous questions, such as "name of the organization," there were others that were obviously written not in English but in "IRS-ese" such as

> 12. If, during any year, your receipts exceed your expenditures, are you required to apply such excess in furtherance of your tax exempt purposes? If yes, please attach any legal document, legislative act or regulation requiring this, to the extent not already provided.

or

> 13. Are the assets of your organization required to be distributed for charitable, educational, scientific, religious or literary purposes upon dissolution? Please attach a copy of any law, government regulation, court decision and/or any legal documents pertaining to this, to the extent not already provided.

The obvious correct answer to these questions is "yes." However, for reasons that I still do not understand, perhaps four out of every five respondents replied to one or the other or both questions in the negative, irrespective of the respondent's nationality and irrespective of the fundamental linguistic differences between Hungarian and the Slavic languages.[23] I then had to explain that a negative reply to question 12 implied that such "profits" could perhaps be pocketed by the rector for his own benefit, or distributed as dividends to powerful professors, definitely gainsaying the not-for-profit status of the institution. Accompanying the questionnaire, the respondent also had to submit the text of the legislative acts or other administrative regulations that established the tax-exemption of the institution, both in English and in the original language, and also had to provide the past four years' income/expense statements and balance sheets of the institution.[24] This also caused immense problems, particularly in the first few years. When the Foundation got ready to make a grant to the Hungarian Academy of Sciences, and I explained on the occasion of a visit to Budapest that we did need these financial statements, I was told in no uncertain terms that that simply could not be done. When I asked why, I was told "we never had to keep books; when we were running low on money, we would just call a certain telephone number, and next day there was money in our bank account."[25] The Academy ultimately did produce the required statements, but only by expending an enormous effort. In addition, the quality of the financial statements was generally atrocious; many financial reports revealed that their accountants could not distinguish between revenues (a "flow" concept) and assets (a "stock" concept) and the general level of accounting skills was dreadful. This alone convinced me that management training with a healthy dose of accounting was a good thing.

I soon felt that I needed to enlarge my circle of correspondents to cover the other countries as well, and circumstances helped a lot. A Polish visitor to the Institute for Advanced Study in Princeton, Dr. Jerzy Kłoczowski, steered me to the university librarian of the Catholic University in Lublin, Dr. Andrzej Paluchowski. A distinguished astrophysicist at the Princeton astrophysics department, Bohdan Paczynski, put me in touch with a Polish colleague of his, Tomasz Hofmokl, who turned out to play an important role in building up electronic connectivity in Poland. Jonathan Fanton, the president of the New School for Social Research, (now the New School University) frequently hosted dinners for distinguished East European visitors. At one of his memorable dinners I met Radim Palouš, the rector of Charles University (Prague); Josef Jařab, the outstanding rector of Palacký University in Olomouc; and Norbert Frištacký, the rector of the Slovak Technical University.[26] I wrote to all of them and many more, but communication with Eastern Europe was very frustrating. Letters took weeks, sometimes many weeks. E-mail was in its infancy and it was in effect only through the efforts of a devoted band of e-mail enthusiasts, as well as the efforts of the Mellon Foundation, that Central and Eastern Europe was able to join EARN, the European equivalent of BITNET, which was the earliest general-purpose electronic network in the United States.[27] The East European telephone systems were in an execrable state, and faxes did not work well: I suffered hours and days of immense frustration as I tried to send a fax five, ten times before it went through. In addition, some East European offices that I dealt with had the odious habit of turning off their fax machines when their working day ended; thus effectively shutting me off from communicating after 11 AM (EST). Telephone systems started to be upgraded after a year or so, but this often led to new telephone numbers being introduced. In Prague, the previous six-digit telephone numbers became seven-digit or eight-digit numbers (and to confound matters, all three types were valid at the same time, so that the number of digits in the telephone number you dialed provided no check of whether you had the right number or not). Five-digit numbers in provincial cities became six-digit numbers. It was not, and is still not, the custom in Eastern Europe routinely to inform your contacts that your telephone number has changed. Thus, I sometimes tried to telephone or fax to the wrong number for days before I discovered that the number had changed.

The final obstacle to making effective contacts was that philanthropy was basically unknown in the region. Some rectors regarded my introductory letter, I believe, with profound suspicion. It just did not seem plausible that, out of the blue, somebody would actually offer them money. I think they could not quite figure out what the catch was. As a result, some of my letters (but fortunately only a minority) were not rewarded with an answer. But sooner or later, contacts did materialize. I realized that I could accomplish only so much via correspondence and began to plan trips to the region. In Hungary, I had no fears of organizing these trips efficiently by myself, although occasionally I had to rely on the good offices of Csekő to arrange an appointment for me. But the nonresponse or very belated response by some people and the intrinsic deficiencies of the telephone system made it clear to me that I would never be able to set up an efficient schedule without a local organizer in Czechoslovakia and Poland. Vladimír Benaček played that role for a number of years in Czechoslovakia and, subsequently, in the Czech Republic. With admirable efficiency and at great sacrifice of his own time, he arranged appointments, transportation, and accommodations. Everything always went like clockwork.

In Poland, the situation was slightly different; because distances in Poland are much greater and the road signs clearly the worst in those parts of Eastern Europe that I am familiar with, I invariable hired a car with a driver when I left Warsaw for some provincial cities—not because I was afraid of getting lost, but because I could not afford the time to get lost. Instead of relying on a friendly economist, I got wind of an export-import company, Wessly Uniservice, the principals of which, Wiesław Rozmyslowicz and Tomasz Jaśkiewicz, were delighted to make the arrangements for me. For many years, they provided me with the same driver, the amusing and reliable Marek Kopeć, who was a biophysicist trying to supplement his meager official income. This arrangement permitted me to do things like leave Lublin in the early morning, get a good day's work done in Sandomierz, and still be in time for dinner in Kraków. While we were driving, I tried to teach him

English and he tried to teach me Polish—a noble undertaking that created much mutual merriment over the years. The first two trips were filled with anxiety about whether all this would work, but it always did, in part due to my compulsive nature which prompted me to start making arrangements for a trip four to five months ahead of time. Many people chuckled good-naturedly, a few snorted with exasperation, and I justly earned the reputation of being the most neurotic American ever to travel in Eastern Europe.

The Status Quo in Higher Education

The history of higher education in Eastern Europe is much more checkered than in North America. Many institutions did not have the continuity that we are used to, because of wars and the not infrequent redrawing of borders. Jagiellonian University in Kraków was founded in an earlier incarnation by King Kazimierz the Great in 1364, was renewed by King Władisław Jagiełło in 1400, and languished from the mid-sixteenth century until 1780, when Hugo Kołłotaj reformed it. After Poland was partitioned in 1795, Jagiellonian University became more German than Polish, but reverted to being a Polish university in 1918 until it was closed by the Nazis in 1939 and finally reopened at the beginning of the Communist era in 1945. Charles University in Prague, the oldest university in the region, was founded by Holy Roman Emperor Charles IV in 1348 and was severely reduced in scope during the Hussite Wars (only the faculty of arts survived). After the Thirty Years War, it was recreated as Charles-Ferdinand University with four faculties. In the eighteenth century, German replaced Latin as the language of the university. In 1882 it was divided into two universities, one German and one Czech, which operated side-by-side until 1939, when they were altogether closed by the Nazis. Charles University reopened in a new incarnation only after the end of World War II.[28] Law schools were an important target for the political forces aiming to control the legal profession, and law schools in Czechoslovakia were closed, first by the Nazis and then by the Communists, and only the law school at Charles University was left.[29] In Estonia, Tartu University was founded in 1632 by Gustavus II Adolphus of Sweden, was closed in 1710 when the Swedish period came to an end, but was reopened in 1802 as the sole German university in the Russian empire.[30] It had the reputation of being the most liberal of the German-speaking universities, and both the Estonian and Latvian national revival started there. This state of affairs lasted until 1919, when it became an Estonian university. In Hungary, one great university that had relatively more stability than the ones described above was the Eötvös Lóránd University, which was founded by Péter Cardinal Pázmány in 1635 in the city of Nagyszombat (now Trnava, Slovakia), was transferred to the capital city of Buda in 1877 and finally to Pest in 1884; it did not have long periods nonexistence.[31]

Libraries were also subject to the vagaries of wars and politics. For example, the wonderful library of engineering and science, dating from the sixteenth through the eighteenth centuries, of the Mining College of Selmecbánya[32] in northern Hungary was spirited out of what had suddenly become Czechoslovakia after World War I and removed to Hungary, where it was eventually divided between the Universities of Miskolc and Sopron. In Poland, the outstanding Ossolineum Library became the repository of books from the university library in Lwów, when that city became a part of Ukraine after World War II. Change, movement, and instability in academic institutions were characteristic of the region.

The Communist period left a profound—we all hope not indelible—mark on higher education and research in Eastern Europe. Prior to this period, the state controlled the universities in significant respects, but there was room for the Humboldtian ideal of quality guaranteed by the professorate and, in spite of the state control, substantial elements of autonomy were present in universities.[33] In the Communist period, political control of the educational and research establishments was all-important and finances were always scarce. The natural sciences and technology were much favored over the humanities and social sciences, and the quality of the latter suffered, much more in some countries than others. Czechoslovakia was effectively shut off from academic contacts in the West and "non-ideological education in the law, humanities, and the social sciences was forced to become historical, precluding analytical, problem-solving approaches."[34] Control required the assiduous application of the principle of *divide et impera*: the functions of universities were separated from those of academies of science, universities themselves were broken up into constituent parts, and new, highly specialized universities were created. Typical of the breakup was the severing of schools of medicine and schools of agriculture from the main body of the university: after 1945, Jagiellonian University lost its medical and agricultural faculties, which became separate universities, as well as its faculties of pharmacology, forestry, and theology.[35] Similar changes occurred in Hungary, where medical schools were uniformly stripped from the body of the universities they had been part of. But other specialized institutions also came into being—so specialized in fact, that one could hardly imagine such an institution being called a university in North America. Thus we find in Slovakia the University of Transportation in Žilina, the University of Horticulture and Food Industry in Budapest, and the University of Chemical Engineering in Veszprém, Hungary.[36] This resulted in enormous fragmentation and inefficiency, and consistent policies were difficult to achieve because different types of universities reported to different ministries: thus, general universities were under the Ministry of Education, agricultural universities under the Ministry of Agriculture, medical universities under the Ministry of Health, and so on. The fragmentation of universities, together with the absence of a credit system, created pernicious inflexibilities in higher education: a student who had a change of heart might have to change institutions and, in effect, start from scratch, because there were no transferable credits.[37]

Universities were viewed as primarily teaching institutions, which is not to say that they conducted no research; but it does mean that academies of science were the prime and much favored research institutions. For example, research expenditures in 1986 in Hungarian universities amounted to 3 billion HUF (Hungarian Forint), whereas research expenditures in research institutions (mostly institutes of the Hungarian Academy of Sciences) amounted in the same year to 9.3 billion HUF.[38] Rivalry between universities and academies of science was a way of life. As a telling milestone in this rivalry, Ernő Zalai (a professor and former vice rector at the Budapest University of Economic Sciences) reported that in 1986, 7,501 researchers in research institutions produced 83 books in foreign languages and published 1,542 papers in foreign journals, whereas 13,814 researchers in universities produced 207 books and 3,203 papers, yielding a book production propensity of 0.011 per researcher in research institutes and a propensity of 0.014 in universities, and scientific paper production propensities of 0.206 and 0.232, respectively, suggesting that university-based scholars were (marginally) more productive, in spite of the advantages enjoyed by the Academy of Sciences. It should be noted that the pre-eminence of academies of science was not completely uniform and the Academy had no monopoly on research in Poland,[39] where the universities remained the main centers of research, although the Academy was better funded.[40] There was also a clear hierarchy of degrees. In Hungary, the lowest form of advanced degree was the university doctorate,[41] followed by the "candidate" degree and the doctor of science degree, both awarded by the Academy of Sciences, as determined by the Scientific Qualification Committee. The latter two degrees were also awarded by the Academy of Sciences in the Czech Republic.[42] In Poland, the degree of doctor was equivalent to the candidate degree in the other countries and was awarded by the faculties of universities. The higher degree was the "Dr hab." (a distant relation of the German Habilitation), which was also awarded by universities, and which required the publication of a significant book of scholarship, some other publications, and a complex process of approval by referees and of a faculty examination committee, a public lecture, and further approval by the Central Qualification Committee.[43] A large percentage of university faculty members did not possess the more advanced degrees. For example, in Hungary, in the late 1980s and early 1990s, of some 17,500 teachers in higher education, only 8,000 had candidate degrees and 1,857 doctor of science degrees. The corresponding numbers in the Czech Republic were 2,779 candidate degrees and 773 doctor of science degrees, out of a total of 12,762. In Slovakia, there were 2,334 candidate degrees and 372 doctor of science degrees, out of a total of 8,103. Many did no research or only poor-quality research, if for no other reason than the fact that academic salaries were so low that many faculty members had to take other jobs and spent much time moonlighting. Moreover, salary structures were extremely rigid and largely prevented better or more productive professors from being paid more generously.[44] The low salaries also started to cause a significant brain drain, not so much to foreign countries, but to commercial enterprises, such as banks, insurance companies, and the like, which were able under the new economic system to pay vastly higher salaries than academic institutions; unfortunately, the ones most in demand tended to be the best people, letting the average quality in the universities deteriorate even further. The low-salary problem is noted as a particular obstacle to recruiting excellent new faculty members at even the more prestigious institutions.[45]

Universities typically provided a five-year curriculum, leading to a diploma; a very small number of students undertook "post-graduate" studies at universities. For example, the number of graduating students in 1991 in Hungary was 23,648, but only 491 were graduate students (Fábry, p. 9). As late as the academic year 1997–98, the number of full-time students at Warsaw University was 24,758, but only 2,824 were post-graduate and 1,142 were doctoral students.[46]

Other types of post-secondary teaching institution existed, as well. These institutions characteristically provided highly specialized and often vocational training in commercial subjects or the arts. They are called colleges in Hungary (more precisely "Főiskola," which translates as "higher school" and more or less corresponds to the German Hochschule or Fachhochschule) and academies in Poland; examples are the College of Finance and Accounting or the College of Fine Arts in Hungary,[47] and the Academies of Economics in Poland (Akademia Ekonomiczna). Training took anywhere from three to five years for a diploma and was more practically oriented than in universities. But teaching methods were vastly different from what we are used to in American universities. Students sat in class some 30 hours per week and were lectured to, or, one might say, "lectured at." The kind of free-wheeling discussion that characterizes teaching in the humanities and social sciences in the United States, paper writing, and independent work were essentially absent from university curricula. Even "seminars" tended to become professorial monologues. Higher education tended to emphasize "training" rather than education and aimed at providing trained rather than educated people for particular slots in the economy (McCrudden et al.).

The Civic Education Project is a not-for-profit organization, originally created at Yale, that provides visiting western lecturers in economics, law, and other social sciences for periods of up to two years to some 70 East European universities. It had conducted an important study of teaching methods at these institutions and the effect on teaching methods of foreign exposure, as measured by the percent of East European teachers who studied in the West, by the number of conferences attended abroad, and by the number of western visitors to the social science departments at these universities.[48] The study examined in particular the effect of foreign influences on the use of western materials, on essay and research paper assignments, and on active methods of teaching, that is, methods that emphasize analytical approaches and reasoning and are based on frequent student–faculty interaction. Some positive influences were found: conference attendance by faculty members and courses taught by foreign visitors at the universities increase the propensity to assign research papers, courses taught by foreigners increase the number of essays assigned, and conference attendance also increases the teaching of cognitive skills and analytic methods. However, on balance, the effect of foreign exposure on teaching methods was limited as of some six years after the beginning of the East European transformation.

The funding of higher education and research was problematical, to say the least. Fábry notes that allocations from government resources were

- *Calculated on the basis of the earlier year's numbers;* therefore, they are independent either from the output of the institutions or from the real financial needs of the particular research programs;
- *Lacking in transparency;* R&D allocations were distributed into several different lines in the national budget;
- *Lacking connection with real costs,* so that no indices of effectiveness were established to restrict or encourage their use;
- *Dependent only on the personalities of the administrators,* without any well-based, long-term concept for development.[49]

Facilities were still seriously lacking in 1996–97, and in the social sciences 37% of Hungarian, 41% of Polish, and 58% of Slovak faculty members had no access to e-mail.[50] Long-term planning in universities was severely hampered by the fact that rectors were typically elected for three-year terms and could hold only two consecutive terms. The problems faced by the higher educational and research sectors in 1989 were truly enormous:

> Trying to reform their educational and research systems, these countries all have to solve comparable dilemmas: *how to grant autonomy and self-government to higher education and research institutions without depriving the state of instruments to enforce its educational and research policies; how to formulate policies avoiding the hitherto central structural steering; how to solve the inherited conflicts between different parts of the of the scientific and educational community linked everywhere with three separate sectors: Higher Education Institutions (HEIs), Academies' Institutes and research institutions subordinate to different ministries; how to decentralize the funding system with only one inadequate source of financing—the state budget; how to get users of research results to support profit-oriented research and development, in the case of (a) low quality of research results, (b) deep economic recession, and (c) lack of private capital; how to introduce the competitive system of financing education and research to a system lacking reliable mechanisms to evaluate institutions and research teams; finally, how to pass and implement new regulations in view of inherited habits of the community, and the conservative attitude of much of it?*[51]

To solve some of the problems, Andrzej Pelczar, former rector of Jagiellonian University, noted that it was important to develop university outreach to the larger community, in particular to industry, which can be a source of funds if universities participate in the solution of applied problems (Pelczar, pp. 8–15). This outreach, which is solid practice in the US going back to the founding of the Land Grant universities, was substantially missing at the beginning of the new era, and in fact was not even very visible among agricultural universities in Poland, where extension services were provided by a state agency rather than universities (see chapter 11). Pelczar also pointed out that it was essential to provide opportunities for student and faculty mobility if they were to change the way in which they thought about the university and their disciplines—an objective that the TEMPUS program has fully recognized and has done much to bring about. Finally, he observed the very significant brain drain that has occurred, a point echoed by numerous observers.[52]

Interestingly, the brain drain from universities to foreign countries, while accounting for a loss of 2,706 scholars between 1981 and 1991,[53] diminished over time, accounting for an annual loss of 302 persons during and shortly after the Martial Law period (1981–84), 231 persons in the period of relaxation (1985–88) and only 191 persons after 1988. At the same time, personnel losses to other sectors in the Polish economy, totaling 4,313 in the same period, increased from 280 per year in 1981–84, to 286 in 1985–88, and to 586 in 1989–91. These losses were primarily due to the higher salaries that could be obtained in the nonacademic sector.[54] The salary differentials between a middle-level academic position at a first-rate university and a comparable position in a commercial or financial firm were unimaginably high in the early 1990s: the commercial/academic salary ratio could be as high as 10. It should also be noted that there is (still) a concurrent brain drain: full-time faculty often work at two or three different jobs because university salaries are so low that average families cannot survive on them. Faculty members spend only half their working time at the university and spend an average of 18% of their time in nonacademic jobs.[55]

The problems of the educational and research establishment were so vast and encompassing that it would not have been reasonable to expect any country to solve them in short order. An incomplete list of issues to be addressed would consist of the following points.[56]

1. How can the disparity between the academies of science and the universities be redressed, the antagonism between the two reduced, the research quality at the universities improved, the considerable talent at the academies deployed in teaching, the system of scientific degrees rationalized, and teaching and research integrated?
2. How can autonomy be granted to universities, which up to this point were entirely subordinated to various ministries, particularly in an environment in which the majority of funding will continue to come from the state?[57]
3. How can competition be introduced into a system where the academic quality in universities was low, funds were allocated to institutions largely without reference to excellence, and academic advancement was mechanical and often depended on holding politically correct views? In other words, how can peer review and accreditation be introduced into such systems?[58]

4. How can inefficiency be reduced and the fragmented structure of universities be improved by suitable re-integration of units?
5. How can the deep-seated conservatism of faculty and administrators and their protective attitude towards the status quo be overcome?
6. How can academic incentives be improved for faculty members whose salaries were too low to provide a decent living, in an environment in which state budgets are miserly and likely to be cut further?
7. How can the infrastructure of universities be improved?
8. How can the quality of teaching be improved?
9. And most obviously, how can the economic position of universities be improved?

In fact, the early years after 1989 witnessed numerous attempts to come to grips with these problems. New regulations concerning higher education were enacted in Czechoslovakia, Hungary, and Poland[59] and various acts were passed to improve the research environment (for example Higher Education Act [1990] in Czechoslovakia, also taken over by Slovakia when it became independent in 1992, the Act on State Support of Scientific Activities and Technological Development [1992] in the Czech Republic, the Law on the Committee of Scientific Research [1990], the Higher Education Act [1990] in Poland, and the Act on Higher Education (1993) and its various amendments in Hungary).[60] One of the greatest early achievements was the creation in Poland in 1991 of the State Committee for Scientific Research (Komitet Badán Nauchowy, KBN), which became a government agency with ministry or near-ministry powers. Given that committee members were typically appointed in the previous regime, it was a considerable achievement that its 60 members were elected. Although KBN was overwhelmingly dominated by scientists, its creation represented a serious step towards introducing academic competition for grants.[61] In 1992, for example, 47% of KBN's budget was spent on statutory grants for the maintenance of research institutes, 20% on peer-reviewed research grants, and the balance on R&D infrastructure and nationally important R&D programs.[62] In Hungary, the National Scientific Research Fund (Országos Tudományos Kutatási Alap, OTKA) was founded as early as 1985. By 1991, its budget was 1.8 billion Hungarian forints; its 1990 budget was divided among the Academy of Sciences (38.3%), institutions of higher education (30.3%) and agricultural and welfare institutions (Glenday). There were also attempts to streamline the relationship, not usually an easy one, between universities and academies of science. It was not really possible to fold academy institutes into university departments, and deep-seated conservatism, coupled with the desire to protect one's own turf, made that otherwise interesting idea a nonstarter (Fábry, pp. 20–1). But several academies of science attempted to reduce the number of institutes and, between 1988 and 1992, drastically curtailed the size of the staff: thus, in the Czech Academy of Science, the staff size went from 10,200 to 8,500,[63] and in the Slovak Academy of Science the change was from 5,941 to 3,943.[64] The change from 1988 to 1992 reported by the Academy is 24%, substantially greater than that reported by Koucký and

Hendrichová. It is also interesting to examine what happened to the number of publications emerging from the Czech Academy of Sciences: total publications, publication of journal articles and monograph publications all increased from 1988 through 1992 and all experienced a drastic decline in 1993 (Kadlecová and Vítek). The reasonable interpretation might be that the publication lag is responsible because of much work "in the pipeline," but this was worked off by 1993, at which point the effect from the much smaller number of scholars working at the Academy became visible.

But it was clear that there were no easy solutions to reforming higher education and research. The funds that the state could allocate to universities were small and decreasing, and what might have seemed a logical step, namely to replace free higher education with a system in which at least partial tuition would be paid by students, was not easy to implement in the short run, partly because students could typically not afford even token payments, and partly because free higher education was guaranteed in some constitutions (for example in Poland).[65] Academic conservatism was rampant. I encountered many an academic who could not conceptualize a drastically different way of doing things. And finally, in spite of an extraordinary amount of good will and desire on the part of foundations, U.S. government agencies, and European Union agencies to assist both the national economic recoveries and the rebuilding of the higher education sector, there was very little coordination among all the agents on the scene. One could fairly ponder whether we might all end up working at cross-purposes, and whether it might not turn out to be the case that "too many cooks would spoil the broth." On the whole, since enormous funds were being channeled into management training and relatively few into library assistance activities, it seemed plausible that duplication and lack of coordination were greater dangers in the former than in the latter area.

4

Building a Program

As I mentioned in the previous chapter, I realized early in the game that I would have to undertake a fairly substantial amount of travel to the region, both to solicit grant applications and to supervise projects once they were funded. Until we had made a sizeable number of grants, which took at least two years, correspondence was inadequate in itself to generate many good applications. This was partly because potential grantees had never heard of the Mellon Foundation and an invitation to prepare a grant application, extended by letter, was not always taken seriously, and partly because they often had only rudimentary ideas about how to prepare a proposal. I aimed at establishing a relatively regular pattern of travel, so that people could form reasonable expectations about when I might visit. Because I was teaching at Princeton until the end of the 1994–95 academic year, I could not very well get away for the requisite time before the middle or end of May. I always scheduled my visits to Czechoslovakia (later the Czech Republic and Slovakia) and Hungary for the end of May and early June, and my visits to Poland so as to overlap the fall term break at Princeton. These trips tended to be two-to-three-week trips and the days were crammed with appointments from early morning until after dinner. In Poland I always engaged a car with a driver to get from city to city, in the Czech Republic and Slovakia I drove myself, and in Hungary I mostly traveled by train. Some of my most delightful car trips took place in the Czech Republic and Slovakia where I was often accompanied by Andrew Lass, professor of anthropology at Mount Holyoke College, who managed our Czech and Slovak library projects. But car travel was also tiring. The road trip from Bratislava to Košice takes about five hours and the trip from Košice to Prague some seven to eight hours, and it is not adequate compensation for the fatigue that the road takes you near the famous battlefield of Austerlitz. So sometimes we went by air and had the dubious pleasure of flying in a Soviet-built Yak 42 jet, which gives one a good understanding of how sardines feel in a can. But getting to the region was easy: Pan American, and later Delta, flew to Warsaw, Prague, and Budapest and permitted me to accumulate large quantities of frequent flyer miles. Later, when Estonia and Latvia were added to our program, my flying became more diversified and slightly more arduous, because to get to the Baltic countries I had to pass through Helsinki or Copenhagen. But on the whole it all became fairly routinized. East European airports were a useful gauge by which one could judge the degree of progress in each country. In 1990, Budapest, Prague, and Warsaw each had old and decrepit airport terminals and one by one, they built new ones; first Budapest, then Warsaw, and finally Prague. As of the fall of 2000, only the Bratislava airport seemed unchanged.

First Thoughts

It was entirely clear from the beginning that we could not deduce the shape of a program-to-be from first principles and that we needed a great deal of input both from East Europeans concerned with the areas we wanted to support and from Americans who wanted to provide some kind of assistance. It was also clear that some project that was needed or that "worked" in one country would not necessarily work in another. For example, since the late 1960s, the dominance of Marxist-Leninist ideology in the various academic disciplines was less pervasive in Hungary and in Poland than in Czechoslovakia. In the latter, Marxist-Leninist doctrine had a much stronger influence than in the other two countries, as a result of which the general state of economic science in Czechoslovakia was noticeably weaker than in Hungary and Poland. An informal name count of Czechoslovak economists revealed that there were perhaps no more than two dozen economists in that country who could claim familiarity with western professional literature.

Many ideas started to percolate and rise to the surface, first and foremost from American organizations that happened to have established some contacts in Eastern Europe and felt that they had something to offer in providing business training. Short courses, of one to three weeks in duration, could easily be organized in some East European country, and by their very brevity they could be specialized in any number of ways: by providing training for small entrepreneurs, for managers in medium-sized businesses, for local government leaders, for those who themselves would become trainers, for the management of export-oriented businesses, or for the effective privatization of enterprises. I was eager to get my feet wet, and the Mellon Foundation funded two short-term training courses for middle-level managers: the Institute for Cooperative Assistance ran such a course in Cambridge, MA, and the Volunteers in Technical Assistance organized one in Hungary. But business and management training could also be thought to mean a one or two-year curriculum leading to an MBA degree within the framework of an established university, and programs of this type were obviously harder to design and much more expensive to implement. My preference generally was for training scenarios that were likely to have lasting effects, and it was not clear what permanent effects would result from a two-week ad hoc course. But easy as it seemed to organize short-term managerial training courses, and straightforward (though more expensive) as it appeared to create genuine MBA programs, nothing was obvious about how rigorous economics could be taught in the region.

Many of the ideas for managerial training came from American sources. But ideas for assisting universities to rebuild infrastructure or to develop programs were not as easy to come by, and it became clear that I would have to utilize my periodic trips to the region both for developing ideas and for supervising existing projects. The many academic contacts I succeeded in establishing during the first few months of 1990, combined with the many more from subsequent trips, pointed to some clear-cut priorities for universities: (1) the purchase of books and periodicals for libraries and investment in library technology (equipment, cataloging and circulation systems, CD-ROMS, microfiches, and, in some cases, the purchase of multiple copies of textbooks); (2) the acquisition of computing equipment and software (mainframes, personal computers, Unix workstations, network components, storage devices) which they needed both for traditional, scientific number-crunching applications and for connecting to the electronic networks then in use in the West (BITNET in the United States and EARN in Western Europe); and (3) funds for faculty and student exchanges.

I had no doubts about the effectiveness of the first two of these spheres of activity, but I had more doubts about the value of exchanges, unless they were particularly well focused and targeted. I had seen too many instances in previous years in which East European visitors to American universities had, through no fault of their own, spent the first few weeks of their stay just getting their bearings. If their visit to North America was scheduled to last only a few months, a nonnegligible portion of it might be spent unproductively. While the experiential learning that this represented was ultimately valuable, it did not seem to be the best use of resources at a time when crash programs were needed. But I felt very strongly that the East Europeans should be able to join the BITNET/EARN networks; that membership in BITNET and EARN was an important step toward integrating East European scholars into the western world of scholarship and information exchange, and that the kind of integration provided by electronic networks contributed to keeping these societies open. In fact, the East Europeans themselves had realized this and they had all applied to BITNET for membership. Unfortunately, things were not that simple. Because of the Cold War, now mercifully over, U.S. Department of Commerce regulations (the "COCOM" regulations) prohibited the export of certain technologies to the Soviet Bloc, and BITNET membership was on the prohibited list.[1] Late in 1989, I contacted an official in the Department of Commerce to inquire about when the prohibition was going to be relaxed. I explained briefly what BITNET was about, and since he was not familiar with it, he asked me just what kind of information could be transmitted from the United States to Eastern Europe via BITNET. Feeling in a somewhat facetious mood, I replied, "Well, I could for example send e-mail to a friend at a university in Budapest and ask him to join me for cocktails at the Budapest Hilton at 5 pm on the following Monday." "In that case," he said, "BITNET membership for Hungary is clearly illegal, because under the regulations only public domain information may be transmitted. But your cocktail plans at the Hilton are obviously private domain information." Fortunately, that view did not become the dominant one, and by April 1990, we had received an Advisory Letter from the U.S. Department of Commerce, which indicated that providing BITNET capabilities in Eastern Europe was now legal.

The third area also seemed important, but not all forms of faculty and student exchanges seemed a priori equally cost-effective. First, bringing students and faculty to North America was clearly much more expensive per capita than sending faculty from the United States to Eastern Europe.[2] Some concern was also expressed by various foundations that bringing people to the United States for a period of a year or more also ran a considerable risk that the individuals in question might want to stay in the United States rather than return to their homeland, which would have been very counterproductive and would have intensified the brain drain. But it was important not to be doctrinaire about such matters, and ultimately the Mellon Foundation did fund a number of programs in which shorter or longer training visits to the United States by East Europeans played a significant part. Interestingly, the student and faculty mobility projects supported by TEMPUS seemed less prone to these dangers perhaps partly because of their scale and visibility, and partly because of better overall control and organization.

Many organizations asked foundations to support short conferences or workshops, lasting two or three days. I was not favorably disposed towards these requests. First, I believed that conferences generally do not produce lasting effects unless they are exceptionally strongly focused on a narrow topic and attended by experts on that topic; second, I had a jaundiced view of American professors making the grand East European tour on which they would recycle the same paper from conference to conference.[3]

Wonderfully articulate and coherent views about what needed to be done were expressed in a letter to the Mellon Foundation on January 15, 1990 by Jonathan F. Fanton, then president of the New School for Social Research (now the New School University). He started by observing,

> The immensely challenging and complicated period of transition from the authoritarian, monopolistic rule of Stalinist governments to successor regimes poses daunting challenges: the reconstruction of competitive, democratic politics; the rebuilding of economic structures; and the reinvigoration of civil society. It is to a particular aspect of the latter—the revival of free intellectual and scholarly life and institutions—that the New School has been directing its efforts for the past half decade, and it is in this domain where we think we can make useful contributions.[4]

He then described the activities that the New School had already undertaken in previous years. First, under the leadership of leading East European intellectuals, such as Adam Michnik in Poland, György Bence in Hungary, and Petr Pithart in Czechoslovakia, the New School had organized seminars on democratic theory and practice. In these, the participants discussed the concept of totalitarianism, the work of Hannah Arendt, Steven Lukes, and Michael Walzer; the relationship of liberalism and Marxism; transitions to democracy in Latin America; and many other subjects. Second, the New School had organized scholarly exchanges and brought distinguished East European scholars to the New School and arranged for New School faculty to give papers in Eastern Europe, including at an underground philosophy seminar in Prague. Third, the New School started a post-doctoral fellowship program for young Polish scholars. Fourth, it provided books and journals for the libraries of several Polish Academy of Science institutes. And last, it had arranged for the publication of social science articles by East European scholars in *Social Research*, a New School journal. Fanton then proposed significant extensions of the New School's programs:

> 1. *Extension of the Democracy Seminar*: The existing democracy seminars in Poland, Hungary, and Czechoslovakia provide an important base from which to build contacts across disciplines within individual countries, between colleagues who have had a clandestine scholarly existence and those who have labored in the official institutions, joining scholars in the different countries of the region, and linking them to people and ideas in Western Europe and North America. . . .
> 2. *Increasing the number and kind of scholarly exchanges*: There is a need to accelerate rapidly contacts between western scholars and those in East Central Europe. . . . We would like to organize an active series of visits by western faculty in the social sciences and humanities. In turn, we would like to make available a series of short visits, of two to six weeks, for academics in East Central Europe to come to the West. . . .
> 3. *Provision of scholarly materials and of a comprehensive bibliography of such materials*: This need is enormous and, for any one university in the region, daunting. . . . it should be possible to develop an inventory of most-needed journals and books, and a plan for their provision as an initial step. We hope that U.S. and West European publishers might be prepared to assist in this venture. . . .
> 4. *Emphasis on student fellowships*: While the longer visits we have been sponsoring by senior figures may now prove anachronistic and not cost-effective, student fellowships, especially at the graduate and post-doctoral levels, may be more important than ever. . . .
> 5. *Further publication opportunities*: We would like to use out contacts and positions on various editorial boards to expand the program we have begun with *Social Research* to insert the best scholarship from East Central Europe into our own scholarly discourses.

While some of these program areas were not entirely consistent with the Mellon Foundation's guidelines for Eastern Europe, and I had particular doubts about the fifth one on the grounds that it might undermine editorial autonomy, the case was most cogently argued and the Foundation did end up making several grants over the years to the New School for some of the above and related purposes.

In the second of the areas mentioned above, the Mellon Foundation ended up instituting a large-scale program of its own, which is still in existence today. Early on, I met Emerita Professor of History Éva Balázs of the Eötvös Lóránd University in Budapest. We had a fruitful discussion of what East European scholars needed, and quickly agreed that scholars in science and technology had had many more opportunities in the previous regime than humanists. At one point in the discussion she asked, "Do you know that there are historians of art in Hungary who have never seen the Louvre or the Uffizi?" This off-hand remark was the origin of a program that would initially send each year 20–25 East European scholars in the humanities on three-month research fellowships in the West and by 1999–2000 as many as 40. We insisted that these fellowships were not for the purpose of some ill-defined improvement of the scholars' general knowledge, but for the purpose of permitting them to undertake or complete a specific piece of research. The program is administered by the Council of American Overseas Research Centers (CAORC), which negotiates an arrangement with selected Western research centers for housing the scholars; they apply directly to the research center of their choice, and the centers in turn decide whether the scholar is qualified or not.[5] Over the years, Mellon funded this program with a total of $8,450,000, and through February 1, 2001, a total of 259 scholars had received fellowships; of these, 26 were from Bulgaria, 40 from the Czech Republic, 77 from Hungary, 85 from Poland, 16 from Romania, and 15 from Slovakia.[6] The program has worked extremely well: the scholars were productive and have produced an impressive array of works, and some performed so well that the participating institutes sometimes extended their stay at the institutes' expense.[7]

We now turn to a discussion of some general rationales for economics and business training and then discuss some particular examples of early Mellon projects that were to create successful—and we hope permanent—institutions in each of the Foundation's areas of primary interest: management training, economics training, computing and networking, and libraries.

Economics and Business Training

It is clear that a substantial difference exists between the disciplines of economics and business or management; at a minimum, this difference is reflected in the fact that economics is strongly oriented towards the scientific method (axioms, theories, hypotheses, empirical testing, and inference), while the field of business or management is generally more practice-oriented.[8] In the post-1989 world, the former state enterprises of Eastern Europe were to be privatized and the countries of the region would adopt a market orientation in their economies. It is fair to say that nobody believed that the managers and economists trained during the Soviet period were up to the challenges lurking around the corner. In the preceding chapter, we have already noted some examples of the near total lack of preparedness by the public to handle even elementary concepts of business and economics. It is also worth noting that while western-trained economists and managers serve different objectives and clienteles, they do tend to look at many problems in similar, and often identical, ways. For the moment, therefore, let us neglect the distinctions between the two approaches (although when it comes to designing concrete programs, they will be very important) and point to some key characteristics of training that had to be provided in order to overcome the inappropriate practices of the pre-1989 regimes.

First of all, any economic or business problem involves choices, and these have to be analyzed in an optimizing framework. Economic and management problems involve the allocation of resources, which involves choices, trade-offs, substitutions, compromises, and the evaluation of what is most effective. But none of this is feasible if we do not know what the facts are. That, in turn, requires measurement, hard data, and solid accounting principles and practices. I have already mentioned one Hungarian institution that kept no books on income and expenditures, because it did not have an effective budget constraint, which clearly contributed to the corruption of accounting practices. With budget constraints becoming hard, it was essential that reliable accounting procedures be introduced.

But facts also permit the estimation of underlying economic relationships and the testing of hypotheses, which require sophistication in statistics, statistical reasoning, and econometrics. A statistical way of thinking about reality was clearly anathema to the economic planners of the Soviet period, and this is one of the reasons that the application of statistical reasoning to economic data was one of the most underdeveloped fields in the region, in spite of the fact that some of the fundamental contributions to probability theory have been made by Russians such as Khinchine, Kolmogorov, Smirnov, and others. The reasons for this are not hard to guess. The application of statistical reasoning presupposes that there is an element of randomness in reality, that there is some element that is unknowable and the effect of which has to be filtered out, so to speak, before the systematic relationships can be revealed. Moreover, applying statistics is tantamount to admitting that we do not know ahead of time what reality really is. Both of these assumptions were clearly objectionable from the point of view of an ideology that claimed precision on theoretical grounds. For this reason, people in the new world also had to learn that there is no dogma, no revealed truth, and that everything may be and should be questioned.

All this argued for providing economics and management training on all levels: academic training in economics, management training at the university (MBA) level, and business training for small and medium entrepreneurs.

Business Training for Small Enterprises

Some small-scale projects could start almost right away. For example, the Industrial Cooperative Association, Inc. in Somerville, MA proposed to hold a three-week training program for seven Hungarian and Czechoslovak participants in corporation finance, dealing with topics such as financial analysis, credit analysis, venture capital and investment banking, and related topics. The audience in this case consisted of fairly prominent persons, such as the deputy general manager of the international banking division of a large bank, the head of the credit department of another bank, a member of the Economics Institute of the Czechoslovak Academy of Sciences, and so on. It was not my favorite format for training, but I felt that we needed to "get on the books" in providing grants. This was an easy way to start, even though in the longer term we stopped these one-shot deals that did not lead to creating permanent institutions. The cost of this program was $30,000, of which travel and subsistence must have accounted for nearly a half, and while the individuals who received training were clearly better off for having had it, I thought that approaches of this type were too unsystematic to have a long-term impact.

It seemed rather important to become more orderly and organized in our efforts and to start building institutions that might become permanent or at least lasting through the economic transition that was beginning. We soon became involved with several organizations that had ambitious plans for Hungary, some of which, not coincidentally, developed joint programs with one another: the SEED Foundation in Budapest, the Talent Foundation in Szeged, the State University of New York in Albany, and the Southwestern Pennsylvania Economic Development District, later called SPEDD, Inc. In Poland, the Mellon Foundation supported the Polish-American Small Business Advisory Foundation (Polsko-Amerykańska Fundacja Doradztwa dla Małych Przedsiębiorstw) and in Czechoslovakia (later only in Slovakia) the Central Europe Institute.

The SEED Foundation.[9] As discussed in chapter 1, in late 1989 I became acquainted with what was at first called the Business Development Foundation and ultimately became the SEED Foundation. I originally thought that it had been U.S. Ambassador Palmer's brain child, but it subsequently turned out that that was not quite the case. In May 1989, Professor Robert Hisrich of the University of Tulsa (later of Case Western Reserve University) was in Budapest as a Fulbright professor and was assisting Zsuzsanna Ránki in getting the International Management Center started and turning it into a recognizably western business school. They met regularly and came to the conclusion that it was important to assist entrepreneurs in small- and medium-sized enterprises. They also determined that no agency of the Hungarian government, nor any organization outside it, was providing the necessary support. They developed the concept of a free-standing, not-for-profit organization, staffed by competent Hungarians, that could provide the appropriate training and counseling for entrepreneurs and could build strategic alliances with other organizations.[10] In due course, a number of government agencies and other organizations signed a charter that established SEED.[11] The charter members made an initial financial contribution toward the operating costs of SEED, but it is noteworthy that since that initial contribution, none of them provided SEED with additional funds.

Through January 1990, I had considerable difficulty in reaching and talking with Dr. Ágnes Tibor, the executive director of SEED. I found out from Ambassador Palmer that Catherine Marshall of the U.S. Small Business Administration (SBA) had been selected to be a co-director of SEED, which seemed like an excellent idea. But when I talked to her, she painted a somewhat bleak picture. In order to be seconded from the SBA to this project, the head of the SBA required that her appointment be requested by Deputy Secretary of State Lawrence Eagleburger, but that had not happened yet. She was nevertheless planning to go to Budapest in a few weeks' time. She was also pessimistic about the completion of the legalities necessary for Mellon support in less than two to three months' time. But by January 18, 1990, I had received a fax from Dr. Tibor in which she informed me that the papers necessary for the legal establishment of SEED had been signed and that she was eager to continue a dialogue with me. In spite of this, things continued to move at a snail's pace. By February 28, Catherine Marshall still had not received permission from the SBA to take up her temporary position at SEED, and the U.S. Department of Commerce was now erecting new roadblocks.

I was definitely nervous about whether this would ever develop to the point where the Mellon Foundation would feel comfortable funding SEED, particularly after I received reports from other observers that the SEED organization was extremely rudimentary, even though the SEED staff had a great deal of enthusiasm. Catherine Marshall finally did receive permission to go to Budapest, met Ágnes Tibor, and reported that she was a very capable person and that SEED, with adequate funding, might well develop into a permanent institution. By March, I had made plans to go to Budapest in late May, and by early April I received a draft of a proposal from the SEED Foundation for funding.

The proposal outlined five objectives for SEED: (1) to provide basic consulting functions to prospective entrepreneurs and to organizations that support entrepreneurship; (2) to provide assistance in developing business plans, marketing plans, and loan applications by prospective entrepreneurs; (3) to establish a resource library; (4) to organize conferences, workshops, seminars, and to help develop curricula; and (5) to provide assistance in establishing international trade contacts. (In later years, when Anikó Soltész became executive director, some additional objectives were embraced, namely (6) to conduct research in areas relevant for enterprise formation; (7) to provide education and capacity building among disadvantaged social groups; and (8) to spread knowledge relevant for assisting Hungary's admission to the European Union.[12]) What was obviously needed was a considerable amount of equipment such as computers, copiers, and fax machines, and training for SEED staff. Also, SEED very rapidly saw the advantages of business incubators; that is to say, organizations that provide space and business services to very small entrepreneurs who, by virtue of their very small size, could not afford some needed equipment. Thus, the incubator might have a fax machine and a copier, which all the small entrepreneurs housed in the incubator could use at a moderate cost.

But the initial versions of the written proposal were inadequate. The fact of the matter was that few people in Eastern Europe at that time knew how to write a convincing proposal. Articulating "proposal writing standards" in the abstract was not always effective. That is, pointing out that a proposal should clearly identify the problems to be solved, name alternative solution methods, analyze the alternatives and give clear reasons for the particular alternative selected, provide a detailed budget, and so on, did not always have the desired effect.[13] Finally, I started to use an example, which I repeated many times during the next ten years, which was something like this: "Look, getting a grant from the Foundation is very much like getting a loan from the bank; the only difference is that you are expected to repay the loan and you are not expected to repay the grant. If you go to the bank and ask for a loan to build a factory, the bank officer will want to know whether you have researched the area where the factory will be built; whether it has adequate energy supply, transportation, and labor force; whether you have researched the market for the goods you will manufacture; and so on. Only if you can give satisfactory answers to such questions are you likely to get the loan. Well, the Foundation works the same way: it does not want to throw money away on losing propositions." While this may have been patronizing, and was perceived as such by some people, it nevertheless did the job.

The SEED Foundation proposal went through several iterations. I received different versions of it on February 19, April 5, July 31, and August 24. In the middle of this period, I found myself in Budapest, actually met Dr. Tibor on May 31 and formed a favorable impression of her, even though the facilities of SEED were minimal. In any event, the last version of the proposal was finally in good shape. The Foundation extended a grant of $151,000 to SEED for (1) sending three staff members to the United States for an intensive two-week training period; (2) identifying and sending to the United States ten Hungarian business consultants for further training; (3) developing, in cooperation with the Portland (Oregon) Community College, a small business international trade program; (4) establishing a small reference library at SEED, consisting of publications and videotapes on free market economics, finance, small business management, marketing, and related subjects; and (5) establishing an international advisory board. I thought it was a risky grant to make: the longer term survival of SEED was by no means assured and we had to take it on faith that all the outputs promised could actually be delivered. But they were, and over the years SEED received a number of additional grants (by this time Dr. Tibor had left and Dr. Anikó Soltész had taken over as executive director): $285,000 in 1991 for the preparation of a 20-episode television series on becoming a small entrepreneur, which was aired on Hungarian television; $85,000 in 1993 for training in how to start a family business and for incubator development; $159,000 in 1994 for developing video cassettes on "best incubator practices"; and $18,000 in 1996 for translating these into other languages. By the end of 1991, SEED had trained hundreds of small entrepreneurs or fledgling entrepreneurs. SEED also established useful connections with various U.S. organizations. One of these was with the Small Business International Trade Program of Portland Community College, which received two modest grants for joint training activities with SEED.

SEED was one of the most cost-effective programs because it always operated on a shoestring. To my knowledge, it never had more than five full-time employees working out of a tiny office; yet it rapidly achieved national renown as a serious organization committed to improving small business practices, as attested to by the fact that the executive director was named special consultant to the Minister of Trade and Industry. SEED has become a permanent institution. In the past four years, that is, in 1997 through 2000, SEED continued to provide short-term training workshops in large numbers throughout Hungary. While some of its activities declined as the demand for them abated, such as training teachers of business, others continued in full force. In certain areas SEED was and still is the pioneer, the most notable of these being the training of women entrepreneurs, advising family enterprises, and providing training for Roma (gypsy) people. Interestingly, training programs restricted to Roma women entrepreneurs caused substantial consternation among the Roma men. In fact, the largest project in 1997 was a joint project with the Center for International Private Enterprise (CIPE) and dealt with family enterprises. SEED has also developed a network of collaborating organizations and has worked together in the Roma training programs with the Autonomy Foundation (also funded by Mellon). Programs became increasingly diversified and, by 1999, SEED provided training for Ukrainian entrepreneurs and ran a workshops to emphasize the importance of innovation and of marketing. Most recently, it has won USAID contracts to provide training for Bulgarians and Romanians engaged in business and local government, thus becoming an important example of east–east assistance.[14] SEED has been able to raise money for its activities by competing successfully for projects and has received project grants from USIS, the World Bank, the Friedrich Ebert Foundation in Germany and others. SEED continues to train vast numbers of people and, in late 1999, had more training and other work assignments than it could comfortably handle. It started the year 2000 with a substantial backlog (Lackó and Soltész). It should be noted that until 1999, SEED's advice to entrepreneurs was offered free of charge, simply because the ability to pay was not there; in 1999, SEED began to charge for its time at the rate of 1,000 HuFt/hour (using the June 1, 1999, exchange rate of 238.87 HuFt/dollar, about $4.19/hour). But acceptance of this practice was, at best, mixed: some clients accepted this without argument, others would just hang up the telephone.

SEED's success has an important moral that could be observed throughout Eastern Europe over and over again: more important than the cleverness of an idea, or the quality of hardware and software provided for some initiative, or the amount of money dumped into a project, is the determination and commitment of the people who implement it. The executive director, Anikó Soltész, and her colleagues, such as Zsuzsanna Lackó, deserve the credit for making this enterprise work. Has SEED been productive and cost-effective? The answer has to be a resounding yes. Will it survive? If it continues to adapt as circumstances change, as it has in the past, and if its clients' ability to pay for advice continues to increase, the chances are very good.

State University of New York–Albany. In February 1990, I was contacted by the Office of International Programs at SUNY–Albany with an inquiry about potential Mellon funding of training programs for small and middle level enterprises in Hungary. There were no concrete proposals as yet and some staff members from SUNY were about to go to Hungary to "feel out" the lay of the land. This was a fairly typical first move: after all, American academics could not be expected to have detailed knowledge of the East European countries and their immediate training needs. However, an initiative ran the risk of superficiality when requests for major commitments of funds were based on a one-week or 10-day trip to the country in question. But there was no way around this problem: needs were obviously great, there was a palpable sense of urgency to get started with economic restructuring and the training that was a prerequisite for it, and the American academic community was eager. Four small proposals emerged, each with a grant request between $48,000 and $50,000.

Two circumstances were entirely favorable. First, the SUNY staff members were aware of how little they knew at the beginning and attempted not to export too many American preconceptions to Hungary. In a report to the Mellon Foundation dated August 7, 1990, the project director, G. Todd Jagerson, said, "we know much more about Hungary today than we did when our proposals were submitted." Second, SUNY immediately realized that none of its plans could be realized without some Hungarian partners. This, in fact, became the pattern throughout the next ten years in all countries: American teams of experts forged alliances with local partners. The danger, however, in seeking partnerships was that East European institutions were deluged with visiting western experts who passed through capital cities with monotonous regularity. This was true for NGOs such as SEED as well as for universities. In particular, the resources of university rectors and academic staff were strained in the extreme. It seemed natural that a rector, offered the possibility of participating in a project funded with a pot of gold, should gladly offer his participation; but it was not clear that behind every such assent there was a genuine willingness as well as ability to commit institutional resources for the project. There were, indeed, several instances in later years in which the initial enthusiasm was not matched by subsequent carry-through.

The first three projects were as follows. First, SUNY proposed a Privatization Assistance Project, in which SUNY collaborated with the State Property Agency (SPA) and the Budapest University of Economic Sciences (BUES, formerly Karl Marx University). The project started with a one-day training seminar, attended by all 35 professional staff members of the SPA, in which each attendee received about 700 pages of printed materials on various issues in privatization. It was followed by a three-day privatization/restructuring conference attended by some 140 persons. The conference dealt with buyouts, capital market issues, valuation, fairness, employee ownership, human resource management, corporate governance, and related matters. The urgency of the privatization issue and the pervasive lack of understanding of its ramifications made the project important, but it is unclear to what extent it actually contributed to promoting privatization in Hungary. Its impact may well have been dissipated by the disorganization that prevailed in the SPA.

Second, SUNY proposed an Export Expansion Assistance Project to respond to Hungary's considerable need to replace its lost exports to the COMECON countries with exports to hard currency areas. SUNY teamed up with the College of Foreign Trade (Külkereskedelmi Főiskola) to identify Hungarian products and firms that had significant export potential, such as software development, autoparts, Terimpex (processed meats), Hungarotex (ready-to-wear textiles, fabrics, table linen, canned food), and Monimpex (wines, goose liver). Contacts were established between the consumer products producers and American counterparts. In the final phase of the project, Hungarian industry representatives were invited to the United States to meet with potential U.S. importers. There is no doubt that this project had salutary effects; the only qualification that needs to be made is that its scale was small—as indeed the scale of many Mellon Foundation projects taken in the context of a country as a whole. Hence, the overall impact had to remain small.

The third project, called The Entrepreneurship/Small Business Support Project, was based on the recognition that ordinary business skills were woefully lacking among Hungarian small entrepreneurs. After a preparation phase in which nine suitable individuals were identified, including Ágnes Tibor of the SEED Foundation, a U.S. tour of 34 days was organized in order to expose the participants to U.S. small business support systems. USIA became interested in the project and offered to underwrite the costs of the tour. Arrangements were made for the participants to experience small business support systems on the federal, state, local, university, and private levels. In July, a two-day conference, chaired by Minister of Trade and Industry Peter Bód, was held at BUES, not so much for a one-way transfer of ideas but for their mutual exchange. The conference included such outstanding Hungarians as György Surányi, later president of the Hungarian National Bank;

Lajos Bokros, subsequently minister of finance; and Mária Augusztinovics, a distinguished member of the Economics Institute of the Hungarian Academy of Sciences. The conference was exceptionally successful and was the last step before beginning more systematic entrepreneurial/small business training. Regrettably, at this early stage, the SEED Foundation was not found to be a suitable partner by the project managers. The project instead selected the Hungarian-American Enterprise Fund. Charles Huebner, who directed SUNY's Export Expansion Project, was named the new director of the Fund. Huebner was much in favor of providing training: he reported that by the time he assumed his duties on July 23, he found 4,321 requests for funding from small Hungarian enterprises. Yet none of these requests contained sufficient information about the firms in question for making an investment decision about them.

The first three projects were obviously one-time events, yet they gave rise to continuing requests for activities from the Hungarian side. To satisfy these requests, SUNY proposed the creation of a Center for Private Enterprise Development, which was to become a permanent Hungarian institution staffed by Hungarians. The Center would provide training services, research services, consulting, and networking. SUNY branches in Albany, Buffalo, Binghamton, and Stony Brook were all going to participate in setting up the Center, as would the BUES, the College of Finance and Accounting, the College of Foreign Trade, the Technical University of Budapest, the SEED Foundation, and others. Interestingly, the original proposal to create the Center, dated March 26, 1990, promised the creation of two MBA programs at two universities; a second proposal aiming at funding three additional years of operation for the Center, dated May 15, 1990, makes no mention of MBA programs, suggesting that realism superseded some of the initial enthusiasm, although in the end, several MBA or MBA-like programs were enriched. In any event, the Center was created and functioned in a useful capacity for almost eight years. It relied heavily on continuing support from USAID. When that agency ended its programs in Hungary, the Center had no choice but to go into dissolution, which occurred at the beginning of 1999.

While it existed, the activities of the Center were vast and affected many aspects of business development and training. By 1992, the Center was operating some 70 two-day training courses per year, reaching about 1,750 persons and a number of one or two-week management courses. It was involved in public education, curriculum development, and similar activities. As recently as 1998, when the ultimate liquidation of the Center was a *fait accompli*, the Center ran an agro-business program in the town of Kiskunhalas and arranged for the publication of the Hungarian translation of Robert C. Camp's *Business Process Benchmarking* and A. Tenner and I. DeToro's *Process Redesign*. It developed a two-stage program for improving the wine industry, attended by 92 wine makers, restaurateurs, consulting firms, viticulture teachers, and tourist office managers. It purchased and distributed computer equipment, and undertook many other activities.[15]

What did the Center actually accomplish and how lasting were its contributions? The marketing department of the Budapest University of Economic Sciences prepared a major assessment of the activities of the Center.[16] The evaluation was based on a complex methodology, involving questionnaires, expert interviews, telephone interviews, and the study of documents of general background.

The crude outputs of the project were (1) the training in short courses of about 5,000 Hungarians from small, medium, and large enterprises; (2) the training of 60 faculty members and managers in three-week programs and of 20 others in eight-week programs; (3) the development of an MBA program at the Technical University of Budapest through faculty retraining; (4) the improvement of teaching and consulting capabilities at seven regional universities.[17] Since many of the activities were based on existing institutions, the evaluators felt that the long-run sustainability of the efforts was good, and also felt that a very large body of "knowledge-stock" had been created through faculty training and other training activities—a prerequisite of effective technology transfer. Indeed, the only way to assure the effective diffusion of technical knowledge is by starting on a significant scale; token efforts may have moderately significant local effects but are unlikely to seep through the relevant sectors of the economy. The training philosophy was based on the principles of Total Quality Management (TQM); the argument for it is illustrated in the following excerpt from the report:

> Management practices in the Hungarian economy, including quality management, have changed dramatically. Quality management, where it existed, was politically influenced in the former regime. Since the changes, TQM practices have been forced by the collapse of the former Comecon markets and the necessity to increase export performance to the European Union. Firms soon came to realize that without quality certification these EU markets would be out of their reach. Quality assurance became a need, which has led to the growth of the ISO 900x market development.[18]

TQM curricula were developed and implemented and six TQM reference libraries were established. This seemed to be a very wise decision. TQM concentrates on a customer focus and process improvement, both of which were lacking in the planned economies of Eastern Europe because anything could be sold in their shortage economies and pleasing the customer was irrelevant. Benchmarking, the search for best practices that lead to superior performance, was unknown, and anyone who traveled to the region before the 1990s can tell tales of horror about the extent to which the customers' desires were ignored.[19] In fact, the beneficial effects of TQM in terms of the number of ISO 900x-certified companies can be seen in table 4-1.

Throughout the training, the participants reported medium-to-high levels of satisfaction, but the report faults both the project and the networking activities of the Center. The report concludes that if networking is deemed important, professional management is essential, and the Center lacked that component. What counterbalanced the lack of professional management was

Table 4-1 ISO 900x: Number of companies certified.

1991	1992	1993	1994	1995	1996	1997	1998
2	14	37	101	209	321	529	550

the strong backing of the entire effort by SUNY-Albany, as a result of which the activities of the Center were largely successful and it can be deemed to have achieved its objectives.

One other interesting note on the short-term training for entrepreneurs: the director of the Center, Árpád Kézdi, and the director of SEED, Anikó Soltész, confirmed in 1992 that there was a great glut of short-term training courses, because too many "single-shot" operations were being organized by German, American, British, Dutch, and French assistance organizations. I concluded that in the future we had to investigate carefully whether a proposed course of this type had positive marginal value.

SPEDD, Inc. Previously known as the Southwestern Pennsylvania Economic Development District, SPEDD has had a distinguished history of supporting business incubators. It started its activities in 1982, and by 1989 it had 15 facilities, over 150 business clients as tenants, and over 200 small businesses as participants in its program. SPEDD made an early appearance in Hungary, when a staff member, Pier Benci, introduced the concept of incubators to Hungarian audiences. Dr. Tibor of SEED responded in a letter,

> The lectures had a great success among the government representatives, experts, local councils, universities, unions, trade associations representatives on the spot. According to your lectures, lots of them realized that our country will need a network like this because an enormous need will come up for such a service. [*sic*][20]

A similar warm response came from Peter Szirmai, cochairman of the Hungarian National Association of Entrepreneurs. On May 9, 1990, SPEDD submitted a proposal for the Mellon Foundation's consideration in which it proposed to undertake, in collaboration with the SEED Foundation, the training of ten persons in the techniques of developing and managing business incubators. SPEDD staff would spend time in Hungary and all ten trainees would spend four weeks at SPEDD centers in the United States, with training activities culminating in the drafting of plans for several business projects. The first grant of $48,800 was quickly followed in 1991 by a smaller one of $13,600. Concentrating on the provinces rather than Budapest, SPEDD succeeded in creating the Hungarian National Incubator Association and developed close cooperation with several small-business advisory and support organizations, such as Regio in Ozd, the Talent Foundation in Szeged, PRIMOM in Nyiregyháza, DN Starter in Tatabánya, Coopsystem and SEED in

Budapest, and the BAZ County Regional Enterprise Agency in Miskolc. Two incubators were in full operation by 1992 and three additional ones had been started; SPEDD also assisted Hungarian companies to find U.S. trading partners, supported a citizens' network for business, and directly assisted Hungarian incubators by providing U.S. internships. A follow-up grant of $50,000 in 1991 provided further training, enhanced the effectiveness of local economic development organizations by improving the communications and service infrastructure, and established organizational relationships between economic development organizations and citizens' associations. All this represented significant achievements, and when a new proposal arrived in 1992 to extend these activities, to provide further on-site training, develop future programs at the incubators (having grown to eight in number), deliver various sorts of technical assistance, and most importantly, to assist the Hungarians in developing a SPEDD-type business database that entrepreneurs could interrogate, the Foundation was pleased to make another grant of $159,000. In 1993, the Talent Foundation in the city of Szeged received a grant $49,000, and in 1995, SPEDD and Talent joined forces for a combined project requiring a grant of $117,000 that would solidify business-to-business interactions between Hungary and the United States, develop greater marketing expertise in Hungary, and develop specific marketing strategies for 50 Hungarian businesses.

The SPEDD strategy was just right: it concentrated on transferring technology and, at the same time, on creating networks and collaborative structures that would become self-reinforcing. SPEDD wisely concentrated on the provinces and selected good partners, and it had a permanent effect on the development of small businesses in Hungary.

Polish-American Small Business Advisory Foundation (PASBAF). The number of small businesses in Poland grew rapidly from a base of some 450,000 firms in 1988 to 860,000 firms in 1989, 1.2 million in 1990, 1.5 million in 1991, 1.6 million in 1992, 1.7 million in 1994, 2 million in 1995, and 2.2 million in 1996, employing a total of 6.3 million persons.[21] This increase was not unanticipated, and in 1990, the U.S. Congress created the Central European Small Business Enterprise Development Commission (CESBED) (Public Law 101–515) for the purpose of fostering small business development in Hungary, Poland, and Czechoslovakia with a budget of $1 million in its first year of operation (1991) and a budget of $1.5 million in the each of the next two years.[22] CESBED contracted with the Scientex Corporation in Washington, D.C. to set up a Polish program, which, in collaboration with the Polish Chamber of Commerce, succeeded in creating PASBAF in 1992. PASBAF staff received training in the United States and three advisory centers were set up, in Warsaw, Gdańsk and Łódż.[23] From the very beginning, PASBAF management (in particular, the redoubtable Mieczysław Bąk) was aiming towards self-sufficiency, because the eventual cessation of government funding and the decline of private funding was never in doubt. The activities of PASBAF consisted of counseling individual businesses and entrepreneurs on problems of management, finance, taxation, and marketing; organizing and running workshops and seminars on various topics for small entrepreneurs; and providing databanks and printed resource materials for them. These objectives were very much in line with the Mellon Foundation's thinking and PASBAF received grants of $194,000 in 1994, $125,000 in 1995, and $92,000 in 1996. By 1994, PASBAF had counseled over 1,100 entrepreneurs and had taught in workshops and seminars some 2,500 small business managers. Because the demand for its services was so great, PASBAF opened additional centers in Bielsko-Biała, Konin and Lublin, but wisely decided to close the Łódż center when it became clear that that particular city had an oversupply of small business advising. The substantive content of PASBAF's efforts changed according to need, and when it was perceived that many small entrepreneurs could profitably engage in foreign trade, appropriate seminars on relevant topics were held. Like the SEED Foundation, from the outset PASBAF did not charge for its counseling services because at that time Polish law prohibited foundations from submitting invoices for services. By 1996, some small charges were being levied, but more importantly, PASBAF was also targeting training activities towards organizations with a greater ability to pay, such as banks and local chambers of commerce. For example, in the 1995–98 period, some 220 credit inspectors from 70 banks were trained in specially targeted workshops. By 1995, 20% and by 1996, 40% of PASBAF's revenues were from such sources. By the latter part of the decade, PASBAF had expertise in the evaluation of investment projects, developing business plans, preparing credit applications, analyzing competition and promotional techniques and marketing plans, and was providing precisely the kind of advice that small entrepreneurs need but cannot obtain on their own in a post-socialist state. In the late 1990s, PASBAF also began to play a significant role in influencing public policy by advising the legislature (Sejm and Senate) and in participating in the evaluation of the budgets for the years 1997–99 PASBAF started to provide advisory services for public authorities on the local (municipality and gmina) level. Most recently, PASBAF has started to provide east–east assistance by exporting its expertise to Ukraine and is working with Ukrainian NGOs to provide counseling for small and medium enterprises there and, to a smaller extent, to Kyrgyzstan, Moldova, Tadjikistan, and Armenia. It also conducts a great deal of applied research on timely issues of economics and business and has received financial support from business firms, USAID and PHARE. One particularly worthwhile project consisted of the preparation of ten case studies in business ethics.[24] At present, PASBAF operates directly only two centers (in Warsaw and Gdynia), with the remaining ones (in Konin, Lublin, and Bielsko Biała) having been spun off to local chambers of commerce. Its principal source of income is donations from banks. Entrepreneurs are not charged for basic or elementary advice, but do pay for more complicated help such as the preparation of business plans, market research, or the development of a distribution network. Its revenues are adequate for its operations and there is no expectation that the contributions by the banks will diminish; hence, sustainability has been achieved.

PASBAF operated on a somewhat larger scale than the SEED Foundation, but they played remarkably similar roles and had remarkably similar effectiveness. Two questionnaire surveys examined the effectiveness of the counseling and seminar/workshop activities of all the Polish centers in 1993 and 1994, largely by measuring client satisfaction (Hoy and Kulawczuk). While the centers did not meet all their clients' expectations, 1994 represented a substantial improvement over 1993, and in no category was the satisfaction rate less than 50%. Counselors understood the clients' needs moderately well, well, or totally in 61% of the cases and the counseling advice was at least moderately useful in 66% of the cases. The effect of counseling on sales revenue and profits was again more favorable in 1994 than in 1993, but even in 1994 less than 50% of clients reported any effect at all, with similar levels of impact from the seminars. On the whole, this is not a remarkable but a respectable showing.

SEED and PASBAF are both examples of technology transfer in which the western influence was almost exclusively in creating the organizations and in providing funds; unlike MBA programs located in universities, there was no massive inflow of western know-how. Even if the effects were less than hundred percent, small size and personal dedication made these programs very cost-effective. The lesson from PASBAF and SEED is that success is contingent on adapting rapidly to changing circumstances and on the organization not becoming bloated.

Other Programs. A few additional, sometimes but not always smaller scale, programs were also funded in those first few years. In 1990 and 1991, the most noteworthy of these was a set of workshops on banking that was conducted by K. J. Walraven, affiliated with Georgetown University, in the city of Lublin (see also the discussion in chapter 12). These were very important because growth in private enterprise activities was not conceivable without an adequate banking system, and because bankers in Poland were typically not familiar with concepts such as bank accounting, risk analysis and credit evaluation, commercial bank management, branch banking, investment banking, and related functions. Walraven affiliated his project with Catholic University Lublin (KUL) and trained dozens of people in the elements of banking. This also cemented a closer relationship between Mellon and KUL, which was to have significant effects later.

In 1993 and 1995, the Mellon Foundation made relatively small grants to the Talent Foundation[25] in Szeged, Hungary, a small business advisory organization not unlike the SEED Foundation. Talent was in fact the creator of the first Hungarian business incubator in 1991 and the first Mellon grant was for supporting this incubator work; the second was to establish contacts between Hungarian entrepreneurs and potential trading partners in the

United States. As indicated above, Talent was another organization that worked closely with SPEDD, Inc.

In Czechoslovakia, and later in Slovakia only, small business consulting was provided by the Central Europe Institute, a small, Washington-based NGO operating on a shoestring. It had done excellent work, but its business advisory operations do not have to be reviewed here in detail. What is very interesting is that this organization, or rather its dynamic president, Peter McFadden, knew exactly how to alter its image and mission as time passed, so as to respond to the local needs. Because this is an instructive story in itself, I shall recount this phase of CEI's operations later, in chapter 12.

Many other programs and organizations came into being that had as their prime objective the conversion of planned economies into market economies, either through training activities or through research or both. Examples are the Adam Smith Research Centre in Warsaw, headed in the early 1990s by the distinguished economist Jan Winiecki; the Research Center for Marketization and Property Reform in Gdańsk directed by Jan Szomburg; the Foundation of Social and Economic Initiatives in Poland; the Foundation for Economic Education (Fundacja Edukacji Ekonomicznej) in Warsaw, led by former Finance Minister Leszek Balcerowicz; the American Independent Journalism Foundation, devoted to promoting independent media in the region; the Center for Economic Development in Bratislava, which aimed at promoting the active participation of citizens in economic activities and their involvement in private enterprises, and enhancing knowledge about the economy; the Márai Foundation in Bratislava devoted to ethnic conflict resolution and confidence building in the Carpathian region, and many others. We considered supporting some of these, and many did more than creditable work. But it was counterproductive to spread ourselves too thin, and none of those mentioned in this paragraph received grants.

Apart from the significant impact of the programs that the Mellon Foundation supported and their high cost-effectiveness, some other lessons emerged from these early activities. First of all, business training and related activities attracted many persons, and many new institutions were created. But most of these could not rely on indigenous funding for their survival. (For example, the president of the National Bank of Hungary patiently listened to a two-hour presentation of what SEED was about, but provided no financial support.) When the western funds disappeared, the activity tended to disappear (SEED and PASBAF are exceptions to this generalization), unless it was centered in an organization with a prior record of stability. Secondly, the persons engaged in learning how to become trainers and participating in the training of others were themselves subject to the pressures of rapidly changing opportunities. Thus, they were extremely mobile, and western funders could not count on personnel continuity. In the previously cited letter, Vidic recounts how a group from Hungary visited SPEDD for the purpose of promoting privatization. They were shown the business cards of over 300 Hungarians with whom SPEDD had had contact in the 1990–95 period; the visitors recognized only a single name among the 300+ shown to them. This is not a circumstance conducive to building permanent relations. As a result of the great mobility of people and the impermanence of most of the institutions in this sector, many requests for funding were declined. I tried to select for support those that seemed to have the best chance for survival.

Computing and Connectivity

All three projects discussed in the previous section were originated, at least in part, by Americans. It was very clear that we could not and should not rely entirely on such projects and that it was important to involve East European agencies directly. It was also clear that I had to make personal contact with many organizations in the region if this was to go forward at all, and particularly so if I wanted to develop assistance programs for universities. Accordingly, I spent two weeks in Hungary and in Czechoslovakia in May/June of 1990, where I met with some 55 different people at universities, institutes of academies of science, national libraries, and other organizations. I had a dual purpose: I needed to get clearer ideas about what it was that Mellon could usefully support, and I had to "introduce" the Mellon Foundation to potential grantees who had never heard about it until they received a letter from me in which I suggested that I come to see them.

It was amply clear that computing technology and electronic connectivity were grossly underdeveloped in the region. In the early computer era (say, the 1950s and 1960s), computers were largely regarded as extremely fast calculating machines, but in the past 20 years, it has become clear that computers are efficient symbol manipulators, and information technology started to significantly raise economic productivity.[27] Hence, this was not an area to be neglected.

In November 1990, I spent two weeks in Poland and visited, among others, Catholic University Lublin (KUL), where I made an obligatory stop at the university computing center. The entire equipment of the computing center was an aging and, for lack of spare parts, nonfunctional PDP-11 (Digital Equipment Corporation) and a couple of IBM PCs; I realized that I had more computing power on my desk at home than the entire university computing center. At other institutions, I had seen makeshift arrangements consisting of a RIAD (Soviet) clone of an IBM mainframe, using a genuine IBM multiplexor and attached to East German tape units; miraculously, this Rube Goldberg machine actually worked. It was not the ingenuity of the people that was the matter, but the scarcity of funds and the COCOM regulations.

Initial interviews were always a challenge. First, I had to convince my hosts that the Mellon Foundation was "for real," that is, that substantial grants were likely to be made if satisfactory proposals were to be submitted. But I also had to be careful not to promise too much; after all, I did not know what the quality of the proposals that we would receive would be. I therefore kept insisting that the trustees of the Foundation have many difficult choices to make and that the process of receiving a grant was anything but automatic. In addition, I also tried to educate my hosts in the art of fundraising in general. This has continued to be a matter of great difficulty to the present day for a variety of historical and cultural reasons. During an early trip to Hungary, I visited the rector of József Attila University, Béla Csákány. At one point I decided to encourage him to seek support from other American and West European Foundations. I urged him to be aggressive in seeking out money for various university projects, told him that universities had to learn the art of proposal writing, that they had to invest some resources in learning about what kinds of projects the various foundations might support, and so on. When I was finished, he lowered his eyes, and said in a soft voice, "Hungarian gentlemen do not beg for money." And it was ever so thereafter; while Mellon and OSI (Soros) did receive a fair number of unsolicited proposals because of the widely held perception of their proactive attitude, I suspect that in the early days few attempts were made by universities to get support from other foundations and hardly any to get support from corporations.

Connectivity. The disparity in Internet use between North America, Western Europe and Eastern Europe, could hardly have been more glaring in 1990. Stephen Ruth cites recent results that show that Internet use per 1,000 persons is 492.6 in North America, 220.5 in Western Europe and Scandinavia, and 37.2 in Eastern Europe.[28] The first question one needs to ask is why it is important to create substantial connectivity. Is it not the case that e-mailing at all hours of the day or night and surfing the web are self-indulgent activities and the time saved by using e-mail is negligible in the greater scheme of things? Those who advocate or demand improvements in connectivity must be motivated by the conviction that connectivity enhances productivity.

While there are arguments on both sides of this question, on balance the productivity enhancing character of connectivity is well established. In a study of Chilean scientific projects, Ruth and Gouet strongly confirm that network users are significantly more likely to produce scholarly output than nonusers.[29] In particular, they confirm that network users will be more productive scientists and will be more familiar with the milieu of scientific networks, and that network impact on professional activity will be perceived to be stronger than other network effects. The most recent evidence about this is in a careful paper by Ruth that analyzes the immediate results, as well as the more distant research and instructional impact, of Internet training conducted at the Romanian Academy of Sciences between 1996 and 1999 with funds provided by a Mellon grant (see also later in this chapter). Some 521 persons received a three-day Internet training course and were asked at the end of this period to complete a detailed questionnaire. The data were analyzed using analysis of variance, and while interesting differences were found between the genders and between scholars in the humanities versus engineering and mathematics, the most important difference for our pur-

> quire an IVL to the Soviet Union. Its classification would depend on the exact nature of the data being sent.
>
> On the basis of the information provided in your letter, the Department of Commerce would probably recommend approval of an export license application to provide BITNET access for the Soviet Union. However, the outcome of such an application could not be predicted, and would depend on the technical details of the application and on the results of the review process.

This, so far, was not particularly encouraging. But many persons were interested in the extension of BITNET/EARN to Eastern Europe and in enhancing the computational power in the region with a view towards improving higher education and research. On January 8, 1990, a seven-page advisory letter from the Department of Commerce somewhat clarified the situation. I excerpt below salient points from the letter of William Clements, director of the Office of Technology and Policy Analysis.

> In the absence of a validated general license, the EAR [Export Administration Regulations] prohibits exports or reexports from the United States of computers, switches, software, and technical data. The term technical data is broadly defined to include virtually all know-how. In addition, the regulations apply to exports in any form, including electronic transmission from the United States and disclosure to foreign nationals in the United States or abroad. . . .
>
> Wide area networks exported to the Soviet Union require an individual validated license. If such validated licenses are issued for exports from the United States of items that will enable the consignee to build a network or establish a gateway for a network, then the EAR does not prohibit or require permission to establish or use such a network so long as prohibited exports are not made over the network . . .
>
> However, I must underscore that this letter does not constitute permission or authority for BITNET or any BITNET member or user to export from the United States technical data or software . . . each individual and each institutional member is responsible for compliance with the EAR.
>
> You have explained that BITNET does not monitor traffic on the network. It is a non-secure network . . . BITNET has taken the position that it is similar to a telephone common carrier, which is not responsible for the illegal export of technical data over its lines by a subscriber, so long as that happens without the knowledge or reason to know of the common carrier. In our opinion, BITNET has a closer relationship to its members than a telephone carrier. . . . BITNET has more opportunity to learn of the export activities of its members than does a telephone common carrier. In exercising its appropriate duty of care under the EAR, BITNET must take into account information provided to it by members. . . .
>
> Your letter refers to user access to BITNET servers and data services. You have informed us . . . that such information is not proprietary and is published electronically without restriction on distribution. . . . However, . . . in the future BITNET may add proprietary data bases that will not

> qualify for General License GTDA. If BITNET becomes the provider of such information in the future. . . . BITNET will require classification of the technical data. . . . If your members export software by file transfer over the network, they must establish or obtain the appropriate general license or validated license.
>
> You have indicated that the network does not enable the user to log onto a remote computer and use the computing capabilities of that remote computer. . . . If the network permitted remote computing by foreign users, that would raise questions for Commerce. For foreign computers, we have long applied a theory we refer to as diversion in place. For example, if a computer is exported to a western country "A" for use in that country, then the stated end-use for the export is breached or violated if a user in an eastern country "B" enters that computer via modem and telephone line in a manner that enables him to use the computing power of the computer. . . . For supercomputer installations. . . . BITNET should also be aware of the basic prohibitions:
>
> 1. Both direct and remote computational access by COCOM-proscribed nationals or work done on their behalf.
> 2. Transfer of technical data or software derived from the use of the supercomputer which relate to COCOM-controlled items.[38]

Various legal opinions were obtained and Edward J. Bergman, CREN corporate counsel, wrote to Ira Fuchs, president of the CREN board of trustees, on February 7, 1990:

> In conclusion, counsel recommends that the proposed connection to entities within Eastern European Countries and The People's Republic of China to the network proceed via entertainment of such applications for processing according to existing membership guidelines. It is further recommended that the General License GTDA Section 2410 of the Export Regulation Act, the Commerce Letter and the Memorandum Opinion of Counsel be disseminated to the membership over the network. All interested parties should be aware that a conservative approach mandates that any questions or gray areas be the subject of inquiries addressed to counsel.

Basically, this meant that we were in business.[39] The East European governments were persuaded to foster the development of national backbones and metropolitan area networks, both of which are essential for the diffusion of connectivity, and highly talented university people in each country played essential roles in the process.[40] A solid step in this direction was the completion in the first half of 1993 of CESNET, the Czechoslovak backbone, which had a bandwidth of 64Kbits/sec.

Grants in Czechoslovakia. Several other influential people were taking an interest in computing and connectivity in the region. Former U.S. Ambassador to Romania, Harry Barnes, who for understandable reasons may have had greatest interest in Romania but was definitely also interested in other countries, was in frequent touch with Stephen Ruth in the department of decision sciences at George Mason University, with James Conklin at BITNET, and with me. Ruth had a tremendous enthusiasm for empowering the East European scholarly community by bringing BITNET/EARN to them. Barnes, with a group from the German Marshall Fund of the United States, visited Prague between February 28 and March 3, 1990, and determined some key facts: that the Institute for Information Theory and Automation of the Czechoslovak Academy of Sciences had an IBM 4631 to which ten of the Academy's institutes were already linked; that there were mainframe computers at Charles University, the Czechoslovak Technical University, and in Bratislava (the latter having been financed by UNESCO); that Charles University had plans for a university-wide computer net; and that Ira Fuchs of Princeton and of CREN had been urging Charles University to try and create a link to EARN/BITNET.[41] By early April, Barnes had put me in touch with Vladimír Kučera, director of the Institute of Information Theory and Automation. I soon had set up an appointment to see him and his network expert, Karel Šmuk, on June 3, when I was to be in Prague. In short order, I also obtained appointments to see Dr. Bohumil Palek of the Charles University Computing Center, and most importantly, Lubomír Ohera and Jan Gruntorád of the Czechoslovak Technical University. By the time I had arrived, the Czechoslovak membership application to EARN had been accepted. It had been agreed that the Czechoslovak Technical University would become the Czechoslovak EARN node (only one node per country was allowed), with Gruntorád as the director of this effort, and all we had to really talk about was how we could assist financially the creation of the appropriate connectivity. The Technical University's needs were modest; all they required were some modems and switches for a total of $28,000, which the Foundation quickly made available. According to Dr. Palek, Charles University also expected to be connected to EARN via a separate link to Nijmegen, the Netherlands, but he emphasized that he was not in competition with the Czechoslovak Technical University for becoming *the* EARN node. This sounded very disingenuous to me, and I thought that the Technical University marshaled more talent for this task anyway.

The Institute for Information Theory and Automation had a larger shopping list.[42] It took a little time to assemble the necessary documentation, but by October 24, I was able to recommend to the president of the Mellon Foundation that a grant of $357,000 be made to the Institute. I said,

> My sense is that if you and the Trustees approve this grant, we shall have made a substantial impact on the Prague computing establishment with our grants to this Institute, to ASCOC [see below] and to the Czechoslovak Technical University. The area where comparable help will be needed to a relatively greater extent *after* this grant is made will be at the provincial universities which are even farther behind.[43]

By June 1992, Vladimír Kučera, director of the Institute, could report that the enhanced computing support made it possible to carry out six new projects, which would not have been possible without it, and that the local network and other equipment acquired through the Mellon grant would increase the efficiency of scientific research, as well as the complexity of the problems to be solved.[44]

The Foundation also made a grant of $246,000 to George Mason University, so that Ruth could provide value-added services for computing and networking. The GMU grant interacted with all the other grants that the Foundation made in the computing/networking area. Ruth spent inordinate amounts of effort to provide modems, computers, disk drives, CD-ROM readers, laptop computers, software, and training to Czechoslovak institutions. In the years that followed, it was practically the case that wherever I appeared, Ruth had already been there, teaching people about e-mail and providing some hardware.[45] Progress during the 1990–92 period was quite rapid. Ruth reported to the Foundation as follows:

> The goals that were established for the first year were specific and measurable and they have all been exceeded. The number of users is twice as high as anticipated . . . [The grant] provided the opportunity to find and utilize a cadre of men and women in CSFR [Czech and Slovak Federal Republic] who are so energetic, capable and influential in their academic environments that they have been able to provide rapid and visible results in a very short time. . . . This current work . . . is fully replicable in many settings and the results are far out of proportion to the investment.[46]

Ruth's project was so successful that the Foundation departed from its geographic guidelines and made another grant to George Mason University in June 1991 in the amount of $330,000, so the he could replicate his activities in Romania. Some of the elementary instructional activities that he was going to undertake there included updating people on network operating procedures, training in information technologies, "hand-holding" with prospective computer users, hands-on classes in using computer systems, help with establishing contacts with colleagues worldwide, working with BITNET/EARN officials to improve service, help with establishing bulletin boards and many similar activities. In general, the assistance that the George Mason University projects provided fell into three categories: small, focused, short-term undertakings like purchasing modems, providing Fidonet training, or providing software tools for improved use of MOSAIC, a predecessor of Netscape; larger, one-time investments, such as the purchase of telecommunications equipment for the Czechoslovak Technical University or the purchase of a large Digital Equipment computer for the Institute of Atomic Physics in Romania; and collaborative projects with other foundations that had already made some investments in connectivity, such as the Soros Foundation and the German Marshall Fund of the United States. One of the greatest successes of this operation was the Romanian Internet Academy, affiliated with the Romanian Academy of Sciences, which provided rigorous training for about 1,000 persons in the past four years (Ruth).

By 1992, 11 EARN nodes were in operation, there were some 800 Eunet users in Czechoslovakia,[47] the first experimental Internet connection came into being in the fall of 1991, and the FESNet project was being organized to provide a multiprotocol infrastructure, which would permit users to take advantage of all major networks, such as EARN, Eunet, and the Internet. The growth of traffic is illustrated by the fact that from October, 1990 to March, 1992, the number of 80-character records transmitted within the country had grown from 6,000 to 11,803,000 and internationally from 560,000 to 10,691,000.[46]

But in spite of excellent general progress, pockets of severely retarded development remained. In June, 1990, I also visited Košice, in Eastern Slovakia, where the main branch of P. J. Šafárik University (UPJS) is located, and where I met the outstanding Rector Lev Bukovský and several of his colleagues. One of these was Professor Zoltan Tomori, a research physician specializing in respiratory physiology who was sufficiently well known for his work to be invited to do research at Dartmouth Medical School. He was carrying out real-time analysis of data generated by physiological experiments on an ancient Commodore computer with 64 Kbyte of RAM and on some IBM XTs! Slovakia established SANET, the Slovak Academic Network, supported by the Ministry of Education, in 1992. The initial connection between the two branches of Šafárika University (in Košice and Prešov) was 9.6 Kbits/sec. By December 1999, SANET provided 2 Mbit/sec connections between Bratislava on the one hand, Trnava, Žilina, Banská Bystrica, and Košice on the other hand, and 384 Kbit/sec connections between Košice and Prešov.[49] Progress throughout the region was generally very substantial; by the year 2000, Estonian connectivity had been upgraded to the point that Tallinn and Helsinki were connected via a 4 Mbit/sec line, Tallinn and Tartu via an 8 Mbit/sec line, and a sizeable number of other city pairs were connected through 2 Mbit/sec, 768 Kbit/sec, or, at worst, 512 Kbit/sec lines.[50] At the same time, in Hungary, a 155 Mbit/sec backbone had become operational.

During the spring of 1990, I also had extensive correspondence with František Janouch, president of the Stockholm branch of Charter 77, the parent organization at that time of the New York and Prague branches. In a letter of February 5, he pointed out two vital needs of higher educational institutions in Czechoslovakia: (1) library books, particularly in social and political science, history, history of art and literature, economics, art, modern fiction and poetry, and (2) modernization of computing facilities. He said,

> I have experienced a certain shock when being told that a PC with 386 processor (like IBM 386 or Compaq 386 with a 100 Mbyte hard-disk) with a laser printer would increase the computing power of the Institute of Physics (the Czechoslovak Academy of Sciences) by a factor of ten or more and solve their problems for many years to come.[51]

It was clear that he was not exaggerating, because similar situations could be encountered elsewhere as noted above.

A many-sided dialogue ensued as a result of the correspondence with Janouch, in which I got to know Dr. J. Nadrchall of the Institute of Physics, and I also put Janouch in touch with Ohera and Gruntorád. Ruth also had extensive conversations with Nadrchall, because he, too, was beginning to be active in the country with the grant from the Mellon Foundation. Out of all this, a proposal was born which promised the establishment of the Advanced Scientific Computer Center (ASCOC) in Prague. The proposal pointed out that the Soviet policy of essentially copying IBM 370 class computers led to the complete failure of the Soviet system to develop modern computing equipment and that, together with the US COCOM regulations, caused all countries in the Eastern Bloc to fall severely behind. ASCOC was to be operated jointly by the Czechoslovak Charta 77 organization and the Academy of Sciences, the main purpose of ASCOC being "to introduce modern computing method into all fields of science and humanities and simultaneously to enhance the access for a broader scientific community to contemporary computer technology and programs."[52] I was concerned whether the humanities and social sciences would get their share of such an establishment, and in an e-mail to Janouch dated April 23, I said,

> On the operations of ASCOC, a question will be raised whether any significant benefit will accrue from ASCOC to the humanities. The Mellon Foundation has supported the humanities in a major way in the past and if one could demonstrate some use for the humanities, this would solve a possible question (e.g., if one could say that there is a scholar of ancient Greek who would like to do a computer concordance of some Greek author or who desperately needs this sophisticated software that has all ancient Greek texts [Thesaurus Linguae Graecae], etc., etc.).

But Nadrchall thought very highly of the proposal and, in the end, a grant of $150,000 was made for the benefit of ASCOC, which decided to locate in a former trade union building that the government made available for the newly created Central European University, the Center for Research and Graduate Education (CERGE) (see chapter 6), and ASCOC. This was a large building but by no means ideal; as late as May 1991, the nine-story building had only three telephone lines. In any event, I do not believe that ASCOC received the broad use that I had hoped for. It ended up buying a STARDENT 3000VS mini-supercomputer, which did not strike me as desirable, except for doing some very high-powered computations. By 1992, I received reports that ASCOC could still not be reached remotely and that it was a woefully underutilized resource. Dr. Jan Koukal, a computer scientist, became the director of ASCOC, but after a while he decided that politics was more to his liking and he was elected mayor of Prague. Eventually, ASCOC was taken over by the Institute of Physics of the Czech Academy of Sciences, where it continues to operate as a special-purpose facility for a handful of physicists and graduate students.

Two other institutions deserve special mention. By 1991, I was busily discussing the computing needs of Comenius University[53] in Bratislava with its prorector (later rector), Ferdinand Devinsky. Comenius was founded in 1918, and in 1991 had about 17,000 students, 5,300 staff members, and an annual budget of about $20 million—a ludicrously small amount compared to western universities of comparable size. But Comenius University was undoutedly the most distinguished institution of higher learning in Slovakia and computing conditions there were intolerable. During the summer of 1991, only the rector's office, the mathematics department, and the natural science departments were connected by a rudimentary electronic network. They proposed to extend the network, install a Vax computer from Digital Equipment Corporation as a server, connect to the Slovak Technical University (the closest gateway to the outside world), all the while making certain that the network would support not only the X.25 protocol but TCP/IP, as well, and that it could accommodate dealing with Novell subnetworks and with IBM mainframes. This was entirely appropriate, particularly in the light of the creation, in April 1991, of SANET, the Slovak Academic Network. Most remarkably, the university was willing to put $170,000 of its own money into the project, paving the way for later matching grants.[54] Fortunately, DEC was as generous with discounts here as it had been in Hungary, (see later in this chapter) and in the fall of 1991 the Mellon Foundation gave Comenius a $266,000 grant.[55] As a result of the rapid progress in building the university network, the demand grew at such a pace that by 1994 the server could no longer handle the traffic, and Comenius requested and received another grant of $60,000 to upgrade the server; by that time, the days of Vaxes had passed and the alpha-chip architecture replaced it. It is noteworthy that seven of the 12 faculties (divisions) of the university were connected to the university backbone and more than 1,000 PCs were on the net. The total amount of its own funds that Comenius University had invested in the project by that time amounted to $728,000. The Ministry of Education had provided a further $245,000. It is fair to say that this project had an enormous impact on the university.

The other university in Czechoslovakia in which the Mellon grants had an enormous impact was Palacký University in Olomouc, the second oldest university in the Czech Republic, founded by the Jesuits in 1573. It nearly disappeared when the Jesuits were expelled from the Austro-Hungarian Empire in the nineteenth century, and it reopened in 1946.[56] By January 1991, the forward-looking rector of the university, Josef Jařab, was using a modem connected to an old PC to send messages over the EARN network.[59] Discussions with the university began in 1990, and when it became clear that what was needed most was the development of an information infrastructure for the university, I suggested that Stephen Ruth become a consultant to the university for purposes of proposal preparation. This was willingly accepted. The proposal identified the key problems, such as the absence of a computer network, the use of old-fashioned machines that were "wrong" for the tasks at hand, the lack of training in the use of computers, and the consequences of early enthusiasms for computing which led to the purchase of wrongly matched pieces of equipment. Palacký University eschewed a piecemeal approach to solving these problems and suggested nothing less than the creation of a University Information Center that would have subcenters for the six academic divisions of the university and one each for the central library, the audio-visual center, the publications center, and the center for computer technology. Phase 1 of the project consisted of designing the details of a university-wide electronic network, the creation of a certain number of ethernets, the purchase of appropriate hardware, and the hiring of relevant personnel. Phase 2 would provide an suitable building for the Information Center, complete the setting up of local area networks at all the divisions of the university and connect them to one another, install a new central computer, establish a regional EARN node, connect terminals to EARN, introduce the CDS/ISIS library automation system, and provide far-ranging CD-ROM services. The university was so certain of its mission, its vision, and its ability to follow through that in the Phase 1 proposal it asserted that if by November 1991 it failed to meet certain benchmarks, it would be willing to forfeit the remaining unspent portion of any grant. This was extremely confidence inspiring and the Mellon Foundation made a first grant in March 1991 in the amount of $200,000, a second grant of $389,000 in December, and a third grant of $32,000 in January 1995. All this made it necessary for me to visit the university several times. In June 1992, Rector Jařab was kind enough to put me up in the university guesthouse, the "Vicarage," which has the distinction that in the 1760s the 11-year-old Mozart stayed there.

Fortunately, additional funds for the project were secured from the Czech Ministry of Education. By the end of 1992, local area networks were functioning in the deans' office, the rector's office, the offices of the faculties of medicine and science, theology, physical education, and pedagogy. An IBM RS/6000 was installed as server for Internet access. In 1993, the local area networks became interconnected at 19.2 Kbits/sec. But as Internet access expanded and the university opted for more centralized network control, bandwidth requirements increased. The university started to cooperate with the City of Olomouc, the Olomouc State Library (ultimately a member of one of the library consortia), Czech Railways and others in order to create a common fiberoptic network. Initially, the system was upgraded to 34 Mbits/sec, and by 1996 work began on upgrading the system to a 155 Mbit/sec ATM network. In the same year, the last of the remote units of the university were connected by 4 Mbit/sec microwave links. All the essential Information Center functions and units envisaged in the Phase 1 proposal came into being and the Information Center has played an important role in turning the university into a research university on the western model (Hladky). It should also be added, but this will be discussed in more detail later, that the changes that were undertaken in these years were a prerequisite for the successful implementation of MOLIN, the Moravian Library Information Network, also supported by the Mellon Foundation (see also chapter 8).

Grants in Hungary. Early in 1990, I began a correspondence with academic administrators at Eötvös Lóránd University (ELTE) in Budapest. I had only modest expectations of how much I could achieve by correspondence, but I was pleasantly surprised when on May 10 I received a request from the faculty of natural sciences for a grant of $136,000.[58] Sums of $60,000 and $26,000 of this total were for library acquisitions in the natural sciences and for developing courses in environmental science. But the balance, $50,000, was earmarked for a Microvax 3400 computer in order to create a widespread e-mail facility throughout the natural sciences. While an X.25 network had existed in Hungary, the university was largely uncabled, although ambitious plans existed for cabling and connecting the university buildings spread widely around the city of Budapest.[59] While the completion of the ELTENET had to wait for another grant of $110,000 in March 1993, of which $75,000 went to ELTE and the balance to the Institute of Computing and Automation (SZTAKI) of the Hungarian Academy of Sciences, which had become the EARN node in Hungary, a server to handle e-mail was deemed essential from the very beginning. The grant was made and, simultaneously, Digital Equipment Corporation decided to grant Hungarian universities a 75% discount from list prices, which enabled the university to purchase a much more powerful machine, namely a Vax 6000–510. As the dean of the faculty of natural sciences expressed in a letter,

> The money from the Mellon Foundation was a crucial and absolutely necessary catalizer [sic] in the—at last—very successful negotiation with DEC. Last week we reached an agreement which means a qualitative [sic] progress in the information system of the whole university: we get a Vax of the 6000 (!) series which would have been impossible without the support of the Mellon Foundation![60]

A curious but perhaps predictable pattern evolved: computer equipment and library software vendors would often not even want to talk to East European universities while they felt that the latter had no money to spend. But as soon as they became aware of grants to the universities from western foundations, they were not only happy to talk to potential clients, but rushed to grant unusually generous discounts. As the dean of the faculty of natural sciences, Ádám Kiss, said in his letter and reiterated for several years whenever we met, the Mellon grant came just at the right time and acted as a major catalyst.[61] This view is further underscored by the fact, reported by Dr. Nadrchall, that prior to the computer and connectivity grants to the various Czech institutions, Digital Equipment had only offered to donate them obsolete PDP-11 computers.

A very similar story can be told about József Attila University in the provincial city of Szeged. I visited the university in May 1990, expecting to repeat what I had said in previous correspondence about what a request for a grant should be like. To my enormous surprise, after I had started to talk about this to the rector and his deputy, they handed me a complete proposal for acquiring a Microvax 3400. At the time, the university mainframe computer was a Russian RIAD built in 1974, with computing power only a fraction of the personal computer I had just acquired for myself. Here, too, Digital Equipment extended the 75% discount and the plans were upgraded to a Vax 6000 Model 410. One other important things happened on the occasion of this visit: I met the University Librarian, Béla Mader, who became a good friend and subsequently played important roles in the development of Hungarian libraries. Progress continued to be rapid and, by the summer of 1991, the university's internal network had grown substantially, so that several departments and administrative offices were linked electronically. The network was growing daily and a fiberoptic ethernet was about to come into service.

But undoubtedly, the most important grant of this type went to the Institute of Computing and Automation. This institute, with a distinguished research record in artificial intelligence, computer graphics, graph theory, combinatorial analysis, operations research, and other fields, collaborated closely with the nearby Technical University of Budapest. The proposal to the Foundation was submitted jointly by the director of the Institute, László Keviczky (later secretary general of the Hungarian Academy of Sciences) and László Zombory, dean of the electrical engineering faculty of the Technical University. At the time, a rudimentary packet-switched X.25 network with a bandwidth of 9.6 Kbits/sec existed in Hungary and connected about 80 institutions of various sorts. Since 1988, an e-mail server (ELLA) was in operation. The network servers were all mainframes; the largest of these was an IBM 4381. In March 1990, the president of EARN, Frode Greisen, informed the Institute that the U.S. Department of Commerce had given formal permission to connect Hungary to EARN and the Institute was designated as the official EARN node. For the time being, Hungary was to use the old IBM spooling system, RSCS, but migration to TCP/IP was envisioned even at this early date. EARN membership was deemed vital for the progress of research and for phasing Hungary into the international network of scholarship. To join Hungary to EARN, some hardware and software was needed, the EARN fees had to be paid, and rent had to be paid for leased lines to the Hungarian Telecommunication Company. Software and hardware costs amounted to about $70,000, the three-year prepayment of EARN fees to $17,500, and the three-year prepayment of leased lines to $135,000, with meeting attendance costs and other minor items accounting for the balance of the $254,000 grant, which was extended in the same year.

Grants in Poland. Two significant grants were made in Poland to enhance computing capabilities as such. The first one was to Catholic University of Lublin (KUL), and I have already alluded to the pitiable state of computing facilities at that institution.

The 1991 request for assistance was extremely modest. KUL requested funds for 34 low-level (286-chip) computers, four 386-chip computers, and some printers, copiers, and word processing and desktop publishing software. All this was easily accomplished with a grant of $182,000.[62] While this was not a grant for cutting-edge computer use, it was a pleasure for me to visit KUL a year later and see students actually sit at computers. KUL had equipped a computer room for students (20 computers) and provided some for the rector's office, the library, the department of experimental psychology and several other units. Most importantly, since the neighboring Marie Curie Skłodowska University, which had become an EARN node, is only a few hundred feet away, it was easy to create a connection between them, and KUL finally had e-mail facilities.

I had first visited Adam Mickiewicz University in Poznań, Poland, in October 1991, where I met Rector Jerzy Fedorowski, University Librarian Artur Jazdon, and numerous other officials. Their most urgent need was for local area networks throughout the university and Unix workstations. The staff members I met with were competent and I was certain that they could produce a good proposal for such a project, which they did and which was funded by the Mellon Foundation in 1992 with a grant of $500,000. When I visited in 1992, I did not expect great progress yet, because the grant had been paid only a few weeks earlier; but there was a great deal of activity, fiberoptic cabling and computers were on order (having received generous discounts from Digital Equipment Corporation), and I was immensely pleased that I could TELNET from Poznań to my computer account at Princeton. Coming years saw the completion of the university network. The rector and faculty reported great satisfaction with the resulting improvement in productivity.

Brief Stocktaking

Even before the first year was over, a number of other initiatives were taking form. But in the initial period, we had made important dents in two areas that clearly needed major assistance. The first of these, the training of entrepreneurs and faculty in how markets and businesses operate, was off to a good start, but the Mellon Foundation was just one of many organizations that felt the need to make contributions in this area. There is no doubt that much technical business knowledge was imparted, but the question of sustainability had not been solved in a definitive way. As we ultimately learned, sprinkling knowledge over the countryside in a more or less random way is by no means useless, because something always "sticks." But it is unlikely to be truly cost-effective unless there are permanent institutions to back up the efforts. Such permanent institutions were the Technical University of Budapest, the newly created SEED Foundation, and PASBAF. An important lesson learned was that the diffusion of managerial technology is considerably enhanced if different organizations succeed in collaborating and building networks among one another, as occurred repeatedly in Hungary.

In the area of computing and networking, the grants fundamentally altered the landscape. During the critical opening moves of the new societies and economies, neither Hungary nor Czechoslovakia could assign enough priority to enhancing connectivity so as to provide the funds that the Foundation did. As Adam Manikowski, later national librarian of Poland, said in a different context, "sooner or later, the automation of Polish libraries would have taken place anyway; however, it is clear that without Mellon Foundation support the automation would have been postponed and would have been driven by an entirely different process."[63]

The same holds true here, too. Eventually, networks would have become upgraded and modernized and faster computers would have been purchased. But it was a key aspect of a successful transformation of the higher education and research establishments that the assistance came at the right time; technology transfer is useless if it is not timely. It also became clear that the Foundation's role was not only to provide funds for particular projects and technology, but to act as a catalyst: donees could often utilize the receipt of a Mellon grant as an incentive for other donors to follow suit. As Ruth had noted, collaboration among donors creates greater leverage. A particularly interesting question is how you can and should measure success. Indicators of success should be monitorable, but can be of two varieties: less active, equipment-centered indicators (for example, the number of PCs delivered), and more active, user-centered metrics (for example, the number of users leveraging training, subjective satisfaction, long-term utility).[64] While in the area of computer connectivity, it may be relatively straightforward to measure the latter types of indicators (say by counting the growth in the number of messages received and sent or by the increase in the number of scientific articles published), in something like small business training, this is not always as easy to accomplish. To be sure, training sessions can and have frequently been evaluated by the trainees, thus providing some measure of subjective satisfaction, but the ultimate test of success here is the extent to which the training has resulted in successful small business formation. While aggregate (national) indicators of small business formation have signaled a lot of success, it is more difficult to attribute this to the diffusion of knowledge provided by any one set of assistance activities.

5

Universities, Priorities, Problems

In all spheres in which western donors were attempting to transfer knowledge, it was important to be sensitive to the needs of the prospective recipients and to avoid patronizing them. This was not always easy to do; in fact, the donors were transferring superior hardware and superior management skills, and such transfers unavoidably required that old ways of doing things be abandoned. Much as one tried to be on one's best behavior, the frustrations of conducting business and coping with the exigencies of everyday life in Eastern Europe added to one's impatience. And frustrations there were plenty.

One May, I spent a week or so in Prague and was unable to reserve rooms at the Intercontinental Hotel, where I always stayed. I ended up in the Atrium, a huge, modern building with a vast interior space, that promised the ultimate in efficiency. The second morning I was there, I decided that I wanted to have the *International Herald Tribune* delivered to my room every day because when I left the hotel in the morning to go about my business, it had not arrived yet and when I returned in the evening, it was all sold out. So I went to the concierge and asked him to arrange for it to be delivered to my room. To my considerable surprise, he asked me "Why?" I explained patiently that I left in the morning before it was delivered to the hotel and when I returned there were no copies left. He understood and sympathized, but informed me that the concierge's desk did not handle such requests and that I would have to speak to the business center about this matter. So I went to the business center, where again I was asked why, and I again explained the reason, perhaps a little less patiently than the first time. The staff person at the business center understood the reason but told me that this was not done by the business center, but by the boutique. So, off I went to the boutique, where I was assured that my request was reasonable, but that the boutique did not do that; I would have to ask the concierge to deliver the paper to my room. So, I returned to the concierge, considerably more frazzled and less patient than 15 minutes earlier, recounted the steps in my odyssey and finally was assured that somehow or other, the newspaper would find its way to my room. I had other unpleasant encounters at the

Atrium, like the time when I had to have the business office make me three or four copies of an eight-page document, which they would not do until I explained why I wanted it done, at which point I lost my temper. Upon departing, I made an appointment to see the manager of the hotel and I explained to him that his employees were acting like Communist apparatchiks who cared nothing about the comfort and convenience of the guest, whereupon he lowered his head and said, with ineffable sadness, "you are right." But we all persevered, and dealing with academics in those years of rapid change was exhilarating and rewarding.

Turmoil in Higher Education

The status quo in higher education was briefly summarized in the last section of chapter 3. It is quite clear from what was said that all countries faced a complex and controversial agenda in a period in which resources were, if anything, diminishing. But substantial forces were being mobilized to assist the development and rationalization of higher education in Central and Eastern Europe. The Institut für die Wissenschaften vom Menschen in Vienna created an Expert Committee to study the problems of higher education and to make recommendations for alleviating them.[1] Numerous agencies supported projects aimed at improving higher education, but not as many American foundations as I would have liked. The World Bank, the European Union and a host of governmental and nongovernmental organizations got into the act. Particularly significant was a $150 million loan that the World Bank had extended to Hungary for education, training, and research; of this sum, $50 million was to be allocated to improving higher education.[2] The World Bank's recommendations for Czechoslovakia included the establishment of government financing criteria for universities, assurances of full autonomy for universities, including full autonomy to commit and use their revenues, and legislation to authorize the levying of tuition.

Growth in Higher Education. The various laws on higher education passed by the countries of the region largely established the autonomy of the universities—but not completely, for they were not generally permitted to dispose of real estate putatively belonging to them. All countries faced the problem that only a small fraction of the population was getting a higher education: in Czechoslovakia, only 7% of the population over 25 held university degrees compared to 17% in West Germany,[3] and 15% of those in the 19–24 age bracket were enrolled in higher educational institutions,[4] while in Hungary, 10% of those in the 18–22 age group were studying at universities, compared to 20–25% in the European Community.[5] In Poland, about 8% of the population over 25 had more than a secondary education in the early 1990s.[6] In general, in East European countries the fraction of the population educated in universities was much smaller than in Western Europe or the United States. The U.S./Polish ratio of this fraction was about

36 in the late 1980s or early 1990s, and the U.S./Polish ratio for Ph.D.s granted per population was approximately 15.[7] A massive push had to be undertaken to increase the student body so as to be able to reach, at least quantitatively, West European standards, at the same time that curricula had to be redesigned, the awkward university structures had to be modernized, and substandard faculty had to be let go. Moreover, since the participation rate in higher education of the young people whose family background was working class or agricultural was substantially lower than that of people whose families belonged to the intelligentsia, the notion of providing equal educational opportunities raised the difficult question of whether admission should be entirely based on merit or whether the system should in some fashion continue the Communist practice of "giving points" to people in whose favor positive discrimination was deemed to be socially desirable.[8]

Fortunately, in a sense, East European universities had excess capacity in personnel, even if substantial portions were useless, either because they were teachers of Marxism-Leninism (compulsory subjects for all students under the earlier regime) or because they had not kept up with their fields. In Czechoslovakia, for example, 15–20% of teaching staff was replaced on the grounds of being unqualified or undesirable (although many teachers had resisted ideological indoctrination and remained teachers of quality).[9] Just as useless teachers were being gotten rid of, so universities needed to get rid of antiquated or doctrinaire syllabi and books. I recall visiting Charles University in Prague in the late spring of 1990; at the turn of a corridor, I and the person accompanying me came upon a pile of books heaped in a large mound, about four to five feet high and with a diameter at the base of perhaps 8 feet. I remarked that this did not seem to be a good way to store books, and was told that they were all books on "scientific socialism" slated for liquidation. I now wish I had asked to keep a few as historical souvenirs!

Precise figures on the growth of student bodies are difficult to obtain because different sources disagree with one another. Thus, for example, *Concept for Higher Education Development in Hungary* (1991) gives the number of university (as opposed to college) students as 41,560, while a publication by the Hungarian Rectors' Conference gives the numbers of full-time students for 1990 as 43,460 and for 1991 as 45,727, and the total numbers of students as 54,659 and 56,497, respectively, without any indications as to the source of the discrepancy.[10] In those same years, the total number of instructors (full and part time) numbered 10,134 and 10,466, respectively, giving a student/faculty ratio of just over 5. In Czechoslovakia, the full-time student body in 1990 is given as 149,000 (96,400 in the Czech Republic, 52,600 in Slovakia), and the student/teacher ratio as 7.7, but ranging from 3.3 to 18.3 in six prestigious universities.[11] The best data on the growth of the student body and the faculty are from Poland, and show that from the academic years 1989/90 to 1997/98, the student body in all institutions of higher learning grew from 378,000 to 1,092,000.[12] University and technical university enrollments increased in this period by a factor of 2.8, similar to the increase in total higher education. The student/faculty ratio increased during this period at universities and technical universities from 6.3 to 15.4—a remarkable change over a nine-year period.[13] But the averages hide a great deal of variation among institutions. Since the number of full professors represents a relatively small fraction of the total faculty, the student/full professor ratio ranged in 1993 at Kraków institutions from 120 at the Academy of Economics to about 32 at the Academy of Mining and Metallurgy, while the same ratio for assistant professors was 25 and 8 respectively.[14]

It is noteworthy that identical types of data are not available for all countries, simply because they collect the data in different ways. Hence, the data are classified in a somewhat different form for Hungary. The number of Hungarian university students grew from 45,500 in 1989 to 103,200 in 1998, while the number of college students grew from 55,400 to 155,100. The number of teachers in the same period increased from 16,300 to 21,300.[15] The number of full-time students as a percentage of the 18–22 age group indicates that the Hungarian educational system had (as did the Polish system) largely succeeded in raising the percentage of the college-age population that receives a tertiary education. Since we cannot separately break out the number of teachers in universities and in colleges, we can only calculate aggregate student/teacher ratios; that ratio increases from 6.2 in 1989 to 7.7 in 1998, a rather modest increase compared to the Polish case. In any event, the increase in the number of students turned out to be quite massive, more so than anticipated by Lord Dahrendorf in 1994 when he wondered whether there would be adequate student demand for university education.[16]

In addition to state institutions, a number of private educational institutions have come into existence. Throughout the region, hundreds were born, largely unregulated by state agencies; many of these attempted to copy the curricula of state institutions, while others tried to find a niche in which they could have some monopoly power.[17] In Poland, while Catholic University Lublin (KUL) had always been private, the number of students in non-state higher educational institutions grew from 14,408 in the academic year 1991–92 to 226,795 in 1997–98; almost a quarter of the number of students enrolled in state institutions. However, the number of teachers in these institutions amounted to only 5,545 in 1997–98, yielding a student/faculty ratio of about 41. The explanation for this is most likely that faculty members at state institutions moonlighted at private institutions and are not counted in the statistics for private institutions. In the Czech Republic, only three private colleges have received state approval by October 1999, namely the Banking Institute, the Prague Hotel College, and the European Polytechnic Institute.[18] In Hungary, there were some 26 church-affiliated institutions of higher learning (four university-level institutions, one of which is Pázmány Péter Catholic University) and six private polytechnic institutions, which accounted for about 10% of the student body in that sector.[19]

In the Czech Republic, student numbers grew from 118,200 in 1990–91 to 165,800 in 1996–97, and total teaching staff grew from 11,500 to 13,200 in the same period.[20] The number of students obviously increased substantially, but not quite as much as in Poland or Hungary; the number of teachers in the professorial ranks was very steady and the total number of teachers at universities increased only slightly, raising the student/faculty ratio from 10.3 at the beginning of the period to 12.6 at the end. The situation was somewhat similar in Slovakia, where full-time students numbered 48,543 in 1988 and 85,742 in 1998, while the faculty size grew from 7,883 to 8,948 in the same period, yielding student/faculty ratios of 6.1 and 9.6, respectively.[21] While one has to be careful about making cross-country comparisons because of the intrinsic uncertainty about the precise meaning of the statistics, by the end of the period Poland had the highest and Hungary the smallest student/faculty ratio. It is worth noting that the general pattern of enrollment increases also holds in recent times in Russia, where the number of students has increased by 50% between 1995 and 1999.[22]

Problems and Agendas. It is instructive to contrast the diagnoses of what ails higher education and the prescriptions of what needs to be done, offered by indigenous observers on the scene, with those of western observers. We will illustrate this first for the case of Hungary, but much of what emerges from the Hungarian example applies, *mutatis mutandis*, to other countries. In general terms, one could argue that all diagnoses and all prescriptions fall into one (or possibly more than one) of three categories: issues dealing with structure, governance, and control; those dealing with instructional and research activities; and, finally, those dealing with financial matters. Early in the process of implementing changes, a pessimistic note was sounded by Tamás Lajos.[23] He claimed that the preconditions for serious reform were unfavorable, because ever since the 1960s, the government had pushed applied research over fundamental research (although some universities had spontaneously started to reform themselves in the 1980s, such as the Technical University of Budapest and the Karl Marx University). The post-1989 emergence of mediocre talent in university administrations and a deep-seated, conservative resistance to genuine reforms did not augur well, in his view, for rapid progress. Indeed, progress was extremely slow, particularly in countries such as the Czech Republic and Slovakia, which can be attributed primarily to four causes: (1) the awkward university structures that placed too much power in the hands of individual deans; (2) the lack of accountability; (3) old habits of fear inherited from the Communist system that prevented even sensible and decent academics from speaking out; and (4) greed, unchecked by a sense of academic ethics. Even in recent years, deans, elected by academic senates, often represented the old guard in the Czech Republic and had enormous power to appoint faculty, assign teaching duties, select the scientific councils that pass on the habilitation of junior faculty and select the individual committees that approve or disapprove doctoral dissertations. These powers can and have been grossly misused. Aviezer Tucker reports the case of Jiři Honajzer at Palacký University, whose doctoral thesis—actually written by another person—was "arranged to be accepted" for the doctoral degree by Dean František Mezihorak.[24] Another well-known case of corruption is the selling of entrance examinations (for $1,500), as well as the answers (for another $1,500), by the law faculty of Charles University, which led to demands in the senate of the university that the dean of the law faculty, as well as the rector of the University, Karel Malý, resign (Tucker, p. 94). Corruption in Ukranian universities appears to be widespread, and professors and administrators are routinely bribed by students; the price for a passing grade is reported to be $20.[25] Most recently, two university rectors in Romania were suspended when it became known that dentistry diplomas were being sold to foreign students.[26] There is reason to believe that corruption is endemic in all spheres of life in the countries that have recently emerged from Soviet domination, which has been abetted by the lack of transparency and accountability in public life.[27] It is important to note here that in many instances the corruption is not episodic or opportunistic; it is what former Šafárika University rector Lev Bukovský has called "programmatic corruption."[28] The fact that the guilty parties often go unpunished just adds to the public cynicism about ethical conduct. In Slovakia, progress was additionally retarded by the autocratic actions of the Meciar government, which attempted to rein in the autonomy granted to universities in the 1990 law and also created several new universities, presumably staffed with Meciar loyalists, without either academic or financial justification, and certainly without the approval of the Accreditation Commission.[29]

Other observers of the scene were willing to take the bull by the horns. Károly Kocsis, former rector of the University of Agricultural Sciences in Gödöllő, Hungary, noted the major problems,[30] that (1) research and education were artificially separated by the creation of liberally financed research institutes; (2) the rights to provide advanced scholarly training and to grant advanced scholarly degrees were denied to universities; (3) curricula were extremely rigid, with a large number of required subjects; and (4) highly specialized universities were created, which made the system extremely inefficient. These and related points were widely echoed throughout Eastern Europe during the post-1989 era: for example, in Poland there were vocational schools representing 300 different specializations, whereas there were only 16 broad specializations in Germany. As Leszek Balczerowicz noted, "This system is producing unemployment—young unemployed people."[31] But he also advocated an ambitious agenda that has the following salient points: (1) universities must have autonomy in education and research; (2) university constitutions should reflect democratic principles;[32] (3) curricula and teaching programs need to become more flexible; (4) requirements for professional degrees have to be raised; (5) standards for degrees must meet international standards; (6) mechanisms must be found to increase the international mobility of students and faculty;[33] (7) education, research, and outreach activities must be effectively combined; and (8) universities must have more financial independence. A very interesting aspect of East European higher education, which illustrates Balcerowicz's point of too much specialized education, is the fact that East European students enrolled in the humanities constituted a much smaller proportion of the total student body than in the West: in the second half of the 1980s, the percentages of students in the humanities, in technical fields, and in law and the social sciences is shown in table 5-1.

Kocsis' agenda is a far-reaching and ambitious one, and it is gratifying that substantial progress has taken place with respect to several items on his agenda. It is important to note that the first of his points speaks to the artificial gulf that the previous regime had created between the universities on the one hand and the Academy of Sciences on the other; attempts to bridge that gulf in Hungary go back to at least December 1990, when a special meeting of the Academy accepted the principle of creating close cooperation between the universities and the Academy within the framework of the Atheneum Association.[34] Points 5 and 6 have been taken very seriously by the EU. The Sorbonne Declaration of May 1999, of the European Ministers of Education underscores the need for a common frame of reference for higher education across national boundaries, and, in spite of some criticisms, it does not require common enrollment standards for students, for example, the 3-5-8 system, by which it takes three years for a bachelor's degree, five for a master's, and eight for a Ph.D. The subsequent Joint Declaration of the European Ministers of Education (Bologna, June 1999) advocated degrees that are easily accredited throughout Europe, a two-tiered system consisting of undergraduate and graduate studies, a uniform system of transferable credits, and increased student and teacher mobility.[35] The Ministers of Education further affirmed the principles of the Bologna process at a May 2001 meeting in Prague.[36]

A somewhat different slant on higher educational issues is taken by Lord Dahrendorf, who notes two major concerns and six other important issues.

Table 5-1 Percentages of students in various fields.[a]

	Humanities	Technical Fields	Law, Social Science
Poland	8.8	17.8	21.6
Bulgaria	7.6	35.8	18.0
Czechoslovakia	1.2	42.1	15.9
East Germany	2.1	32.3	16.3
Hungary	2.5	17.3	16.9
Denmark	14.2	16.0	24.3
West Germany	12.7	17.6	28.0
Sweden	14.9	19.3	29.1

Note:

a. Table abridged from Hryniewicz et al., *The Brain Drain in Poland*, Warsaw: University of Warsaw, p. 24.

The major concerns are how to guarantee the autonomy of research and teaching in a state system in which the purse strings would always remain largely, if not wholly, in the hands of the state, and how to fund the universities at adequate levels. The other six problems noted were (1) how to redefine and regularize the relationship between universities and research institutes; (2) what to do about overstaffing; (3) what to do if student demand for higher education fails to materialize (a concern that did not turn out to be justified); (4) how to introduce "demand-led" subjects, that is, subjects that may be nontraditional and the demand for which may come from the students; (5) to determine where innovation will come from; and (6) how to create a "common market of the mind," that is, an international community of shared scholarship and education (Dahrendorf). Some of these problems were actively being worked on by 1994 (for example, the setting up of European chairs that would attract first-rate people for extended periods), and some turned out to be less serious than anticipated; but on the whole, Dahrendorf listed serious problems to be dealt with.

Efforts to Fund Higher Education and NGOs. Kocsis' last point is intrinsically bound up with the entire budgeting system of universities and with the flow of resources from state budgets to universities. Budgets have declined in real terms: in Poland, for example, state expenditures on higher education amounted 2.5% of the state budget in 1993, 2.3% in 1994, and, in real terms the 1994 expenditures were only 60% of what they had been in 1990.[37] Total budgets were orders of magnitude smaller than those of American universities of comparable size. For example, the 1993 budget of Jagiellonian University (16,900 students, 280 full professors) was $22.4 million (of which 73% was from the state, 15% from the State Committee for Scientific Research [KBN] and the balance from other sources, including summer school tuition, room rentals, and conference fees),[38] while in the academic year 1992–93, the budget of Princeton University (with about 6,000 students) was $395 million,[39] a discrepancy that is huge even if we allow for differences in the purchasing power of dollars in the two countries. The disparity is also reflected by per student spending at public universities: in 1997, this figure was $5,430 in Hungary, $5,351 in the Czech Republic, and $4,395 in Poland, while it was $17,466 in the United States.[40]

A significant problem in financing universities has been the lack of strategic planning and the decoupling of academic decisions from the planning to provide long-term resources for the maintenance of the activity contemplated. Information systems were poorly developed and commitments were made without regard to long-term consequences. Thus, for example, McCrudden et al. point out that at József Attila University in Szeged, Hungary, the decision to introduce Serbian language study, to admit 1,000 new Ph.D. candidates, and to start a college of business were made without long-range financial planning and without setting institutional priorities.[41] The same report also points out that the system of university overhead charges on research grants was typically inadequate to finance the ancillary costs of research projects, such as the extra power consumption created by some large piece of equipment.

But another aspect of financing, for which there is essentially no precedent in Eastern Europe, needs to be thought of, namely financing received from alumni, other private donors, and corporations. As early as 1990, I had started to try to persuade university rectors that they must cast their net wide and think of raising funds from a multiplicity of sources. I illustrated in chapter 4 how much this went against the grain, by citing a rector who embarrassedly likened university fundraising to panhandling. But beyond the emotional content of not conforming to the mores, ignorance about how one raises money from private sources was extremely widespread. There was no appreciation of the fact that fundraising was a professional activity for which, to be sure, one had to have an aptitude, but the skills for which could be learned. Even as late as October 2000, it did not occur to many academics and university administrators that, for example, a western manufacturer of computer peripherals located in their town might find it advantageous to donate its own equipment, such as printers, to the university as a marketing strategy for the future. An important exception to the general myopia about fundraising was Rector Andrzej Wróblewski of Warsaw University, who reported in October 1991 that he had approached 300 business firms and asked them for a donation to the university. It is a depressing characteristic of the times and the region that he was successful in only four cases. Another interesting attempt to fill the perceived needs was the creation of the Friends of Hungarian Higher Education Foundation (Felsőoktatás Baráti Alapitvány), created in early 1998 with the help of the U.S. Treasury Department's Office of Technical Assistance, simultaneously as a Hungarian NGO and also as a 501(c)3 organization in the United States. Its purpose is precisely to streamline private-source fundraising by the higher education sector in Hungary, by helping universities establish development offices, train persons in fundraising, cultivate alumni, etc. While it was totally clear that FHHEF would face an uphill battle, the Mellon Foundation provided a $40,000 grant for its activities in 1998 and another grant of $20,000 in 2000. Some other foundations, such as the Rockefeller Brothers Fund and the Mott Foundation, also contributed significantly to its operations. But the problems that FHHEF faced in its early existence, arising from the fact that administratively it operates within the framework of the Association of Higher Educational Conferences (Felsőoktatási Konferenciák Szövetsége) and shares office space with its secretariat, give one pause about the ease with which this type of nonhardware-based technology can be transferred into an environment that is, at best, skeptical and at times uncooperative. The initial Hungarian Board (there was also a separate American Board) was largely inactive, which gives some credence to the views of Tamás Lajos, cited earlier, that a lot of mediocre persons surfaced in the post-1989 period. In the last year improvements had taken place on all fronts. The new executive director, Ágnes Soós Nagy, had taken steps to restore normal working relations with the Secretariat, the Board, and even the Ministry of Culture; the first steps had been taken to establish development offices in several universities; and this raised some hope that the FHHEF would be able to deliver what it was meant to. Unfortunately, Soós Nagy ultimately also found it difficult to work with the Secretariat and resigned in 2001. The future of FHHEF is now under a cloud.

An interesting experiment was undertaken in Hungary in the "1% Law" passed by the Hungarian Parliament in late 1996.[42] The law stipulates that taxpayers may designate 1% of their (paid-up) tax liability for the benefit of a not-for-profit organization, provided that the total tax liability exceeds 10,000 HUF (Hungarian Forint, about $35 in mid-2000), that the tax is paid in a timely fashion, and that the designated NGO is appropriately identified. Potential designations not actually or validly described result in the corresponding sums being retained by the Treasury for general governmental purposes.

The Act has been criticized on the grounds that it creates incentives for the Treasury to invalidate designations and also creates incentives for NGOs to lobby the public so they get designated on tax returns, thus diverting their energies from raising new funds from the public. In fact, the fraction of valid NGO designations did not exceed 40% of the returns in the first year of the law's existence (1997) and has not increased much since that time. The total number of NGOs receiving support from the government increased from 6,788 in 1996 to 14,438 in 1997; education in that year represented 40.6% of the "1% designations."[43] The significance of the program to individual institutions can be illuminated slightly by examining "thank-you" notices published as classified advertisements in the Budapest newspaper *Népszabadság*.[44] Relatively few of these donations are directed towards educational institutions and most of them are very small; in hardly any case does the amount earmarked for an institution *from all tax returns* exceed $5,000 in a given year. But most importantly, there does not appear a crowding-out effect. Total NGO income in 1997 increased over 1996 by 45.3 billion HUF (about $246.2 million), representing increases in every category of income (support from the central government, local governments, corporate donations, individual donations, etc.) except for foreign donations. However, to date, the total of "1% donations" is very small even on a Hungarian scale: $1.8 billion HUF ($9.8 million) out of total NGO sector income of 284.4 billion HUF ($1.54 billion) (Vajda and Kuti, pp. 175, 217). While the impact of the Act should not be exaggerated, it is a step in the right direction.[45]

Some Detailed Recommendations. Just two years after the Kocsis paper and shortly before the appearance of the Rocki, Koj, and Handke papers and the Campbell-Dahrendorf volume, McCrudden et al., produced a document commenting on the desirable course for József Attila University in Szeged, Hungary. They made no bones about looking at what is desirable in a university from a very American point of view, and while their views derived directly from an analysis of Hungarian institutions, their diagnoses have wide applicability to the entire region. Their analysis included the view that "many societies have recognized that schools both reflect and transmit a society's core values. They are, in short, a powerful instrument for social change. This is especially that case with schools in democratic societies" (p. 2). They then went on to speak explicitly about universities' institutional mission, a term that is less frequently encountered in the writings of East Europeans, and dealt with issues of faculty and institutional autonomy. Concerning the curriculum, they finally noted that "democratic societies manage to embed democratic values in the curriculum itself" (p. 4). It is interesting that the teaching of political or moral values was no stranger to East European curricula; after all, the endless courses on Marxism-Leninism before 1989 were precisely that. But soon after the regime change, a reaction appears to have set in, and some or perhaps many universities did not wish to set up compulsory courses on democratic institutions or business ethics or even market economics. Nearly ten years later, at least Czech universities are beginning to sense that moral education of the students does lie within their sphere of responsibility and are more attracted to the Anglo-American pattern in which universities care about students' and faculty members' moral welfare.[46]

The McCrudden et al. report then discussed in detail the areas where urgent action is needed.

1. Long-range planning is essential for universities and involves, among other things, the establishment of priorities for resource allocation. The report noted that the short term of university rectors (typically three years, once renewable) makes this difficult to achieve. It is interesting to note that this point, which I have raised in conversations with university rectors in each country, typically fell on infertile ground: rectors tended to support the short term of rectors on the grounds that (a) rectors in Eastern Europe are not professional managers but scholars, and must be permitted to return to their academic activities; and (b) rectors are often succeeded by their vice rectors (also called prorectors), and thus continuity is ensured anyway.

2. Institutional missions must be more finely articulated; staffing patterns must rely less on historical precedent than on demonstrated manpower needs; more attention needs to be given to undergraduate education; and outreach to government and business must be strengthened.

3. Student/faculty ratios need to be increased, but that will not be sufficient for improving university finances, because new students also bring with them new costs by increasing demands on library and computer services and space in general.

4. A credit system must be introduced so as to permit a substantial increase in student mobility through the transfer of credits.[47]

5. Institutional governance must be changed; specifically, outside boards of trustees should be established. However, the idea of outside boards is a foreign concept to East European universities and, never having had them, they do not generally understand how useful it can be to have outside per-

ceptions and input, and how valuable boards of trustees can be in mobilizing support within society at large for the mission of the university.

6. Information and financial systems need to be drastically improved. In 1998, I was speaking with the rector of a university about the possibility of developing a fundraising activity that would target alumni—it has to be stressed immediately that this, in itself, is a totally foreign concept in the region, but one whose time may have come. I asked whether the university kept any records of who the living alumni were and where they were, and the answer was absolutely not; in fact the university had no idea whatsoever how to contact any alumnus. Deficient information and weak financial and management systems were clearly very general problem in the region: Andrzej Karwacki reported the following as the principal defects in university management: lack of a clearly developed university development strategy, lack of a clear conception of a management system, disintegration of programs and applications, inadequate personnel training and employee fear of computing equipment, and instability in the principles of university operations and in the university environment.[48]

7. A tuition system has to be introduced or regularized in some way, preferably combined with a system of financial aid for needy students. The fact is that free higher education was guaranteed under the previous Communist governments and experiments with tuition have been few and not wholly successful. Hungary introduced a modest tuition for university students in 1996 (10,000 Hungarian Forint per semester; less than $50 at 1999 foreign exchange rates), the newly elected government of Hungary abolished this tuition charge. The most recent higher education act of the Czech Republic has banned tuition at all state institutions.[49] Poland has constitutionally prohibited the charging of tuition for regular full-time students in state institutions. But Handke strongly argued for a tuition system and for subsidizing students rather than institutions: he argued that that is the only way to introduce true competition into the higher educational system, by allowing students to "vote with their feet."[50]

8. East European universities typically do not employ the American system of coordinating university research and the funding of research through an administrative office that ensures that proposed research is compatible with the mission of the university, that private donors are not inundated with so many proposals as to alienate them, that research budgets contain a charge for overhead so as to reimburse the university for the additional costs it may incur as a result of research undertaken, and that potential donors are assured that a research proposal has passed some internal scrutiny. A particular feature of the American system is that professors submitting research proposals to outside organizations often budget for themselves summer salaries (typically two-ninths or three-ninths of the academic year salary), which is a powerful incentive to prepare and submit viable research proposals.

9. Faculty and staff development, particularly in the area of teaching,

has to be organized on a significant scale. We have already seen in chapter 3 that some six years after the beginning of the changes in higher educational institutions, the degree of change in teaching methods was very modest; but even at the end of 1999, teaching methods still emphasize rote learning and memorization (Burton). There was reason to believe, particularly in the area of the humanities and social sciences, that faculty members were unable to keep up with western developments, if for no other reason than the artificial scarcity of library materials (see chapter 7).

10. While the educational system as a whole tended to be highly centralized—being run in effect by decrees issued by government departments[51]—university functions were excessively decentralized. Not only were universities divided into separate institutions by fields—a fact we have discussed before—but even within universities, functions were excessively decentralized: deans had exceptional powers compared to the American model, and service functions (for example, janitorial service) were often independently organized by the several divisions within the institution. The power of rectors was correspondingly diminished and they often had no say in the hiring or firing of faculty members (Jařab).

University Integration. The decentralization of universities and university functions is not an unavoidable consequence of authoritarian and transitional societies, as can be seen through a comparison with the universities in South Africa. While support services in the latter may have been as inefficient as in Eastern Europe, the academic side of the enterprise has been much more centralized, and it is just now that South African universities are undertaking major efforts to decentralize; that is, to devolve power onto divisional deans, who will have substantial financial and academic independence. It is therefore interesting that while South African universities are designing plans for decentralization, Hungarian universities are doing the opposite, that is, implementing plans for centralization (perhaps one should say recentralization). The basic idea for integration reflected the desire for greater efficiency and the increasing exposure to western models of higher education in which special and single-purpose institutions are the exception.[52] In three academic centers (Szeged, Debrecen, Pécs), in particular, an institution called "universitas" has replaced the aggregate of several institutions of higher learning. These newly created institutions are now called Debrecen University, Szeged University, and Pécs University. In each of these three cities, there was a university representing primarily the arts and sciences, a medical university, and some assorted other institutions, such as an agricultural university and a Protestant theological college in Debrecen, a school of music and a technical college in Szeged, and so on. In each city these have now become one institution with one rector. The Gödöllő Agricultural University has become the lead institution of another universitas, now called St. Stephen University, which also comprises the University of Veterinary Medicine, the University of Horticulture and Food Industry, a teacher training college, and a technical college.[53] In some cases, an existing university with geographically dispersed campuses was dismantled and its constituent parts recombined with different other universities. The Budapest University of Economic Sciences has absorbed the College of Public Administration. And three college-level institutions, the College of Finance and Accounting, the College of Foreign Trade and the College of Commerce and Hospitality Management have fused to become the Budapest Business College. While integration was much less extensive in Budapest than in the provinces (the Medical University in Budapest did absorb two other medical institutions and the Eötvös Lóránd University remained untouched), in the aggregate, 41 university-level institutions have been or are being merged into 17 universities and 22 college-level institutions into 13 colleges.[54] While the idea of the universitas is not new and is mentioned as early as 1991,[55] it is clear that the World Bank's support for the integration of institutions, motivated by the desire for greater efficiency and avoidance of duplication, hastened the process. But opponents of this idea were not wanting. It was suggested that various government ministries would oppose the formation of such full-scale universities because they would lose control of the special-purpose institutions that reported to them, and university administrators would object because they stood to lose their jobs.[56] Integration would almost certainly increase the power and prestige of provincial centers at the expense of Budapest universities, and the latter tended to be skeptical about the benefits of integration (Morgan). Even supporters of integration warned that it must not be at the expense of competition among institutions.[57]

The Law on Higher Education specified that the potential integration partners would have to decide for themselves whether they wanted to integrate, but once they decided in favor, they would have to accomplish it within two years. Different cases present different degrees of smoothness in the transition. A case study of what difficulties can arise in such a process is provided by Debrecen. In that city, the largest institution is the multipurpose Kossuth Lajos University (KLTE) with divisions of arts, sciences, economics and management, and law. The other institutions are single-purpose institutions such as the Medical University, the Agricultural University, the Protestant Theological College, a teacher training college and an institute of the Academy of Sciences. The leadership of KLTE committed some strategic errors when it negotiated the number of votes KLTE would have on the newly constituted University Council (which elects the new rector of the integrated institution), and also when it accepted the condition that KLTE would be represented on the Council not as a single large unit, but separately through each of its faculties. This latter provision made it particularly difficult to enforce "party discipline" in voting for a rector, and allowed some faculty representatives to ally themselves with the Medical University and the Agricultural University. The latter two felt that they needed to retain power in the new organization, because unlike KLTE, they had cost centers with substantial revenues that were outside the normative educational funding of universities by the state: the Medical University has a hospital and the Agricultural University a research center with revenues independent of the Ministry of Education. They feared that in the integration these centers would be folded into the general budget of the university and that they would consequently lose control of these funds. Because of the voting scheme put in place, a member of the Medical University became the new rector, another person from the same institution the chief financial officer, and a person from the Agricultural University became the president of the University Council (Muraközy). Many members of (the former) KLTE have felt threatened by these developments and have felt quite properly that they have lost out in the power struggle. It would have been a gesture for social peace if the newly appointed university librarian for the integrated university had been chosen from among the KLTE librarians, but here too, the Medical University won out. The only semblance of fairness was the decision to rotate the rectorship among the three principal institutions in the following manner: the first rector (from the Medical University) would serve for a year and a half; he would be succeeded by a person from KLTE for one year; then that person would be succeeded by a person from Agriculture for one year; and that person in turn would finally be succeeded for a year by another KLTE person (with the further proviso that if the first KLTE person was from the humanities, the second would have to be from the natural sciences, and conversely). This is an absurd arrangement that will maximize inexperience in the top position, as well as the time wasted in campaigning for election.

It took almost ten years to carry the idea of integration to the point of realization and now that it is on the verge of full realization, not everybody is happy with it. While some of the arguments in the preceding pages, arguing for greater rationality in allocating resources, strongly underscore the need for larger academic units, which are perceived to be more flexible from the students' point of view, the execution of the mergers is far from trivial and raises questions of financing, governance, and administrative detail. Some rectors have lost their positions and may become vice rectors in the new university, and some university librarians may become directors of branch libraries. Questions arose concerning which of the libraries joining in the universitas should become the main university library. There is no question that substantial frictions arose between the medical faculty and the other faculties over this and other questions. In particular, in Debrecen, the plan for the new university does not designate a central library, even though the Kossuth Lajos University library is by far the largest among the participating libraries. There is some danger that the new university leadership may not maintain KLTE's previous commitment as the second National Library of the country for collecting Hungarian materials on an exhaustive basis.[58]

Only time will tell whether the Hungarian experiment will succeed. It is interesting that it has not been emulated in the other countries except in isolated circumstances; to wit, the Medical College of Kraków rejoined Jagiellonian University as one of its divisions on May 12, 1993 (the anniversary of the founding of the Kraków Academy in 1364). It is not too bold or romantic a prediction that ultimately these integrative moves in Hungary will succeed; after all, many of the parts that are currently uniting are merely reuniting and have been traditionally part of the same university in other contexts. But the transition is obviously painful and will be so for several years. A clear lesson seems to be that integration tends to be more friction-free if one large institutions absorbs very small units, but that it is fraught with peril if the union is one among relatively more equal partners.

Structural Problems of Management. The new laws on higher education essentially restored academic rights and freedoms to universities and vested the power to elect university rectors in the university senate (with the power of final appointment reserved to a government agency, such as the president of the country). But the power of university rectors to shape university policy and long-term development was limited by one or more of the following factors. (1) University rectors are typically elected for a period of three years and can be reelected only once. If one assumes, as is reasonable, that university rectors are characteristically selected from among the faculty and not from a pool of professional administrators, and that it takes a newly elected rector a year or two to find his or her bearings, there is not much time left for implementing long-term policy changes. From this point of view, the Debrecen rotation discussed above is particularly pernicious. (2) In some instances, rectors and deans are only advisory members of the university senate. (3) Deans and other officials in charge of particular schools or colleges within a university have immense powers compared to their U.S. counterparts; both the university and its constituent colleges are legal entities, and their presiding officers, say deans, have full powers to determine relevant policies with respect to personnel, curriculum, and other matters. Since there may well exist a conflict between the interests of a school or department and the university that it is a part of, university management is problematic and sometimes extremely inefficient.[59]

Promoting Academic Reforms. The ferment of academic reform was powerful enough by 1991 that the Mellon Foundation felt it desirable to participate in modest ways in promoting structural changes in universities and in improving managerial standards. The structural problems discussed above were matched by poor management of the details: bad or useless accounting systems, intermingling of accounts that destroyed transparency, and a budgeting system that created no incentives for efficient allocation of resources. In some countries, budgets were handed down in three categories: a salary budget, an operating budget, and an investment budget; and savings in one category could not be used to cover deficits in another. This is the case in Slovakia, and any surplus remaining in an account at the end of the year has to be returned to the central pool, killing all incentives to economize. In some countries, such as Slovakia, income from service generated by a unit of the university, such as would accrue when a library had commercial sales of a CD-ROM it produced, could not be kept by that unit, but would have to be turned over to the state budget. A notable exception to such poor budgeting practice was the case of the rector of the Kaunas (Lithuania) Technical University, who told me in 1995 that one winter he drastically reduced the heat in his university and spent the savings on buying 80 computers.

Once again, the State University of New York–Albany, was the standard bearer in improving the situation. With a $50,000 grant, SUNY organized three higher education management seminars in Hungary, which took place in early 1992; two of these were directed toward university rectors and senior university management, and one was aimed toward the principals and senior management at colleges. While the implementation of the seminars was not without difficulties, a total of 50–60 persons attended each series.

Overall, the seminars were very successful in presenting the basic concepts of university autonomy, accountability, finance, and financial management. As I indicated before, Hungarian universities and colleges were in the middle of substantial reappraisals and restructuring, the topics were "hot," and these seminars could not have come at a better time.[60]

In spite of the ultimate success of the seminars, it did not make sense to repeat such general introductions in Poland and the Czech Republic, because they had already had similar presentations under the auspices of USIA, OECD, and other organizations; but it did make sense to tailor a set of seminars to their particular remaining needs. This was done, and the Polish seminars were primarily devoted to alternative financing and cost containment, organization of the curriculum, and enhancing interinstitutional cooperation. The seminars to be held in the Czech Republic addressed a particular Czech problem, namely the absence of a middle tier of institutions between high school and university, and were thus directed to the senior management of a rapidly growing group of post-secondary institutions below the university level.

I should say that the topic of enhancing interinstitutional cooperation, which was one of the topics for the Polish seminar series, was an extremely important one. The Communist system did not encourage cooperation among like-placed units in society, because alliances among such units are contrary to the spirit of direct control from the center. This was so deeply ingrained, that it often took me major efforts to bring about collaboration. I recall visiting the rector of a university one day in 1990, who described the problems of his university to me. The next day, I was visiting the rector of another university, who happened to be grappling with exactly the same problem. I remarked that it seemed useful for the two rectors to pool their information about their problems, and by doing so they might find a faster and more cost-effective solution to both their problems. The second rector drew himself up and said, "What, you want me to tell *that* guy *my* secrets?"

By 1994, the need for institutional cooperation was broadly appreciated. East European academics started to puzzle out the seeming paradox that competition had become a general feature of higher education, because research grants were increasingly awarded by a competitive process (by KBN in Poland and OTKA in Hungary). Yet the quality of an institution and the ability of its members to move freely among a wide array of institutions depended on cooperation among institutions and on formal institutional links, often with institutions in foreign countries. Particular regional arrangements for cooperation have worked exceptionally well: a case in point is the Rectors' Conference of the higher educational institutions in Kraków.[61] One of the successful cooperative projects of this group is the Kraków Library Consortium, funded by and created at the insistence of the Mellon Foundation, which we shall discuss in more detail in chapter 9.

In connection with interinstitutional collaboration, we should also mention briefly the state of the academies of science. We have already noted in chapter 3 that during the Communist period, the academies of science were the much favored institutions and that the post-1989 restructuring that was undertaken severely reduced both the number of academy institutes and the number of researches at them. Yet it is clear that the academies continue to have valuable "plant and equipment," and employ many persons who represent an important human resource. The most recent and promising development is that there are some signs of collaboration between academy and university staffs for creating political support for research. In both Slovakia and Hungary there has been a definite rapprochement: in the former, nonacademy personnel may compete for research funds controlled by the Academy, and in the latter, nonuniversity personnel may do so for funds largely controlled by university representatives.[62]

The Hungarian Higher Education Project. HHEP started, indirectly, with a visit to Hungary in October 1991 by Derek Bok, emeritus president of Harvard, Henry Rosovsky, former dean of the faculty at Harvard, and myself at the invitation of Árpád Göncz, the president of Hungary. The visit was organized and took place under the general auspices of the Citizens Democracy Corps (CDC) and was intended to begin a systematic study of higher education in Hungary. This led to a second visit by Bok, William F. Massy of Stanford, and Robert Zemsky of the University of Pennsylvania in October 1992, after which Bok prepared a review of the Hungarian parliament's proposal to reform Hungary's higher education system.[63] A 1993 conference sponsored by CDC, the Institute of International Education, and the Higher Education Support Program (HESP) of the Soros Foundation caused various groups of American educators to work on financial and administrative assessments of two major Hungarian universities. By 1994, there was an active group of professional university managers working on various academic reforms with support from the Mellon Foundation, the Carnegie Corporation, USAID, CDC, the Hungarian Ministry of Education, the Hungarian Rectors' Conference, and the firm of Benko and Associates.[64] They made detailed recommendations for creating early retirement incentives to induce staff attrition, for curriculum review in order to eliminate wasteful duplication, for evaluating new teaching methods that might be more economical, and for the introduction of modest tuition fees.[65]

In early 1995, the Mellon Foundation received an ambitious proposal from the University of Utah to undertake a program of reform in higher education. The three topics of the proposal that the Foundation decided to fund in the amount of $100,000 (with a required 4:1 match with outside funds) were (1) accreditation; (2) financial management; (3) post-secondary education. Each of these topics appeared to be of considerable importance. These efforts dovetailed very well with the major reform activities spurred by the $150 million loan from the World Bank in 1997 (and matched by $100 million from the Hungarian government) to improve content as well as efficiency in Hungarian higher education (Morgan).

Accreditation. I had been arguing with Hungarian, Polish, and Czech academics for years that accreditation was an essential feature of a system of higher education, and I felt disappointed that these suggestions were not accepted with more alacrity. But the argument against accreditation and any form of peer review was always the same: "this is a small country, everybody knows everybody else, and peer review will just lead to bad feelings." Fortunately, the new Hungarian Law on Higher Education (July 13, 1993) specifically assigned to the newly established Accreditation Council the task of building specific models of accreditation. The Council selected a distinguished advisory board and established 24 professional committees that were actually in operation by the middle of 1993.[66]

Analogous developments took place in Slovakia and Poland. The Czechoslovak Law on Higher Education established an Accreditation Commission and stipulated that, "faculties shall be merged, split and dissolved by the rector with the consent of the academic senate of the higher education institution, after a prior opinion of the Accreditation Commission and an agreement of the Ministry of Education,"[67] and that the Commission would have an important function in the financing of higher education. "The amount to be granted from the state budget shall be decided by the Minister following the opinion of the Accreditation Commission and the Higher Education Council."[68] But there is no evidence that the Czech accreditation commission has imposed strict standards and Tucker refers to it as "indulgent" (Tucker, p. 97). Meciar, who was prime minister of Slovakia after the split with the Czech Republic on January 1, 1993 until his final defeat in 1998 (except for a nine-month period in 1994 when he was briefly sacked by the Parliament), turned increasing hostile to the universities because many of them represented a vocal opposition to his increasingly autocratic, racist, and xenophobic policies.[69] During his reign, the Accreditation Commission became largely a political tool and was used to "punish" nonconforming universities. During this period, the method of accreditation relied heavily on citation indices and similar measures of scholarly output, but reliance on these measures was so complete and ignored the differences among fields to such an extent as to border on the absurd. Annual evaluations of the faculty were required, but the criteria for the Commission's recommendations to the Ministry of Education were not made public. In any event, the outcome of the evaluations affects the universities' right to grant habilitations and recommend professorial appointments. Clearly, during the unfortunate Meciar period, accreditation in Slovakia left a great deal to be desired. After the election of Mikuláš Dzurinda, a new Accreditation Commission was appointed, chaired by the outstanding former rector of Šafárika University, Lev Bukovský, and there is now considerable hope that the new Commission will function in a responsible and objective fashion.[70]

Poland progressed toward accreditation in a more leisurely fashion, and while a state accreditation body, the National Accreditation Commission, is on the drawing boards, it does not yet function as of August 2001. However, as a result of the private initiative of 16 universities or university-level pontifical academies, the rectors of these institution established on January 31, 1998, the University Accreditation Commission (UAC) which now operates on an entirely voluntary basis and has accredited some 85 units (faculties, departments) as of August 2001, with another 80–90 expected to be accredited before the end of 2002. The governing principles of the UAC state that (1) participation is voluntary; (2) accreditation is for or by fields and not institutions; (3) the outcome of accreditation deliberations is binary (accredited or not accredited) rather than one with a many-point scale; (4) criteria for accreditation should be developed by experts in the relevant fields. Accreditation is an approval given for two or five years and is based on quite standard criteria, including, most interestingly, the condition that student evaluations of the teaching be solicited by a department seeking accreditation. The UAC is in operation and it is hoped that, if and when a state-run National Accreditation Commission (NAC) begins to operate (probably by early 2002), the two commissions will coexist happily. While the existence of two commissions might complicate things more than necessary, the beginnings of accreditation in Poland are promising.[71]

HHEP Continued. The new law also changed the universities' system of finance by making state support respond to the teaching and research outputs of universities. Hence, it made sense to try to devise schemes that would produce the appropriate incentives. Also, there was no model of post-secondary education in Hungary comparable to the American community colleges, and it seemed very desirable to provide such education. New funders joined in the project (for example, the Hewlett Foundation and, most importantly, the World Bank) and HHEP successfully organized training for accreditation in Western Europe and in the United States for representatives of nine Hungarian institutions. Eventually, the work of the HHEP team spilled over into activities aiming at the realization of the universitas concept, and finally, when it became evident that project funds had not been exhausted, into a subproject that would attempt to create lay (that is, nonprofessional) boards of trustees for Hungarian universities.[72]

The social-political environment into which Derek Bok and his colleagues attempted to plant the seeds of lay boards was one characterized by substantial power in the hands of professors, substantial power and control in the hands of the Ministry of Education, and lack of experience with lay boards throughout the system. The institution of boards of trustees in universities is probably best developed in the United States and in Great Britain, and the usefulness of this institution is widely accepted. As William G. Bowen said,

> Both for-profit and non-profit entities operate in inherently complex settings, where matters are rarely cut and dried. The exercise of collective responsibility through a board can slow down some kinds of decision making, but it can also dampen the enthusiasm of the aspiring autocrat. It provides checks and balances by adding layers of judgment and protection against abuse of power and some forms of self-dealing, self-promoting, and favoritism. The existence of a board also encourages the development of an institutional sense of purpose and continuity. . . . My short answer, then, to the question of whether boards matter is that they *can* and *should* matter a great deal, and not only when a crisis is at hand.[73]

Boards are needed for discharging the specific functions of (1) selecting and evaluating the chief executive of the organization; (2) reviewing and adopting long-run strategic objectives; (3) ensuring the availability of the necessary resources for achieving the objectives; (4) monitoring management; (5) ensuring the responsible and effective operation of the institution; and (6) nominating suitable candidates for the board (Bowen, pp. 18–20). Bowen's view, which is the dominant view in the United States today, provides a salutary contrast to the views of Thorstein Veblen expressed in a chapter filled with diatribes against businessmen.[74]

It was to be expected that the long-established structures in which higher education was embedded would make the introduction of lay boards an uphill fight, even though they would serve the dual functions of making institutions more responsive to changing economic and social conditions and protect them against unwanted political interference (Morgan, p. 24). The autonomy recently granted to universities and the Ministry's lack of trust in the administrative competence of professionally dominated boards made the Ministry reluctant to share its powers. This affected the Ministry's attitude towards both of the hypothetical alternative realizations of the idea of boards, namely a single national board or institutionally based boards. Yet the message that boards can play important functions by becoming buffers between the Ministry and the institution and can play important roles by carrying the message of the university to the wider society and promoting university outreach was not entirely lost. Active discussions ensued between HHEP staff, officials of the Ministry and the Debrecen and Szeged integrated universities. But difficulties continued to surface even at the institution level, partly because the institutions had difficulties in understanding what exactly the role of boards would be, and partly because powers were already vested in university senates and other bodies. It was not clear what they might have to give up if boards with powers to govern were to come into being. An important socio-political reality is pointed out by Morgan and Bergerson in noting that "many Hungarians do not trust laws and organizational structures. What they do trust is long-established, personal networks" (pp. 9–10).

A national board, the Higher Education and Scientific Council (FTT), Felsőoktatási és Tudományos Tanács, had been created by the Act of 1993, and it became rapidly clear that governing (as opposed to advisory) boards were not compatible with the Hungarian political and higher educational environments. An immediate issue was the composition of FTT. It also became clear that an entirely lay board was unthinkable; that is, some memberships had to be reserved for experts on higher education. Its scope was overall strategic planning, refinement of funding and allocation formulas, review of institutional development plans, giving advice on new degrees and programs, and commenting on broad policy issues. The 21-member board that was created had representatives of clearly defined constitutencies, such as the Conference of Rectors, the Conference of College Principals, the Ministry of Education, the Hungarian Academy of Sciences, the Mayor of Budapest and others. It was generally regarded with suspicion, and alternately viewed as usurping the Ministry's powers and representing a bottleneck in getting academic programs accepted. Even university rectors had some doubts about this new institution because they viewed it as interfering with the usual channels of communication (Morgan and Bergerson, pp. 11–14).

Even though "supervisory councils" (later changed to "advisory councils") had been mandated for institutions merging into a single university, as in Szeged and Debrecen, their creation was no easier and no more successful. The HHEP project team had drafted bylaws for the Debrecen Universitas, but misunderstandings arose quickly concerning both the functions and the composition of the council. Businessmen were reluctant to serve if the council was large or dominated by academics, universitas leaders had doubts about the willingness of business leaders to become council members, and yet it seemed clear that business leaders on the council would give it political credibility. The 1999 Amendment to the Law made the councils optional and required that at least four of its members be elected from among outstanding representatives of economic, financial, and scientific public life. These institution-level councils were not perceived by the Ministry as being particularly threatening to its own authority and, except for Szeged and Debrecen, institutional understanding of these councils was feeble. The attitude on the part of universities is still that oversight of universities is best left to academics, and the government elected in 1998 has taken steps to weaken the influence of the national board, FTT. The difficulties of introducing lay boards in Hungary illustrates the obstacles to effective technology transfer when the technology being transferred is not of the nuts-and-bolts variety, but rather presents a significant change in cultural and social norms and in accepted modes of thinking about political institutions.

Private Universities

Private institutions of higher learning have never been the norm in Eastern Europe, with the exception of theological seminaries. Since 1989, literally hundreds of institutions have come into being that were or claimed to be devoted to higher education, many of them with the single mission of teaching business and management. It is also the case that many of the new institutions are of extremely low quality and nothing would be gained by a detailed review of these. For the sake of completeness, I shall briefly describe three of the distinguished institutions.

The Catholic University of Lublin (KUL).[75] KUL, in the Polish city of Lublin, was founded in 1918 as a denominational institution concentrating on the humanities and social sciences. Remarkably, it remained independent even during the Communist period after World War II and played a significant role in maintaining an anti-Communist mentality. It has master's or doctoral programs in the divisions of humanities, social sciences, philosophy, theology, law, and mathematics and natural science, although the last of these consists of only mathematics and environmental protection and offers only master's degrees. It is the largest of the private institutions in the region, with approximately 15,000 students and 850 faculty members.[76]

Pázmány Péter Catholic University.[77] Located in Budapest, as well as in the small Hungarian town of Piliscsaba, the university's department of theology is viewed as the intellectual heir of the university founded by Péter Cardinal Pázmány in 1635. It has only divisions of theology, humanities, law, and information technology. It was recognized by the government of Hungary in 1993 and its departments of theology and canon law operate under "ecclesiastic rules concerning ecclesiastic faculties."[78] Its humanities division had some 1,700 students in 1997.

Central European University (CEU).[79] CEU is the most important academic institution created after the regime-change in 1989–90. The creation of a new university was the idea of George Soros, and it proved to be a most ambitious undertaking. It began on a very small scale, with about 100 students, in Prague and was originally intended to have both Prague and Budapest campuses. Indeed, for a short period of time, CEU shared a building with the Center for Economic Research and Graduate Education (CERGE) in Prague (see chapter 6). But the Prague operation was not to last long. The building in which CEU had space in Prague was owned by a trade union and in 1993 it decided to triple the rent charged to CEU. Soros expected the Czech government to absorb the increase in view of the fact that his foundation was already paying several million dollars a year for the Prague operation. The Ministry of Education refused the request and Soros decided to move CEU to Budapest, which seemed all the more reasonable in the light of the rumor that he and Prime Minister Václav Klaus were not on the best of terms anyway.[80]

By January 1995, the Hungarian Ministry of Culture and Education formally recognized CEU as an educational institution. In September 1996, the Board of Regents of the State of New York granted CEU an absolute charter. In August 1998, the Board of the Association of European Universities admitted CEU to full membership, and in June 1999, it was granted the status of "candidate for accreditation" by the Commission on Higher Education of the Middle States Association of Colleges and Schools.[81]

In the academic year 1999/2000, CEU enrolled 788 students in master's and Ph.D. programs. Its programs are in the humanities and social sciences, law, and business administration, with a single program in the sciences (environmental science and policy). In several of the programs, the emphasis is on Central and Eastern Europe.[82] Its budget for 1999–2000 was $14.6 million. It is interesting to compare this to the Jagiellonian University budget of $22.4 in 1993 mentioned earlier. While these budget years are six years apart, Jagiellonian has more than 20 times as many students as CEU and teaches the full range of scientific disciplines, which are much costlier to maintain than the humanities and social sciences.

Initially, I was skeptical whether it is possible to create in a short period of time a university *ex nihilo*. While it is true that enormous resources have been invested in CEU (see table 2-3), these resources have been well used. CEU has become a quality institution at which solid education is provided and a substantial amount of research is conducted.[83] Most recently, there has been some turmoil in the institution because the new rector, Yehuda Elkana, has been perceived by some of the faculty to act in an autocratic manner; but I have no doubt that the institution will survive this episode.[84]

Concluding Comments

We should note at the outset that not only have some private universities of quality come into being, but other kinds of innovative institutions have been created, as well, many of them within the framework of existing universities or colleges. Darvas[85] identifies eight innovative institutional characteristics and associates a particular institution with each of these: (1) building entirely new institutions (AI, and later AIN; see chapter 6); (2) restructuring disciplines (CERGE-EI, see chapter 6); (3) creating new foundations for high level knowledge (The Center for Theoretical Studies at Charles University, Prague); (4) systematizing doctoral studies (The Graduate School for Social Research, Polish Academy of Sciences); (5) promoting excellence in student development (The Invisible College, Budapest); (6) replacing old hierarchies of knowledge (The Institute of Fundamental Learning, Charles University); (7) diversifying undergraduate programs: The Program of European Studies (Berzsenyi Dániel College, Hungary); and (8) relying on membership and self-governance (Rajk László College at the Budapest University of Economic Sciences). All of these were unthought of and probably unthinkable during the previous regimes and their creation represents solid achievement.

Some cautious generalizations are in order. First, the four countries faced essentially identical problems, which were the legacy of the Communist period: central control of universities, inadequate funding, antiquated teaching methods and inadequate facilities, poor management structures, division of responsibilities between universities and research institutes, fragmentation of higher education, lack of student mobility. All of them set out to grapple with these problems through new laws on higher education, often amended several times, that established university autonomy, introduced accreditation in some cases, coped with the fragmentation issue in others, and generally moved (but slowly) in the direction of resolving the conflict between universities and research institutes by rationalizing the structure of and the authority to grant degrees and permitting competition for the same research funds. Western assistance did much to improve management structures or at least to create an awareness of the management problems that needed to be solved, but by no means to the same extent in the various countries. Remaining unfinished are the tasks of solving the funding problem of universities (in spite of the introduction of normative financing) and of carrying out broad-based reforms of teaching methods.

6

Economics Education

We have already discussed in broad terms the differences between economics education and management training. Western Ph.D. programs in economics attempt to turn out graduates who are capable of independent research and who will become faculty members at other institutions, or researchers or executives at consulting firms or government agencies such as the Federal Reserve Board or the Antitrust Division of the Department of Justice, or at international agencies such as the International Monetary Fund or the World Bank—people who will formulate new problems on a conceptual level, who are able to abstract from idiosyncratic elements of problems and concentrate on the essential characteristics, and who will invent new methods of measurement and quantitative analysis. And while I hasten to say that there are many distinguished professors in business schools who do exactly the same, the bulk of the thrust of MBA programs is towards individuals who will occupy positions in business enterprises, where they will have to solve problems in their gory complexity, without abstracting from the details.

The differences in orientation are reflected in notable differences in the required training. To get an MBA usually takes two years, while a Ph.D. normally takes a minimum of four, and more usually five or six years. The requirements for a Ph.D. include a dissertation, which is a substantial piece of independent research, while the MBA degree has no such requirement. It is highly desirable for the faculty that teaches Ph.D. students to have—in addition to a specialty—a general understanding of the economy as a whole, and this desideratum is expressed by the common requirement that Ph.D. candidates qualify in microtheory, macrotheory, econometrics, and usually two additional fields. It seems less important for teachers of MBA students to have as broad a command, and one could have very adequate teachers of, say, finance who do not know much about marketing or accounting; although in western MBA programs broad command over material is common. All these considerations convinced me that setting up MBA programs would generally be easier and quicker than developing Ph.D. programs in economics.

Of course, not all economics education programs are Ph.D. programs, like the program of the Center for Economic Research and Graduate Education (CERGE) in Prague. Some may be master's level programs or programs corresponding to the first year or two of a Ph.D. program (Academia Istropolitana, in Svätý Jur, near Bratislava), some may be undergraduate programs (economics honors program at Warsaw University), and some may be purely post-graduate research programs (Central and East European Economics Research Center at Warsaw University). Some may even be ad hoc programs of a few weeks duration. By their very nature, the latter tend to be irregular and not sustainable, and for these reasons I was skeptical about their long-term value. But in any event, the Mellon Foundation did fund a few of these, some of which were organized by the Institute of International Education or by the Foundation for Teaching Economics. They were addressed to varying audiences, from economists to high school teachers. Their usefulness was relatively higher in the early years, when there were very few people in Eastern Europe with even a rudimentary understanding of market economics, and inevitably declined in the later years.

The Case of CERGE

As early as January 18, 1990, I was aware of an inchoate effort to start some economics education in Czechoslovakia. Earlier that month, at a meeting of various funding organizations held at the Ford Foundation, George Soros had stressed the importance of economics education. I shared his views, but it seemed fairly clear that there were only two institutions in Czechoslovakia that had a staff that could be useful in such an effort, namely the Czechoslovak Academy of Sciences' Institute of Economics (Ekonomický ústav) and Institute of Forecasting (Prognostický ústav). In particular, at the latter I had known the director, Waltr Komárek, who had become deputy prime minister, Václav Klaus, the newly appointed minister of finance, and Vladimir Dlouhý, also a deputy prime minister and later minister of trade and industry. Dlouhý, in particular, was familiar with technical econometrics, and in 1988, on a brief visit to Prague, we did some statistical computations together. At the Ford Foundation meeting, I had a brief conversation with Jan Vanous, the publisher of *PlanEcon*, the best source of economics data covering the former Soviet Bloc, and he told me that he and Jan Svejnar of the University of Pittsburgh were thinking of starting some systematic economics education and were, for the time being, focusing on the Institute of Economics. Svejnar and I had known each other for a long time; he had received his Ph.D. in economics from Princeton in 1979, had been a student of mine, and had become in the meantime a very distinguished economist with numerous books and articles to his credit. I immediately called him and encouraged him to develop this idea further.

By the end of February 1990, great progress had taken place, at least conceptually. The department of economics at Pittsburgh and Charles University had agreed to create a graduate-level economics department at

Charles University. They envisioned a professional staff of about 20 or more people, of whom about one-half would be western (that is, West European or North American) economists. Svejnar anticipated strong support for this from the Czechoslovak government and thought that the "Czechoslovak side" would pay for a building and would also pay the salary costs of Czechoslovak personnel. By April, Josef Zieleniec, later foreign minister of the Czech Republic, was playing a significant role in the formulation of the plans and I was expecting a formal proposal in the near future. I had hopes that by October of that year a recommendation for significant funding could be submitted to the trustees of the Mellon Foundation. When I visited Prague in May, I made a point of meeting with Zieleniec and his colleague, Vladimír Benaček, to discuss progress in creating what by then had been named the Center for Economic Research and Graduate Education.[1] Almost all important details had been decided, except where CERGE would be housed. One of the most important decisions had been to place the governance of CERGE into the hands of an Executive and Supervisory Committee (ESC) that would have both Czechoslovak and western members.

As far as housing CERGE was concerned, there were five potential sites at that moment, of which I was able to inspect four, and each of those appeared to be suitable. A sixth site would have been in the building owned by the Institute of Economics, but at the time that did not appear to be as desirable an alternative as the others, because the Institute had been denuded of its best people who had been called into government service. (This was rather ironic, because several years later CERGE and the Institute of Economics joined anyway to create CERGE-EI.) But the issue of housing was not only a matter of physical location. When I met with Dlouhý a day or so later, he mentioned that George Soros had hinted to him that CERGE really ought to become part of the Central European University (CEU), which Soros was in the process of creating. I did not know at that time how seriously to take this suggestion. But I did not think that it was a good idea, at least not from CERGE's point of view, since it was not yet clear that CEU itself would ultimately be a success.

The creation of a new department *ex nihilo* is never an easy task. As a minimum, it requires (1) decisions about program content, program requirements, and a course sequence; (2) the mobilization of financial resources and equipping it with a library and technical equipment; (3) the recruitment of a teaching and administrative staff; (4) the establishment of rules of governance; (5) a definition of the ground rules for the relationship between CERGE and the rest of Charles University. The formal proposal to create CERGE was dated July 2, 1990. It described the mission of CERGE as follows:

> The first mission will involve retraining part of the present generation of economists whose university and graduate education were confined within the narrow limits of Marxist analysis. It will also entail training the first and subsequent generations of economists whose education will have transcended the Marxist paradigm. The economists trained or retrained at the

> Center will, by assuming positions at Charles University as well as other universities, colleges, and institutes, in turn educate undergraduate students and undertake professional activities needed by the Czechoslovak and other East and Central European economies.
>
> The second mission will entail the launching of a significant research program that will generate high quality theoretical and empirical studies. In particular, the Center will set and maintain standards consistent with those of the leading western universities and research institutes. Achieving this end will require the introduction of modern (and now virtually non-existent) computer and library facilities, the definition (or at least a major conceptual reorientation) of the research agenda and allocation of a significant portion of faculty time to this endeavor. In view of the importance and complex nature of the economic transformation in Central and Eastern Europe, the main research agenda will be oriented toward this topic.
>
> The third mission will consist of disseminating the research findings and making the human capital of CERGE available more broadly to government, business, and academic professionals. The goal of this mission will be to provide externalities by (a) stimulating research in other Czechoslovak institutions, (b) evaluating and assisting in the formulation of government economic programs, and (c) interacting with businessmen and officials from international organizations. These goals will be accomplished by circulating working papers and organizing briefings, seminars, symposia and conferences on specific topics.[2]

The intention was to organize CERGE as a western-style graduate department, along the lines of the Center for Operations Research and Econometrics (CORE) at the Université Catholique de Louvain, Louvain-la-Neuve, Belgium. The faculty size would be about 25, and faculty members would be selected by the Executive and Supervisory Committee (ESC) by the usual criteria of academic excellence. CERGE would enroll about 20 students annually, initially mostly from Czechoslovakia, but eventually from other East European countries, as well, and aim at a four-year program. Admission would be based, as it is in the United States, on the applicant's undergraduate record, on test scores on the TOEFL and GRE examinations,[3] and on letters of recommendation. Successful applicants would have to demonstrate thorough knowledge of calculus, linear algebra, and elementary economics. Students would spend two to three years taking courses and would have to pass a comprehensive examination in micro-and macroeconomics, statistics, and econometrics, and would have to pass two field examinations from the usual list of fields.[4] All this conformed accurately to U.S. practice. Faculty members would spend about one-third of their time teaching courses and two-thirds doing research and supervising Ph.D. dissertations. The initial members of the ESC were Jan Svejnar (chairman), Josef Zieleniec, Vladimír Dlouhý, and Kevin Sontheimer, at that time the chairman of the department of economics at the University of Pittsburgh; in the spring of 1991, the ESC was also joined by Philipe Aghion, a distinguished economist who was director of research of the European Bank for Reconstruction and Development. Pittsburgh was to maintain a close supportive relationship with CERGE for the foreseeable future and operations were to begin in earnest in the spring of 1991.[5]

This was exactly what was needed, and on July 3, I recommended to Bowen that he ask the trustees at their September meeting to approve a grant of $900,000 jointly to the University of Pittsburgh and Charles University for the creation of CERGE. I summarized the proposal and pointed to three possible risk factors: first, that Charles University or the Czech government would renege on various promises that had been made; second, that CEU would succeed in incorporating or taking over CERGE, which would then probably lose its distinctive character; and third, that further fundraising efforts would fail. Fortunately, as soon as one of these risk factors appeared on the horizon, we managed to avert disaster.

CERGE was duly launched in March 1991, and the first incoming class of students registered on June 24, 1991, for a preparatory summer term. The program took shape as intended, the regular faculty began to stabilize (Jan Svejnar and Randall K. Filer of Hunter College, City University New York, became regular faculty members for the time being). CERGE was able to attract a star-studded cast of shorter term visitors from Western Europe and the United States. By 1997, the student body had grown to 120 and represented not only the Czech Republic (30%), but also Poland, Hungary, and Slovakia (12%); Russia (13%); Romania (14%); Ukraine (11%); and other countries in the former Soviet Bloc (20%). The CERGE library had some 70,000 volumes and subscribed to 250 periodicals.[6] For a number of years I made annual visits and by 1993 or 1994 the students were sounding very much like American graduate students: they knew the material at the cutting edge of the discipline and they were sassy and unafraid to speak their minds.[7] But problems continued to surface and vigilance was needed lest they get out of hand.

Problems with Academic Programs and Management. Many of the problems could be attributed to the newness of the institution, the thinness of its staffing, and the unfamiliarity of both local staff and students with what a Ph.D. program was really supposed to be about. By the summer of 1994, František Turnovec had been slated to become the new CERGE director, although some doubts were voiced about his effectiveness. However, there was not much to choose from: under the rules, the head of CERGE had to be a tenured professor of Charles University and, for obvious reasons, it was desirable the he or she be Czech or Slovak. The core courses were already being adequately staffed by Czech personnel, but thesis supervision was inadequate because there were not enough experienced senior Czech professors and thesis supervision truly requires experienced researchers.[8] Most students started the program with inadequate preparation in economics. By 1994, the Economics Institute[9] had merged with CERGE. Its members felt that their salaries should reflect the fact that they were able to get outside grants for EI and should, hence, be permitted to increase their salaries by more than the 10% that the rules specified. Indeed, they would have liked to increase their salaries by a stated percentage of the grant. I explained that that is not the way it works; at Princeton, for example, a faculty member who gets an outside grant may pay him/herself out of the grant up to two-ninth of his/her academic year salary during the summer, but never a fixed percentage of the grant.[10] A member of the faculty of Simon Frazier University had spent the academic year 1994–95 at CERGE and prepared an "exit report"; he felt that a third of the CERGE students had no business being in a Ph.D. program. Many students, he argued, did not know what economics is about and did not know what they liked or wanted. While some of these findings were disputed by other temporary visitors, the litany of findings indicates that nothing could be taken for granted and that the CERGE management had its hands full all the time.

Space and Relations with CEU. Initially, CERGE was given space in a building owned by a trade union in a complicated deal that involved CEU. The largest part of the building had been ceded to CEU, and CERGE was essentially a sublessee of CEU, with an allocation of eight offices and shared classroom space. While I had a written guarantee from Minister Dlouhý that CERGE would receive adequate housing,[11] I had something better in mind than what materialized in the short run, and in any event, CERGE (as well as ASCOC) had been told that they would be bona fide tenants, rather than subtenants. CEU was running its own economics (master's level) program in Prague, and while CERGE's total space allocation had been increased to 18 offices by the spring of 1991, this was still not adequate. Frictions between CEU and CERGE continued over this matter. It turned out that Petr Pajas, the director of CEU in Prague, had promised too much and CEU appeared not to be able to deliver on his promises without jeopardizing its own operations. On May 24, 1991, Jan Svejnar wrote to Charles University Rector Radim Palouš:

> The western sponsors insisted on a clear commitment by the Czechoslovak side to these and other conditions as a prerequisite to releasing funding for CERGE. In the spring of 1990, Vladimír Dlouhý, the Czechoslovak Deputy Prime Minister, and Radim Palouš, the Rector of Charles University, wrote letters to the Andrew W. Mellon Foundation confirming the acceptability of these conditions. Subsequently it was agreed that CERGE would be located in the building at Olšanske Square and this was supposed to be included in the letter of transfer of the building from the Czech Government to CEU. The initial understanding between Petr Pajas and Josef Zieleniec was that CERGE operations within the building would take up one-fourth of the square footage of the operations of CEU.[12]

It was small consolation that Pajas was eventually dismissed, because it became increasingly clear that there was no solution for all the parties in the building at Olšanske Square. The proximity of CEU and CERGE also created repeated feelers, if not strong efforts, on the part of CEU to effect some kind of merger between the two institutions. CEU argued that CERGE was not viable, because a four-year or five-year program in Prague would lose students to other operations that provided opportunities in Western Europe (as did CEU). I agreed that it would be useful for CERGE students to get exposure to West European universities and I encouraged the development of collaborative initiatives between CERGE and CEU; but I strongly objected to a merger of the institutions because I felt that CEU would emerge as the dominant partner and eventually influence programs and faculty appointments to meet its own objectives. CEU, which had been contributing to the stipends of non-Czech students at CERGE, felt that it could continue to do so and could also contribute more to the jointly operated CEU-CERGE library if some collaborative ventures were in place and encouraged the Mellon Foundation to take a more proactive attitude in promoting such collaboration. My attitude was that we would welcome collaboration, but it was for the management of CERGE and CEU to draft the plan. In spite of these problems, personal relations between CEU and CERGE were excellent, and CEU was clearly sympathetic towards the objectives of CERGE, but there appeared to be no way of solving the space problem in the building that the two institutions occupied. In fact, the problem caused a fair amount of anxiety, because during my May 1991 trip to Prague, William Newton-Smith of the Soros Foundation (also Fellow of Balliol College, Oxford) indicated that within a year's time CEU might have to evict CERGE from its quarters.

But even the space problem had a solution, and perhaps an unexpected one. The management of CERGE and the president of the Academy of Sciences, Rudolf Zahradnik, decided that there were considerable externalities to be realized if the Academy's Institute of Economics and CERGE were to forge an alliance. This was accomplished in 1994 and the two institutions became known as CERGE-EI. Concomitant with the institutional change, CERGE moved into the building owned by the Economics Institute, a quite splendid old building originally owned by a bank, which immediately solved the space problem. Jan Svejnar became the director of the Institute, while remaining the chairman of the ESC of CERGE, and there is no doubt that the two institutions together represent more than the sum of their parts. Eventually, the CERGE Foundation, a 501(c)3 organization formed for the purpose of fundraising for CERGE in the United States, changed its name to the CERGE-EI Foundation and continues to this day under that name.

Relations with Charles University. Charles University is not only a large institution with many faculties and libraries, but an ossified one in which tradition and privilege play a significant role. Placing a new institution such as CERGE in this environment should have led us to expect recurrent problems.

The first of these was the decision to make CERGE a part of the Faculty (School) of Social Sciences, in intellectual (if not physical) proximity to the School of Journalism. Václav Klaus wrote to Rector Palouš:

> I feel that locating those subjects of study in one department is not appropriate. It is a problem inherited from the past, that the Department of Journalism was an extremely ideologized institution, and this may affect the new Department; if this occurred, it could prevent the exploitation of unusual possibilities for building the new institution.[13]

But CERGE stayed for the time being in the Faculty of Social Sciences and this led to fears that CERGE would be "taken over" and funds intended for it would in some way be appropriated by other administrative units, for example, by the Charles University Economics Institute directed by Boleslav Urban. In fact, funds made available for CERGE did not appear in CERGE's account for some time, which caused me to write to Rector Palouš on March 14, 1991. But administrative difficulties continued to mount, and on March 26, 1993, I had to write to Rector Palouš again:

> I am relieved to know that you are adhering to the idea of creating and maintaining CERGE as a special center of excellence and that you are taking steps to move CERGE directly under the Rectorate. The Andrew W. Mellon Foundation and other sponsors of CERGE have been extremely concerned about the uncertain situation of CERGE within the Faculty of Social Sciences and have also come to the conclusion that a priority project such as CERGE needs to become an autonomous degree granting institution directly under the Rectorate. . . . I would like to urge you to affect the transfer of CERGE to the Rectorate as soon as possible. Apart from reducing the uncertainty and ensuring the smooth functioning of CERGE, a fast action on your part is also desirable for practical reasons. The June 7, 1993 meeting of the Board of Trustees of the Mellon Foundation will be considering the proposal for a three-year continuation of the grant to CERGE and may not approve it if the Trustees cannot receive full assurance that CERGE will be immune from pressures from the Faculty of Social Sciences.[14]

CERGE, in fact, did achieve autonomy, and it was agreed that it would report directly to the rector of the University, but new clouds were not far from the horizon. By 1996, František Turnovec was the director of CERGE, and Karel Malý had become the new rector of Charles University. In February, 1996, Turnovec informed the ESC that it had no formal authority over decisions at CERGE. Serious disagreements arose between Svejnar and Turnovec and governance became increasingly difficult. Rector Malý appointed a prorector, and the president of the Academy appointed a vice president to work with the ESC to help resolve the situation. Rector Malý proposed the creation of a "joint workplace" for CERGE and EI and the establishment of new rules of governance, by which ESC could make binding recommendations to the CERGE director concerning curriculum, standards, and hiring, with the rector arbitrating in case of disagreement. But the ESC and Turnovec continued to disagree, whereupon Turnovec resigned, but Malý did not accept his resignation. A draft for the new joint workplace was prepared and the ESC in effect issued an ultimatum to the effect that it would resign if the

Academy and the University failed to sign it by the end of January 1999. The Academy had actually agreed to sign the agreement in 1998, but the University was still hesitating, perhaps because of the potential impact of a new law on higher education that was in the works. Therefore, on January 25, 1999, I wrote to the rector and after recapitulating the important milestones since 1996, I said:

> I think that the situation is critical for the future of CERGE. The importance of the joint workplace derives from (1) the fact that it would underscore the integration of teaching and research, (2) the signal it provides that Charles University adheres to the original conditions stipulated when CERGE was started, and (3) the fact that the joint workplace would provide the ESC with more legal authority than under the present situation. You should be aware of the fact that CERGE has achieved a reputation that is unparalleled in Central and Eastern Europe. It is a remarkable PhD program, the like of which does not exist anywhere in the region. It is crucial for the continued excellence of CERGE that it be able to attract Western funding for the foreseeable future. One of the essential elements in CERGE's ability to attract such funds is the guarantee of quality that is provided by the supervisory functions of ESC and the respect accorded world-wide to the members of ESC. If the members of ESC resign because of the reluctance of the Director of CERGE to accept ESC supervisory functions and because of your unwillingness to sign the agreement on a joint workplace, CERGE will not disappear overnight, but its standing in the world of economics will be seriously impaired and within a very short period of time it will lose the credibility that it has built up so laboriously over a number of years. . . . You have an asset in CERGE that may be wasted at a rapid rate if you do not nurture it. While I would regret the waste of $1.5 million that we have funneled to CERGE, I would much more regret the unfavorable judgment of the western world on the ability of the Czech academic establishment to maintain such an effort and the very likely permanent drying up of western sources of funds for Charles University.[15]

The agreement on the joint workplace was signed, Turnovec departed as director, and normalcy returned to CERGE. It was a tumultuous first nine years, and the substantive successes of CERGE would not have taken place without the dedication and persistence of Jan Svejnar, Randall Filer, the members of the ESC, and most importantly, the dedicated work of the Czech and western faculty. The cost in terms of human wear-and-tear was high, as was the dollar cost. The Mellon Foundation alone had provided (in four separate grants through 1997) a total of $2,050,000 for CERGE. Within the first three years of its existence, CERGE also raised $1,590,000 from USAID, $300,000 from the Pew Charitable Trusts, $104,000 from the Sarah Scaife Foundation, $131,000 (in kind) from Digital Equipment Corporation, $71,000 from the Soros Foundation, 67,000 ECU from the European Community's ACE Project, 187,000 ECU from TEMPUS, and $98,000 from the Center for International Private Enterprise. It continues to raise moneys successfully, which is illustrated by the fact that between May 1998 and August

1999, the CERGE-EI Foundation has raised some $618,000 from private and corporate foundations, the World Bank, and other organizations. Most extraordinary is the fact that Charles University now covers approximately 40% of CERGE's annual budget, just over $2.5 million, which is an unparalleled level of support by the University.

In terms of academic achievements, CERGE has surpassed all reasonable expectations in some areas and is solidly on track in some others. It was designated by the European Community's ACE program as its only "Recognized Center of Excellence in Ph.D. Studies in Economics" in 1992 and 1994. In 1993 and 1994, USAID recognized it as a Center of Excellence in Economics Education in Central and Eastern Europe. In 1995, the European Economic Association selected CERGE-EI as the host of its annual meeting. In 1999, CERGE-EI was selected as the regional coordinating institution for the World Bank's Global Development Network. Its faculty currently has 16 permanent members, several of whom have Ph.D.s from Princeton University, Massachusetts Institute of Technology, the University of Chicago, the University of Minnesota, the University of Pittsburgh, and New York University. The staff has produced some 140 working papers, 90 discussion papers, and hundreds of other studies. CERGE-EI faculty has also had considerable success in getting research grants: in the most recent PHARE/ACE competition of the European Union, CERGE-EI had submitted four proposals; three of these were among the 38 that PHARE/ACE funded out of the total of 253 applications. At the 1998 meeting of the European Economic Association, CERGE-EI staff gave 11 of the 183 papers presented there; this achievement was exceeded only by the Stockholm School of Economics and the University of Toulouse. Finally, it is noteworthy that the graduates of CERGE are beginning to be found in important positions: several are employed in Czech ministries and at the European Bank for Reconstruction and Development. One graduate recently left the position of chief economist at the Czech Ministry of Finance in order to guide long-range planning and privatization at the largest Czech bank. Another graduate from Romania has accepted a position as research director at a group of western agencies attempting to advance economic research in the former USSR.[16]

But in other areas, improvement is desirable.[17] The drop-out rate for graduate students is too high, the degree completion rate is too low (only 15 students have actually been awarded Ph.D.s by the summer of 1999), and the length of the program has crept from the targeted four years to over six years. The financial support that CERGE-EI is able to provide students is not adequate for the foreign students, and the lack of sufficient experienced senior faculty has forced students beyond their second year to work more on their own than is desirable (although even this problem appears to be nearing a solution). And finally, in spite of the truly impressive number of research reports produced by CERGE-EI faculty, the productivity has not yet translated itself into the appearance of a string of publications in the most prestigious economics journals. It is fairly clear that solving some of these problems needs more time. The western presence is still needed, but as Czech staff members gain experience and become bolder in tackling the prestigious journals, the publication record will improve. If CERGE-EI keeps up its good record of raising funds, it will be possible to continue to maintain an adequate level of western presence. The achievements of CERGE have recently been recognized by a distinguished committee of economists who have strongly recommended to the New York State Education Department, Bureau of Doctoral Program Review, that it register CERGE as a doctoral program, and in the late summer of 2000, New York State did so register CERGE.[18] Quite recently, CERGE-EI has embarked on a campaign to raise $10 million for a permanent endowment and Citigroup Foundation has made the first contribution of $2.5 million to it in June 2000, which the Mellon Foundation followed with a grant of $1 million in September. While cutting the umbilical cord will have taken longer than originally anticipated, CERGE-EI is the only western-style Ph.D. program in the region, and one with solid achievements and a promising future.[19]

Academia Istropolitana Nova

The story of Academia Istropolitana Nova[20] (AIN) and of its predecessor organization, Academia Istropolitana (AI), is a fascinating story of technology transfer achieved with high levels of competence by dedicated people through the collaboration of several western donors in the midst of governmental incompetence, treachery, and evil.

Because the oversight over educational establishments remained in many ways strongly centralized in the post-socialist countries, political events had, unavoidably, a larger role in what happened than was desirable. By 1992, the Czechoslovak Civic Forum had split into factions. In the elections of 1992, Václav Klaus' faction, the Civic Democratic Party (ODS), won handsomely in the Czech part of the Federal Republic, and Klaus became prime minister of the Federal government. In Slovakia, the winner was the Movement for a Democratic Slovakia, and its leader, Vladimír Meciar, became the prime minister of Slovakia. Klaus and Meciar could not agree either on economic reforms or on the future shape of the federation and agreed to split the Federal Republic into two countries,[21] which took effect on January 1, 1993. Slovakia, the less developed part of the federation and the part that faced more severe economic problems, was gotten rid of by Klaus in one stroke. On the whole it was a boon to the Czech lands, but Slovakia was saddled with the nationalist, xenophobic, and authoritarian Meciar.

The name of the institution is part of history. King Mathias Corvinus (1458–1490) of Hungary petitioned Pope Paul II to authorize the creation of a university, and the petition was granted in 1465. The university was located in Pozsony (Bratislava today) and started to operate in 1467.[22] But like two previous attempts to create universities in Hungary, the effort was short-lived. Partly because of the participation of the university's strongest patrons, János Vitéz and Janus Pannonius, in a conspiracy against the king, and partly because of the ascendance of the nearby University of Vienna, Academia Istropolitana ceased to exist in 1485.[23]

After the momentous events of 1989, a new institution was created in Bratislava under the name of Academia Istropolitana. Its origins go back to May 1990, when a preparatory committee was set up to consider the establishment of AI. From the beginning, Alena Brunovska, then deputy minister of education, and Ján Pišut, minister of education, played a significant part. The original conception was that AI would become a Slovak state university. In fact, for the first several years, the AI state budget component was listed among the state university budgets. Subsequently, some consideration was given to a plan for AI to become allied with CEU which was to have a branch in Bratislava. However, this did not materialize, but it so happened that a large USAID grant was given to the University of Pittsburgh for fostering economics and management education in Czechoslovakia. CERGE was the obvious beneficiary in the Czech lands, and Kevin Sontheimer of the University of Pittsburgh economics department tried to use some of the USAID grant to support economics education at AI. I first heard about AI when I received a letter from the (then) executive director, Kamil Sarka, in May 1991, in which he inquired about the possibility of Mellon support for the new institution. I replied, but did not hear from AI again until March 8, 1993, when the (then) acting director Alena Brunovska wrote to me. Soon thereafter, I met her for the first time when I visited Bratislava in June of that year. At that time, AI was offering a two-year curriculum in public administration. For the academic year 1993–94 it had increased its scope to offering programs in architectural and urban heritage conservation and in environmental planning and management. Sontheimer was unable to persuade USAID to allow him to use funds at AI, since the head of the USAID mission in Bratislava, Patricia Lerner, was fixated on using the funds only at the Bratislava Economics University—a decision that ultimately turned out to have been deeply flawed. Lerner also strenuously objected to letting AI evolve into the "younger brother" of CERGE. Sontheimer felt strongly that Slovakia desperately needed well-trained professional economists and that AI was the right place to train them. Hence, by the spring of 1994, the University of Pittsburgh submitted a proposal to the Mellon Foundation for a substantial grant to establish at AI the Professional Program in Applied Economics (PPAE), intended to be a two-year, master's-level program, to accommodate 30 students. The program was well designed, but I was concerned that Sontheimer, who was also on the CERGE executive committee, would be stretched too thin, that Brunovska was still only acting director, and that the future of funding for the institution as a whole was not clear. The first of these concerns was allayed by Sontheimer's decision to concentrate his efforts entirely on AI, the second by the AI management's assurance that the delay in Brunovska's appointment was only caused by temporary disarray in the Slovak Ministry of Education. The third concern could not be solved immediately, although AI qualified (by 1995) as a "semibudgetary institution" under the Ministry of Education; this meant that it did receive some Slovak government funds (but not for student stipends) and had more administrative latitude than a university; but as a semibudgetary institution, it also failed to qualify for accreditation. Sontheimer supported the request for a grant by noting that certain other donors were very likely to provide assistance.[24] In any event, the times were auspicious, because the Meciar government was replaced in March 1994 (alas, too briefly) and in June 1994, the Mellon Trustees approved a grant of $566,000, which made Mellon the largest funder of AI through 1996.[25]

By the 1994–95 academic year, AI had four programs, its permanent staff had grown to 12, and it utilized the services of some 50 visitors from abroad. Another program, in European studies, was added, and in spite of the generosity of many donors, it was clear that more money would be needed; moreover, while the current inflow of funds would momentarily keep the wolf from the door, long-term sustainability was by no means assured.

In order to improve the odds for long-term sustainability, in 1995 I made an attempt to unlock the coffers of USAID. On May 25, 1995, Kevin Sontheimer and I went to see Patricia Lerner. She wanted to know what the Mellon Foundation was doing in Slovakia, which I readily explained. She noted with satisfaction that Mellon was also supporting a program at the Agricultural University in Nitra, (see later in this chapter) which USAID felt was important. But on the issue of some funding for AI, she remained unbending. I pointed out that I was not arguing for a major USAID grant, but rather a token grant of, say, $25,000, which could have important symbolic effects on other potential donors. She countered by saying that USAID had no money for this. Then I had what I thought was a rather inspired idea. I said that the Mellon Foundation would make a $25,000 grant to USAID, which USAID could then regrant to AI. I thought she was irritated that I had called her bluff, but said she would not be part of such a scheme anyway.[26] Thus, USAID never came through for AI, which I consider one of its most dismal misjudgments in Eastern Europe.

After Meciar's ouster, AI experienced a brief period of tranquillity. By the summer of 1994, Alena Brunovska had been appointed director of AI, and support from western institutions, such as the Institute of Advanced Study in Vienna and the University of Wuppertal, Germany, had solidified. AI had an outstanding board, which included among others the excellent Lev Bukovský, the rector of Šafárika University. But other clouds were gathering, an order of magnitude more serious. After the return of Meciar to power, subtle and not-so-subtle attacks on AI were initiated by the Ministry of Education. By March 1995, the Ministry had decided on an extensive audit of AI to make sure that students were paid stipends in accordance with regulations. The problem was that originally AI received state funds for the stipends of the public administration students, only; subsequently, funds had been provided for other students, as well, but the ministry's record-keeping was imperfect in this matter and Brunovska was then easily accused of using state funds for unauthorized purposes.[27] Shortly thereafter, the ministry announced that no funds for student stipends would be made available and that salaries for faculty and staff would continue to be paid by the ministry until the end of 1995. The ministry went public with its attacks and the objectives of its continual interference appeared to be to rewrite the constitution of AI, to establish a new board of directors and to fire Alena Brunovska. The ministry also reclassified AI as a "semibudgetary institution," which meant that AI would have to raise its own operating expenses. On April 13, 1995 I wrote to the Slovak ambassador in Washington, Branislav Lichardus, and implored him to intercede on behalf of AI, but all was in vain.[28]

The intensifying attacks by the Ministry of Education were now widely reported in the press. Sontheimer asked that I attend, in late May, a meeting of the AI board, at which representatives of the ministry would also be present. We gathered in Bratislava, and present were the board chairman, Mikulaš Luptačík, who was a Slovak national and a professor at the Technical University of Vienna; Andreas Wörgötter of the Vienna Institute of Advanced Study; a professor from the Slovak Technical University, a representative of a German foundation that had supported AI; the Chargé d'Affaires of the Slovak Embassy in Vienna, Kevin Sontheimer; and three representatives of the Ministry of Education. The discussions were conducted in Slovak, and an AI staff member provided a running translation for me. The ministry officials struck me as dim-witted or disingenuous in the extreme. Their basic argument was that there were "irregularities" in the conduct of AI's operations because the founding document of AI defines it as an educational institution, but it had come to their attention that AI staff not only taught but also conducted research. Since the founding document said nothing about research, AI's operations were illegal. They had the graciousness to apologize for the newspaper attacks on AI, and the meeting ended rather inconclusively, with some vague promise of partial funding from the government. Sontheimer felt that AI could survive another year and there was some hope that the Soros Foundation would make a substantial contribution towards student stipends. I asked each of the three trustees present what their interpretation of the outcome was: one was optimistic, one was noncommittal, and one interpreted the proceedings to signal the start of the final attack to liquidate AI.

Sontheimer felt that my attendance was salutory because it made the ministry officials understand that considerable outside attention was being paid to this case; but ultimately none of this mattered. In early February 1996, Sontheimer reported to me that Brunovska had been fired as director as of January 22, and that the Ministry of Education had rewritten the rules so that AI's board of trustees would henceforth be appointed by the ministry. He feared that the continued hostility of the ministry would make it difficult or impossible for the institution to provide quality education to a highly capable student body. Soon thereafter, Andreas Wörgötter resigned from the board of trustees and said in his letter of resignation; "The decision of the Minister to remove Alena Brunovska from her position as Director of AI without any attempt to consider communication with the Board is a clear and unacceptable break of rules [*sic*] for managing academic institutions."[29] That was the point at which informal planning began for saving what had been achieved. One thought was to fold AI into the Economics University. Another thought was to meld it into Comenius University. Yet another, much bolder but also preferable, plan was to privatize AI in some fashion. But it would take some time before plans could be fleshed out.

I returned to Bratislava in October 1996, and gave an address at Comenius University. I spoke about academic freedom and the importance of preserving the autonomy of institutions of higher learning. While Rector Juraj Švec thanked me for my appropriately timed remarks, I was quite certain that if they were heard by any of the Meciar faithful, they would go in one ear and out the other. Later during my visit, Sontheimer and I went to call on the new director of AI, Petr Škultety and his deputy, Josef Sivak. Škultety spoke no language other than Slovak, but fortunately, Sivak was quite fluent in French, and so he became our translator. After some opening pleasantries, we got down to business. I asked Škultety to outline his vision for AI's future. It became painfully clear within minutes that he had no idea of what AI did or what it was about. He did say that the government had an interest only in the public administration and environmental planning and management programs; these would be transformed into training programs for civil servants and the programs will lose all academic independence effective immediately. He also claimed that he had no interest in interfering with the other programs (a good thing, since he knew nothing of what was going on in these) and would respect the faculty's academic freedoms—an assertion that I outright disbelieved, since he favored the proposal for the university law before the Parliament on the grounds that "there is altogether too much academic freedom." It was a depressing meeting after which Sontheimer and I agreed that AI was, to all intents and purposes, finished as an independent institution.

But all was not over. During the preceding months, many discussions had taken place between Brunovska, Sontheimer, Joseph Schull of the Ford Foundation, and Peter Darvas of the Open Society Institute (OSI), a central organ of the Soros Foundations, with respect to the creation of a new entity, Academia Istropolitana Nova, which would be private and independent. Plans for its operations had been drawn up and it had actually come into being as a new not-for-profit organization by October 1996. Public administration would be dropped, because the demand for it had been decreasing, but a new field in banking and finance would be added. A very large percentage of the staff would resign from AI and join AIN at the end of the academic year (not before, because of the staff's moral obligations to the students currently enrolled in AI). The situation was immensely complex. In all fairness to the AI students, their stipends would have to be paid through the end of the academic year; and while OSI had been paying these through the end of October, OSI's attitude was that it would continue to do so only if requested by the director of AI (Škultety), who was entirely indifferent toward the question of student stipends. If AIN was to open its doors in September 1997, student recruiting would have to begin forthwith; but because of the conflict of interest situation, a faculty member could not simultaneously work at AI and recruit for AIN. In any event, AIN would have to have an office and start recruiting with a skeleton staff no later than January 1997, but had no funds to do so. In addition, AIN would eventually need space for carrying out instructional and research activities, and in October 1996, there was nothing on the horizon. Last, but emphatically not least, there was a substantial momentum among the faculty to create AIN and to get on with this enterprise. Both Sontheimer and I feared that delays at this point would dampen the enthusiasm and the momentum would be lost: if the faculty could not be assured that by June 1997 they would have a paying position at AIN, they would start drifting away and accept other positions, in which case it would become impossible to recapture them for AIN.

AI had done first-rate work. Its graduates had obtained important and responsible jobs upon graduation. The excellence of its program had been recognized by the fact that AI was a finalist for the prestigious Hannah Arendt Prize awarded by the Institut für die Wissenschaften vom Menschen. I was convinced that it must be given a chance to reestablish itself. It appeared that AIN would have to receive infusions of funds of about $1,385,000 to keep it afloat through the year 2000. If the Mellon Foundation, Ford, and OSI could provide this amount of funding within the next six months, the outlook would be fairly positive. But immediate help was needed for the groundwork to be laid. On November 13, 1996, I recommended that a grant of $45,000 be made to AIN. After reviewing the history of AI and the fledgling AIN, I said,

> Numerous discussions with Alena Brunovska and Kevin Sontheimer . . . have convinced me that no program will remain safe at AI and that we either need to terminate the current grant to the University of Pittsburgh or find an alternative to AI. This latter has been accomplished in principle with the creation of Academia Istropolitana Nova, a civic and tax-exempt organization that proposes to assume the teaching function in the applied economics, environmental management, urban heritage and conservation, and European studies sections of AI; it also proposes to add a professional program in finance. Some 15 members of the AI staff have given signed undertakings to move to AIN either as of January 1997 or as of the end of the current academic year . . .
>
> During this [initial] 3-month period, the following particular tasks will be accomplished: (1) The senior management team will be reestablished, and they will prepare for the transfer of the academic programs, (2) a new Board of Trustees will be established . . . , (3) While AIN has a set of byelaws and rules, . . . these will have to be revised . . . , (4) A new physical plant will have to be established . . . , (5) The recruitment of a new class will have to be begun . . .
>
> This recommendation is for a very risky project. If either of the other two potential funders [Ford and OSI] decide not to provide grants to AIN, it is very likely that AIN will not be able to get started and the Foundation's grant will be, in that event, completely wasted.[30]

This grant was duly authorized and by January 15, 1997, I received a full-fledged proposal from AIN for more substantial funding. In addition to the tasks enumerated in my letter of November 13, the new proposal concentrated on faculty recruiting, short- and long-term fundraising activities and the finance program. While the project remained risky, I was counting on transferring to AIN the balance of the original grant to the University of Pittsburgh (which was calculated to amount to about $200,000 by June 1997) and on grants from OSI and the Ford Foundation totaling $873,500. With a further $400,000 from the Mellon Foundation, AIN would have a reserve of $1.52 million, enough for three to four years of operation. I was pleased that this grant was soon approved by the trustees of the Mellon Foundation.

AIN did get established and did attract a new class for the opening of term in October 1997. It had succeeded in finding a potentially excellent piece of real estate: the former Pálffy Castle in Svätý Jur, some 10 miles outside of Bratislava. It was in terrible condition when AIN signed a long-term lease for its use, but renovations were undertaken with dispatch, and it became a truly wonderful space. The board of trustees acquired an outstanding new chair in the person of Donna Culpepper, onetime attaché at the United States Embassy in Budapest and currently president of the Civic Education Project (see chapter 7). But not all problems were solved equally easily. My attempt to interest Neil Cohen of George Washington University in organizing and creating a program in finance, banking, and insurance failed.[31] Brunovska was insisting on a flat salary structure for AIN staff, which was too low to attract top-quality faculty in economics, because economics teachers had much greater employment opportunities than teachers in the other fields. Sontheimer justifiably felt that tuition payments could not be exacted in the East European environment unless teaching was of the highest quality, but that Brunovska's flat salary plan would interfere with achieving that objective. Worst of all blows was that Brunovska was diagnosed with cancer and died by the end of the summer 1998.

A search for a successor was undertaken right away and, once again, AIN had good fortune: the extremely energetic Katarina Vajdova was willing to accept the position and was appointed in February 1999. Funding was at reasonable levels; in addition to the $445,000 provided by the Mellon Foundation, Ford contributed $390,163, and the various Soros sources $918,866. HESP (Higher Education Support Program; an arm of the Soros Foundations) seemed likely to provide institutional support for another three years. This was a clear example of how western foundations working together could lever each other's effort to great advantage. A new program in journalism was started in January 2000 as a cooperative venture with the Center for Independent Journalism. The program in finance was approved by the board of trustees of AIN on December 9, 1999, and is under development. There was some hope that the Comenius University Department of Mathematics and Physics might help to provide faculty training in economics and finance.[32] The finance program is definitely planned to be one for which a tuition charge will be levied. By mid-2000, it became evident that there was a renewed need in Slovakia for training in public administration as the country inched closer to EU accession, and AIN devised an imaginative new program that would also provide training in NGO management and administration. In September 2000, Mellon made a final $350,000 grant to AIN to support this project. The aggregate student body had suffered some decline as a result of the uncertainties in the years 1996–98 but appears to have stabilized during the current academic year at 76 (16 in architectural conservation, 17 in European studies, 21 in PPAE, 7 in environmental management, and 15 in journalism), with another 5 expected in journalism in the near future.[33] AIN has taken advantage of the World Wide Web to make itself visible and now publishes an online student journal.[34] The graduates are of high quality and much in demand in Slovakia and the countries from which they come.[35] There is reason for cautious optimism that this high-quality operation will prove itself to be sustainable. The legacy of Alena Brunovska is an impressive one, indeed.

Economics Education in Warsaw

Columbia University was among the earliest institutions to spearhead major economics projects in Eastern Europe. All of these focused on Poland, in no small measure because Columbia Professor of Economics Stanisław Wellisz was Polish by birth. The first of these was the Study Group for Polish Reform, funded by the National Endowment for Democracy, which brought together six western economists to discuss stabilization and privatization issues. The second project, funded by the Ford Foundation, assisted the Polish Policy Research Group at Warsaw University, which conducted empirical research on policy issues. The third one, also funded by Ford, was the Polish Finance Ministry Advisory Project, which sent Wellisz to advise the Polish Government on privatization and economic restructuring.[36]

After several false starts between the summer and November 1991, and many discussions with Professors Stanisław Wellisz, Deborah Duff Milenkovitch, director of Columbia's Institute on East Central Europe, and Kevin Quigley of the Pew Charitable Trusts, I received the final version of a proposal from Columbia University to establish an economics program in Warsaw.

The new proposal was very promising. It planned to create an honors program in the economics department at Warsaw University and promised that at least two western economists would be in residence each year. Students would enter the program in their second year of university studies; the program would follow the American pattern of concentrating on microeconomics, macroeconomics, statistics, and econometrics, with some elective special fields and a master's thesis in the fourth and fifth years of study. Graduating students would be equipped to enter western graduate schools, and the economics faculty at Warsaw University would be strengthened. The program would be overseen in Warsaw by Professor Wojciech Maciejewski, a distinguished Polish economist and chairman of the department.

It appeared very likely that the Pew Charitable Trusts would fund this effort with a grant of $200,000, and there appeared a good likelihood that the Ford Foundation would co-fund it as well. This again promised a significant degree of collaboration among U.S. foundations. Since the need in Poland for well-trained economists was well-established, and since we already had a solid program under way in Czechoslovakia (CERGE), it seemed reasonable to proceed. In April 1992, the trustees approved a grant of $230,000 to Columbia University for the project. It is noteworthy that there was no plan from the beginning to make this program permanent; in a letter to me, Wellisz said, "I cannot, at this time, say whether the program will fully accomplish its purpose in three years or whether further extension for, say, two more years will be desirable. I am certain, however, that much can be done in three years and that, regardless of further funding, the effects of the program are bound to be very positive."[37]

The possibility of additional funding for the project, as well as some complications, emerged during the summer of 1992. Wellisz had tried to enlist additional help from the Soros Foundation. George Soros seemed definitely interested and had announced publicly by June 23 that he would provide $500,000 for the project over the next five years. While Ford did not participate in this particular project (although it would do so in a closely related project to be described below), the funding for the project seemed to be adequate for the time being.

During the next two years, the project made considerable progress. Wellisz himself spent the academic year 1992–93 full-time in Warsaw and undertook to spend half his time there during the next two academic years. Several western academics were hired to teach in the program, a solid library was established, and most importantly, my friend and erstwhile coauthor of several papers, Wojciech Charemza of the University of Leicester, had also agreed to teach in the program. He confirmed what Wellisz himself had said, that the quality of teaching was high and the students were exceptionally able and enthusiastic.[38] But, he pointed out, while the Polish faculty members were themselves learning at a rapid rate, the program could not stand on its own feet as yet, and more western visitors were required. This signaled to me that a follow-up grant would be needed sooner or later.

By the middle of 1993, Wellisz had another idea, namely to establish a Central and East European Economics Research Center (CEEERC) under the auspices of Warsaw University, for the purpose of raising the level of economics training among Polish economists and enhancing their ability to do applied economics research using well-established western methods of econometric and economic analysis. He noted that the poor pay that academic economists received in Poland forced them to moonlight on jobs that had little or nothing to do with academic pursuits, so that little genuine research was being accomplished. CEEERC would have resident senior western scholars for guiding the research of the fellows invited to the Center and the fellows would receive a stipend sufficiently generous so that they could, with a clear conscience, promise that during the tenure of their fellowship they would not engage in remunerative work other than teaching. The final proposal to the Mellon Foundation said,

> Economic research in Central and Eastern Europe falls short of Western standards. Few of the region's scholars have the technical ability that matches that of their Western colleagues. There is also little incentive to do rigorous work. Government research units have to generate quick answers to problems of immediate urgency. University departments lack the means to sponsor major studies. Academic standards are lax, and promotion is often unrelated to the quality of work. . . . The Center will promote the formation of a community of scholars and create a milieu in which high-level performance will be expected. Central and East European economists will be given the opportunity to collaborate with Western scholars and acquire new skills.[39]

During the preceding year, intensive negotiations took place between the Mellon Foundation, the Pew Charitable Trusts, the Ford Foundation, and the Soros Foundation.[40] The key question was no longer whether this was a desirable project, but rather which organization was going to contribute how much to it and under what circumstances—that is to say, under what funding scenarios by the other foundations. Ultimately, I recommended to the trustees of Mellon a contribution of $300,000, which was approved; the Ford Foundation contributed $250,000, the Pew Charitable Trusts $300,000, and HESP another $150,000, making it a round $1 million, which seemed adequate for a start. Because of the protracted negotiations, the funding did not become available until the end of 1994 and early 1995.

By the end of 1994, it was also clear that the honors program at Warsaw University would run out of money in the not-too-distant future, Mellon therefore approved another grant of $250,000 for that program and was joined by Pew with another $250,000 and by HESP with $500,000, giving the honors program another round $1 million. All of these contributions were coordinated and we communicated with one another continually to make sure that everybody was on board with the amounts expected by the colleagues at the other foundations. This was another triumph of collaboration among foundations, and it took four large foundations to ensure the fiscal health of the two projects.

The honors program was not intended to become permanent, and after six years of operations it went out of business. As time passed, the effectiveness of the program increased: faculty members became more aware of the capabilities as well as the limitations of Polish students and the attitudes of the students changed for the better as they began to realize that learning was the key to success in the new world.[41] The program also provided and enjoyed positive externalities from the proximity of CEEERC: the program faculty gave advice to the young scholars at CEEERC and often participated in research projects with them. The program faculty also engaged in useful interchanges with the regular faculty members at Warsaw University and with those at the Warsaw School of Economics (Szkoła Główna Handlowa). The teaching program included, in addition to the usual theoretical and statistical subjects, fields such as decision-making under uncertainty, industrial organization, labor economics, international finance, public finance, environmental economics, and economics of health and education. Most importantly, the program filled an important void that characterized much of the East European university education based on rote: it made participation in a seminar on research and writing compulsory. In this seminar, students were taught how to select a topic, how to formulate a hypothesis, how to select the appropriate research methodology, and how to present a seminar. The research undertaken by students in the 1995–98 period was on a high level and dealt with a wide variety of subjects.[42] Some 77 certificates of completion were given during the second three-year cycle and no new entrants were admitted in the fall of 1997. But in addition to the regular students, many doctoral candidates from the Warsaw School of Economics also attended; the total of such extraneous attendees was 717 during the three-year cycle. Of the students who graduated, six were subsequently admitted to do graduate work at distinguished institutions in the United States, three were admitted in England, and two in Belgium. Many others have future plans to attend graduate school or have received good positions in research organizations.

The scale of the program was small. It produced some 20 people who could go on to advanced training in western economics departments, and another 130 or so who obtained good positions where they would function at a much higher level of competence than could otherwise have been the case. This was accomplished at a cost of about $1.5 million, roughly $10,000 per person. This is obviously much less than what comparable training delivered in the United States would have cost. Considering the very large effort that is needed to create master's-level economists, the cost does not seem to have been excessive. But this may be too narrow a basis for calculating; the fact is that as a result of the honors program, the Warsaw University Economics Department is now able to offer highly adequate instruction up to the middle-undergraduate level, which is much more than it could six years earlier.

CEEERC could not start full-fledged operations until 1996. It took most of 1995 to renovate the space it would occupy, which was worth waiting for, because Warsaw University provided handsome and functional quarters for the Center. It was designed to accommodate 10 fellows at any one time, and a massive campaign of advertising was begun to attract candidates. Information about CEEERC was sent to the chairmen of thirty U.S. and Canadian economics departments in order to alert potential East Europeans studying in the United States; this campaign netted a single applicant. Letters were sent to the rectors of all the universities and deans of all the departments of economics in the region, as well as to the 1,200 graduates of the

Batory Foundation (the Soros "country foundation" in Poland) summer school in economics. All this yielded a total of 70 applications, of which six were accepted, yielding a first-year cohort of seven fellows. By the end of 1998, 20 fellows had been accepted of whom 11 were Poles and the rest from Belarus, Romania, Russia, and Ukraine. The low rate of acceptable applications was due in large measure to the inability of many applicants to write a coherent research proposal; the management of CEEERC therefore decided to make the fellowship even more of a training program than had been intended originally. In fact, CEEERC provided some written guidelines for proposal preparation; even today, these guidelines are a model of what one would want to hand out to graduate students. The fellows were under obligation to present papers on their research, and many of those turned out to be of fully professional quality by exacting standards. The senior fellows, such Wojciech Charemza, Staniław Wellisz, Marian Wiśniewski, Wojciech Maciejewski, and others provided research guidance, gave lectures and even taught courses. Short-term visiting lecturers contributed to the program regularly. By the end of 1999, 20 working papers had been produced by the fellows and half-a-dozen publications in refereed journals or books. After four years of full operation, CEEERC has clearly raised the standard of economic research in the region, but has terminated its activities as grant funds were used up. The real question here is again one of cost-effectiveness. This is difficult to come to grips with, because we do not know what multiplier effects will be due to the fellows' return to their home environment. Will the lone Romanian or the lone Belorussian or even the three Russians have a further impact in their own countries? The most likely impact is likely to be in Poland itself, which has had 12 fellows through 1999, and the overall impact of the honors program together with CEEERC is likely to be non-negligible. But in retrospect, it might have been more productive to be less cautious and to conceive CEEERC on a grander scale.

The basic problem was that economists lacked systematic graduate training. This was to be addressed by the creation of a Joint Graduate Instruction Center by Warsaw University and the Warsaw School of Economics. It was planned to correspond to the first two years of a doctoral program, but was not intended to become a complete doctoral program, because it would have competed too effectively with other Polish universities:

> This solution is not acceptable to the Polish universities which insist on the right to grant individual degrees. Centralization, they argue, would deprive departments of doctoral students who conduct exercises (these are all important given the large size of the typical lecture course) and have other important teaching duties. Moreover, the promotion to professorial rank in a teaching institution is contingent upon the candidate having supervised at least one successful doctoral dissertation. The end of the right to grant doctorates would spell the end of internal promotions.[43]

Wellisz proposed a scheme by which Warsaw University, jointly with the Warsaw School of Economics, would organize a common course of formal graduate study, with the other universities retaining the right to supervise dissertations and grant degrees. This idea certainly had some merit, and if it had been presented five years earlier, we might have been more interested in it. As it is, it came late in the Mellon Foundation's involvement in the region and it clearly required a major, multi-year financial commitment, not to speak of other problems, such as my doubts about the individual universities' ability to provide effective dissertation supervision. Whether the decision was right or wrong, we did not pursue this avenue and it is doubtful whether such a program will actually be created.

Economics at the University of Nitra

As early as November 1991, I contacted the rector of the Agricultural University in Nitra, Slovakia, Ladislav Kabat, with a view towards exploring his potential interest in Mellon support for a managerial (not economics!) training program. He replied most graciously some weeks later and informed me that there already was a management program in place, jointly with the University of Iowa and funded by USAID. But he stressed the importance of economics training, which was not being provided. I visited Nitra in June 1992, a small town in the foothills of the Carpathians, and was much impressed by the qualities of Kabat. I encouraged him to develop a proposal, and in short order he produced a document that outlined the operations of a program in policy analysis and a graduate- (master's) level program in agricultural and applied economics. The main planks of the proposed project were (1) establishment of a research program in agricultural and rural policy analysis; (2) establishment of a one-year M.A. program implemented by U.S. institutions; (3) a four-year faculty development project at western universities; and (4) attainment of self-sufficiency by the end of the fourth year.

This seemed reasonable, and by the last quarter of that year I was casting about for U.S. partner institutions. The University of Iowa was one obvious possibility, but what ultimately seemed to gel was a partnership with the College of Agriculture and Life Sciences at Cornell University. As early as July, Cornell and Nitra signed a memorandum of understanding to the effect that they would combine efforts to strengthen teaching, research, and extension in food science, industry, and related fields and to broaden the international experiences of the staff of both institutions. At the same time, Slovakia and Slovak universities were being swept up in the uncertainties of the separation of the Czech Republic and Slovakia, which was scheduled for January 1, 1993. There was much concern that the vote against President Havel in the Slovak Republic would bring old-time Communists to the fore, there was some fear that the new Slovak government had more urgent matters to attend to than worry about university budgets. There were ominous signs that Meciar was increasingly replacing competent people in universities and other organizations with people who were faithful to him. Kabat's term was expiring. As of September, there was considerable uncertainty as to who the next rector would be; nevertheless, Larry Zuidema and Olan Forker of Cornell decided to go to Nitra for a three-week visit, from November 19 to December 12, 1992. This turned out to be a very productive visit, and by April 12, 1993, I could recommend to the Foundation that it make a grant of $491,000 to Cornell for a two-year project which would create an Institute of Economic Studies at Nitra for the purpose of teaching master's-level agricultural economics. The program would admit annually 30 students who had already finished the first four years of study and subject them to intensive work in microtheory, farm and business management, food industry management, agricultural marketing, financial management, international trade, environmental economics, regional development, and agricultural pricing. The program was thus a mixture of economics and management subjects. The program was to be taught by both Cornell and local faculty in five-week segments (so that Cornell faculty would not have to be absent from the Cornell campus for, say, a whole term). In addition, two Nitra faculty members would spend five months at Cornell each year and two Cornell students would spend a comparable amount of time at Nitra. The overall program was envisaged as a five-year project, but Cornell was seeking funding at that time for only the first two years.

There was no doubt that the graduates of this program would be much sought after in Slovak agricultural businesses and government agencies. The risks were the uncertain state of the Slovak environment, and, although mercifully Rector Kabat's term had been extended to the end of 1993 and he felt sure that his successor would honor any agreements made concerning this program, I did not feel entirely confident.

The grant was duly awarded and optimism reigned because Kabat was still rector in the summer of 1994. Meciar had been forced out of power in March 1994, and the Cornell efforts stared with a bang. The spring of 1994 represented a pilot phase, with the full program starting in September, and eight five-week segments were to be taught by two Cornell professors each, which would bring a total of 16 Cornell staff members to Nitra in the first full year (although actually only 14 came). Kabat was quite enthusiastic in June 1994. The project also had some modest support from the Soros Foundation ($30,000). But there were some problems: it is interesting to note what burden even little things represented for such a university. For example, it was quite clear that Cornell faculty members should have access to telephones in their apartments and that they occasionally needed to be driven from here to there; yet even these small extra costs were perceived as nontrivial burdens on the already overburdened university budget.

In 1995, the Mellon Foundation made a second grant in the amount of $70,000 to Cornell, specifically for doctoral training of Nitra students at Cornell. By January 1996, the original two-year project was coming to an end and Cornell requested a third grant for a somewhat reoriented effort at Nitra. Loren Tauer was slated to take over from Olan Forker as project director for the second phase. Cornell faculty would continue to teach Nitra faculty members and graduate students in short workshops, but they would no longer teach undergraduates; the Cornell staff came to the conclusion that additional faculty development was the basket into which the eggs should be placed, and if this were done well, the teaching of undergraduates could be left in the hands of the Nitra faculty. They particularly wanted to strengthen the research component in teaching, encourage student participation in teaching, and create research contacts between Cornell and Nitra faculty members. All this was sensible and would capitalize on the achievements of the first phase; among the impressive outputs of the first phase of the project was a published book of students' final research papers; a thoroughly professional effort.[44] But there were renewed risks, because Meciar was back in power, and he had proposed extremely restrictive legislation, much of which fortunately did not pass in its most restrictive form.[45] I was much concerned about these risk factors and sought the advice of George Soros, who emphatically recommended that we shoulder these risks, because Slovakia needed programs such as the Cornell program at Nitra. After due deliberation, the Foundation awarded a second grant, of $217,000 for continuation of the program. Among the important outputs expected from it were an enhanced capability for Nitra faculty to do research on the urgent economic problems of Slovakia and to incorporate the appropriate technical materials into their courses, thus handing down these skills to new generations.

The program continued on schedule. In the academic year 1998–99, three workshops were held by Cornell faculty at Nitra and five Nitra faculty members visited Cornell for several months each. Three additional faculty members were also selected to visit Cornell in 1999–2000, and the consensus was that the interaction between the two institutions was useful and productive. The visitors to Cornell could observe alternative teaching approaches and participate in research, and the workshops at Nitra provided opportunities for all Nitra faculty to participate in professional development.[46] The funds used for bringing Nitra students to Cornell also had an impressive payoff: in addition to several working papers, one student had been coauthor of two papers published in refereed journals,[47] and another obtained her Ph.D. degree from Cornell with an excellent and technically demanding dissertation.[48] In all, 14 Nitra faculty members made visits to Cornell, spending anywhere from eight weeks to a whole semester, and eight workshops were held at Nitra on topics such as agricultural policy, welfare economics and trade, marketing, environmental economics, demographic analysis in marketing, econometrics, and numerous other topics.[49]

But what about the overall impact? In one significant respect, expectations were disappointed. One of the objectives of the project was to educate young faculty to the point at which they could enter Ph.D. programs and then return to Nitra with academic qualifications substantially in excess of the norm. Unfortunately, most of the young faculty who went through the program received such lucrative job offers in commerce and industry that they declined to pursue Ph.D. degrees (Kabat).

In other areas, the project was much more successful, although not necessarily immediately. Ĺubica Bartová, a faculty member at Nitra, noted that the initial phase of the project had not much effect on the teaching techniques employed by Slovak professors.[50] Faculty in Slovakia have, by western standards, unreasonably heavy teaching loads and unreasonably low salaries; the result is that whatever time is left over from seeing to their teaching obligations is largely spent on extracurricular remunerative activities. The second phase of the project was more successful in this respect and the changed program structure allowed local faculty to integrate new materials and techniques into their courses, as well as delve more deeply into research projects. That is precisely the area in which Cornell project director Loren Tauer felt that the greatest breakthroughs were made:

> In the workshops we instilled in the participants the process of research problem identification, selecting appropriate analysis techniques and objective reporting (i.e., how to do research). Many have been introduced to techniques that they had seen in textbooks but with very little knowledge of application of these techniques. In the early years they did not fully understand economic concepts. I think that our work has helped change that. The participants, either in the workshops or by visiting Cornell observed economic concepts being used to analyze economic problems. I think we will see more of their research change from descriptive reporting to analysis and conclusions. The teaching will benefit from this research because they will have concrete example to use in courses.[51]

This is exactly right. How one does research in the social sciences, and in particular in economics, is a crucial issue; how one takes theoretical concepts and applies them to concrete research questions, and then marries the resulting analysis with appropriate statistical or econometric techniques to allow one to test hypotheses and draw inferences is the key that opens the door to useful applied research.[52] And it is precisely these two elements that tended to be missing among economists trained in the previous regime: how does one determine which theoretical concepts are appropriate for a concrete problem at hand, and how does one tell whether the data support one hypothesis or another or a third? Tauer is right in believing that a breakthrough in this area will undoubtedly spill over into teaching and improve it substantially.

Final Reflections on Economics Programs

Even very specific-sounding programs, such as economics programs, may serve a variety of objectives. Some, such as CERGE and CEEERC, may be primarily directed towards honing research abilities and teaching competencies. Others, like AIN and the Nitra program, may be directed primarily towards producing well-trained applied economists who can function in the demanding worlds of government service or business enterprises. Still others, such as the Warsaw University undergraduate honors program, may be viewed as serving the latter function but also representing a preparatory step towards ultimate attainment of an advanced degree. Nor were these programs established with common expectations concerning their longevity. The Warsaw honors program and the Nitra program were not intended to last forever, although there is reason to believe that their impact will not disappear. On the other hand, CERGE and AIN were definitely viewed by their creators as permanent institutions, with CEEERC perhaps falling in the middle between these two extremes.

While it might be foolhardy to claim victory and declare that CERGE and AIN have achieved permanence, both of them have scored significant successes in that direction. More importantly for the present evaluation, all these institutions have succeeded in their formally stated missions to train and educate students for particular roles. But we must also ask how significant the outputs of these institutions will be for the region. CERGE, AIN, CEEERC, and the continuing efforts of the Nitra faculty can produce annually in the steady state perhaps 70–80 individuals trained for some particular role in the economy. The majority of these will be concentrated in Poland, the Czech Republic, and Slovakia, with a sprinkling in other countries such as Bulgaria, Romania, and the countries of the NIS.[53] Given the sizes of these countries, this may be a sufficiently large number to make a significant difference over a period as short as a decade.

7

Working with Libraries: The Beginning

I am fortunate to have spent my professional life at a university with a great library, and until I encountered East European libraries, I never dreamed how much damage 45 years of neglect could do. Since the Mellon Foundation has had a distinguished record of assisting research libraries in the United States, including the great independent research libraries such as the Folger, the Huntington, the Newberry, and the Morgan, it did not take much persuasion to get the green light from the Foundation to investigate what we could do for them in Eastern Europe. In 1990, the first year of Mellon activities in the region, there were so many small business development, computing and connectivity projects that I had only limited time to deal with libraries. But I did make some important contacts, many of which led to lasting relations between Mellon and some East European libraries, and made many friends in the library world. Even in that first year, I met in Hungary Béla Mader of the József Attila University in Szeged, Olga Gomba of Kossuth Lajos University in Debrecen, Miklós Fogarassy of the National Széchényi Library, and Gábor Vályi of the Parliamentary Library in Budapest. In Czechoslovakia, I met Vojtěch Balík, National Librarian of Czechoslovakia and his deputies, Martin Svoboda and Adolf Knoll, and in Poland, Jadwiga Krajewska of the Warsaw University Library and Jan Pirożyński of the Jagiellonian University Library. While we did start to make some grants directly to East European libraries, much of the first year, and even some of the second, was spent in learning about library systems, needs, and agendas. But our work in the three (and later four) countries did not move in parallel, and the first grants of significant size in Hungary were in 1990, in Poland in 1991, and in Czechoslovakia in 1992. There were three principal types of library projects that the foundation funded: improvement of collections, training of librarians, and library automation. By the end, the third category had received by far the largest commitment of funds.

My first of many visits to Jagiellonian University in Kraków was particularly memorable. In many libraries, I was shown some of the greatest treasures and rarities they possessed, and so, at the Jagiellonian library, I recall being shown a late fifteenth-century volume in which the names of all the matriculating students had been inscribed and looking at the page on which Copernicus' name appears; I remember being upset that that particular page was almost completely faded and illegible because it was obviously the page to which posterity turned most frequently. I also recall meeting the excellent Professor Stanisław Waltoś, director of the university's museum, housed in Collegium Maius, a large and wonderful fifteenth-century structure. He took me on a tour of the museum, including rooms that were not normally open to the public, and when we entered one very large room, perhaps 30 feet square, he pointed to a sizeable earthenware stove standing in a far corner, decorated with glazed tiles, and asked me whether I thought that the motif of the decorations was western or rather eastern in character. I am no expert, but I seemed to discern some arabesques that reminded me of moorish motifs and I replied that it seemed eastern in origin. He then took me right up to the stove and repeated the question. Looking at it close up, it did appear to be different than at a distance, and I (feeling acutely conscious that I might be making a fool of myself) said that I now thought it looked more western in origin. "Ah, you see," he said, "it is just like Poland. If you look at it from a distance, it looks eastern, but when you look at it from close up, you realize that its character is western."

While I have always liked libraries, during the ensuing years I learned to love them. These libraries desperately needed library materials, training, and technology, and I would have to figure out how to fill their needs in the most effective manner. I shall start by describing some particular libraries and their treasures, then characterize East European libraries' acquisitions and personnel policies. Then I shall deal with projects that directly addressed the size and variety of collections, and only afterwards turn to library automation in chapters 8–10.

Some Great Libraries

The purpose of this section is merely to give a descriptive flavor of some of the great libraries in Eastern Europe. It is not intended to make the reader an instant expert on East European libraries, but simply illustrate the range of the libraries in the region, many of which are largely unknown in the west.

Poland. While there can be no satisfactory conclusion to a debate about which libraries are the greatest or most important in a country, my own short list of five includes the National Library (Biblioteka Narodowa), the Warsaw University Library, the Jagiellonian University Library, the Wrocław University Library, and the Ossolineum (Ossoliński National Institute) Library.

The National Library, housed in a relatively modern building, with an even more recent addition, as well as in a palace on Krasiński Square, containing its special collections, has a large collection of some 7 million items, which include 1.9 million monographs since 1801, 680,000 items of serials, documents of social life, musical scores and sound documents numbering some 1.9 million, and special collections amounting to some 620,000 items, including over 23,000 manuscripts. It is a legal deposit library (one of two in Poland, the other being the Jagiellonian University Library) and is charged with the preservation of Poland's cultural heritage and the publication of the Polish national bibliography. Its past, unfortunately, has been checkered in the extreme. Based on the original Załuski Library, which opened its doors to the public in 1747,[1] it suffered a series of indignities. In the last decade of the eighteenth century, the Załuski Library was removed as war booty to St. Petersburg. The National Library was formally founded in 1928 and some of the treasures of the Załuski Library were recovered; however, in the Warsaw uprising during World War II many of its treasures were lost again. But it has been building its collections steadily since the end of World War II and plays a very significant role in the Polish world of libraries.[2]

The Jagiellonian Library, originally housed in Collegium Maius, traces its origins to the early years of the university; a separate library building was built for it in 1517 and at present it holds about 3.3 million volumes. It serves a large reading public, since Jagiellonian University has about 30,000 students, and is a major research asset. It holds 3,250 incunables and 100,000 other treasures and works printed before 1800; in particular, it holds the manuscript of Copernicus' *De Revolutionibus Orbium Cœlestium,* the famous *Behem Codex*[3] (1505) and the *Boguridzica* manuscript (1407).[4]

The Warsaw University Library, with 1,520,000 books, 603,000 volumes of periodicals, and a total of 2.4 million volumes is one of the great libraries with extensive special collections. An early acquisition by the library was the collection of King Stanisław August Poniatowski, the last Polish King of Poland (1763–1813); particularly noteworthy in that collection from an American point of view are five mezzotints and engravings of George Washington.[5]

The University of Wrocław (Breslau in German) was created in 1945 and may be considered to be a successor of a former German University at that location. Its staff came mostly from the cities of Lwów (now Lviv, Ukraine, having been ceded to the Soviet Union after World War II) and Vilnius (Lithuania). It has a library with about 3 million volumes, manuscripts, and other items, with the largest collection of early imprints in Poland, some 200,000, including 3,200 incunables. Among the rarities in the library are John Presbyter's *De ritu et moribus indorum*, printed by Speyer, Columbus' *Epistola de insulis nuper inventis* (Paris, 1493), Marco Polo's *Liber de consuetudinis orientalium regionum* (Antwerp, 1484), and one of the three surviving copies of Martin Luther's "Theses," printed in Wittenberg in 1517. Secular works are represented by the first dated edition of Erasmus' *Moriae Encomium* (Strasbourg, 1511), the first Basel edition of Vesalius' *De humani corporis fabrica* (1543), and many others.[6]

The fifth Polish library I treasure is the Ossolineum Library. The Ossolineum National Institute was created by Count Józef Maksymilian Ossolinski in 1817 and was in the city of Lwów from 1827 to 1945. It served the Jan Kazimierz University in that city; after the war, part of the collection was transferred to Wrocław. It is estimated that 60% of the collection, including autographs, diplomas, prints, photographs, coins, and part of the manuscript collection remained in Lwów. It currently has some 1.5 million items, of which nearly 700,000 are nineteenth and twentieth-century books; its collection of nineteenth-century Polish books is unmatched. It is overwhelmingly a collection of materials of interest to humanists and has some 30,000 readers.

While Poland has such great library treasures that one could go on and on, I shall mention just one more, relatively obscure, library, namely the Hieronim Łopaciński Voivodship Public Library in the city of Lublin. It has a significant collection of manuscripts (although some of these were moved to the State Archive in Lublin), including the "Łopaciński Fragment" dating from the end of the fourteenth century; the *Dzierzwa Chronicle* from 1508; a collection of 57 incunables, including Jakub Locher's *Sand Parysa krolowica Troiańskiego* (a carnival comedy in three parts), 1542, the first play ever to be written in Polish.[7]

Hungary. Arguably, the great research libraries in Hungary are the National Széchényi Library (Országos Széchényi Könyvtár, OSZK), the Library of the Hungarian Academy of Sciences, and some of the great university libraries, such as those of Kossuth Lajos University and József Attila University. The National Library, with about 7 million library units, was founded in 1802. Its mission is to collect all works published in Hungary, in the Hungarian language, by Hungarians, as well as foreign materials dealing with Hungary. Its special collections contain some 8,500 works published before 1711, such as 1,700 incunables, including the first book published in Hungary, *Chronica Hungarorum* (1473).[8] The Academy Library, founded in 1831 with a donation of 30,000 books by Count József Teleki, its first president, has grown to hold 1.8 million items, of which about 1 million are books. The library collects in all fields of science, social science and humanities; very significant are certain special collections, such as the medieval manuscript collection,[9] the Szilágyi Collection, containing some 500 Turkish and Persian manuscripts, the Csoma Collection (named after Alexander Kőrösy Csoma, the founder of tibetology) of some 3,000 Tibetan items, and the Kaufmann Collection of Hebraica consisting of codices and manuscripts, Genizah fragments, and printed books.[10]

Two Hungarian libraries are special collections that are unusual in several respects. The first of these is the Selmec Museum Library, three quarters of which is housed at the University of Miskolc and the remainder at the University of Sopron. It was the library of the former College of Mining and Forestry in Selmecbánya (1735–1918), in what used to be northern Hungary and is now part of Slovakia. It is a collection of books on science, mining, engineering, and forestry. The total collection amounts to 45,000 volumes and contains gems such as Agricola's *De re metallica* (1557), Ubaldo's *Me-*

canicorum liber (1616), and complete runs of *Annalen der Physik* (from 1790) and *Annales des Mines* (from 1794). The Miskolc branch is one of the very few libraries I know in the region that is dehumidified and air conditioned. The other very special library is the library of the Reformed Church College in Debrecen. It has played an extremely important role in Hungarian education; two very great poets, Mihály Csokonai Vitéz (1773–1805) and János Arany (1817–1882) and countless other important personalities were among its students.[11] It has over 600,000 volumes, which makes it the largest ecclesiatical collection in Hungary, and has been in continuous existence since 1538.[12] Its initial collection was primarily devoted to philosophy, theology, classical and biblical languages; over the centuries, law, medicine, and science collections came into existence. The library's stock contains numerous unusual items such as Péter Méliusz Juhász' *Herbarium* (1578), the first text on botany written in Hungarian.[13]

Czech Republic and Slovakia. The arguably most important five libraries in the two countries are the National Library in Prague (Národní knihovna v Praze), housed in the college built by the Jesuits in 1566 and called the Klementinum, the State Scientific Library in Brno (Moravská zemská knihovna), the State Scientific Library in Olomouc (Státni vědecká knihovna v Olomouci), the University Library in Bratislava (Universitná knižnica v Bratislave), and the Slovak National Library in Martin. *Primus inter pares* is the National Library in Prague, which after 1622 also functioned as the library of Charles University. Its collection amounts to about 6 million volumes. Most of the historical collections consist of Bohemica, and the core of the manuscript collection is a set of codices donated by Charles IV; the collection contains, among others, the *Vyšehrad Codex* (1085), the *Passional of the Abbess Kunhuta* (1312), the *Velislav Pictorial Bible* (fourteenth century), part of the library of Tycho Brahe, and numerous other important manuscripts.[14] It is particularly noteworthy that, under the direction of Alfred Knoll, the National Library has done important work for scholarship and preservation by digitizing materials and producing outstanding CD-ROMS.[15]

While the resources in all these countries are immense, using these libraries in the 1980s was neither easy nor pleasant. Although we shall encounter each of these points in more detail in various contexts later, let us simply note some of the crucial deficits.

1. Collection development since 1945 suffered greatly as a result of limited library budgets and particularly because the Communist ideological restrictions strongly discriminated against acquisitions in the humanities and social sciences.
2. The physical plant of libraries tended to be hopelessly inadequate. Part of the reason was that many great libraries were (and are) housed in buildings not designed for libraries; thus the National Széchényi Library is housed in the former royal palace, the Bratislava University Library in the former Hungar-

ian Royal Chamber Palace, which was the seat of the Hungarian Parliament in the years 1802–48,[16] the Czech National Library in a former college, the Warsaw University Library (until recently) in a palace, the Latvian National Library in some eight or nine buildings scattered all over Riga, and so on.[17] Space was often inadequate and books were frequently kept in substandard conditions. As Andrew Lass stated about Czech libraries, "The overflow of volumes in current facilities has led to troubled shelving, storing volumes in plastic crates, and housing folio sized volumes (many of considerable antiquarian value) vertically on floors (some 600,000 volumes in the Klementinum alone, others stored in off-the-shelf conditions around Bohemia). Clearly, completion of a new building is pressing."[18]

3. Libraries were generally user-unfriendly in the extreme; partly because the Communist system did not set much store by encouraging library use and treated libraries as mere repositories of knowledge, and partly because library systems were antiquated and not designed for effective information retrieval. Essentially, no library was "open stack" as in the United States; moreover, classification systems made browsing impossible, even if one had been allowed in the stacks. Books were often assigned shelf numbers by size category, and within size category, by the order of the books' arrival at the acquisitions department. Thus, in such a system, a shelf number might be J.II.1756, indicating that the item is a journal in size category II and is the 1756th item received in that category. If such a library has multiple copies of a given book, it is entirely possible, even likely, that every one of the copies has a different shelf number and is located in a different part of the stacks. I myself have browsed in the stacks of one such library and came upon a book on cardiology next to a novel by Thomas Mann.[19] Union catalogs were typically nonexistent, and thus a person seeking a particular work might have had to visit a number of geographically dispersed branch libraries.
4. Library standards and protocols were not unified or codified and automation was severely underdeveloped if not entirely absent. Work was largely done manually and inefficiency was rampant.
5. The decentralized nature of universities (recall that universities were centrally controlled by the state, but the faculties and chairs in the universities themselves had a great deal of autonomy in some areas) led to the creation of scores of departmental libraries; some reasonably well looked after, others in an execrable state (such as the social science library at Charles University, which has some 16 "central libraries" and over 450 "specialist" libraries[20]).

In the next section we shall provide some brief characterization of the acquisitions policies of libraries; here we conclude with noting that libraries had some wonderful assets but were woefully neglected and needed modernization in a number of areas.

Statistical Characterizations

Libraries differ among themselves along a number of dimensions, and ideally one would like to characterize the production function of libraries—that is, the processes by which inputs get converted into outputs. If one had a good handle on the production function, it would be possible to compare libraries along an efficiency scale. This can, indeed, be done with the aid of linear programming; an important study accomplishing this with 1995–96 ARL (Association of Research Libraries) data is contained in a paper by Shim and Kantor, employing a technique called "envelopment analysis."[21] Shim and Kantor's approach requires observations on input and output variables; in their study, the input variables were total volumes held, net volumes added during the period, monographs purchased, total current serials, number of professional staff, number of support staff, number of full-time equivalent hourly student employees (a largely unknown concept in Eastern Europe), full-time student enrollment, total graduate student enrollment, and full-time instructional faculty. Their output variables were total interlibrary lending, total interlibrary borrowing, number of people who participated in group presentations or instruction, reference transactions, and circulation (initial as well as renewal). In fact, we may think of the input variables as "producing" the observed outputs, and the question that the Shim and Kantor paper asks is how efficiently a library converts the inputs into outputs.

In this chapter, I will content myself with something more modest than a full data envelopment analysis of East European libraries. I shall first discuss some aspects of library acquisition policies in selected libraries and then turn to some questions of staffing.

Acquisition Levels. It is generally believed that the acquisition of library materials, but particularly of western materials, suffered greatly during the Communist period and deteriorated further in the 1980s and early 1990s as the financial position of the East European countries worsened. In its heyday, in 1972, the Polish libraries imported from the West 9,238 periodical titles and a total of 188,808 copies (that is, they imported on average just over 20 copies of each periodical title). By 1993, the number of titles had increased slightly to 10,088, but the total number of copies had declined to 29,823.[22] At the Warsaw University library, the annual acquisition of Polish monographs fell from 9,983 in 1989 to 2,036 in 1994, while the acquisition of foreign monographs fell between those years from 2,055 to 14![23] My erstwhile coauthor, Wojciech Charemza, told me that, in the early 1980s, while he was still teaching at the University of Gdańsk, Poland subscribed to a single copy of *Econometrica*, a very prestigious journal in economics. If he needed to read an article in it, he had to travel for several hours from Gdańsk to Warsaw and back to gain access to it!

In the present section, I discuss some acquisitions data obtained from several libraries. The libraries that I have selected for more detailed analysis can, by no stretch of the imagination, be called a random sample. The principle of sample selection was to attempt to include some large and some small libraries and to obtain some representation from different countries; but the main principle of sample selection was the practical one of data availability. Even so, unlike the case of the ARL libraries for which exhaustive data are available on the World Wide Web,[24] I would not have had access to the data had it not been for some extremely helpful librarians, who provided me with annual data from 1989 to 1998 of the number of books purchased, total books acquired, journals subscribed to, and the cost of these acquisitions.[25]

A large and important library is that of Kossuth Lajos University (KLTE) in the eastern, economically relatively depressed part of Hungary (now part of Debrecen University—see chapter 5). It has faculties of arts, sciences and architecture, with about 60 departments; the current stock of books is 1.26 million, and the library has 163,000 periodical volumes and over 1.5 million other documents. In 1989 it purchased 2,212 books. But in most years out of the next ten, the purchases were less than 1,600 and returned to the 1989 level only in 1998 (with 2,388 books). But the volume of total books acquired was very much larger (ranging from 17,912 at the beginning of the period to 19,336 at the end). A significant disparity between books bought and acquired is characteristic of most libraries in the region. The reason is that acquisitions through exchanges and gifts are extremely common in the region. The disparity is particularly large at KLTE because its library is one of two legal deposit libraries in Hungary (officially, the library is the second national library[26]). It is noteworthy that the dollar value of purchased book acquisitions in exceptionally low by western standards: total acquisitions (including serials) amounted to $471,366 in 1998. For the sake of illustration, the corresponding figures at Princeton University, the University of North Carolina and McGill University were $7.98 million, $6.95 million, and $4.70 million in the same year.[27] Equally noteworthy are the frequently observed facts that the periodical budget is much larger than the monograph budget and that periodicals are gaining at the expense of monographs.[28] Since total subscriptions declined from 996 to 732, the gain of periodicals in the share of costs is clearly driven by their relative price increase. József Attila University (JATE) library in Southern Hungary acquires about half as many books but twice as many periodicals as KLTE; it is therefore curious that its expenditures of books and periodicals are only about a third of those at KLTE. We finally note that the general pattern that total book acquisitions exceed purchases by a sizeable margin, and that periodical subscriptions drop over time and account for the lion's share of the acquisitions budget, also holds true at the University of Horticulture and Food Industry (KEE). It has a much smaller library than the other two and the volume of acquisitions reflects that. It is clear that beginning with 1996, its library budget suffered a substantial diminution; in fact, this institution is the only one in which the trendline of acquisitions has a negative slope. But it continues to be the case that periodicals consume the bulk of the acquisitions budget.

For Polish libraries, I obtained similar data for Adam Mickiewicz University (AMU), for the University of Gdańsk, and the Technical University of Łódż. It is true for all three of these libraries that the amounts spent on periodicals are vastly greater than those spent on monographs. Where data are available, it is also the case that total book acquisitions are larger, or in one case much larger, than book purchases. The total acquisitions budgets for the Polish libraries are of the same order of magnitude as that of KLTE: less than half-a-million dollars a year in most years.

But one of the most interesting aspects of acquisitions, for both the Hungarian and Polish libraries, is the extreme year-to-year variability of purchases. Even superficial examination of the annual data for the North American libraries reveals that the acquisition figures change only slowly year by year; this is true for Princeton, a large, private university library; North Carolina, a large state university library; and McGill, a large Canadian library. More formally, we can get an insight into the relative variability as follows. We divide for each library the book acquisitions in any one year by the acquisitions in the first year for which data are available; this procedure generates a set of index numbers for acquisitions. We then regress the acquisitions index on a variable indicating the year (for example, 1989 = 1, 1990 = 2, etc.). In other words we calculate a linear trend and then obtain the standard error of estimate, that is, the square-root of the residual variance, which is a reasonable measure of variability around the trend. The resulting figures are 0.31 for KLTE, 0.46 for KEE, 3.27 for AMU, and 0.59 for Łódź. The corresponding standard error of estimate for Princeton is 0.20. The difference is largely explainable in terms of the relative turmoil in which East European institutions found themselves after 1989. Budgets were uncertain and changing, institutional changes, such as mergers of units, were occurring with some frequency, and the East European libraries had not yet had time to adjust toward an equilibrium in handling the explosion in the prices of library materials. It is likely that systematic collection development will not improve until budgetary stability is achieved and consistent expectations can be formed by librarians concerning future funding levels.

Staffing Levels and Requirements. Technology transfer in libraries crucially depends on the size and general quality of the library staff, on the organization of the staff and work-flows, and on the specific training that library staff have received in the light of the changes in library services to be provided. These issues have been analyzed in an illuminating manner by Robert M. Hayes.[29] Hayes has obtained data for a sample of 12 libraries in the region, constructed a hypothetical "composite library" out of these data, and compared the individual libraries, as well as the composite library, to a U.S. comparison library with the aid of his library staffing model. He first investigated five specific questions, and reached the following conclusions.

1. Patterns of use between the U.S. composite library and the East European libraries are substantially different; the use of services in Eastern Europe is, on the average, half of what it is in the United States, and particular services show substantial differences; thus, within-library use of materials relative to outside circulation is about 2 to 1 in the United States but more like 3.5 to 1 in Eastern Europe.
2. The patterns of acquisition of materials and the resulting distribution of workloads are again substantially different, with the U.S. library acquiring

monographs with firm orders and microforms on approval more frequently than the East European counterpart.
3. Overall patterns of staffing in United States and East European libraries are substantially similar as far as size and functional distribution are concerned.
4. Staffing patterns within the group of East European libraries are substantially similar.
5. The type of staff employed in the U.S. and East European libraries are substantially different: in the U.S. comparison library, a total staff of 114 consists of 31 professional staff members, 53 support staff members, and 30 hourly employees; in the East European counterpart library, a total staff of 94 consists of 72 professional and 22 support staff members, with no hourly employees.

Among the policy implications emerging from this is the desirability of a drastic reduction in the cataloging staff and the use of copy cataloging on a much broader scale than before. Some of the redundant catalogers would presumably be reassigned to provide user services, which is increasingly necessary in a world in which electronic library services, with which many users may be unfamiliar, become dominant. This, in turn, requires new and possibly higher levels of technical qualification and almost certainly new management abilities capable of dealing with the changing functions and complexities of libraries.

A slightly different picture emerges from a (nonrandom) sample of (13 Polish, 2 Hungarian, 4 Czech and Slovak, and 6 Estonian) libraries I investigated and for which I obtained the total number of monograph volumes, the number of periodical volumes, and the numbers of registered readers and professional and support staff in 1999. Computing the percentage of professional staff to total staff shows that the median percentage is 0.58 and the mean percentage 0.64, with a standard deviation of 0.15 if we exclude the Estonian libraries; this is to be compared with Hayes' mean figure of 0.77, which is roughly a standard deviation greater than what we find. This may well be due to differences in the sample and also to the fact that these data are three or four years more recent than those on which Hayes' calculation was based. But if we include the Estonian libraries, the mean percentage rises to 0.70 (with a standard deviation of 0.18), which is much closer to Hayes' figure. The plausible explanation is that the Estonian libraries were relative latecomers to modernizing the staff ratios along western lines, whereas the libraries in the Visegrad countries had succeeded in taking important steps in this direction by 1999.

The second interesting finding emerges when we analyze the same data with a simple regression analysis. We shall attempt to predict the input, the sum of professional and support staff denoted by y, on the basis of the "outputs" to be generated, that is, the number of volumes (denoted by v), the number of periodicals (denoted by p), and the number of readers (denoted by r); however, we use v and p not separately but only as their sum. Thus, this regression is

$$y_i = \alpha + \beta(v_i + p_i) + \gamma r_i + u_i$$

where u denotes the unobservable error term and the subscript i indexes observations. A second regression is identical to the first with the exception that we include three dummy variables; the first of these, denoted by *dpl*, has value equal to 1 if the institution is in Poland and 0 otherwise; the second, denoted by *dcs*, has value equal to 1 if the institution is Czech or Slovak and 0 otherwise; and the third one, denoted by *das*, has value 1 if the institution is a university of arts and sciences and 0 otherwise. The second equation is

$$y_i = \alpha + \beta\ (v_i + p_i) + \gamma r_i + \gamma_{dpl} dpl_i + \gamma_{dcs} dcs_i + \gamma_{das} das_i + u_i$$

The results of ordinary least squares estimation are in table 7-1. Both regressions are highly significant as a whole, as shown by the *F*-values. In both regressions the output variables have positive coefficients, as one would expect, and the sum of volumes and periodical volumes is highly significant. But the number of registered readers is not significant in either regression. The country dummies in the second regression are not significant, but indicate that both Polish and Czech and Slovak libraries use more staff than correspondingly sized Hungarian and Estonian libraries would. Moreover, the coefficient of the arts-and-sciences dummy variable weakly indicates that that general arts and sciences universities use fewer library staff members than correspondingly sized specialized institutions would. The regression does not change appreciably if it is rerun without the Estonian libraries. All

Table 7-1 Regression results.

Variable	Coefficient (t-value)	Coefficient (t-value)
Constant	−0.894D+0 (0.043)	−0.220D+2 (−0.461)
v+*p*	0.114D−3 (7.134)	0.115D−3 (7.345)
r	0.822D−3 (0.972)	0.754−5 (0.896)
dpl	—	0.302D+2 (0.624)
dcs	—	0.324D+2 (0.588)
das	—	−0.225D+2 (−0.738)
r-squared	0.72	0.74
F	28.84	10.64

these results, of course, have to be taken with a grain of salt, because the sample is small and, in particular, the number of libraries falling into any particular category is extremely small. A further critique of the approach is that we are really estimating here a production function in reverse, that is, estimating the input on the basis of the outputs, and that some vital output functions are missing—namely, the volume of circulation and the number of interlibrary loans, just to mention two obvious ones. But the results indicate that, on average, the addition of about 10,000 volumes to a library results in the addition of one staff person.

Substantial attention has been paid recently to the required educational attainments of library professionals. Many have noted that library schools in Eastern Europe have not been in the vanguard of preparing their students for the electronic age in libraries. Maria Śliwińska argued that in the 1990s Polish library staffs were not at all prepared for coping with library automation. In her questionnaire survey of 39 libraries, the respondents stressed the importance of training in computer skills and in foreign languages.[30] Jela Steinerová examined future manpower needs in Slovak research libraries, with particular emphasis on the required educational background of persons who will be information analysts or information managers,[31] and in a survey found overwhelming agreement on the need for familiarity with information resources, information analysis, the Internet, document processing, text analysis, information management, and related areas. Ewa Krysiak, in another survey, found high variability among libraries in the amount of training that they offered staff and also in the range of attitudes concerning the training they received.[32] She also argued convincingly that "technostress" is prevalent and described the responses to questions about stress as indicating "fear associated with the introduction of new standards and methods," "fear of PCs in general and of radiation," "fear by elderly persons of having to learn a system by heart," "fear of losing data stored only in electronic form," "fear of changing the work environment, of additional tasks, and of becoming redundant," and many other similar attitudes.

As the libraries of Eastern Europe embraced the new opportunities provided by foundation support, it became necessary to redefine the functions of the various categories of library staff, assess staffing needs, and deal explicitly with training, human resource management, and management in general. For the first time, perhaps, library directors discovered that they had to deal explicitly with issues of staff motivation and job satisfaction.[33] An interesting study undertook an opinion survey of 23 carefully selected experts on libraries in Eastern Europe, the Baltic countries, Western Europe, the United States and South Africa.[34] The respondents generally agreed that future librarians would have to be extremely versatile, and would require solid interpersonal and communication skills; they would have to have a customer orientation and at the same time be managers, be able to problem-solve, work in teams, deal with computers, and be skilled in a variety of other functions.

It is not clear that foundations active in providing assistance to libraries, including the Mellon Foundation, were fully aware of the deficiencies in training and the psychological aversions to dealing with new equipment and methodologies. Of course, foundation officers such as myself dealt primarily with representatives of libraries who were the most highly skilled, the best trained, and the most thoughtful about what libraries needed as they emerged from the socialist period. Although even the very best-trained librarians had an attitude that betrayed sadness at their plight and hopelessness about the future, they were well informed and tended to create the impression that all was well throughout the ranks. This was emphatically not the case, and in retrospect I believe that we should have concentrated more effort on providing training—beyond the standard training that vendors of library software provide as part of the sales contract. In fact, the training that the Mellon Foundation provided, although often crucial because it provided training at just the right time and place, was small scale and episodic: Mellon made 16 grants for librarian training purposes between 1991 and 2000, ranging in size from $2,000 to $48,000 and scattered over five countries. All of these grants were basically ad hoc, in response to particular needs at particular libraries, with the exception of three grants totaling $60,000 for the International Center for Information Management, Services and Systems (ICIMSS) at Copernicus University in Toruń, Poland, a new school of library science intending to teach librarians throughout the region the most modern aspects of information management. ICIMSS, under the devoted leadership of Maria Śliwińska, collected an outstanding board of advisors and obtained good teaching space at the university. But it was unable to garner sufficient funds for continued operations, and by late in 2000 it seemed that it might have to suspend operations. In retrospect, I now see that the transfer of library technology might have been more seamless if we had concentrated on this type of activity as early as 1990 and on a significant scale.

Book and Journal Donation Projects

I have noted earlier that one of the consequences, most damaging to effective teaching and research, of the long socialist winter was the extent to which university and other research libraries were deprived of scholarly materials, particularly in the humanities and social sciences. When I visited university libraries, which I did regularly from 1990 on, I made it a habit to check the card catalog for some commonly used books; much more often than not I was disappointed. It was not obvious what the best strategy was for overcoming this problem, nor was it obvious how far it was economically sensible to proceed with any kind of book or journal donation program; but it was totally clear to me that without some careful planning the outcome was bound to be waste, confusion, and the provision of masses of unwanted books to East European libraries. One early horror story recounted the receipt by a library of some kindly donation which included titles such as the "Widener Library Shelflist of Books on Protestant Missionaries in China."[35] In any event, it was important to provide books that were needed, and in principle, this could be accomplished either by dealing directly with the end users or with intermediary organizations. Relying heavily on the former avenue of distribution would prove arduous and time-consuming for us, but in particular cases it was important to respond quickly to the needs of the end users; hence, we ultimately availed ourselves of both approaches.

One should note right at the outset that there were many donors and many methods of conveying library materials to recipients; so much so, that in 1993, the American Council of Learned Societies produced a manual for international book and journal donations.[36] Substantial book donation programs were carried out by British agencies, such as the British Council, the academic library sector, the public library sector, and charitable foundations, and by German organizations, such as the Federal Ministry for Education and Science, and several German foundations.[37]

Grants to End Users. The needs were immediate and I was not sure at the beginning how effective the donation programs through intermediaries would become. As a result, the bulk of the Mellon Foundation grants made directly to end users were in 1990 and 1991. The list of principal grants is shown in table 7-2. End users had very different desires. The first Eötvös Lóránd University grant was for a five-year subscription to various science journals. The second grant and both Jagiellonian University grants were for their respective American Studies Center libraries. The National Library of Poland grant was divided between acquisitions of American literature and covering the costs of designing a library automation plan. The grant to Janus Pannonius University was for miscellaneous book acquisitions, and the grant to the Budapest University of Economic Sciences was mostly for the acqui-

Table 7-2 Library acquisition grants to end users.

Year	Grantee	Amount
1990	Janus Pannonius University, Hungary	$100,000
1990	Budapest University of Economic Sciences	$75,000
1990	Eötvös Lóránd University, Hungary	$60,000
1991	National Library, Poland	$100,000
1991	Jagiellonian University, Poland	$50,000
1991	Gödöllő University of Agriculture, Hungary	$120,000
1992	Eötvös Lóránd University, Hungary	$30,000
1998	Jagiellonian University, Poland	$25,000
1990–91	Kossuth Lajos University, Hungary	$115,000

sition of journals and CD-ROM databases. Many of the grantees had submitted detailed lists of books to be acquired, which provided a nice check that the prospective grantees had done their homework.[38]

In addition to some large grants of this type, smaller assistance was provided to a number of other institutions such as the College of Finance and Accounting[39] and the College of Foreign Trade[40] in Budapest. I have no doubt that these grants, small or large, were very useful, but it was also clear that we could not persist with them too long: there were too many universities and too many fields to make that a practical proposition.

Grants to Intermediary Organizations. The first contact with what was to become an intermediary organization was a letter from Professor William A. Hunt of St. Lawrence University on March 2, 1990. He wrote

> I am writing to propose the establishment of an informal network of American Scholars in Solidarity with Poland. . . . As you know, the collapse of the Polish economy has had devastating consequences for academic life. The victory of Solidarity has brought full intellectual freedom, but Polish universities remain cruelly deficient in the material resources needed for the full creative exercise of that freedom. . . . My sense is that even quite modest efforts on the part of a few American academics could bring measurable, and immediate, benefits to Poland's intellectual life . . . Many American libraries routinely discard books and journals that would be highly valued by Polish universities.[41]

By the time he wrote to me, he had already been in touch with Vice Rector (later Rector) Włodzimierz Siwiński of Warsaw University, who warmly embraced the idea of supplying books and journals to the university. Hunt's efforts were targeted, first and foremost, at the Warsaw University Library. The director, Jadwiga Krajewska, wrote to him on April 10 and said that the library needed monographs in the humanities and social sciences, many periodicals, and was looking forward to receiving lists of periodicals to be discarded by American libraries. She also indicated quite emphatically that the library was not interested in receiving textbooks, because of the differences between Polish and American curricula. She also submitted a list of 63 particular monographs ranging over library science, mathematics, political science, economics, literature and criticism, and philosophy.[42] This appeared to be a well-organized and low-cost effort and we started out by making an $8,000 grant in 1990, followed up by another $9,000 before the end of the year. By November 1990, Hunt had contacted the publishers of some 200 periodicals for individual issues missing from the Warsaw University Library collection and had shipped 5,000 books to Warsaw, some through the courtesy of the Peace Corps and some paid for by the Foundation grant. By March 1991, he had established contact with the Sabre Foundation (see below), which he consulted for advice and help, had shipped a total of 7,800 volumes, and I was convinced that that he was extremely sensitive to the specific needs of the recipient. He had collected another grant in the amount of $25,000 from the Leo Koerner Charitable Trust and expanded his scope to Romania (4,000 volumes), and various medical schools in Poland, as well as Jagiellonian University, Catholic University Lublin, and Charles University in Czechoslovakia. He was doing so well that the Foundation made another grant of $25,000 in 1991. By March 1992, he could write to me,

> many intellectuals and academic administrators, especially at the major universities, now have far easier access to the western academic world, and hence to western sources of financial support. For that reason, it no longer seems appropriate for a small enterprise like the Solidarity Project to expend its limited resources on the general needs of major institutions, such as Warsaw University, or the Jagiellonian University. . . . We also plan to disengage from support for the teaching of English, which has hitherto been one of our three major concerns (along with medicine and the environment). . . . The Solidarity project will henceforth seek to support smaller institutions and organizations that combine the following attributes: 1) demonstrated competence and integrity; 2) creativity in addressing especially critical socio-cultural needs; 3) ability to employ relatively small contributions so as to produce the maximum multiplier effect.[43]

The organizations that he selected for support were the Helsinki Watch Committee in Poland, the Polish Ecological Club, the Łódż Medical Academy, the Budapest University of Economic Sciences, and the Ministry of the Environment of the Czech Republic. He also noted that many "energetic and competent scholars in Eastern Europe are simply unaware of the existence of important western works in their field," and that "money spent on providing multiyear subscriptions to [appropriate] journals would do far more to promote the intellectual integration of Eastern Europe with the West than an equivalent sum spent on individual books." This is, of course, what it was all about: the objective of providing materials was not just to enhance the holdings of East European libraries, but to provide the means for such an intellectual integration. The case was well made, and the Foundation made a final $25,000 grant to St. Lawrence University's Solidarity project, which was a model of what energy and dedication could accomplish.

In the latter part of 1990, a group of academics including Wesley Posvar, then the president of the University of Pittsburgh; Joseph Kohn, professor of mathematics at Princeton University; and Theodore Rabb, professor of history at Princeton created a not-for-profit organization, the American Czech-and-Slovak Education Fund (ACSEF) with an ambitious program that promised support for CERGE and the Czechoslovak Management Center, planned to facilitate student exchanges between the United States and Czechoslovakia, promote the flow of Czechoslovak research assistants to the United States, provide business training at a hotel school in the High Tatras, and engage in the wide-ranging rehabilitation of libraries in Czechoslovakia. During the first half of 1991, the focus of ACSEF mercifully sharpened. By

October I received a proposal from it which correctly analyzed the problems of Czechoslovak libraries (inadequate physical space, inappropriate repositories for historical manuscripts that were rapidly deteriorating, huge gaps in collections, little technical equipment or automation, and inadequate standards of librarianship) and proposed a two-pronged approach: (1) survey U.S. efforts to assist libraries and conduct a survey, combined with a field trip, to determine the needs of Czechoslovak libraries, and (2) design a plan for rebuilding the collections of the libraries. This latter step involved the computerization of the 150 "request forms" that had been distributed to and received from libraries during the earlier part of the year, the engagement of consultants with field-specific knowledge, and the identification of donors of library materials and of publishers who would provide deeply discounted subscriptions. I sought the advice of Professor Andrew Lass of Mount Holyoke College, who was an expert on Czechoslovakia, and he confirmed my suspicion that there was not much further value in surveying either U.S. library donation efforts or the needs of Czechoslovak libraries. However, he felt, and I agreed, that much could be gained from intelligent collection rebuilding efforts. Accordingly, I recommended to the Mellon Foundation that a grant of $50,000 be made to ACSEF. In my letter to the Foundation's president I wrote,

> Since this is an untried organization, there is some risk in making them grant. However, their approach to the acquisition program is rational and unified and corresponds entirely to the overwhelming conclusion of the October 18th meeting[44] that unified approaches should be sought to solving the problems of Czechoslovak libraries; thus, for example, the present approach will be able to control the number of copies that they acquire of a given work and will be able to suggest to some of the Czechoslovak libraries that the work has been provided to some *other* library from which they should strive to obtain it on interlibrary loan.[45]

ACSEF rapidly moved into high gear and accomplished much. It managed to purchase important reference books and professional journals at substantial discounts, and to generate substantial donations of books; it enlisted the help of the Sabre Foundation in shipping books, convinced Czechoslovak Airlines to transport 220 lbs (later raised to 440 lbs) of books free of charge as luggage accompanying a traveler, managed to raise some additional funds ($10,000 from the AT&T Foundation) and provided some travel grants for Czech university administrators and scientists. ACSEF was turning into an effective provider of library materials, which was due in no small measure to the organizational talents of its executive director, Irina Rybacek. As was entirely appropriate, the Foundation approved another grant, of $30,000, in July 1992. In the next six months, the activities of ACSEF continued to grow and it turned into a significant player in its niche. It also started to cooperate with the New School Journal Donation Program (see later in this chapter) and, by December of 1992, the estimated market value of books and journals ACSEF had supplied was on the order of $250,000–$500,000. It seemed entirely reasonable to continue this effort, and while ACSEF would have liked a five-year commitment of much larger funds, in my recommendation for another grant of $100,000 I wrote as follows:

> Although I believe that this has been a very successful operation filling a genuine need, I think that it is entirely reasonable for the Foundation not to consider funding this project beyond the $100,000 recommended here. Such a sum is consistent with our own priorities and it would be incautious in any event to commit funds for more than one year, since the supply-demand situation in Eastern Europe may well change in the coming years. As more and more books are shipped, eventually, the marginal benefit from the additional books and journals that can be obtained at low cost may well decline. The Czech and Slovak Governments may be able, in time, to assume a larger share of the responsibility of acquiring books. The increased efficiency that ACSEF will have by virtue of its experience may be increasingly offset by the rising marginal cost of acquiring books.[46]

The grant was made in March 1993 and the book and journal program continued to function throughout 1993 as anticipated. But in late 1993, Irina Rybacek decided to return to her original homeland, the Czech Republic, and a new executive director was named. The funds that had been provided by Mellon were being consumed by the book program and new funds were not being raised. At that point, ACSEF attempted a few desperation strategies, such as an alliance with the Czechoslovak Management Center (working closely with the University of Pittsburgh), the development of an academic exchange program, the creation of a university administrator training program, and an ambitious but basically flawed plan to turn the Charles University library system into a "model library." This latter project, in particular, seemed to me misguided, partly because of the disorganization of the Charles University library system with its myriad branch libraries, and partly because the director of the Charles University Computer Center, Dr. Pavel Krbec, was the local distributor for TINLIB, an integrated library software product produced by IME in England.[47] This would have created an intolerable conflict-of-interest situation that we could not be part of. ACSEF very much wanted further Mellon support, but in December 1993 I wrote to Bowen,

> With considerable regrets I told them [ACSEF representatives] that we would be unable to provide such support. The unfortunate fact is that they no longer seem clear about their programmatic mission, and while they do want to keep on supplying books and journals to research libraries, this mission seems somehow less well articulated at present than it used to be. . . . Unfortunately, I could not justify giving them hope without a clearer vision on their part as to where they were going.[48]

During 1994, ACSEF stopped operations and went out of business. The lesson here is threefold. First, the board of directors failed its executive director by not aggressively providing additional sources of funds. Second, an organization with essentially a single source of funds is almost certainly doomed. Finally, if the desperate attempt to stay alive induces the organization to engage in scattershot program development, its mission becomes diluted, which jeopardizes the support of the one donor interested in a single mission. But all who worked with ACSEF can take solace from the fact that for three years the organization did good and important work.

The third, initially small-scale but enduring, effort in this area was the Journal Donation Project of the New School for Social Research. It was the brain child of Arien Mack, professor of psychology at the New School and editor of *Social Research*, who submitted a proposal in the fall of 1991. The proposed activity was to supply scholarly journals to research libraries throughout the region—in other words, not restricting the project to Hungary, Poland, and Czechoslovakia. The key idea was to persuade publishers of scholarly journals to donate 15 free subscriptions for a period of three years and, whenever possible, to further persuade the publishers to absorb the cost of shipping them. By the time the proposal was submitted, she had commitments for 108 journals.

From the very beginning, the Journal Donation Project attempted to maintain close relationships with the recipient libraries, partly to ascertain that they received journals that were truly needed and wanted, and partly to make sure that journal deliveries were timely and that the recipient libraries cataloged them and made them available to the reading public. In fact, full descriptive material about the journals was sent to each library, which was then asked whether it had ever subscribed to the journal in question (in order to avoid the possibility that a previous regular subscriber substituted a free or subsidized subscription for a regular one), whether the title was relevant for curricula and research activities, and how the library would advertise the item's availability if it were placed on the recipient list.[49]

A sampling of the initial group of journals includes such stalwart publications as *The Public Interest*, the *American Journal of Comparative Law*, the *Harvard Business Review, The New Republic*, the *Economic Journal*, the *Journal of Political Economy, Neural Computation*, the *Theatre Journal*, and the *American Sociological Review*. In each country, the project engaged local "agents" to help determine the best targets.[50] By November 1992, Mack could report that 330 journal titles were being donated to libraries in Bulgaria, Croatia, Czech Republic, Slovakia, Latvia, Lithuania, Poland, Romania, Albania, Estonia, and Hungary. Since most of these titles were being provided in 15 copies, the total annual shipments of journal issues amounted to about 5,000. In January 1993, the project reported that the number of journals was up to 339 and the market value of the journals shipped over a three-year period was estimated at $1.2 million.[51] The number of recipient organizations had increased to 149, of which 67 were in the Visegrad countries. As the scale of the project increased, so did the administrative burdens of negotiating with publishers, arranging for shipping, following up in the recipient countries, and so on. While fortunately much of the shipping was taken care of by the donor publishers, for the rest the Sabre Foundation (see below) once again stepped into the breech and helped out a kindred organization. But more funds were needed and the Mellon Foundation made another grant to the project in the amount of $75,000. I also called the project to the attention of Anthony Richter of the Soros Foundation in the hope of generating more funding for it.

A key aspect of the program was that no journal was sent to a library unless that library specifically requested it. But even so, more and more copies were needed and the project management realized that publishers would not supply free copies in arbitrary numbers and had to resort to obtaining journals that were not free but at least deeply discounted. The project caught the attention of the American Council of Learned Societies which used its influence with affiliated learned societies to generate more free subscriptions.[52] By September 1993, the number of titles being donated had increased to 580, the number of recipient libraries to 254, and funds raised from other organization amounted to $74,369 from USIA and $66,000 (of which $25,000 was a pledge at that time) from the Soros Foundation. By March 2000, the total raised by the project amounted to over $1.5 million, of which Mellon had provided $160,000, USIA $272,000, Open Society Institute in New York $113,376, and Open Society Institute in Budapest $659,197.[53] This was clearly an incredibly cost-effective operation, due in large measure to the effectiveness and persuasiveness of Arien Mack, who took painstaking care to ensure that the right journals got to a library where they would be used. By 2000, the project had largely ceased operations in Estonia, Hungary, Poland, and the Czech Republic, but it carries on vigorously in 27 former Soviet Bloc countries in which it continues to supply some 400 libraries with 10–20 subscriptions to 667 journals. In February, 1998, Béla Mader of the József Attila University Library wrote to Arien Mack,

> In our experience, your project has been the only journal donation scheme to date that has worked. In the past decade, there have been many such programs, but the plans usually came to nothing. We have received a considerable number of books, but no periodicals on a continuing basis apart from the ones received from you. I think this was mainly due to the fact that you have had a direct and friendly relationship with both the publishers and the editors who have been providing the periodicals and the beneficiaries. We particularly appreciate that you did not deal with us by proxy, but have established direct and friendly contact with us, on a standing basis, and has [*sic*] been interested in our needs, our culture, our scholarship and opinion during our years of cooperation.[54]

In several places in this section I had occasion to refer to the Sabre Foundation, the largest and most ambitious organization in the book and journal donation business. The Sabre Foundation had established its Scientific Assistance Project in order to help East European institutions through technical assistance as early as 1986.[55] The Mellon Foundation had actually given it a $50,000 grant in the fall of 1989, shortly before I joined the Foundation. In these first three years, Sabre had provided over 250,000 volumes with a market value of about $10,000,000 to organizations in Hungary, Poland, and Czechoslovakia.[56] Serious discussions between Sabre and Mellon began in September 1990. The Sabre program planned to concentrate on three approaches. The first one was to solicit tax-advantaged donations from for-profit publishers under Section 170(e)3 of the U.S. Internal Revenue Code, which permits publishers a deduction of up to twice the marginal cost of production for certain inventory donations.[57] The second approach was to solicit donations from noncommercial and non-tax-advantaged sources, such as the libraries of individual scholars, library deaccessions of books, review copies, academic presses, and the like. Josiah Lee Auspitz of the Sabre Foundation was already aware of the efforts of Arien Mack in the New School Journal Donation Project and planned to assist her activities. The third approach was to create a Scientific and Cultural Assistance Fund that would, through a reduced-cost purchase program, provide additional materials, make matching grants to stimulate further institution building efforts on the part of United States and East European groups, and provide staff support for building a consortium of groups that would provide material support. Initially, I was somewhat concerned about the extent to which Sabre could tailor donations to actual needs. Auspitz replied that,

> We try to treat overseas partner groups as customers rather than donees and to assure that every book sent is, in effect, ordered by them. . . . First a general needs profile is developed in the field with each overseas partner group. . . . Second, as inventory donations become available, the actual lists of proposed donations are sent by fax to the recipient group for selection. . . . Books not selected are not sent. Third, where lists of titles are not sufficiently descriptive, we provide further descriptive materials based on professional evaluations, library research, and materials from publishers' catalogues.[58]

In Hungary, the partner organization was Sabre Hungary, headed up by Miklós Fogarassy of the National Széchényi Library; in Poland, Sabre operated in a more disaggregated fashion with independent programs in Warsaw, Lublin, and Kraków. In Czechoslovakia it had been suggested to Sabre that it should locate its activities in the local IREX office. All this was reasonably reassuring, and in October I recommended that Mellon give a $230,000 grant to the Sabre Foundation.

Progress was good because Sabre was anything but lackadaisical. After three months of negotiations, in May 1992, Sabre acquired a collection of 81,133 volumes of new science and technology books from Elsevier, with a list value of $11.4 million, which immediately increased Sabre's warehousing requirements. In addition, Elsevier and Pergamon Press (ultimately merger partners) gave a 50% discount on all of their combined 1,200 journal titles for three years. Gábor Vályi, the head of the Hungarian Librarians' Association and former director of the Parliamentary Library, estimated that Hungary alone could usefully absorb the entire amount made available by these publishers. In any event, Sabre's needs increased as a result of this and other acts of largesse, and it needed to engage a consultant in inventory management, increase warehousing space, provide open shelving for single-copy retrieval, obtain some equipment, and provide for warehouse staffing. By the end of the summer, Mellon made another grant of $100,000 to Sabre to assist in these matters. By the end of 1992, Sabre was beginning to try to wean the recipients from donation to purchase, tap EC funds, establish a regional distribution center in Budapest, and provide a better distribution system for Poland. For the next two to three years, Sabre was going to spend 40% of its efforts on Hungary, Poland, the Czech Republic, and Slovakia, another 40% on the NIS, and 20% on other countries. Another $250,000 grant followed in 1993 and two grants totaling $155,000 in 1995, for a total of $785,000 over six years.

Cooperation with other donation programs was excellent (such as the New School Journal Donation Project and the American Czech-and-Slovak Education Fund), and by the middle of 1994, Sabre had distributed in the four Visegrad countries 147,159 volumes.[59] By 1995, Sabre had taken the first steps towards introducing a document delivery service at the Prague Institute for Advanced Study for scientific articles. It was actually implemented in November 1996; documents were typically provided within five working days and cost the end user about $4 per document. Documents were scanned and then sent to the end user by FTP (File Transfer Protocol) and then printed. It was an effective system that ultimately had to be discontinued (in 1998) because no funding could be found to pay the full cost of the service. The cumulative total of books shipped by Sabre through 1998 was 847,471 books with a total list value of $28.3 million, to which we must add the list value of $2.6 million in reduced-cost subscriptions for three years to 918 scientific and medical journals. It was a massive effort, well organized and effective.

There were some other organizations that sought Mellon funding, but I could not convince myself that their activities were well matched to our objectives. The Company/College Gifts-In-Kind Clearing House was interested in providing "materials" produced by American industry, which were supposedly in demand in Hungary, such as electronic products used in manufacturing, metal and plastic parts for production engineering, and so on. I certainly did not feel that I had the expertise to evaluate what was needed and to what extent the materials supplied matched requirements. The Brother's Brother foundation proposed to supply massive shipments of textbooks, and again I was not certain that it was worth investing in an effort that measured the amount shipped by the pound weight. A particularly harebrained request was for financing the operational expenses of a collection of some five million volumes of journals, reported to be mostly disbound, which had been acquired from the *Science Citation Index*. The collection was modestly referred to as Bibliotheca Universalis and was to be housed in a building in the middle of nowhere in the Hungarian countryside (in a former county office building in the town of Szentes). None of these efforts were funded by Mellon, but I was content to believe that the important ones were.

Evaluations of Book and Journal Donation Programs. While the various book and journal donation programs were under way (in other words, long before glowing testimonials could be received), I was very concerned about how effective these programs were, particularly because the grapevine was filled with rumors that the "wrong" books were being sent to Eastern Europe, that the books were not reaching the designated target audience, and so on. For that reason, the Mellon Foundation commissioned a study of the effectiveness of these programs, which was carried out by the Civic Education Project (CEP).

CEP was founded in 1990 by two Yale graduate students, Stephen Grand and William Antholis, for the purpose of providing western instructors to East European universities for one to two academic terms, predominantly in the social sciences; fields included economics, law and legal theory, political science and international relations, public administration, and public policy. Its creation was assisted by Yale, Charter 77 Foundation, and the Soros Foundation. The instructors taught, on the average, two courses that were deemed relevant by their host university. In the academic year 1991–92, CEP placed 14 graduate students and two professors in eight Czech and Slovak universities. By 1992–93, it could send 82 instructors to 42 universities in nine countries; this grew to 110 instructors in 50 universities in 11 countries in 1993–94.[60] In 1995, the western instructors' activities were supplemented by the Eastern Scholars Program, partly to assist in CEP's instructional efforts, and partly to encourage young people from the region to choose academic careers. As of January 2000, total current and previous Eastern Scholars numbered 191, representing operations in 20 countries.[61]

Since CEP had a cadre of instructors widely dispersed in the region, it seemed reasonable to ask CEP to take charge of a study of the kind I had in mind. CEP planned to contact university rectors and deans in order to study whether donation programs were responsive to demand, whether donated books and journals were available to the intended audience, and many related questions; local CEP staff could supervise the gathering of data and also use local students and faculty for more detailed information. In 1994, Mellon gave a $50,000 grant to CEP to carry out such a study.[62]

The resulting small book was highly critical of donation programs. While certain organizations, such as the New School Journal Donation Project, the Sabre Foundation, the International Book Bank, were only infrequently accused of specific misdeeds in the study, its overall thrust was largely negative. Its principal chapters dealt with book donations, the conduct of in-country partner organizations, and journal donations.

In book donations, the most common complaint appears to have been about the quality and age of the books sent, and while this complaint was most pointed in the case of the bulk shippers, those who worked with approval lists were not immune from the charge either. Certain kinds of important books (for example, computer books) were, apparently, severely undersupplied. Donors were criticized for providing incomplete sets of books, foreign language books in basic subjects the readers of which could not be presumed to know foreign languages, and most significantly, for not ensuring that the books were accessible. A sampling of libraries determined that 70–100% of the books sent to particular libraries could be located in only 25% of them; in fully 55% of the libraries either less than 30% of the books could be located or the collection was not accessible at all. The conjecture was that many of the books ended up in the private collections of professors or university administrators, and the study expressed the wish that books appear in open stacks rather than in restricted areas, such as departmental libraries. While the study admitted that actual usage was difficult to estimate, it found some evidence that the availability of the books was not being brought adequately to the reading public's attention.

Local partner organizations were faulted in several ways. They were primarily held responsible for the poor awareness on the part of the reading public of the availability of certain books. When partner organizations did not have adequate warehousing facilities, distribution suffered. The books did not even get to the partner organizations in a timely fashion if they were being held hostage by customs officials, which had happened on more than one occasion. Partners failed to see to it that donations were used effectively and often did not provide adequate feedback to donors.

In the journal donation area, the study saw the same quantity versus quality dilemma that had beset the book donation programs. It singled out the Journal Donation Project for not making it easier for libraries to select the right journals (a direct contradiction of what the Journal Donation Project had reported), although it was praised for providing at least three-year subscriptions (since it was thought that subscriptions for a shorter period were essentially useless; a charge that held for the International Science Foundation, which had provided 20–25 two-year subscriptions to 107 titles, with a market value of about $7 million [Becker, p. 40]). A particularly good point was that since there had been relatively little coordination among donors, duplicate titles were delivered to some institutions.[63] Journal donations were also claimed to be poorly publicized, donated journals were said to have a disappearance rate of 25%, and the donations were overly concentrated on current journals at the expense of archival copies.[64]

The report evoked a strong reaction from the Sabre Foundation and others associated with it. In a long document, Josiah Lee Auspitz took the report to task for a substantial portion of its analysis and conclusions.[65] In addition to criticizing the overall research design, the looseness and inaccuracies in the quantitative analysis, he pointed out that the CEP study suffered from a significant cultural bias by not realizing that (1) books may be difficult to locate in highly disaggregated library systems lacking union catalogs; (2) open stack systems are essentially unknown in the region; (3) keeping books under lock and key is the safest way of ensuring their continued availability and may in fact comply with the requirements of Section 170(e)3 of the Internal Revenue Code that prohibits resale of donated tax-advantaged books. He argued that the study ignored the realities of the supply side and that redonated books (that is, provided originally by publishers from their inventories) and procured books (low-cost purchases for purposes of donation) simply could not provide what might optimally be desired. He severely criticized the anecdotal evidence, which he referred to as "snide" and "offensive," and faulted the study for ignoring questions of cost-effectiveness. While he did praise the report for its orientation toward the interests of the end user, for suggesting end-user surveys of donated materials and distribution audits, and for its general understanding of the library situation in Eastern Europe and its general evaluations, he concluded that,

> The most disappointing and self-damaging feature of the Report is the use of terms suggesting ineptitude and favoritism, where honest and intelligent strategies better explain the same facts. This, in turn, stems from deeper conceptual failings: the use of a fictionalized version of the American open-stack library as a criterion for Eastern Europe, the failure to distinguish practices specific to donated books from general library protocols, the denigration of non-library holdings, inexperience in the practicalities and evaluative literature of librarianship, inadequate grounding in the basics of the supply side of donation, neglect of cost considerations in formulating recommendations, and a condescension in interpreting anecdotal material that can and should be corrected in orientation sessions for future volunteers. (Auspitz, p. 14)

This was by no means the end of it. The Sabre Foundation organized a conference in Budapest, November 6–8, 1995, entitled Dialogue of East European Partners, which was devoted to a discussion of donation programs. It adopted the Budapest Declaration, which made recommendations to donors and stated the commitments of the East European partners.[66] One paper presented at the conference took a sample of its own among libraries and largely rejected the findings of nonavailability reported by CEP.[67] But the quantitative techniques employed by this survey and the sophistication of the survey instrument itself was certainly no better than those employed by CEP. The participants were largely people sympathetic to and grateful for Sabre's efforts. The executive director of CEP, Kerry S. McNamara, had been invited to attend, but did not do so; instead he replied to Auspitz's withering rebuttal by a letter of his own.[68] He first pointed out that the CEP findings of book unavailability was compatible with the other finding, namely that some 90% of professors had read some donated book, simply because so many books had been donated that 90% was entirely plausible, even though many books might have been missing. He denied arguing in favor of open stacks but insisted on better record-keeping by libraries of the books that passed through their hands, in part so as to track books that had been redistributed to other end users by the libraries that originally received them. He pointed out that the information received from Sabre was not sufficient to track any books in the Czech Republic and no audits were performed at all in that country. He reiterated the validity of some of the anecdotal information.

This provoked what was to be the *dernier cri* from Auspitz.[69] Auspitz renewed his charges of shoddy research design and implementation; he argued that the statistics of unavailability were totally unreliable because the four Czech libraries in the sample were not audited at all and yet were counted in the missing or inaccessible category. The letter analyzed the statistical inferences that could be drawn from standard audits versus spot audits, pointed to lapses in the use of CEP's own file materials, claimed that librarians had found the study impractical, naive, and arrogant, and stated that the CEP statistics presented were, when judged by other surveys, incorrect by a factor of 10. It reemphasized the importance of cost considerations, and generally castigated McNamara for his "misunderstanding at a still more basic level." It gave a prominent role to the opinions of one CEP lecturer who described local library practices, according to which it was reasonable and productive to keep books in faculty offices.

Little did we think when we commissioned the original study that nearly two years' worth of heat (and some light) would be generated as a result. There is little doubt that Sabre had done a very good job in the region. An evaluation of its activities in early 1994 revealed that its partner organization in the Czech Republic, the Center for Democracy and Free Enterprise, appeared to be efficient, but that perceptions about the quality of books were difficult to elicit; Hungary was regarded as the role model, with quality regarded as good; in Poland, Sabre was doing at least as well as other organizations.[70] The CEP study was by no means evaluating primarily Sabre, but its analysis clearly touched a raw nerve. But what are we to make of all this?

Cutting through the trees for a better glimpse of the woods, let us admit that the CEP study did not qualify as a proper statistical study by academic standards. Whether because of pressure of time or shortage of staff or confused thinking (the options among which Auspitz's criticisms oscillate), the findings of the CEP study were not conclusive and alternative explanations, some properly drawn from a dispassionate contemplation of the local social and library context, were entirely appropriate. Moreover, anecdotal explanations are always suspect to some extent: what am I to make of the fact that in the Tartu University library I did see a copy of a journal supplied by the Journal Donation Project, but I failed to see a copy of another journal that I know was sent to that library? Yet anecdotal evidence has some value. Hoarding of books and hoarding of information was endemic in the pre-1989 societies and it is unlikely that this disappeared overnight. Yes, cataloging standards and library protocols differed from ours, and the existence of scores of small, inefficient departmental libraries in large university settings made record-keeping and auditing difficult; but the fact is that if it was difficult for a CEP auditor to locate a book, then it was difficult for a faculty member or student at that university to locate it, as well. We must therefore not necessarily blame the donation programs at all for unavailability, but the systems in place; if anything, the donation programs should have exerted more influence in changing prevailing systems, in the same way in which library automation programs changed prevailing systems. But we had not anticipated that it might be important to budget for mechanisms to transform cultural norms and deep-seated library customs. And, in fact, not only did grants fail to make provision for such activities, but the intermediaries in the donation programs themselves were far from equipped to handle this dimension. It is also true that the donation programs evolved over time; my own sense is that both Sabre and the New School Journal Donation project became more efficient with the passage of time and increasingly sensitive to the importance of responding to users. On balance, the donation programs had a massive influence, so massive in fact, that with a few exceptions they have run their course in the Visegrad countries.

The areas in which even those countries will continue to need donations are in current and archival journals. The current journal need will be much alleviated by the Open Society Institute's decision to subscribe to EBSCO, which provides over 3,900 current journals in electronic form in 39 countries; in the year 2000, some 1.4 million full-text articles were downloaded through this program.[71] The archival need may be met in selective areas by JSTOR, which, through a Mellon grant, has recently been made available to a few institutions in the Visegrad and Baltic countries at a small fraction of the cost paid by U.S. or West European institutions.[72] It is interesting to note how few of the East European universities have been able to muster the funds needed for such reduced price subscriptions to JSTOR: as of May 18, 2001 only the Czech Academy of Sciences, Palacký University, Masaryk University, Silesian University in Opavá, Comenius University, Tartu University, Kossuth Lajos University, and the Library of the Hungarian Central Statistical Office were subscribers. It is well known that in the past several decades, the prices of periodicals have increased substantially faster than university budgets, particularly in the areas of medicine, science, and technology.[73] And as the current journal prices keep increasing faster than the East Europeans' ability to pay, the need for journals in whatever format will become more intense. This is well documented by the rapid acceptance of JSTOR in those few institutions in Eastern Europe that have become subscribers: in the year 2000 (up to December 20), in Hungary, 5,856 pages from 2,392 articles were viewed and total page accesses were 13,410; in the Czech Republic 20,026 pages from 6,775 articles were viewed and total page accesses were 46,372. In the Czech Republic, in particular, the growth of usage has been rapid: in the last three months of 2000 usage was over five times as great as in the last three months of 1999.[74]

An Electronic Resource Consortium: HUSLONET

In early 1992, Miklós Fogarassy of the Hungarian National Széchényi Library (OSZK) had the bright idea of creating a consortium of several Hungarian university libraries and one Slovak university (Šafárika) library for an as yet undetermined purpose. The institutions that seemed to be potential beneficiaries of some kind of collaboration were three institutions in Debrecen, Hungary: the Kossuth Lajos University, the Medical University, and the Agricultural University in that city; the University of Miskolc; and, finally, Šafárika University in Košice, Slovakia (later also joined by the Technical University in that city). The common bond among them was that they were all situated in a reasonably compact geographical area in the northeast of Hungary and the eastern portion of Slovakia; to me the particular charm of such a consortium was that it would be the first international library consortium in the region (except for CASLIN, which became international by an accident of politics); thus, if these institutions could find a common problem that was amenable to a consortial solution, the consortium would directly respond to George Soros' often reiterated call for "regional solutions," with which I wholeheartedly agreed. The librarians of these universities started to meet in order to define their common needs and eventually concluded that the purpose they could best serve through a consortium was the joint acquisition and sharing over a wide area network of a number of expensive CD-ROM-based electronic databases. Working out the details of this took the better part of 1992, and the first draft of a proposal arrived at the end of February 1993.

The realization of such a plan had three major components: (1) How could CD-ROM databases be shared over an electronic network? The libraries were largely DOS users, and it was not possible to remotely use the CD-ROM retrieval programs in a DOS environment. (2) Would the wide area network have enough bandwidth for satisfactory remote use of large amounts of data? (3) Would it be possible to negotiate with CD-ROM database vendors wide area network licenses which would make it less expensive for the five institutions to obtain a database than it would for five individual licenses? The solution proposed for the first of these problems was the acquisition of additional software and hardware—specifically, LogiCraft Gateways costing some $35,000. Complete solutions to the second and third problems could not be given until serious attempts had been made to send data over the network and negotiate with vendors. But all of us feared that this was not going to be easy; there was no direct link between Debrecen and Košice and signals would have to travel from Debrecen to Budapest to Vienna to Bratislava to Košice—certainly a roundabout way. Among the CD-ROM databases that the universities wanted to acquire were titles such as *Books in Print with Book Reviews, Ulrich's Plus, Chemical Abstracts, Science Citation Index, Biological Abstracts,* MEDLINE, *Social Science Citation Index, Oxford English Dictionary*, and many others. In any event, the proposal presented a "minimal" plan and a "maximal" plan, costing $325,000 and $866,000, respectively. OSZK was going to exercise oversight and be responsible for financial management, as in two previous consortial projects (see chapter 8). In March, I recommended that we fund this project at the level of $425,000; the minimal plan seemed to be too small to be viable, but without further experience I was not inclined to risk more. At the same time, since the project took a regional approach, I sent the proposal to George Soros in the hope that he, too, would find it appealing and that once again, we might co-fund a project. It turned out that he was interested; he asked Richard Ayres, the outgoing librarian of the Central European University in Prague, to evaluate it. The latter made a number of useful comments and asked several searching questions. After these were satisfactorily answered, it seemed that there was no further obstacle to a Soros contribution. In fact, OSI did not contribute funds to the acquisition of databases, but made an aggregate contribution of $55,000 for other related purposes, such as enhancing the interlibrary loan connections between HUSLONET members and establishing a network for delivering requested materials electronically.[75]

Negotiations with vendors were hard and not nearly as successful as had been hoped. In particular, SilverPlatter was very adamant and refused any wide-area discount, although other vendors did provide one, ranging from 20 to 30%. It turned out that some of the original conceptions had to be revised, since the consortium received better terms for ordering duplicate copies of certain databases than sharing a single one over the WAN.[76] More recently, some vendors seemed to prefer consortial pricing, and discounts even higher than 30% were obtainable.[77] In spite of a noble attempt by OSZK to circumvent the problem of Slovak import duties and VAT by acquiring all the equipment itself and then donating the appropriate share to UPJS, a lot of delays ensued because of border problems. Furthermore, the vendor of the equipment, a Hungarian firm Dataware, did not provide adequate service on the other side of the border. I also had the sense that the Slovak librarians felt somewhat outnumbered by their Hungarian counterparts in spite of the general amiability that characterized the operations of the consortium, and there was some lack of agreement among Hungarians and Slovaks in the evaluation of bids. But finally, after long delays, HUSLONET was officially opened in a public ceremony in Miskolc in October 1995. Soon thereafter, it became clear that OSZK had completed its assignment and there was no further reason for its continued involvement. Accordingly it withdrew from the project. HUSLONET became an independent formal consortium with Kossuth Lajos University as the manager. In 1996, the Mellon Foundation made a supplemental grant of $80,000 to the university for the purposes of HUSLONET.

HUSLONET has been in use through 1999, but the intercity traffic was much less than had been anticipated, at least in part because of the somewhat orthogonal interests on the part of the institutions. The principal databases that became reachable from any of the sites were the *Science Citation Index, Metadex, Biological Abstracts, EconLit,* MEDLINE, CAB *Abstracts*, the *Hungarian National Bibliography*, the *Bibliography of Hungarian Periodicals* and *Swetscan*.[78] There is some disagreement within HUSLONET about the usage of databases in Hungary: according to one source, 20% of the databases are used by only one university, 40% by two, and 40% by three universities,[79] whereas according to another, 25% are used by two, 25% by three and 50% by more than three universities.[80] In Slovakia, 63% are used by only one library, 27% by two, 8% by three, and 2% by four.[81] The Hungarian part is currently funded by the Hungarian Research Fund, but the Slovak part of HUSLONET is moribund because the lack of continuing financing prevents the acquisition of databases (see also the discussion of KOLIN in chapter 8). A critical but early assessment of HUSLONET lays blame on the untested technology, the lack of a clear agenda and timeline for the project, and the absence of a project manager.[82] But whatever the shortcomings of the project were, it established two important principles: first that international cooperation can be made to work, if only imperfectly, and that WAN access to electronic databases is worth the attempt. The improvement in bandwidth in the past several years has been quite impressive, with the Košice–Bratislava link ranging from 0.5 to 2 Mbits/sec, the Bratislava to Vienna and Vienna to Budapest link at 34 Mbits/sec, and Debrecen to Budapest and Budapest to Miskolc at 43 Mbits/sec, soon to be upgraded to 155Mbits/sec.

Vendors need to think about the pricing of CD-ROMs and similar databases. If it is the case, as seems plausible, that (1) the demand curve for such products is downward sloping, and (2) that the average costs for providing the product are also declining (because there is a large fixed cost of creating a database and a constant marginal cost for distributing it), then marginal cost pricing is irrelevant. Setting a price that provides a fixed margin over average cost does not seem like an optimal policy from the firm's point of view, because there are additional potential buyers who would buy the product at a somewhat lower price. Setting a price for a consortium with n institutions that is less than n times the unit price will generate more sales and allow the seller to, in effect, move along the demand curve, thus behaving like a perfectly discriminating monopolist, for higher aggregate profit, and great benefit to buyers.

8

Library Automation: Loose Consortia

It was fairly clear that there was only so much one could do with book and journal donation programs without running into diminishing returns, and from the very beginning I thought that much good would come from introducing modern, integrated automation systems in the libraries of the region. I started out on this mission with only two handicaps: I knew next to nothing about library automation systems and I knew a comparable amount about how East European librarians actually dealt with books; how they cataloged them, circulated them, and so on.

In every country, I made a point of visiting the national library, and at every university I visited, I asked to be taken to the library. I talked to librarians, asked them to describe what systems they had in place, and encouraged them to think about automation using the most up-to-date technologies. I found some strange things, such as the immense proliferation of branch libraries, functioning essentially as departmental libraries, many of which had a very small staff and collections that have grown in a topsy-turvy fashion. It seemed clear that in the first instance automation had to deal with the universities' main libraries, if for no other reason that, unlike the pattern of American campuses, East European urban universities tended to be scattered over the city and building a university-wide local area network was not in the cards in the short run.

In spite of my general ignorance, I was aware of some East European automation systems that were more or less integrated,[1] but I did not think that these were good candidates for the regions's libraries, in spite of the fact that many of them were actually in use. It struck me that if the objective was to integrate East European libraries into the world of western scholarship, one would want to use systems that had had substantial international exposure and that had been created by sufficiently large companies so that their long-term survival was, if not assured, at least plausible, and that they were not "dead-end systems," that is, that they would continue to be developed and improved. There were, in fact, many international companies selling library software; through 1991, CTB and IME (TINLIB) were leading with 1,250 and 1,350 systems sold, with some other well-known vendors represented with smaller numbers, such as Ameritech (78), CLSI (309), Dynix (743), Geac (290), Innovative Interfaces (251), Notis (159), Sirsi (144), and VTLS (214).[2]

There was plenty to choose from, but in some places there was definite resistance to selecting a western system. Indeed, the first time I talked about this with Rector Zbigniew Grzonka of the University of Gdańsk, he said something to the effect that he did not need to *buy* an automation system; he would just lock ten of his best programmers in a room for a year, and when they emerged, they would have a full-blown library automation system ready! Considerable resistance developed in Latvia to replacing the home-grown system ALISE with a western one (see chapter 10), and as recently as March 2000, I was told by the deputy minister of culture of Estonia, Margus Allikmaa, that Estonia should not have wasted money on an expensive system such as INNOPAC,[3] because Estonian programmers are exceedingly gifted and could have developed such a system from scratch; he illustrated his point by noting that a group of Estonian programmers had recently devised a very effective theater-seat reservation system. Wisdom kept me from arguing about the relative complexities of theater reservation and library automation systems. In any event, technical universities and universities of arts and sciences seemed to have a predilection for seeking government funds for domestic software products, on the grounds that they were cheaper and had the desirable quality of simplicity.[4]

The provision of automation software and hardware for the libraries of the region became a ten-year adventure and an immense learning experience, and it substantially transformed the ways in which the East European libraries functioned.

The Status Quo in Automation

The creation of MARC (Machine-Readable Cataloging) records revolutionized the way in which libraries and library patrons could deal with bibliographic information. As Béla Mader pointed out, "The key element of library automation was the recognition that once a bibliographic record has been created in machine-readable form, that record can be used repeatedly and for different purposes."[5] This was undoubtedly one of the key innovations that permitted the modernization of library processes, and to the extent that library automation forced some libraries in Eastern Europe to adopt MARC formats, it played a vital role in technology transfer. Mader distinguished four periods of library automation in the western countries, starting with the introduction of first-generation cataloging systems in the late 1960s, followed in the mid-1970s by systems utilizing standard bibliographic records, and then in the 1980s by systems based on much more powerful hardware utilizing open architecture operating systems (Unix). The final stage, the one in which we find ourselves at present, is characterized by the explosive growth of the Internet and the globalization of information.

In a sample of 70 East European libraries, Christine Borgman examined the importance to librarians of five reasons for automation.[6] The most important of these was *cooperation*: automation encourages the sharing of bibliographic data, reduces original cataloging, leads to union catalogs, etc. The second most important reason was *access*: online catalogs have vastly increased librarians' and readers' ability to tap into library information.[7] The third most important reason was *standards*: in Borgman's words, "Conformance to standards is the bond that unites information infrastructures and enables interoperability among networked information systems." The last two were *mission*, the ability of automation to allow libraries better to serve their communities, and the impact of automation to reform and improve library *management*.

Eastern Europe was clearly lagging behind. While a few libraries expressed some interest in automation as early as the 1970s, serious work did not start until the mid-1980s. Libraries typically did not have computers, and personal computers started to make an appearance around 1984–86. An exhibition of software took place in January 1986 at the premises of the Applied Computing Corporation (Számitástechnika-alkalmazási Vállalat) in Hungary. Nine of the 17 systems exhibited were for Commodore-64 computers—machines that would have been considered antiques in the United States at that time.[8] But clearly some efforts were being made to modernize the information infrastructure. Beginning in 1980, some libraries were permitted to work online and to acquire data-tapes from the West, and in 1986 a formal program was initiated in Hungary to develop the information infrastructure (IIF) in order to create systems that would be compatible with western networks.[9] But development was not very rapid, and into the 1990s, the predominant PCs were IBM PCs, IBM XTs, and IBM ATs or their equivalent. I had visited quite a few libraries in the early 1990s where everything was still being done by hand. Most library software was not integrated but performed only one or at most two functions; perhaps cataloging or circulation. Some early (western) integrated library software ran on mainframe computers; examples are DOBIS/LIBIS and NOTIS. A case in point is Hungary, where the National Széchényi Library (OSZK) obtained a computer of its own and adapted ("localized") DOBIS/LIBIS for its own purposes, namely for the maintenance of the Hungarian National Bibliography, which by May 1992 contained some 400,000 machine-readable records.[10] This was followed in short order by the adoption of the same system by the Hungarian Academy of Sciences.

Local area networks in Hungary were fairly rare in the 1986–90 period and the bandwidth through which a library might be connected to a backbone was often no more than 2,400–9,600 bits/sec. Moreover, in a vast majority of institutions, only one PC was connected to the outside world (Tolnai). Software was of many different flavors, which made the transfer of skills difficult, reduced the mobility of librarians, and probably added to a pervasive fear of computers. Finally, foreign software was regarded as very expensive, and the cost of converting data made automation even more so (Füredi).

Several countries attempted to develop their own software, and quite a few domestically built systems were in use as late as the end of 1994: Christine Borgman reports that in her sample of 70 East European libraries 15 (21.4%) used such systems (Borgman). In Hungary, OMIKK (Országos Műszaki Információs Központ és Könyvtár, National Center and Library for Technical Information) developed a computer-based cataloging tool, OSZKÁR, which ran on an IBM mainframe. In Poland, in 1987, the National Library started to develop MAK for personal computers, and by 1991 it was able to carry out several library functions, among others cataloging the Polish National Bibliography.[11] But efforts to automate actually started some years earlier; the problem with these early efforts was that records were not in MARC format, and their ultimate conversion to MARC format created substantial difficulties. In any event, the DOS-based MAK system became enormously popular; by 1992 it was being used by about 50 Polish libraries and by 1997 by about 400. At the National Library, some 100 users could use it simultaneously over a Novell network, where it was used for cataloging the national bibliography, foreign books, Polish serials, and special collections. MAK was a reasonably flexible system and could accommodate various MARC formats, and users could modify the structure of databases (Sadowska, p. 106).

In (then) Czechoslovakia, two efforts were undertaken to develop an automated system: MAKS in the Czech Republic and IKIS in Slovakia. MAKS and IKIS were intended to promote communications from the "center" to the peripheral libraries, as was consistent with a highly centralized political system.[12] MAKS employed standardized data structures and its core contained an exchange format with a strong resemblance to UNIMARC.[13] The automation of the Slovak National Bibliography started as early as 1972 at the Slovak National Library and the project lasted till 1985, after which a new software, IKIS, was developed in the 1985–92 period. Its main purpose was to create a library computer network, a union catalog and shared cataloging.[14] Perhaps its most important effect was the introduction of a MARC format, developed in cooperation with the National Library in Prague. While the development of IKIS permitted the distribution of IKIS records on diskettes and the discontinuance of printing catalog cards, ultimately IKIS could not function adequately as a national network solution with the existing hardware, and the Slovak libraries turned to the software CDS/ISIS.

CDS/ISIS clearly paved the way for the ultimate introduction of more ambitious software packages. It is distributed free of charge by UNESCO on an "as is" basis and it has more than 20,000 registered users worldwide.[15] It clearly represented a major advance in Eastern Europe over some earlier home-grown packages in that it employed variable-length records with subfields, permitted repeatable fields, provided a search engine with Boolean and proximity operators, and enabled data exchange by virtue of being based on the ISO 2709 standard. Later versions of CDS/ISIS were compatible with MARC formats. Of course, when Slovak libraries adopted CDS/ISIS, it became necessary to convert the existing 350,000 record database to CDS/ISIS format, which proved to be a difficult task, even though IKIS also employed the ISO 2709 exchange format. Among the early achievements with CDS/ISIS is the creation in 1977 of an online catalog at the library of the József Attila University in Szeged, Hungary, with 113,000 records. But as late as the end of 1994, some 20% of Borgman's sample of 70 libraries had achieved no catalog conversion, 27.1% had converted only the last two years, 21.4% had converted up to the last five years, and only 14.3% had converted more than five years of data.[16]

The Bratislava University Library began to process foreign serials on IBM 7040 and IBM 360/20 computers in the early 1970s, but this effort did not get very far. In 1981, it started to store bibliographic records on an RPP16S computer originally designed for industrial process control. Finally, in 1989, the University Library introduced the Magic system produced in Israel, which necessitated another major conversion of records.[17]

One of the greatest problems the libraries of the region faced was the absence of commonly agreed upon standards, such as the MARC standard. This standard provides a precise format for the bibliographic description of library items, and while there is more than just one MARC format, they exhibit strong resemblances to one another. While USMARC, originally developed at the Library of Congress in 1965–66,[18] is probably the most widely used MARC format,[19] others have developed over time, such as UNIMARC, which was created in order to provide an "exchange format," that is, a format that would facilitate the conversion from one national MARC to another. In any event, without agreement on common or nearly common MARC formats it was not practical for libraries even in a single country, let alone across international borders, to share bibliographic information, and it is clear that a great variety of formats was in use.[20] A similar profusion of different standards was being applied to character sets.[21] The variety of bibliographic formats and of character sets also made it extremely difficult to develop "shared cataloging" approaches, by which a given item would be cataloged only once, so as to avoid the burden of each library having to catalog the same item over and over. As a result of this, original cataloging amounted to 83.6% of items cataloged in Borgman's sample of 70 libraries. (Borgman, p. 18)

As the new age dawned in 1989, the libraries of Eastern Europe had to overcome numerous handicaps:

- They were poorly equipped with hardware;
- As a result, librarians' familiarity with computers was weak and many librarians were subject to "fear of computers";
- Their wide area networks and their international gateways had low bandwidth;

- They largely relied on a more or less random mixture of nonintegrated library software, much of it home-grown;
- Library protocols and standards were far from uniform and were largely not compatible with western models;
- Anything that needed to be done was expensive; a very special handicap in countries that had to exercise strict financial self-restraint;
- They had to face up to the task of developing collaborative approaches in order to diminish the amount of original cataloging done by the libraries and thus increase their reliance on copy cataloging.

I shall dwell at some length on the library automation initiatives that were funded by the Foundation. I want to examine, to the extent possible, (1) what was done; (2) what was the impact of automation on library protocols and procedures; (3) what was the role of the "human factors;" (4) what was the influence of various management structures; (5) what role library automation played in the evolving relationship between the libraries and the respective governments; and (6) what was the overall impact. It is interesting in retrospect, although I did not fully realize this at the time, that the course of library automation was quite different in Hungary, Poland, and the Czech and Slovak Republics, and had its own individual characteristics in the Baltic republics. These differences reflect to some extent fortuitous circumstances, but they are also explained in part by where I was at any one time on my own personal learning curve.

Automation in Hungary

Mellon Projects. As I had indicated before, I had written to numerous libraries in the early months of 1990. My letter to the National Széchényi Library was answered by Miklós Fogarassy, of the library's Center for Library Science and Methodology.[22] A lively correspondence grew up between us during the months of April through August. But I was unsure about how to generate quick proposals from several other libraries. I had already started discussions about libraries with Kossuth Lajos University, Budapest University of Economic Sciences, Janus Pannonius University, and Eötvös Lóránd University and was reasonably well advanced in discussions about computer support with József Attila University (see Chapter 4), but it was not clear to me how to widen the reach of my net. Fogarassy produced the brilliant idea that OSZK would act as an "aggregator": he, as project manager, would canvas the various libraries, determine their needs and desires, and produce a single proposal for a number of institutions. My meeting with him and other librarians at the end of May confirmed my hypothesis that library technology was the single greatest need that Hungarian libraries had, and also confirmed that it would make good sense for several libraries to make a joint proposal for a grant under the leadership of OSZK. By July 31, 1990,

I could write in my recommendation to the Foundation that "the proposal is the outcome of this joint effort and it is nothing short of miraculous that nine libraries have gotten their act together in so short a time and made individual proposals. I think that this greatly simplifies life for us, since the National Library is willing to administer a grant for the benefit of all the participants."[23]

Nine libraries had submitted proposals in this package for a total request of $252,722, which the Foundation rounded to $260,000.[24]

All the proposals dealt with introducing or further developing library technology. All the libraries were actively involved in building databases for online searches, establishing networks inside the library for administrative purposes and general library management and connecting them to the outside world for enabling user access, and establishing databases for bibliographic purposes. The specific requests ranged from PCs to network components, Novell software, CD-ROM readers, and in some cases to acquisition of databases and funds to repair some rare books. Among the proposed database acquisitions of, say, the József Attila University library we find *The Bible Library, The New Grolier Electronic Encyclopedia, The Serials Directory, The General Science Index, The Social Science Index, The Humanities Index*, and *ERIC*. Implementation of these mini-projects did not pose great challenges, and all the libraries succeeded in installing the equipment and software. The libraries were extremely satisfied with the improvement in their operations resulting from this, although none of these changes was revolutionary in character. On the strength of this success, OSZK put together another group of libraries in 1992; this second group proposal was very similar to the first and included 12 libraries for a grant of $295,000.[25] There are two noteworthy things about this proposal and grant. First, the original proposal included 32 libraries, of which all but nine were not university or college libraries but municipal or county public libraries. All but one of the public libraries were cut from the grant because the Foundation viewed its mandate in Eastern Europe as primarily fostering higher education. However, the Szabó Erwin Library, also a municipal public library, was retained as a grantee because that library is a major research library in Budapest. Second, once again, the grants were evolutionary rather than revolutionary in character and the grantees had no difficulty in accomplishing their objectives. But it is also the case that these grants enhanced the human capital of librarians only modestly, and the technology that they transferred was small scale compared to some other grants.

One may be tempted to characterize these two grants as grants to "consortia," and in a sense they were; but they were consortia in name only, not held together by a common purpose or mission. For this reason, they have to be sharply distinguished from "true consortia" about which I shall speak later in this and also in subsequent chapters.

Two multipurpose grants to Kossuth Lajos University in 1990 and 1991 contained a total of $170,000 for improving the state of library automation. The university had not yet decided exactly how to automate its library, and it was also the beneficiary of a World Bank loan to Hungary for higher education; part of the university's share of this loan would also be used for automation. Ultimately, after a brief flirtation with Aleph, produced by Ex Libris, the university acquired the Voyager system, which has been in full operation for a number of years.

The only other institution in Hungary at which the Foundation was the sole or principal contributor to library automation was the University of Agricultural Sciences at Gödöllő (GATE). In October 1991, the time that I recommended a grant of $191,000 for this institution (which also included a sizeable component for book acquisitions, with $70,000 earmarked for library technology), the library had four IBM AT-type computers and no online capabilities at all. After a brief flirtation with a Hungarian system called HUNGALIS, which was being recommended by the Chamber of Hungarian Libraries, the university leadership wisely chose TINLIB, a widely distributed software package produced by IME in the United Kingdom. In retrospect, it would have been a bad decision to adopt software not backed by a large company with international exposure. TINLIB, which appeared on the market in 1986, seemed particularly well-suited to the needs of small libraries that did not need complete functionalities in an integrated system. Since it was substantially less expensive than some other, more ambitious systems, it made good sense for smaller libraries, such as the libraries of GATE, of the University of Veterinary Medicine and of the University of Horticulture and Food Industry to acquire it. TINLIB had both DOS and Unix versions, but only the latter could be reached remotely by using TELNET. A practical size limit suggested for effective use of TINLIB was 50 concurrent users and 500,000 volumes.[26] (TINLIB is now considered a "dead-end" system because the vendor decided not to develop it further.)[27]

Dealing with the Hungarian libraries was not complicated. We were not asked to help design a major restructuring of library processes and did not face difficult-to-solve issues of how and when to automate. TINLIB was widely regarded as a user-friendly system, and no great problems of training emerged. While all libraries suffered from inadequate resources, there did not seem to be great political issues involved in questions of which library received what kind of assistance. While the Foundation's support to libraries was important and solid, it was not of the kind that would cause massive transfusions of new technology or would revolutionize the ways in which libraries went about doing their business (although many libraries did outstanding work in particular areas such as retroconversion[28] or subject headings and authority control[29]). The Hungarians seemed to know what they wanted and could specify their needs in unproblematic fashion, so that we could deliver the needed help. The situation could not have been more different in Czechoslovakia (later the Czech Republic and Slovakia) and in Poland.

The Hungarian Union Catalog. It seems appropriate to report briefly on an important project that came into being without any Mellon assistance.

Union catalogs are in many ways the culmination of integrated library systems; they permit shared cataloging, create huge databases that allow for comprehensive searches, and facilitate nationwide interlibrary loans. An important starting point for the attempts to create a unified Hungarian approach was a report occasioned by the World Bank's support for higher education in Hungary.[30] The report clearly outlined the options for a union catalog and the economies in cataloging that would ensue from it, and stressed the importance of (1) coordination and agreement between automation systems and international library standards; (2) UDC (universal decimal classification) as a way of solving the classification problem; and (3) the importance of properly handling diacritical characters. All subsequent Hungarian efforts in this area have reflected the content of Borgman's insights.

Two more-or-less parallel efforts came into being in Hungary. The first, called VOCAL, is based on the approximately 20 libraries that use the Hungarian version of Voyager, called Corvina. A union catalog was created by 1997, and its creation and usage rest on the following principles: (1) the catalog continuously reflects the state of the members' individual catalogs; (2) the members can download bibliographic and authority records from the central catalog; (3) the central catalog provides holding and status information; (4) the integrity of the members' own catalogs is safeguarded; (5) in order to promote the automatic growth of the VOCAL central catalog, the local "save to OPAC" option was replaced by "save to OPAC and VOCAL" option.[31] VOCAL has the important feature that other integrated systems can communicate with it through the Z39.50 protocol and that authority records can be browsed and downloaded.[32] VOCAL is not a nationwide system, although some of the most important libraries are members of it, such as the libraries of Kossuth Lajos University, Janus Pannonius University, and József Attila University. As of June 29, 2000, it contained 571,314 bibliographic and 723,695 local authority records.[33]

MOKKA (Magyar Országos Közös Katalógus) is intended to be a Hungarian shared catalog—union catalog—covering enough libraries so that the catalog's coverage would amount, to all intents and purposes, to 100% of Hungarian holdings. Work on MOKKA has been financed by the Ministry of Education, with some support from OSI. As of November 2000, MOKKA was not as well advanced as VOCAL, although many of its principles follow the VOCAL model: members build their own databases and upload records to the central database. It follows the principle that a given document should be cataloged only once and that any reader anywhere should be able to determine where, if at all, a particular item is available in the country. Unlike VOCAL, which is based on USMARC, MOKKA intends to employ HUNMARC as the bibliographic format. Some have thought that VOCAL and MOKKA could fruitfully coexist, and that VOCAL could be a useful consortial member of MOKKA, providing it with a large and uniform database (Bakonyi). Others strongly disagree with this proposition on the grounds that a number of crucial issues have been left unresolved: (1) when VOCAL members upload records, do they upload to the VOCAL catalog, the MOKKA catalog or both;

(2) when VOCAL and MOKKA rules differ, how do they get reconciled; (3) if a record already exists both in VOCAL and in MOKKA, where does a member download it from?[34]

The actual history of MOKKA has been uneven to date. A tender was announced for the software that would run MOKKA, and the expert committee charged with evaluating the bids recommended Voyager/Corvina. However, the majority of the members of MOKKA chose Dynix, and the chairman of the MOKKA group, Béla Mader, resigned after this decision. MOKKA actually contracted with Dynix, but it eventually became clear that Dynix could not deliver what was demanded, and the MOKKA group canceled the contract with Dynix. It now appears that MOKKA will return to the originally recommended decision and contract for the Corvina system. MOKKA now seems back on track.[35] The Hungarian librarians deserve credit for having realized the importance of a union catalog and for developing the concepts of VOCAL and MOKKA. But the events underscore the perils of dual approaches and the importance of forging a consortium that speaks with a single voice, as we tried to do in Poland (see chapter 9).

Czech and Slovak Libraries: CASLIN

First Steps. I did not know Czechoslovakia nearly as well as I knew Hungary, and it was not obvious to me how best to proceed. In June 1990, I visited the National Library in Prague and met with the director, Vojtěch Balík, as well as his deputies, Martin Svoboda and Alfred Knoll. We had a pleasant conversation and I tried to tell them that the Mellon Foundation was interested in providing support for library automation, but after our meeting I was not yet convinced that their automation plans were going to advance vigorously. Fortunately, I had already gotten in touch with Andrew Lass who had very special qualifications to be of assistance to me. By a quirk of fate—his parents had been journalists and had been stationed in Prague from November 1948 to early 1973—he spent his formative years in Prague, including his university education at Charles University, and was bilingual in Czech and English. When Lass and his parents were expelled from Czechoslovakia (before he could complete his degree at Charles University), he enrolled at the University of Massachusetts in Amherst, MA, from which he subsequently obtained a Ph.D. degree in anthropology. He had been a professor at Mount Holyoke College ever since. It is amusing as well as gratifying that in 1990 he was "politically rehabilitated" by Charles University, which belatedly awarded him the "magister" degree.[36]

Lass knew many people in Czechoslovakia, and had an abiding love for libraries. He had heard that I was directing the Mellon Foundation's program in Eastern Europe and had written to me; I wrote him back in a highly encouraging tone, and by June 6, 1990, he replied,

> Yes, I am certainly interested in taking the lead as I try to organize support for the Clementinum Library in Prague. As I may have mentioned, I am

> leaving for Prague to speak to the new leadership of the library on the 11th of this month. I plan to listen and to look carefully and assess their needs and, basically, give them some, at this point vague, idea of what the possibilities may be.[37]

Little did I know at the time that our chance correspondence would mature into a close collaboration and an enduring friendship.

I was clearly impatient and perhaps hoped that Lass' visit to the National Library would produce instant proposals for a radical overhaul of the library system of all of Czechoslovakia. In any event, I wrote to Dr. Balík on August 25, and pointedly informed him that I had already received a library automation proposal from the National Széchényi Library in Budapest—hoping that he would not want to be outdone by his Hungarian colleagues. But nothing much happened until November of that year, when I was contacted by Dr. Elizabeth Kennan, the president of Mount Holyoke. She felt that a fact-finding trip to the National Library by a group of experts was the first important step to be taken. She was to lead this group and had received commitments to participate by Warren Haas, president of the Council of Library Resources; Lee Jones, president of the Midatlantic Preservation Service; Robert Hillier, president of the Hillier Group; and, of course, Andrew Lass. The committee was to assess how to computerize various phases of library operations, review space requirements and preservation needs, and determine the needs for library networking. The visit to Prague was to take place in January 1991, and would be followed by a return visit by Dr. Balík in the fall of 1991, after which a full-fledged proposal might be submitted by the National Library. This seemed like a very good plan, and the Foundation promptly extended a grant of $8,000 to Mount Holyoke College to cover the expenses of the trip. Soon thereafter, I received a letter from Dr. Balík, in which he cautiously explored the possibility of submitting a proposal for automation by using the GEAC system; but I decided that we should not jump the gun at this point and should make no commitments until we had the experts' evaluation in hand.

The Kennan-team duly visited the National Library, but came back with only two firm findings: first—and this would have no effect on the Foundation's funding program because of its policy not to finance construction projects—that new space was desperately needed and that the National Library was eminently sensible to plan for a new building "which would warehouse much of the collection now located in several woefully neglected sites,"[38] and second, that before rational programmatic recommendations could be made, a team of Czech librarians would have to visit the United States for about three weeks to study the workings of a variety of libraries. This, as well as a proposed second site visit to the Klementinum, seemed to be very sound proposals and led to two further Mellon grants to Mount Holyoke in the amounts of $48,000 and $9,500 respectively.[39]

But as the spring and the summer progressed, and while the Mount Holyoke team was both arranging the U.S. visit of Czech librarians and planning for its return visit to Prague, matters became more complicated. It gradually became clear that a lot of organizations had a strong interest in the National Library. The American Czech-and-Slovak Education Fund (see chapter 7) wanted to become the coordinator of all the activities surrounding the National Library. The president of the Fund for Arts and Culture of Central and Eastern Europe, Jillian Poole, wanted to get involved. Norman Ross Publishing was already involved with the National Library through a large microfilming project. Wendy Luers, the president of Charter 77 Foundation, also had an interest,[40] as did the New York Public Library and the Library of Congress. I had a panicky feeling that among all these cooks the broth might end up spoiled, as each organization presented its own agenda to the National Library and perhaps obtained its agreement to implement it.[41] This would diffuse the National Library's energies and our relatively well-defined agenda might get submerged in a maelstrom of other projects, not to mention the fact that the Mellon Foundation clearly had much larger resources to invest in the libraries of Czechoslovakia than any of the other players.

The Birth of CASLIN. In the early summer of 1991, Wendy Luers, Andrew Lass, and I met for dinner to sort out the available options, and it occurred to us that it would be very useful to hold a small conference at the premises of the Mellon Foundation in New York, to which I would invite the most important librarians in Czechoslovakia, representatives of libraries in the United States, and a variety of people from the foundation world. As the summer progressed, Lass and I talked more and more frequently about this idea, and it seemed to us that the Czech and Slovak librarians ought to be given the opportunity of explaining to their American counterparts and to potentially interested donors their perceptions of the library situation and their needs, and to find out first-hand how western experts assessed their situation. I was also beginning to have a glimmer of the potential advantages of cooperation among libraries and had some hope of persuading the Czech and Slovak librarians to adopt a cooperative approach, even though I could not wholly articulate those advantages until I started to urge Polish librarians to form consortia.

As soon as I became aware of the desirability of meeting with a group of librarians, I started to organize the conference and set the date for October 18, 1991. I had determined, with the assistance of Lass, that arguably the four greatest and most important libraries in Czechoslovakia were the National Library in Prague, the State Research Library in Brno, the Slovak National Library in Martin, and the University Library of Bratislava. Also attached to the National Library in Prague is the Slavonic Library (Slovanská knihovna), which is a research library specializing in the politics, history, literature, and cultural life of all Slavic countries, with holdings of about 750,000 volumes.[42] Another reason for choosing these libraries was that their directors had, soon after 1989, already come to an agreement on cooperation. The Slovak National Library had a somewhat anomalous position in that it was technically subordinated to and received its budget from an organization known as Matica Slovenská, founded in 1863 as a Slovak cultural institution for the promotion of nationalist and patriotic activities.[43] Since the purposes of the National Library and the Matica were quite different, the relationship between them was not an easy one; but an entirely happy and promising resolution of the strains took place on June 15, 2000, when a new law on libraries was passed by the Slovak Parliament. The Library Act established the Slovak National Library as an independent agency overseen directly by the Ministry of Culture and further permitted it to absorb certain former Matica departments such as the Literary Archive and the Slovak National Literary Museum and Biographic Institute.[44]

From Czechoslovakia, I had invited Vojtěch Balík, director of the National Library; Milena Klímová of the Slavonic Library; Miroslav Bielik, director of the Slovak National Library in Martin; Jaromír Kubíček, director of the State Research Library in Brno; and Emil Vontorčík, director of the Bratislava University Library. On the American side, I had invited experts with various backgrounds: librarians, computer experts, and foundation representatives.[45] Together with the American contingent, it was a star-studded cast and I had high expectations.

The presentations of the Czechoslovak library directors and the subsequent discussion were extremely far-ranging. The following paragraphs represent a composite view of the librarians and are taken from the report of the rapporteur, Kristen Willard:

> The current crisis in Czech and Slovak national libraries is the product of years of neglect and misdirected resources, brought on by historical and political circumstance. Under the previous regime, libraries came under the responsibilities of the Ministry of Culture, not the Ministry of Education, even though regional libraries were expected to support university research activities. Libraries were used as repositories of knowledge more than sources of information; inaccessible to the general public, they focused on gross accumulation of materials rather than "reader services." As a result, collections have over-run storage space. In many areas, books are simply stacked on concrete floors.[46] Many are uncataloged. More are inaccessible. Almost all are in severe danger of irreparable damage from pests, water, humidity, etc. Preservation . . . must be a primary goal of any relief effort . . .
>
> In addition, despite the emphasis on accumulation, the collections have some conspicuous gaps. Dissident and ex-patriot literature after 1945 is almost entirely missing. Western periodicals, particularly academic journals and newspapers either have gaps or were not collected at all, either for political reasons or due to limited foreign exchange. This problem is particularly critical at those libraries which are charged with providing research support for universities . . .
>
> Because the public views libraries more as cultural museums than as research tools, there is only limited awareness of the problems that the Czech and Slovak libraries are facing and, consequently, no popular support for

> the massive expenditures which would be necessary to rehabilitate the library system . . .
>
> The breadth and depth of the national libraries' problem will require national effort to preserve and conserve the collection; to increase and improve storage facilities; to move towards a more usable system, with enhanced reader services and current scientific research capabilities.[47]

The particular goals that the librarians felt needed to be pursued were

- public relations
- interlibrary cooperation
- staff training
- conservation and preservation
- technological modernization, including automation
- physical improvement of library space
- acquisitions

While conservation and preservation and physical improvement of space were not items that fell into the Foundation agenda, and while we were already doing something about acquisitions, it seemed that there was much on this agenda that the Foundation could help with. In particular, in discussing automation, a fairly strong consensus developed that there was much to be gained for several or many libraries to automate, using identical software and identical hardware. While all software has some limitations, the general view was that the advantages of cooperative solutions far outweighed the shortcomings of any particular software. In addition, it was deemed absolutely essential by the western experts that all libraries adopt the same MARC standard. Many other issues were discussed, and in retrospect I was amazed how much was crammed into a single day.[48]

Everybody had learnt a great deal from the meeting and felt that it was a success; but I did not know just how much of a success until a few weeks later, when I received a letter under the signatures of Balík, Bielik, Kubíček, and Vontorčík, which said,

> We have agreed that we will jointly prepare a unified project of automation of Czech and Slovak libraries and jointly submit the proposal to the Foundation.
>
> The basic idea of the project is to lay a solid foundation of a national library network. In this stage all the four libraries will use the same system of software and hardware; due to high standardization the network will be open to any further additions, and will be able to communicate with the majority of other systems on both national and international levels. We are proposing to install fully integrated modular systems; their main task would be a substantial improvement in the access of scholar, student and general public to the wealth of our stocks; at the same time improving the economies of our libraries by the benefits of computer supported ac-

> quisitions and circulation and by the supply of reliable managerial information.
>
> With respect to the administrative structure of the ČSFR [Czechoslovak Federal Republic] the network will have two main centers in the National Library in Prague and the Slovak National Library in Martin. Each of the centers will serve dual purpose [sic]:complete bibliographic coverage of the nation's document production; and gradual build-up of the Union Catalogue of all the participating libraries. To ensure backup . . . all the changes and additions in one data base will be mirrored in the other one.[49]

The letter was accompanied by a formal letter of intent in which the library directors agreed upon their common intention "to lay solid foundation of a nationwide library network providing both home and foreign users with easy, fast, and unrestricted access to the information stored in or mediated by the libraries and information centers."[50] The letter of intent went on to describe the goals of such a library network as being a powerful public service orientation, substantially improved access to catalog records, strengthened cooperation and labor sharing, and coordination in acquisitions and other activities. This was, indeed, a remarkable breakthrough and seemed as if, in one stroke, a whole new way of thinking about libraries had been miraculously transferred to the Czech and Slovak librarians. As Esko Häkli, the National Librarian of Finland put it: "In my opinion it [the letter of intent] is a unique document and could be used as a model in developing policy decisions also in other former socialist countries."[51] And thus the Czech and Slovak Library Information Network (CASLIN) was born.[52]

Czechoslovak librarians were to address themselves to defining their automation needs, with a view toward submitting a draft proposal to Andrew Lass (ultimately to be submitted to the Mellon Foundation) for his scrutiny by the spring of 1992. But Murphy's Law works in Czechoslovakia, too, and something did go wrong, and for a while it seemed that it went horribly wrong. On February 13, 1992, Minister of Culture Milan Uhde, accused the National Library at a hearing of the Parliamentary Committee for Science, Education, and Culture of embezzling 13 million Czech crowns (about $465,000 at that time). If the charges had been true, it would have clearly dampened the Mellon Foundation's enthusiasm for proceeding with a large automation grant to a consortium in which the National Library was the pivotal member. Lass immediately telephoned Balík and Eva Kantůrková, a member of Parliament and of the Committee, who was clearly sympathetic towards the library. It became clear that the accusations had generated a tremendous amount of publicity. By February 20, Kantůrková replied in writing and pointed out that the minister had not presented any evidence for his accusation and that she regarded it as unfounded. It also turned out that the National Library had been subjected to an audit during the last three months of 1991, which had revealed no criminal activity, although it did reveal irregularities in bookkeeping and in the payment of invoices for construction. On February 22, I wrote to Kantůrková and said, among other things,

> Apparently there has been some concern, by all parties involved, that this negative report in the press might cause us to reconsider our interest in assisting the library. In view of this, we think it is important that you know the [Mellon] Foundation's position on this matter. First, let me assure you that nothing of what we have learned so far leads us to conclude that we should withdraw our initiative—especially since the allegations against the library are not based on a complete investigation. At the same time, we have an obligation to examine (through a regular process) the financial status of any organization to which we might make a grant.
>
> It is standard procedure in our country to have audits conducted by an independent audit firm. This is thought to be important in order to guarantee objectivity, a point of particular importance in the case of an institution whose primary function is to serve the public and that must therefore be accountable. It is not our place to suggest how you should proceed, but if you were interested in pursuing an approach of this kind, engaging an independent accounting firm with an international reputation, I would be prepared to recommend that the Foundation assist in meeting the attendant costs, provided that the auditing firm selected meets the standards of the Foundation.[53]

Lass, as well as Elizabeth Kennan, had thought that the Foundation's attention to and intentions for the National Library had made it a "political football," and Minister Uhde may have thought that by discrediting the director of the library, which might have led to his dismissal, he could capture the project and enhance his own standing. Lass also believed that the mere offer of an outside audit would still the turbulent waters. By the middle of March, it became clear that no embezzlement was involved, although some moneys may have been spent unwisely. Richard Kneeley, a doctoral student at Columbia University, who had done research at the Slovanská knihovna in Prague, thought that that was an issue about which reasonable men could well disagree.[54] By April 1, the Ministry of Culture formally dropped the charges of embezzlement, and by April 13, Balík reported to me[55] that the sum of money expended in irregular fashion was actually only 5.8 million Czech crowns (about $207,000) and that the nature of the irregularity consisted mostly in spending it in a fiscal year other than the one for which it had been appropriated. As a result, two persons in the National Library's accounting office were dismissed. On April 28, Kantůrková wrote to me again and informed me that the audit was actually being carried out by the Ministries of Finance and of Control and expressed her thanks for my and the Mellon Foundation's assistance.[56] Elizabeth Kennan's team had made a second visit to Prague around this time and focused considerable interest on the National Library. Minister Uhde decided that at this point there was more political capital to be made by being friendly toward the library than by attacking it, and offered his help for securing a large building for it on the outskirts of Prague which could be used to house overflow materials. Lass, as well as Kantůrková, thought that there was actually some benefit from the four-month ordeal, not only from coming close to securing a new building, but also from the additional publicity that the library received from its exoneration.[57]

It was a salutary lesson for me in byzantine politics and taught me that philanthropy and technology transfer may well get wrapped up in the nefarious Macchiavellian schemes of people looking out for their own interests. But offsetting these negatives was the great interest that the Czech and Slovak libraries had elicited in the West and the wonderful responsiveness shown by the librarians.

The CASLIN Proposal. Since 1989, the Czech Republic and Slovakia moved on divergent paths. Václav Klaus had decided to halt much of the arms production, most of which was located in Slovakia. Unemployment levels in that country were high and were further unfavorably affected by the halt in armaments production; indeed, they were much higher than in the Czech Republic (see table 1-2, chapter 1). In the elections of 1992, in the Czech Republic the winners were the conservative and liberal parties, whereas in Slovakia they were the nationalists and the intellectual heirs of the Communists (Stroehlein, et al.). Klaus, and the Slovak winner, Meciar, agreed to divide the country, and in July 1992, the Slovak National Council accepted the Declaration of Sovereignty of the Slovak Republic. I was elated, on the one hand, that we had unwittingly created an international consortium of libraries, but was also concerned that the division of Czechoslovakia might interfere with the effectiveness of CASLIN. However, Miroslav Bielik wrote to me on July 21,

> I can assure you that between the Slovak National Library (the same holds also for the University Library and other libraries in Slovakia) and the National Library in Prague (and other libraries in the Czech Republic) there will continue an immediate and unrestricted cooperative relationship in accordance with the joint goal of the Council of the Main Libraries' Directors in the ČSFR established on November 11, 1991. Regardless of whatever constitutional changes, we will continue to work closely with the National Library in Prague on the information network of libraries with a view towards opening access to other libraries and close contact with the information system of Hungary (good cooperation with the Országos Széchényi Könyvtár [OSZK], Poland (Biblioteka Narodowa), and other countries of Central Europe.[58]

Similar thoughts were being expressed by other librarians, and I was reassured.

Frantic activity characterized the first eight months of 1992.[59] The Czech and Slovak librarians were busy assessing their requirements and finding technical ways of meeting them. Lass was assisting them all the way and took on the task of preparing the final draft of a proposal to the Foundation. He and I were in constant contact about details. But lest anyone think that all this occurred in an atmosphere of total harmony, it is important to note that during the summer I received an independent proposal for library automation from Dr. Kubíček of the Brno State Library, which I of course declined on the grounds that the four libraries had to submit a single proposal and that there was no room for individual adventurism. Lass thought this was a symptom of potentially deeper troubles and wrote perceptively,

> Here is a problem that may be difficult to tackle . . . Yet, delicate diplomacy aside, I think it is essential that even the slightest quivers that could be symptomatic of potentially serious troubles down the road, should be addressed. This one goes as follows: In the "real" world, the relationship between the four institutions that constitute CASLIN is far from perfect. It appears that the cooperation between Prague and Martin (the two designated centers) is and has been excellent both at the director's level and in terms of the automation project itself; the relationship with the two satellites (Brno, Bratislava) is full of quirks. I gather, there is more to it than just my own impressions of the two directors. The complaint I have heard repeatedly over the Summer—and most clearly this past week when I called Prague to bring them up to date—has Bratislava's director passive (but not obstructive) and Brno's just the opposite; Not very cooperative, impatient, threatening to do "his own thing" (software, hardware). . . . I don't get the sense that he wants out, but rather that the others . . . find the director's negative charisma counterproductive and perhaps a bit worrisome.[60]

Nor was the Brno library director the only one to worry us. To leap for a moment to the future, while in June 1993, Igor Prokop of the Slovak National Library reported that collaboration among the libraries was exceptionally good,[61] by the early summer of 1995, Emil Vontorčík, director of the Bratislava University Library, had managed to earn from me the epithet "rogue librarian." While he had completely agreed in the planning stage that the National Libraries should receive a slightly higher share of the grant than the satellite libraries, he now saw fit to complain to the Slovak minister of culture that the grant was not treating him fairly. As a practical matter, this was preventing or at least delaying the creation of a united front by the two Slovak libraries for pressing the Slovak Ministry of Culture for more government resources.

But in spite of all the ups and downs, an excellent proposal arrived in September 1992, and it was complete except for one thing: the integrated software package had yet to be selected. But the principles of the consortium were clear: the two National Libraries would be designated primary centers, with the Brno State Library and the Bratislava University Library becoming "satellites;" a distinction that was reflected in the division of resources to be provided for CASLIN. The libraries would prepare to automate using identical hardware and software, and would commit themselves to creating a union catalog. During the summer I had spent a lot of time worrying about the management of this project. Given periodic outbursts of pettiness, I was very much afraid that if we designated a Czech library to be the project manager, the Slovak librarians would be offended, and conversely if we designated a Slovak library. I therefore decided, with Lass' and Kennan's willing consent, that Mount Holyoke College, through the person of Lass, would be project manager. The total amount requested in the proposal was $1,188,000; of this sum, $88,000 was to be used for the Mount Holyoke management costs, $900,000 for hardware and software, and $200,000 for a major upgrade of the two National Libraries' microfiming/reprographic capabilities. The search for an appropriate software system was to be undertaken at once, and Lass proposed to engage an outside library consultant, Johan van Halm & Associates in Amersfoort, Netherlands. The funds for hardware and software were to be divided as follows: each center would receive $112,500 for software, $150,000 for hardware, and $66,000 for training, and each satellite library would receive $112,500 for software, $75,000 for hardware, and $34,000 for training. Digital Equipment Corporation had been giving substantial discounts in Eastern Europe and their 200 MHz alpha-chip workstations were cutting edge, and the thinking was that CASLIN's servers would be DEC machines.[62] All this seemed entirely satisfactory, and on December 14, 1992, the Mellon Foundation awarded the grant to Mount Holyoke. Soon thereafter, the Pew Charitable Trusts came through with a grant of $200,000, as we had hoped.[63]

Lass believed that the software decision could be made in six months. In fact, on January 23, 1993, he participated in Brno in the final deliberations and decisions concerning the choice of a software vendor, and he was able to tell me on March 4 that the Aleph software, produced by the Ex Libris company (Israel) had been selected. He also used the occasion of his visit to Prague to publicize the CASLIN project and received the promise of both the ministers of education and of culture that they would support the project. Lass' role in directing the government's sympathetic attention to CASLIN cannot be overemphasized: as late as 1994, the then Minister of Culture, Pavel Tigrid, was unaware of the existence of CASLIN because his staff did not bother to tell him about it, and might have remained so if Lass had not given him complete information.[64]

On May 31, an agreement was signed with Ex Libris and the project received excellent coverage in the press then and in later years; it was truly launched.[65] Acceptance No. 1 (covering OPAC, cataloging, and acquisitions) was signed and DEC servers were installed in August 1994, database structures and tables definitions were set in January 1995; by February all books were being cataloged in Aleph. The Magic system at the Bratislava University Library was closed in January 1996 and WWW access to the database was enabled in May of that year. Finally, the able Gabriela Krčmařová used a Mellon grant to upgrade the National Library server and purchase an Oracle license for creating a union catalog, which was deliberately called the CASLIN

Union Catalog in order to preserve and foster CASLIN-like cooperation. Nearly two dozen Czech libraries were brought together under its auspices, and thus the Czech side of the CASLIN Union Catalog became reality.[66] Most recently, CASLIN has started to migrate to Aleph 500 and the Czech National Library has successfully completed this step.

Czech and Slovak Libraries: KOLIN, LINCA, MOLIN

KOLIN (Košice Library Information Network), LINCA (Library Network of the Czech Academy), and MOLIN (Moravian Library Information Network) were the three additional library aggregates that joined CASLIN to form what was ultimately called CASLIN+. LINCA was the first significant extension of CASLIN and took place in 1994; KOLIN followed in 1995 and MOLIN in 1996. However, since my contacts with the principal institution of KOLIN, Šafárika University (Univerzita P. J. Šafárika v Košiciach), predate by far any serious contacts with the others, I will deal with it first.

KOLIN. Šafárika University, also referred to as UPJS, was the most significant institution of higher education in eastern Slovakia in the early 1990s. I was eager to make contact there, following my ground rule not to concentrate excessively on capital cities. I wrote to then Rector Rudolf Korec on August 15, 1990 to call the Mellon Foundation's existence to his attention, and a useful dialogue began in which I tried to clarify for him what kinds of things the Foundation might do for UPJS. By February 1991, he had been replaced by Lev Bukovský, a distinguished mathematician and outstanding university administrator. It was clear that I would have to visit Košice, which I did during a visit to Czechoslovakia in the first two weeks of June, 1991.[67]

I flew from Prague to Košice and was picked up at the airport by Vice Rector František Longauer, who surprised me by addressing me in fluent Hungarian. I probably should not have been surprised, because the Košice region has a substantial Hungarian minority, but I still do not know how he found out that I could speak Hungarian. In fact, most of the faculty and administrators I met on that occasion did speak Hungarian. The university librarian, whom Bukovský had already decided to replace as of January 1992, did not strike me as sympathetic toward substantial library reforms, the staff appeared to be completely untrained for computers and modern systems, and the library had nothing even resembling automation. Bukovský asked me whether the Mellon Foundation would consider funding study/training trips to the United States for a few of his librarians. I had been thinking about such a project anyway and had broached the idea with Donald Koepp, the Princeton University Librarian, and Princeton seemed willing to host a few East European librarians for two-month training visits. I encouraged Bukovský to flesh out this idea, and by midsummer 1991 I had received a preliminary proposal for sending two or three librarians to Princeton for two months each. The proposal was refined, and in November 1991, the Foundation approved a modest $17,500 grant to UPJS for this purpose. The first incumbent was the new library director, Darina Kožuchová, who spent February and March 1992 in Princeton, and the second, a year later, Ĺubomíra Šoltésová, her deputy and also head of the Medical School library. My idea was that they would learn about modern library systems and management as much as possible and after their return to Košice, they would have the wherewithall to prepare a sensible library automation plan. These two librarians directed the activities of KOLIN for a number of years (Kožuchová still does) and Lass and I agreed that they were, arguably, among the very best librarians in all of Slovakia.

On January 18, 1993, Bukovský sent me a proposal requesting $270,000 for library automation. UPJS had two campuses; one in Košice, where the natural sciences, law, and medicine were, and one about 20 miles away in Prešov (Eperjes in Hungarian), where the humanities, teacher training, and divinity faculties were located. The proposal was reasonable and dealt with automating both campuses (the library's books being divided about equally between the two branches); the only feature that gave me pause was their choice of TINLIB. But in any event, I could no longer encourage him in this plan, because I had begun to think "consortially." While I subsequently refined my thoughts about the advantages of consortia, I wrote to him that we would not fund this proposal because

> it has been the increasing conviction of the Foundation that it is not cost-effective to fund library automation proposals that come from *individual* libraries; rather we are interested in funding proposals that come from *consortia of libraries*. The reasons for this are the following: (1) If several libraries cooperate in installing a unified system of library automation, it may be possible to save on hardware costs; i.e., it is possible that two or more universities can share the same server. (2) There are almost certainly economies in training and operating the system, because the number of librarians who become familiar with the system is larger than would otherwise be the case. (3) A joint approach may allow the creation of an *interuniversity* public access catalog and may even lead to a coordinated acquisitions strategy.(4) Finally, and perhaps most importantly, confronting the potential vendors of both hardware and software may induce them to give you better discounts. [Emphases in the originals]

Fortunately, declining this proposal had no permanent effects, because KOLIN was just around the corner. KOLIN was to consist of UPJS, the Technical University of Košice (TU) and the State Scientific Library of Košice (SSL), the largest library in the city. Kožuchová was the chief executive of this consortium and Ĺubomírá the project manager. An excellent proposal for $476,000 was submitted by Mount Holyoke and funded in 1995. The proposal anticipated the acquisition of the Aleph software, and as a result of some brilliant negotiations by Lass, the eventual contract with Ex Libris stipulated that

KOLIN could upgrade to Aleph 500, a much more advanced, web-oriented system, and Oracle based software, at a very low additional cost. But the negotiations with Ex Libris were horrendous and took a toll on everybody. The KOLIN view was that the vendor failed in various ways: questions posed by KOLIN about the draft contract were not answered, contrary to prior agreement, a Slovak version of the draft contract was not made available, points made by KOLIN negotiators were dismissed as irrelevant or unacceptable, and so on. Ex Libris negotiators considered some of the KOLIN requirements unreasonable and were offended by the fact that during the contract negotiations some points that had been agreed to by KOLIN staff appeared to have been disavowed by next morning.[68] Mutual recriminations so enraged Ex Libris that it presented KOLIN with an ultimatum: accept the contract as is by July 15 or the deal is off. It took all of Lass' negotiating skills to put the almost derailed project back in order. Finally, the agreement was signed. Hardware was delivered and software installed in the first half of 1997. Training of librarians was begun. I was quietly optimistic.

But that was not the end of the story. By November 1996, the government had decided to turn the State Scientific Library into Regional Scientific Library, which superficially may seem unimportant, but gave the government the excuse to remove the excellent library director, Daniela Dzuganová, and to replace her in November 1998 with a purely political appointee, Josef Mižák, who was not even a librarian. Mižák was taking his cues from Dr. Alojz Androvič, a deputy director of the Bratislava University Library, who had risen to greater prominence because the director, Dr. Vontorčík, was fired in 1997, and his successor, Josef Gerboc, knew nothing about libraries.[69] Androvič wanted to replace Aleph at that library with some other system.[70] When I visited the library in June 1998, I could not meet with Mižák because, as I was told, he absented himself from Košice on purpose; but I told his deputy that if the Regional Scientific Library abandoned Aleph, I would demand the return of the all the computing equipment that had been purchased with Mellon money.[71] The point of this was that such a step would have delayed even more the automation process at the Regional Scientific Library and would have made it more difficult for it to cooperate with the other libraries. Be that as it may, the library has consistently lagged behind the other KOLIN members, even after the defeat of Meciar and his government in the September 25–26, 1998 elections,[72] although the library reverted to its original status and is once again the State Scientific Library (since May 1, 2000). Fortunately, the uncooperative Josef Mižák was himself replaced by Dr. Ján Gašpar in June 1999, who appears to be much more committed to make a success of Aleph.

But even worse was to come. As of 1997, UPJS was carved up into two universities: the Košice branch became UPJS, while the Prešov branch became the University of Prešov (UP). The rector appointed to head up UP was Karol Feč, and I saw him in June, literally a week or two after his installation. He was, unsurprisingly, not well informed about KOLIN and what it was all about. UPJS also had a new rector, Dušan Podhradský, a scientist from whom, for some strange reason, I expected a great deal. My hopes were not to be realized. By the summer of 1998, the complaints of Kožuchová and Ĺubomírá were so persistent that something had to be done. They complained of substantial lack of interest in KOLIN on the part of the rectors, lack of financial support for elementary supplies such as toner cartridges for printers, paper, etc., and of the total inaccessibility of Podhradský. In 1999, the qvestor (chief financial officer) of UPJS even refused to pay the annual software maintenance fee to Ex Libris, with corresponding loss of credibility for the library. This refusal occurred in spite of the fact that in June 1998, Lass and I had managed to arrange a one-and-a-half-hour meeting with Podhradský, Feč, and Miloš Somora, the rector of the Technical University of Košice, in order to make them more favorably disposed toward the libraries and KOLIN. I started out at that meeting by telling them about the importance of KOLIN for the academic community, of the role that information technology could and must play in their universities, and the many ways in which their lack of support for KOLIN hampered the attainment of these objectives. Somora immediately grasped the point and Feč was much better informed and much more enthusiastic about KOLIN that I had been led to expect a year earlier. Podhradský just sat there. At Lass' and my urging, Kožuchová opined that she would need about $15,000 a year to cover those petty expenses that she was given no budget for; I then turned to the rectors and told them that if they provided such a sum, the Mellon Foundation would match it. Somora and Feč immediately said that they would make available to KOLIN their share of $15,000, but Podhradský still did not seem to get it. We finally adjourned with the understanding that Mount Holyoke would prepare a proposal to the Foundation for $15,000 and would disburse this sum as matching contributions by the participating institutions were made.

Hardly had we left Košice that a meeting of rectors and vice rectors took place, the minutes of which reveal—in addition to the laudable decision to commit themselves to further cooperation for the benefit of KOLIN—a number of absurdities, such as the formation of a commission to develop regional automated library systems, the appointment of the Regional Scientific Library to coordinate purchases of periodicals, and its appointment to coordinate the KOLIN project—all this without a mention of the fact that KOLIN already had an able management team headed by Kožuchová. We prepared a detailed answer, to which we received a reply on August 24 from Rector Somora, in which he apologized for the "misunderstanding" and stated that the original objectives of the project had not been changed. He admitted that it had been a mistake not to have the library directors present at that meeting and that no radical changes in KOLIN were contemplated. He strongly affirmed the rectors' intention to continue with KOLIN according to the original conception.[73]

As of this writing, the KOLIN libraries have mostly migrated to Aleph 500, the librarians are continuing to do the best they can, and Aleph 500 is slowly coming online. While relations between Ex Libris and some of the libraries are still not good, much has been achieved, as can be seen from table 8-1, showing the modules that are in operation as of October 2000. Rector Feč has been extremely responsive to his library's needs and technical progress at all four university-related libraries has been good.

But at UPJS, Rector Podhradský still indolently ignores the needs of KOLIN and of his own library, which has a zero acquisitions budget for monographs in the academic year 2000–01. The university does acquire some monographs, but these are all paid for by faculty- (department-) based research funds; while they are cataloged, they are inaccessible, except for the professor in whose office a particular book may be kept.[74] This is another example of the enormous power that deans of faculties exert and of the lack of recognition that the library must play the role of a central university service.

The continued ill-treatment of the library by Rector Podhradský prompted me to write him one final letter (with a copy to Vice Rector Leonard Siegfried, who had just assumed his position a few weeks before I visited in October 2000 and who seemed much more favorably disposed toward the library). After a general discussion of KOLIN, and of the importance of libraries and of information technology in a modern university, I said,

> "the Director of the Library is as much an expert in the acquisition of materials and in delivering them to patrons as the Kvestor [chief financial officer] is in financial matters and the Vice Rectors and Deans are in their own portfolios; it is therefore essential that the university take advantage of the Library Director's expertise in formulating the university's vision and strategic plan for information systems in the broadest sense. The simplest device that the university can use to utilize the knowledge of the Library Director is to include the Library Director in the Rector's Council, as has been done at the University of Prešov and is routinely done at American universities, where the director of the library (system) frequently has the title of "Dean of Libraries" . . . It is clear that the library budget at Šafárika University has been completely inadequate to fulfil the various obligations that a library must fulfil in a university aspiring to academic

Table 8-1 Aleph 500 operations.

	UPJS– Law and Science	UPJS– Medical	TU	UP	SSL
OPAC	Yes	Yes	Yes	Yes	Yes
Cataloging	Yes	Yes	Yes	Yes	No
Acquisitions	Yes[a]	Yes	Yes	Yes	No
Circulation	Yes	Yes	No	Yes[b]	No
Serials	No	No	No	No	No

Notes:

a. Except for some minor functionalities.

b. As of November 1.

(teaching and research) excellence, although the library staff and the Director have done an outstanding job with the meager resources available to them. . . . In fact, in the past year, the university failed to pay the software maintenance fee agreed to with the Ex Libris company; thus not only bringing Šafárika University into disrepute, but jeopardizing the entire KOLIN undertaking. . . . It is essential that the university leadership publicly support its library; a *desideratum* that has yet to materialize. As a result of these failures, the morale in the university library is low and the University is risking the loss of some of its very best people to other universities or commercial organizations. It is difficult to escape the conclusion that the Šafárika University Library has been systematically neglected, and if this neglect continues, not only will the University's mission to teach and do research be severely undermined, but so will the future of Slovak university and research libraries. It will also become an international embarrassment to you personally.[75]

As of August 17, 2001, I have received no reply to this letter and I expect it to have no effect on the conditions under which the library has to operate.

As if it were not enough that KOLIN does not enjoy the support of one of the rectors, all is not well within KOLIN itself. On the occasion of my October 2000 visit to Košice, I met with the library directors of all four institutions participating in KOLIN. Apart from the usual complaints about inadequate software services by Ex Libris, all but Kožuchová were rather upbeat about the level of funding they had received (from the respective university administrations in case of the Technical University and the University of Prešov, and a state agency in the case of the Regional Scientific Library). But as I alluded to above, the Šafárika University library had its back to the wall. I proposed to Kožuchová that we bail her out by providing her library with a grant to pay three years' worth of annual software maintenance fees (a total of $12,000), which was the expected duration of Rector Podhradský's tenure as rector, and also provide each of the four institutions with an extra $2,000 of discretionary funds for their library needs. While such a grant would have tended to provide the wrong incentives for the already unhelpful Šafárika administration, I felt that it was more important in the short run to provide the necessary bail-out.

After I left, Kožuchová informed her director colleagues of what I had suggested, and they were dismayed that they would receive smaller bail-out grants than Šafárika University. Why should a Mellon grant pay that institution's annual maintenance fees but not those of the others? How could this be fair? It occurred to none of the others that the concept of fairness was not well defined,[76] and that in the short run the most important thing was to keep the consortium alive in the face of an uncooperative rector, that we had to economize overall resources, which might well mean that a more enlightened administration would end up paying more out of its own pockets than a less enlightened one. The directors' attitude seemed to represent sheer envy and rent-seeking; it suggested that librarians who are envious do not care much for productivity (Porter, p. 21). In game theoretic terms, the behavior exhibited by these librarians is not "subgame perfect": if the librarians and I had been debating how best to solve the financial problem that the Šafárika library was facing, it would have been entirely rational for them to argue for some split of a potential grant other than what I had proposed; however, in the light of, effectively, a *fait accompli*, the rational behavior would have been for the other libraries to happily accept the net increment of $2,000 in their spendable resources and be done with it. I pointed this out to them in a letter on November 22, 2000, and invited them to make countersuggestions, but have received no reply as of August 17, 2001. They (collectively) managed to cut off their nose to spite their face.

Linca. Mount Holyoke received a grant of $350,000 in 1995 for automating the library of the Czech Academy of Sciences, which was the first new member of CASLIN. The funds were for the usual components: a local area network, a server, software, etc. The only surprise was that at the last moment the Academy decided that it did not like Aleph after all, and selected BIBIS, a system designed by a Dutch firm. But this was a small matter and the library is still able to communicate effectively with the other CASLIN libraries.[77] The grant was an occasion for the library to undertake a comprehensive renovation of its informational system: the library created a network within the library and gradually connected it to most of the individual academy institute libraries. The LINCA aggregate catalog was made available in 1997 and contained 160,000 entries from 21 institute databases; in addition, card catalogs were scanned and the scanned catalog contains 850,000 items. By 1999, the library was creating a catalog of electronic resources and implementing Z39.50 under a Ministry of Education program.[78] But currently, the library does not intend to upgrade from BIBIS-97 to BIBIS-98, because the latter is reported to be incompatible with UNIMARC; the library is contemplating the adoption of an untried system still under development, which is worrisome.

Molin. The last subconsortium to join CASLIN was the Moravian Library Information Network, consisting of Masaryk University in Brno, Palacký University in Olomouc, the State Scientific Library[79] in that same city, and the State Scientific Library in Brno, which was a founding member of CASLIN. Mount Holyoke was again the manager of a $350,000 grant for automating MOLIN. While the extent of MOLIN is much larger than LINCA, three of the libraries already had automation systems: the State Scientific Library in Brno had Aleph from the original CASLIN grant, and Masaryk and Palacký Universities already had TINLIB. Only the State Scientific Library in Olomouc needed new software and substantial hardware, and it chose Aleph. The only anxiety I had in this arrangement was over the potential communication problems between Aleph and TINLIB libraries. We specified that they had to prepare an explicit interface that would permit seamless communications between the two. In fact, the problem was solved not by preparing an explicit interface, but a two-step process. The first step in the preparation of the interface was the creation of a new TINLIB cataloging instruction that substantially improved TINLIB's ability to convert to and from UNIMARIC, as a result of which TINLIB libraries could begin to contribute to the CASLIN Union Catalog. The second step was the development of a WWW viewer which can send a query to both Aleph and TINLIB library catalogs and merge the results for the user's inspection.[80] As a result, MOLIN has become an organic part of CASLIN+.

Implementation Issues

Because the library projects in Hungary were of smaller scale and simpler, I had no inkling of the range of implementation problems that could surface in more complex cases. I interpret the term "implementation problem" in the broadest sense; I do not only mean the narrow issues of getting a piece of software running on a set of computers and having librarians use them in some fashion. Rather, I mean all those issues that have to be faced as a library system converts its entire operations from an older to a newer system. The principal (and quite disparate) areas are (1) relations with the vendor, which means primarily with the software vendor, because it is the software that has to be adapted ("tuned") to the user's circumstances; (2) training and, more generally, the "human factor;" (3) conversion of databases and retroconversion of catalog cards; (4) standards and protocols; (5) Value Added Taxes; and (6) issues of national politics and budgets. To a greater or lesser extent, each of these six areas played an important part in every large-scale automation effort I was involved in, whether in the Czech Republic and Slovakia, or Poland, Estonia, or Latvia.

Vendor Relations. The very first issues are the language of the contract and what the acceptance criteria are with respect to the software; in other words, when does the software perform sufficiently well from the user's point of view so as to trigger contractual payments. CASLIN insisted that acceptance would have to be conditional on its staff receiving adequate training, without which they would not be able to be verify whether certain functionalities were performing as required.[81] Furthermore, CASLIN insisted that functionalities had to be checked both against Aleph specifications and against Ex Libris' replies to CASLIN's Request for Proposal. Ex Libris' view was that all that was required for acceptance was a demonstration of what Aleph could do.[82] In another context, Ex Libris also suggested that acceptance would occur automatically at the end of a certain period or whenever the user libraries turned to production, whichever occurred earlier.[83] In any event, determining whether a functionality is present or not, and whether it works as claimed by the vendor or not, is a subtle issue, and it is fair to say that East European librarians did not have experience in the nuances of contract language. And then there is large room for disagreements in interpreting just how well a functionality works. In one instance, the library staff claimed that some of the functionalities were "not working *properly*." The interpretation of such remarks has to be that the functionality was working all right, it was just not quite working the way the users ideally wanted it to work. In any event, the state of acceptance procedures in the fall of 1993 in the Czech Republic left a fair amount to be desired, but thanks to Lass' carefulness in dealing with the wording of the contract between CASLIN and Ex Libris, the various parts of Aleph were ultimately accepted, with some objections but no damage to the libraries.

A second aspect of relations with the vendor is software maintenance. A complicated software system like Aleph is not like a toaster which, if it does not work when one brings it home from the store, one either exchanges it for another one or returns it for a refund. Software requires continual adjustments by the vendor, partly to eliminate "bugs" and partly to add enhancements. It is well known that eliminating bugs or otherwise modifying the system can inadvertently introduce new bugs. Software failure is not a well-defined concept except in some gross instances; the failure may be replicable in some contexts but not replicable in others and some failures permit "work-around," that is, improvisation by the user that basically avoids confronting the failing functionality head-on. In any event, frequent contact between the user and the vendor is necessary.

This leads to a whole host of problems. The first and conceptually the most trivial is one of language. While many East Europeans are conversationally fluent in English (and Ex Libris staff are typically bilingual in English and Hebrew), discussions about software performance require a degree of linguistic precision that is often lacking in the region. This clearly created some problems at first, and did not improve for a while even after Ex Libris established a branch office in Prague. In other instances, Ex Libris engaged a local firm to be its representative, which had its own problems: the staff of the local firm tended to be less well trained than Ex Libris staff and the local firm had other sets of obligations to its own customers as well.[84] Software maintenance is an extremely complex task and many times a patch will be loaded overnight via the Internet; to determine exactly which problem a patch is supposed to fix may be a nontrivial task. All these difficulties created frictions between the vendor and the libraries, reduced the effectiveness, and retarded the pace of the technology transfer that was to take place.

Training and the Human Factor. Vendors differ in how they conduct training and in what depth they provide it. Ex Libris provided some initial training in Israel, which was followed by further training seminars in the Czech and Slovak Republics. The training component of the contract with Ex Libris was funded by a Pew grant, which funded four subsequent Pew Seminars.[85] In fact, the original Pew Seminars were so successful and were deemed by all concerned to be so important, that a several-day-long seminar has become a regular, annual event and has not only led to the creation and dissemination of important technical knowledge, but has also become an instrument for welding the libraries closer together.[86] While the original transfer of library technology may have been primarily of the point-to-point variety (Williams and Gibson, p. 44), the Pew Seminars became the vehicle for genuine diffusion. The general approach always tends to be to train the trainers; that is, train some people in sufficient depth so that they can then train others in their libraries. But what is sufficient depth? Some training is clearly on an elementary level and concerns daily functions that librarians have to carry out ("if I enter such-and-such in this field and then click on this button or press that key, the consequences are . . ."). But other training needs to deal with the core of the system, with matters such as "table-setup" and with setting values for system parameters, all of which determine how efficiently and well the system is working as a whole. This is particularly important in complex systems such as Aleph 500, the innards of which are less transparent to the user and may rely on third-party software, such as Oracle, for database management. It was, in fact, the case that many librarians in Eastern Europe had only modest exposure to computers and automated systems, and many had considerable anxiety about using them—we noted in chapter 7 some of the specific aspects of "technostress." At the same time that librarians had to try to overcome their anxieties and learn new skills, they also had to adapt to new managerial modes of operation, characterized by greater reliance on communication, problem-solving, planning, teamwork, and conflict management (Virágos), because the earlier management structures were largely hierarchical, in fact occasionally referred to as dictatorial.[87] On top of all these problems, the economic opportunities outside the library world were sufficiently more attractive that libraries tended to lose personnel; most often, the best trained people. As Řepišová put it,

> The biggest problem of the [Bratislava] University Library was that it underestimated the human aspects of the CASLIN Project's implementation. From 1994 to 1997, the personnel in the key positions of systems librarian, systems programmer, and project manager changed ten times. It is important to stress that at no time in the library's history of Aleph's implementation was there a full-time programmer employed by the library. In addition, five of the so-called superusers who underwent training provided by the software vendor Ex Libris, and were intended to be future trainers for the whole library staff, have left the library, with no provision having been made for their successors. A very long time has passed since trained staff members had the opportunity for the kinds of hand-on experience with the system that would allow them to extend their knowledge. This has led to a reluctant and distrustful attitude toward even the smallest system changes off the beaten track. The main driving force, the feeling of commitment to the idea of a far better automated future that worked perfectly from the beginning, has faded away. Communication among departments, library management, and the implementation group was not easy and straightforward, and problems arose because solutions were often ad hoc ones rather than systematic ones. All this confirms the earlier observation that it is important to remember that automation is done by humans and not by computers. (Řepišová, p. 214)

One cannot stress too highly how important it is that all participants have a strong commitment to the common goals. In June 1994, Lass and I visited some sites in Slovakia together. We carefully arranged the time when we could visit the Bratislava University Library and alerted Dr. Vontorčík that we intended to make a site visit to have serious talks about technical issues and to find out what progress had been made at the library. Vontorčík received us in his immense office, but was not inclined to engage in a technical conversation. His principal contribution to our immediate welfare was his announcement that we were to attend a major luncheon at the Ministry of Culture that would be attended by the Minister himself. We were hoping for a quick lunch, but it soon became obvious that this was to be a two-and-a-half-hour nightmare. The minister did not show up, although his deputy did, and the time was completely wasted. After that, he told us, we were to be received at the Parliament; but by the time we got there, nearly everybody had gone home. In the end, upon returning to the library, we had perhaps half an hour with a systems librarian—better than nothing, but far from enough for understanding the complexities of progress. The fact is that Vontorčík did not care at all about Aleph, library systems, or progress; all he cared about was to parade us (distinguished visitors and donors to *his* library) before his superiors and thereby enhance his own status.

In the light of the problems surrounding vendor relations, training, and the human factor, it may be a miracle that implementation ultimately succeeded, that CASLIN and CASLIN+ "work," and that the library directors' goals have been largely achieved.

Conversion and Retroconversion. Most large libraries had fairly substantial machine-readable databases built with various predecessor systems, such as MAKS, IKIS, Magic in the Czech Republic and Slovakia, MAK in Poland, and CDS/ISIS in nearly every country. Since it would have been extremely wasteful to recreate these records in the new system, the old databases, in whatever form they were recorded, had to be converted to the new system; Aleph in the present case. In general, there are three ways of going about this: in-house conversion, contract with the software vendor to carry out conversion, or outsource this task to other companies.[88] In the Czech and Slovak Republics the conversion was carried out by Ex Libris.

Retroconversion can be a monumental task, because there may be hundreds of thousands or millions of catalog cards to which there does not (yet) correspond a computerized record. How is one to convert these into computerized MARC records? Two very different approaches were selected by the National Library in Prague and by Kossuth Lajos University Library in Debrecen, Hungary.

The National Library, together with Comdat, a Prague firm specializing in conversion, developed a methodology by which catalog cards were first scanned to TIFF images, the scanned images were then converted to ASCII, and finally the unstructured ASCII files were tagged or structured into UNIMARC fields. The second step of this process was likely to fail in the case of very old, handwritten catalog cards, but on the whole, the methodology was extremely successful. The Mellon Foundation made a $150,000 grant to Mount Holyoke College to permit this work to go forward, and by the end of 1998, some 230,000 records had been added to the bibliographic catalog of Czech books.[89] This project, called RETROCON, was part of the broader CASLIN+ initiative and was extremely successful: the Czech Bibliography of the twentieth century is now available on CD-ROM.[90] The general catalog, with some three million entries, has been scanned, but its conversion will be demand driven. The cost for converting a full bibliographic record amounted to $1.50.[91]

The Kossuth Lajos University Library and the libraries of other higher educational institutions in Debrecen[92] elected to convert catalog cards by hand. The choices were outsourcing, in-house conversion, or derivation of records from some existing database, such as OCLC (Online Computer Library Center). But for Czech and Hungarian material, the "hits" in OCLC were not likely to be very numerous, and the decision was in favor of in-house conversion, using students of the department of library science, as well as librarians on overtime (which they welcomed because of their low salaries). The costs per record ranged from a low of $0.97 (Institute for Nuclear Research) to $2.20 at Kossuth Lajos University.[93] By the end of the first phase of the conversion process (1994–95), 117,183 records had been converted (Gomba).

Still other methods needed to be employed in retroconverting the American Studies Center Library of Warsaw University. Because of inconsistent catalog records of the original ASC collection, retroconversion required physical examination of each item. The portion of the collection that was a gift from USIS was converted by comparing the original computer records of the donated items with OCLC's CAT CD450 cataloging system. The total cost per record ranged from $0.61 to $0.90.[94]

While we can have no assurance that the detailed steps in the conversion covered the same ground in the Czech, Hungarian, and Polish cases, it is noteworthy that the costs are of roughly comparable magnitude. Outsourcing the conversion would have increased the cost by about 31% in the Hungarian case and would have more than doubled it in the Czech case.[95]

Standards. The staff of the Czech National Library realized in the 1980s that it was imperative to start to explore modern western standards which had been slow to make their way to Eastern Europe. The questions that were posed were (1) what standards to accept, and (2) once accepted, would they be used without modification, or did they require localization (just as integrated automated software systems did).[96]

Several standards were examined in some depth. ISBD (International Standard Bibliographic Description) has been in use without modification since 1989. UNIMARC was initially difficult to understand for Czech librarians and was incompatible with the earlier versions of CDS/ISIS and instead, in the pre-CASLIN days, the National Library staff improved the Czech Exchange

Format; but while some of these improvements aided the ultimate conversion to UNIMARC, others delayed it. There was a strong initial bias against AACR2 (Anglo-American Cataloging Rules-2), precisely because of their Anglo-American bias.[97] Library of Congress Subject Headings (LCSH) were also initially rejected because of their North American bias; however, UDC (Universal Decimal Classification) has been used widely in Czech Libraries (Stoklasová, pp. 220–21). But as automation arrived and CDS/ISIS was abandoned, it became clear that all these standards were important. Czech translations were prepared for ISBD formats and AACR2 was accepted without modification. UNIMARC was also translated into Czech. As far as LCSH are concerned, the National Library abandoned the use of keywords for subject access and the library decided to use LCSH in the areas relevant for the collections. The limitations of older approaches, such as Magic at the Bratislava University Library, are lucidly discussed by Řepišová (Řepišová, pp. 205–8), and these limitations make it understandable that the various western standards were ultimately accepted. What that fact does not reveal is the depth of the soul-searching that had to take place and the immense amount of work that had to be performed before such decisions could be taken with confidence. The National Library became the first East European library to supply records to OCLC, and they are now available in OCLC's WorldCat database (Stoklasová, p. 223).

Value Added Taxes. Lass made the unpleasant discovery that Value Added Taxes (VAT), which became effective as of January 1, 1993, would be applied to CASLIN's purchases of hardware as well as software. Since the VAT rate on hardware was 23% and on software (as well as books) 5%, this threatened to become a major drain of funds, not to mention the fact that not-for-profit organizations in the United States are exempt from sales (and income) taxes. I brought this problem to the attention of Karel Dyba, minister of the economy, when I met with him in the early summer of 1993, and he said that legislation for relief was under preparation, but he could give me no indication of when such legislation might be passed.

According to the existing law, any legal entity had to pay VAT on goods that it received or purchased *from abroad*, which sharply differentiates the treatment from that accorded to goods obtained domestically: a Czech company could donate goods to a not-for-profit organization, receive a tax deduction, and the recipient would not be charged VAT. Some of the basic provisions for levying VAT on goods obtained from abroad were as follows:[98]

- all equipment, whether bought or received as a gift, was subject to VAT of 23%;
- all book gifts were subject to VAT of 5%;
- all books acquired via exchanges were subject to VAT;
- books returning via interlibrary loan were subject to VAT;
- publications sent by publishers to scholars as review copies were subject to VAT.

The regulations seemed capricious to the extent of being absurd, but I feared that we would be stuck with them. Fortunately, a solution existed in the form of a loophole in the law. Donations could be made to foundations without incurring VAT, and fortunately, the National Library had created the Foundation in Support of the National Library; such a foundation could then pass on the goods to its beneficiary, again without incurring VAT. This arrangement lasted for about a year, after which some new legislation exempted universities and libraries from VAT, provided the gift was accompanied by a letter certifying that it was a gift. This fortunately solved a very vexing problem, which I encountered numerous times in other countries as well.[99]

National, Political, and Economic Issues. I have already discussed how in the early days, CASLIN seemed to be at the mercy of the minister of culture, who was motivated in his attack by an agenda of his own. In Slovakia, during the Meciar years, political correctness (that is, professed support for Meciar's policies) by university rectors was a necessary condition for budgetary largesse. I noted, in 1996, that the "loyalist universities" (Bánska Bystrica, Žilina, Zvolen) were flush with money, the "oppositionist" universities (Comenius, Šafárika, Trnava) were hard-pressed, and the ones that were not particularly vocal (Slovak Technical University, Nitra Agricultural University, Economics University of Bratislava, Veterinary University) were in-between.[100] Vice Rector Ferdinand Devinsky of Comenius University (subsequently elected rector for two terms), an outstanding and courageous man, seriously feared that either quality would have to deteriorate or enrollments would have to be curtailed at his university. Moreover, the three-year limitation on a rector's term of office and the rule that a rector could succeed himself only once, made it likely that Meciar would try to replace good rectors with his cronies. While none of the libraries in CASLIN was a university library,[101] the danger of political influence was ever-present, and as I showed in the previous section, we were soon to deal with genuine university libraries in Slovakia as well.

But there were "political" issues within the library (or university) world; this is what Lass has aptly named the "micropolitics of libraries." Some of the political tensions arose because of the sheer weight of the capital city relative to provincial ones; these tensions were probably less important in Poland than in the other countries. I noted, for example during my visit to the Czech Republic in June 1992, that the leadership of the Olomouc State Scientific Library regarded the National Library as excessively "Prague-centric" and therefore not to be trusted. Provincial university rectors in Hungary (more so than library directors) tended to regard universities in Budapest with envy and suspicion.

On the economic front, absurdities reached new high points. In late 1994, the Czech government undertook drastic budget cuts, which severely affected libraries. Not only was the budget of the Czech National Library cut, but so were the budgets of other cultural institutions. Some of the reasons for the cuts were undoubtedly macroeconomic and originated with the Ministry of Finance.[102] But there were other reasons as well. Balík, a frugal manager, had saved part of his budget during the previous fiscal year for certain expenses that he knew would be incurred in the current one. The government auditors concluded that since he did not spend all his money, he obviously did not need it, and hence it was cut from the next year's budget. On a related matter, he expected to receive funds for building renovations in the autumn; but renovations can really be done only in the spring, summer and early fall because of the weather. But if he were not to spend the renovation budget by December 31, it would be concluded that he did not need the money.[103] When I asked why he did not prepay the renovation expenses, he told me that by law one may prepay only up to 10% of such expenses. Fortunately, in this case the cuts were more likely to affect materials and binding, salaries, and building maintenance than the activities of CASLIN.

Summary and Evaluation

There could hardly be two automation experiences more disparate than those in Hungary and in the Czech Republic and Slovakia. The former was largely unproblematic and scored solid but unspectacular advances. The latter, while it had quiet phases as in the cases of LINCA and MOLIN, was often plunged in the midst of roiling political struggles, personal ambitions, poor governance, and hostility between vendors and library. In spite of these obstacles that at times seemed insuperable, CASLIN and CASLIN+ have become a functioning system and must count as extremely successful steps in technology transfer and library modernization.

None of this would have succeeded if it had not been for the constant attention lavished on CASLIN by Andrew Lass. The trust he commanded on the part of the Czech and Slovak librarians was a well-deserved reward for his immense work, his sensitivity, his understanding of complex issues and his sensible and effective problem-solving. It is even more remarkable that CASLIN "works" in the light of the fact that it is not even a legal entity; is merely a loose aggregation of libraries held together by the original letter of intent signed by the four founding librarians. It is, therefore, not a true consortium but a voluntary association held together merely by the libraries' perceptions that they have common objectives: as Vojtěch Balík said, "there is no institution or body bearing the name of CASLIN at the present time."[104] Balík further stresses that the participating institutions were not willing to enter a tighter bond with one another and regarded CASLIN with some ambivalence, formed primarily to achieve a concrete objective supported by a grant (or by now, several grants). Lass puts it thus:

> Sadly, the Consortium of Czech and Slovak Libraries was never established as promised. Unclear developments in legislation pertaining to associations of state organizations and now involving two countries created a major

> stumbling block, and there seemed to be no collective will to find a common solution to the problem and follow through on the initial intent. As a result, CASLIN remained an informal organization held together as much by momentum and the fact that some of the project's management was located in the United States . . . as by a common vision or its own simple organizational rules.[105]

Balík noted that "there is still no governmental project designed to formulate an information policy for the Czech Republic" (Balík, p. 128). This is underscored by a statement by the Office of the Czech Republic that is a masterpiece of opaqueness and is certainly more one of hopes than achievements.[106]

On the level of the individual subconsortia, it is painfully clear from the example of KOLIN that a great deal remains to be learned and fully internalized in the realm of values and attitudes. But in spite of the difficulties, CASLIN and CASLIN+ have achieved a number of important milestones.

1. They have raised the awareness of the government and the public and have succeeded in mustering important government support: (a) the Czech Ministry of Culture awarded CASLIN a matching fund to meet the challenge of the Mellon grant; (b) it provided funds for internet access for the first two years; (c) the CASLIN principles have been used as a model by the Ministry of Culture for the development of a library network; (d) other funds have been made available from other governmental programs.
2. The intellectual capital transferred as a result of CASLIN has resulted in a major modernization of Czech and Slovak library standards, protocols, and practices.
3. Very large transfers of skills have taken place, primarily to librarians but also to library users.
4. The tangible output, namely shared cataloging and the union catalog, has made a substantial difference to users and has done much to reintegrate Czech and Slovak libraries into the world of western librarianship and scholarship.

In fact, union catalog work in the Czech Republic has raised renewed hopes of forging a tighter CASLIN consortium and the major Slovak libraries are now seriously considering retroconversion as a first step toward a Slovak union catalog. In particular, the appointment of Tibor Trgiňa as the new director of the Bratislava University Library and of Dušan Katuščák as the director of the Slovak National Library has created a welcome *rapprochement* between these two traditional rivals, to the extent that they will now collaborate with one another and with the Košice State Library on Slovak retroconversion; an effort supported by a $50,000 grant from the Mellon Foundation in early 2001. This new development raises significant hopes that the Slovak union catalogue development will get a substantial boost. Similarly, the creation of a partial union catalog in Hungary (VOCAL) and the hope for a full union catalog (MOKKA) promises major improvements in the functioning of Hungarian libraries. But the efforts of both countries underscore the immense problems that have to be solved in embarking on such an enterprise.

9

Library Automation: Tight Consortia

My first trip to Poland under Mellon auspices was in October 1990. In Hungary, I most often traveled by train because of the excellent intercity trains that connect Budapest with the larger provincial cities. In the Czech Republic and Slovakia, I sometimes drove a rented car or took a plane. In Poland, the distances are large, my itinerary often made trains impractical, and on most trips I rented a car with a driver for a geographically clockwise tour that would start in Warsaw and carry me to some or all of Lublin, Kraków, Wrocław, Poznań, Gdańsk, Toruń, and back to Warsaw.

As I started to meet with university as well as government officials, I soon realized that there was a substantial style difference between the Hungarians and the Poles on the one hand, and the Czechs and the Slovaks on the other. The Hungarians and Poles tended to be informal: discussions took place around a table over coffee, and they had the air of conversations. In the Czech Republic and in Slovakia, comparable events were often (but by no means always) very formal. The "locals" would sit on one side of a long rectangular table and I (or I and a colleague if I was accompanied) would sit along the other side; occasionally our side of the table would fly a small American flag while the other side would fly the appropriate local flag. There would be no conversation on such occasions, but speeches would be made: an opening sally by the home team, a reply by me, and so on; this might end with a ritual glass of brandy, even if it was 10 o'clock in the morning.

There was one issue associated with making library automation grants that I did not really have to face in Hungary, the Czech Republic, and Slovakia, namely what difference it makes whether the software system is or is not already selected by the library at the time that the grant is awarded. In Hungary, I did not have to deal with integrated automated library systems, and in the other two countries the oversight provided by Andrew Lass insured that the preparatory work was scrupulously careful and detailed, even if the final decision had not yet been made. The advantages of having the software decision made early was that the cost of the system could be known with greater precision and that the library would have had to do all its homework before the grant was awarded. However, the difficulty with requiring a prior software selection was that many software vendors would not even talk to libraries until they were assured that they had the financial resources to buy a system. Not insisting on such a prior decision avoided this latter difficulty, but made the project more open-ended than I would have liked. In the end, both types of approaches were employed at different times.

There was another issue that I never even thought of until much later. Software packages are priced partly on the basis of what modules the library buys (for example, purchasing the OPAC module will be less expensive than purchasing the OPAC module *and* the cataloging module *and* the acquisitions module) and partly on the basis of how many "licenses" are purchased—that is, how many concurrent users of the system there may be. The number of licenses would be arrived at by negotiations between the vendor and the buyer, but for a long time I did not ask the question whether the number actually agreed upon was the right number in some sense. It subsequently occurred to me that there is a trade-off: the more licenses are purchased, the costlier the package, but at the same time the lower the probability that a user (librarian or library patron) will want to use the system and cannot for lack of an adequate number of licenses. This clearly suggests that there might be an optimal number of licenses; one that offsets the cost of additional licenses against the (implied) cost of a librarian not being able to do his or her work. It turned out that this is a conceptually fairly easy problem which I was able to solve, but the solution method did not dawn on me until about 1998.[1]

The automation of the Polish libraries turned out to be significantly different from the automation of Hungarian, Czech, and Slovak libraries. In Hungary, we dealt with (roughly) two dozen libraries that (predominantly) made marginal improvements in their library processes. In the Czech Republic and Slovakia, we dealt with a dozen libraries organized into one loose consortium (with several subconsortia) that overwhelmingly used a single software system. In the case of Poland, we dealt with about 40 major libraries, using three different major automation systems and several tight consortia organized in drastically different ways. This made dealing with Polish libraries somewhat more complicated. It was fortunate that Professor Jurand Czermińsky of the University of Gdańsk took on the task of coordinating much of the work of, at first, the so-called VTLS-libraries, that is, the libraries that purchased the software of VTLS, Inc.,[2] and then later, more broadly, the work of libraries using other systems as well. His role was not exactly analogous to that of Lass in the Czech Republic and Slovakia; he did not participate in writing the libraries' proposals to the Mellon Foundation or administer the grant funds. But he did much to facilitate the work of the libraries, participated in the planning of longer term strategies and in resolving differences of opinion, and he interceded for the libraries with government agencies such as the State Committee for Scientific Research (KBN). His contributions have been of lasting value to the Polish library scene.

All the libraries funded by the Mellon Foundation in Poland ended up choosing one of three integrated automated systems: VTLS, Dynix/Horizon,[3] and INNOPAC.[4] We discuss them in order.

The VTLS Libraries

Warsaw, Kraków, Gdańsk, Wrocław. The VTLS libraries did not start out as a consortium, nor did they necessarily start out with the intention of acquiring the VTLS software. I had begun an intensive correspondence with rectors and librarians at Warsaw University, Jagiellonian University, and the University of Gdańsk in the early summer of 1990, in preparation for a visit I had planned for October of that year. Warsaw and Jagiellonian were obvious choices as the two most prestigious universities in Poland. I think I must have picked Gdańsk to be in the first group of libraries to be contacted because my old friend, Wojciech Charemza, had taught there before he moved to England. In fact, I did not have time in October to visit Gdańsk but succeeded in deepening the contact between us by meeting with Prorector Zygmunt Kraśucki, who kindly came to Warsaw to meet with me.

In Warsaw, I met Rector Andrzej Wróblewski; Prorector Zbigniew Koczorowski; the dean of the International Business School, Andrzej Koźminsky; and importantly, University Librarian Jadwiga Krajewska. I was, of course, taken to the library, which I found cramped and unsatisfactory, although it had vast holdings and contained great treasures.[5] I gained the sense that Polish universities were in some ways even more impoverished than their Czech, Slovak, and Hungarian counterparts; I later wrote to William G. Bowen that "it is common, for example, to have four to six faculty members share a small room as an office and it is typical that not every person has his or her own desk in that office."[6] In that same letter I recommended a grant of $240,500 to Warsaw University, of which a small portion, $20,000, was to pay for the training of a few librarians at Stanford University, but the bulk of which was for library automation.[7] It is noteworthy that Ms. Krajewska's proposal and my recommendation made no mention of the system to be acquired, nor did it specify a fully integrated system. Up to this point, the library was using CDS/ISIS, but had minimal databases, although its work was obviously respected, since it was partnering with the British Library in work on the *Eighteenth Century Short Title Catalog*. The library's immediate objectives were to create databases of newly acquired library items, of Polish and foreign periodicals, and of local catalogs in departmental libraries; the secondary objective was to retroconvert older materials into machine-readable format and create an online catalog. The hardware request was for some 27 personal computers, including some with Intel 286, 386, and 486-chip processors, and ancillary equipment such as printers and CD-ROM readers; the software request was for unspecified appropriate software. I think I knew I was taking a risk, but I had confidence in the people I had met and I believed that the rector would support library automation.

Visiting Jagiellonian University in Kraków, the capital of Poland until

1609, when King Sigismund III moved his seat to Warsaw,[8] is a feast for the eyes. St. Mary's Church in the marketplace (Rynek), Wawel Castle, the ancient buildings of the university are just few of the gems that awaited me. But my first visit not only pleased my senses but was productive as well. Rector Aleksander Koj (soon to be replaced by Andrzej Pelczar, whom he in turn replaced three years later) was abroad and I met with Prorector Jerzy Wyrozumski; Stanisław Waltoś, the director of the University Museum; Jan Gawlik of the Jagiellonian University Foundation; Jan Pirożinsky, the director of the Library, and Renata Dobrowolska, who ultimately became the deputy of the vice rector for international affairs and has helped me immensely over the years. In contrast to many other institutions, the Jagiellonian University officials had firm ideas about the kind of support they needed, including some that did not fit the Mellon Foundation's priorities, such as converting the heating plant of a student dormitory from coal to gas. The final version of the Jagiellonian proposal contained a few small items not related to library automation and a grant of $560,000 was approved in June 1991.[9] By April 1991, I had also received an automation proposal from the University of Gdańsk, and that institution received a grant of $540,000.

These three grants were largely fortuitous, brought about more or less by my possibly peculiar institutional priorities. None of the institutions had decided on the software system at the time of the application for a grant, but soon after they received the funds and addressed themselves to the process of selecting a system, something momentous happened: all three selected VTLS. What is more, the Academy of Mining and Metallurgy in Kraków (Akademia Górniczo-Hutnicza)[10] also decided to purchase VTLS, entirely with its own funds. What was created, more or less by accident, was the core of the VTLS Group. By April 1991, Jurand Czermiński had the title of plenipotentiary (for library automation) of the rector of the University of Gdańsk. By October 1992, the Warsaw University library reported that Warsaw University (WU), the University of Gdańsk (UG), and Jagiellonian University (JU) had created a Joint Library Computerization Project, soon to be joined by the Academy of Mining and Metallurgy (AGH), in order to improve and facilitate the development of the individual local projects. The project was open to all Polish not-for-profit institutions on two levels: (1) participation in the program for authority control and format development, which was to have immense consequences later, and (2) VTLS software implementation.[11] The universities had, in fact, decided to form a consortium with an informal board consisting of library directors, heads of project teams, and advisors. It was decided that the board would meet once a month, and it was headed by Anna Paluszkiewicz of Warsaw University, the foremost expert on authority control in Poland. It was very important for the future that the Ministry of Education rapidly recognized the importance of this organization and provided funds so that the project could start librarian training in formats for bibliographic description, authority control, and in implementation strategies for automation. The close cooperation that developed among the four institutions led to an important division of labor:

Warsaw would take charge of authority files, Gdańsk would specialize in serials, and Jagiellonian would concentrate on subject headings. By 1993, there was a formal division of responsibilities as shown in table 9-1.

Wrocław. Quite independently, and long before I had thought of library consortia in Poland, I had started a correspondence with Vice Rector Józef Ziółkowski of the University of Wrocław, a sizeable institution with almost 12,000 students in 1988.[12] I had visited there in the autumn of 1991 and had met him and University Librarian Andrzej Ładomirski, and by August 1992 it was clear that they were definitely thinking of VTLS as the automation system, to be run on a Hewlett-Packard platform. Wrocław desperately needed to automate: almost everything was being done by hand and it was not unusual for a reader to have to wait an hour or more for a book to be retrieved from the stacks. The university did not even have a union catalog of all its library sites. No online searches could be conducted at the university and finding a book was a nightmare in itself: the library employed the system described in chapter 7.

A proposal arrived on September 14, 1992, and by that time the university was actively exploring VTLS implementation questions with the more experienced Warsaw University and Gdańsk libraries. In the covering letter, Vice Rector Ziółkowski indicated that it would be extremely useful if the

Table 9-1 Task Assignments.

	Task	Institution
Provide instructions for:	USMARC bibliographic data	WU
	USMARC authority files	WU
	USMARC for holdings	UG
	OPAC (for books)	JU
	OPAC (for manuscripts)	WU
	Acquisitions module	JU
	Cataloging enhancer	AGH
	Keyword and Boolean searches	WU
	General information	AGH
Provide translations for:	Messages	WU
	Menu driven search system	UG
	Intelligent workstation	UG
	Cataloging enhancer	UG
	Help subsystem	AGH
Prepare rules for authority files:	For names	WU
	For subject headings	WU
	For serials	JU
Central authority file creation		WU, JU, UG, AGH

Ossolineum library, also in Wrocław, could take part in automation using VTLS—a suggestion that was realized in 1996, when the Mellon Foundation made a $100,000 grant to that institution. The university indicated that it would fund the creation of local area networks necessary for automating its two principal buildings and numerous branch libraries out of its own resources and additional resources it hoped to receive from KBN; all other aspects of the proposal were straightforward, and the Foundation gave it a grant of $465,000 in 1993, followed by a supplemental grant of $201,000 in 1996. Thus, the VTLS "consortium" had grown to five members.

In general, progress in Wrocław was satisfactory. Contracts with vendors were signed in 1993, training of library staff began and was undertaken much more expeditiously than would have been the case without the active participation and help of the "charter members" of the VTLS consortium. But some software problems had surfaced early, and while the library was already using the cataloging module of VTLS in early 1995, the librarians' annoyance began to be noticeable. When I visited the library in the fall of 1994, they already had told me of some problems. In February 1995, I received a letter from Dr. Ładomirski, in which he listed a number of complaints (not all about the VTLS software):

> There is no possibility of copying records between VTLS92 and VTLS94.[13] We cannot copy records from the other libraries which means we cannot use the VTLS system. The system is useless for us because the authority control records which are common for all Polish libraries implementing the VTLS system are in the VTLS92 standard. Those records are necessary to have our database coherent with the other databases. The other libraries do not want to change the version of the system because of the bugs found in the VTLS94 version.
>
> For several months we didn't have the Polish character set. . . .
>
> AFAZ [actually, AFAS; Acquisitions and Funds Accounting System] still are not installed. . . .
>
> We received Easy Pack software without any communication protocols. . . .
>
> Command Matrix wasn't initialized in clas02 account during installation performed by VTLS. Because of it we couldn't access this account.
>
> We still cannot set the parameter 1.13 to have the Polish version of the system. We receive Error STRING_CODE=01666000.[14]

Some of these were clearly transient problems, although the character sets tended to create problems irrespective of country and software vendor.[15] The problem, of course, was not only that diacritical characters must look the way they are supposed to look in the language in question, but that sorting into alphabetic order must work properly, as well; thus, for example, L precedes Ł in the Polish alphabetic sequence, just as r precedes ř in Czech.[16]

In any event, I immediately transmitted the complaints to Vinod Chachra, president of VTLS, with a request that he look into these matters, but by April I had received renewed complaints from Wrocław. Eventually all these deficiencies were fixed, but the bugs and various other vendor failures were a constant source of irritation for the otherwise overworked and beleaguered librarians.

Lublin. The autumn of 1992 seemed to provide an appropriate moment to revisit Catholic University of Lublin, which had received a sizeable grant for computerization a year and a half earlier. I met with the rector, the Rev. Stanisław Wielgus, and the deputy librarian, Rev. Tadeusz Stolc. Somewhat to my surprise, the latter was desperately hoping for a grant so that KUL could acquire VTLS and join the Polish VTLS Consortium, preferably in association with Marie Curie Skłodowska University (MCSU), a large state institution practically next door. When I visited MCSU the next day and met with Rector Kazimierz Goebel and University Librarian Teresa Gaworczyk, they actually presented me with a proposal for library automation, which they had prepared in anticipation of my visit. The KUL and MCSU libraries were large (over 1 million and 2 million volumes, respectively) and largely complementary, with substantial humanities and social science collections in the former and science collections in the latter. I then found out that the Agricultural Academy (AR)[17] was also quite near, and when I visited that institution later in the day, I put the bee in their bonnet about a potential Lublin Consortium.

After this I never thought about automating individual libraries again, except in unavoidable cases. I was confident that consortia could obtain more favorable unit prices from vendors. I was certain that some savings could be realized by having fewer servers than consortium members. I also thought that the training of librarians and patrons could be streamlined and made more efficient in a consortial framework. I firmly believed that a great deal of duplication could be avoided by "localizing" the software system only once rather than separately for each institution: it seemed a sheer waste to translate manuals, system messages, and error messages over and over for each institution. I certainly hoped that the presence of a large number of users might make it reasonable for the software vendor to establish local representation. And finally, and this was a dream that I actively supported for many years, many institutions using the same system might make it easier to create citywide or even broader union catalogs.[18]

Intense correspondence started after that by mail as well as e-mail. Zbigniew Skorżyński at MCSU was put in charge of computing and networking details, and by March 1993, the Lublin Polytechnic University (PL) also decided to join the Lublin Consortium. The configuration of institutions can best be seen in table 9-2.

The Lublin institutions had Internet connections since 1992, with TCP/IP, TELNET, FTP fully functional, which I verified by telneting from three of the institutions to my computer account at Princeton University. The Lublin area metropolitan network was fully functional, although the backbone to Warsaw had a bandwidth of only 9.6 Kbit/sec as late as August 1993; but it was to be upgraded to 64 Kbits/sec within a month. The initial offer by VTLS

Table 9-2 Characteristics of Lublin institutions.

Institution	Library "Units"	Full-time Students	Part-time Students
KUL	1,525,000	8,000	NA
MCSU	2,085,000	11,880	7,913
AR	360,000	4,040	696
PL	302,000	3,379	873

was for the version VTLS92; but at that point Skorżyński made the sensible-sounding, but ultimately fateful decision to acquire the brand-new Unix version of VTLS, as had the University of Wrocław, but with even more dire results. In any event, it had been decided that two servers would be sufficient, of which one would be at KUL and the other at MCSU, serving that institution and AR as well as PL. It was also decided to keep the KUL database separate from that of the remaining institution; while this was not deemed to be the most convenient arrangement, KUL insisted on a degree of separateness.[19] The original VTLS offer contained a 10% discount and required five years of maintenance to be prepaid for an actual seven years of maintenance provided; I managed to persuade VTLS to increase the discount to 20% and to change the maintenance terms to four years prepaid for six years of actual maintenance. The price of the project was becoming reasonable, but I did not feel that we could make a provision for the cost of retrospective conversion; so I sent the proposal and my analysis of it to George Soros and suggested that he might be interested in joining the Mellon Foundation with a grant of his own in the amount of, perhaps, $200,000. Soros, in turn, once again asked for the expert advice of his (outgoing) CEU librarian, Richard Ayres, who had several useful comments and questions; but all these were satisfactorily answered and Soros ultimately did provide the sum I had suggested, which was a net addition to the $951,000 that Mellon gave to MCSU in October 1993, the lead institution and financial manager of the project.

I was satisfied with the librarians' enthusiasm for starting a new system, although a few were noticeably apprehensive. Their choice of VTLS was strongly motivated by the fact that Warsaw, Gdańsk, Jagiellonian, and AGH had selected the system and that the VTLS Coordinating Group those institutions had created was performing useful services. I believed that they seriously underestimated the retroconversion costs they would face (their estimate was $0.33/item). But all in all, I believed that in about 15 months from the time of the grant, librarians and patrons could make serious use of the new system.

There was only one more hurdle to be scaled. As I was contemplating and analyzing the final version of the proposal in July 1993, and was about to write a recommendation to the Foundation for funding it, I realized that there was nothing in the proposal about the management of the project. I was assailed by panic that the Lublin people did not really know how complex an enterprise it is to automate four libraries at once. Almost immediately, I got on an airplane and went to Lublin for two days, for some intensive discussions of the management aspects of the project. I insisted that they had to create a management team and suggested that (1) the team should have a librarian, a computer person, and an administrator on it (although a single person might qualify under more than one rubric); (2) the team be empowered by the rectors to assign tasks to other experts as needed; (3) the members of the team be relieved of a sufficient volume of regular duties so that they could do justice to their assignment; (4) the team be responsible for day-to-day implementation and supervision of the project; (5) the team be the first level of conflict resolution; (6) in the event of the team's inability to resolve conflicts, the matter be referred to the rectors or prorectors; (7) the rectors agree to a mechanism by which their own potential conflicts would be resolved; and (8) the team be ready to operate by early October. These suggestions were accepted, and a five-person task force was established in November, with Professor Grzegorz Nowak of MCSU as the task force leader. The integration of the Lublin Consortium into the broader VTLS Consortium in Poland moved forward significantly when, in early 1994, the rector of MCSU was asked to sign the Agreement of VTLS System Users, previously signed by the rectors of WU, JU, UG, and AGH, and when he asked Jurand Czermiński, the general factotum and troubleshooter of the VTLS libraries, to become an official advisor to the Lublin project.[20]

During the year 1994, the usual sequence of events occurred: Hewlett-Packard servers were installed, LANs were built, training courses were conducted both by VTLS and by the staff of other VTLS libraries, VTLS92 was installed (at first), and bugs in VTLS92 were discovered and called to the attention of VTLS, just to be told that they would not be fixed until VTLS94 became available. The cabling of the local area networks began and by December, the LAN in the MCSU main library was complete, and VTLS94 (Unix) was installed.

The next 18 months were a period of nearly endless discoveries of software bugs or of missing functionalities. An abbreviated chronology follows (Nowak, pp. 8–17):

February 1995: VTLS94 Version 1.7 was installed. Its *EasyPac* module was unable to display Polish diacritical characters correctly. VTLS promised to fix this in V. 1.8.

March 1995: V. 1.8 was installed and numerous diacritical character defects were found.

May–June 1995: V. 1.9 was installed, together with *EasyPac* V. 1.31. Some bugs in the previous version had been fixed, but most of the diacritical character problems remained, in spite of efforts by VTLS as well as the local teams. VTLS promised to fix them in V. 1.10.

July 1995: V. 1.10 was installed, along with *Cataloging Client* V. 2.75.

The Polish characters still did not display correctly and VTLS promised to fix this in V. 1.11. In October, I was handed a list of malfunctions and promised personally to intervene with VTLS.

October 1995: V. 1.11 was installed, along with *Cataloging Client* V. 2.76 and *EasyPac* V. 1.32. Both of the latter are cataloging tools; *Cataloging Client* for a DOS environment, *EasyPac* for a Windows environment. The latter of these still could not display Polish characters properly. VTLS promised a fix in V. 1.12.

January 1996: EasyPac V. 1.40 and *EasyCat* V. 1.40 were installed; both were able to generate the Polish characters.

February 1996: V. 1.12 was installed and additional errors were found. V. 1.12F was then installed and was the first version that permitted efficient cataloging.

April 1996: SSHD V. 1.2.13 (used to ensure secure access by only authorized personnel) was installed. But under certain circumstances the software permitted access to the system by unauthorized personnel.

July 1996: SSHD V. 1.2.14 was installed and this problem was fixed.

October 1996: VTLS94 V. 1.13 was installed, which fixed many previous problems; however, when used in conjunction with *EasyPac* V. 1.42, a serious memory management problem arose, which prevented simultaneous cataloging sessions on several terminals. At the end of the month, V. 1.13 was removed and V. 1.12 was reinstalled. The bugs that remained in the system at that time were quite horrendous. When I visited Marie Curie Skłodowska University that month, I had a demonstration of how the password-protected command "process" could be circumvented.[21]

November 1996: VTLS94 V. 1.13 was reinstalled; together with *EasyPac* 1.42b, *EasyCat* 1.42 and *Cataloging Client* 2.77, the system finally became fully operational.

The recitation of software battles and frustrations tends to obscure the very substantial achievements during this period. Librarians were trained to use the system, in spite of some resistance because of inadequate relief from their ordinary duties, and a substantial amount of retroconversion was carried out, thanks to grants from the Open Society Institute and KBN. Nowak had the important insight that training that is not customized to the circumstances of the country and the libraries in question is, more or less, useless, as the VTLS-provided training was.[22] Truly old-fashioned librarians, several of them library directors, who were unable to adapt to the new system were dismissed or retired and their place was taken by younger people with a more modern outlook. In 1996, an additional Lublin library, the Łopaciński Library, received a grant of $210,000 from the Mellon Foundation so that it too could join the Lublin Consortium.[23] As of December 31, 1999, the server for MCSU, AR, and PL held 100,795 bibliographic records, 120,086 copy records, 3,798 serials records, and 94,930 authority records; a substantial achievement (Nowak, p. 27). The amount of effort that went into automating the Lublin libraries is staggering: the number of person-hours is displayed in table 9-3:

Table 9-3 Person-hours of labor in automation.[a]

	1994	1995	1996	1997	1998	1999
Librarians	1,560	18,000	18,900	45,500	24,800	12,000
IT specialists	1,540	3,600	4,050	5,100	3,600	1,200
Task Group	1,290	1,360	1,470	1,800	630	450
Total	4,390	22,960	24,420	52,400	29,030	13,650

Note:
a. Personal communication from Grzegorz Nowak. No other consortium that I know of has kept track so meticulously of effort expended.

The grand total of all the person hours is 146,850, which is approximately 70 person-years.

The management structure in Lublin was much less elaborate than in some other consortia—to wit the Kraków Library Group (see later in this chapter). KUL was always a bit separate from the other institutions and the implementation did not make use of standing subteams to deal with specific issues. But the Lublin institutions worked very closely with the Centre for Formats and Authority Control, formally established by the Ministry of Education at the Warsaw University Library in 1996, which was the successor to the Joint Library Computerization Project.[24] In the sense that all VTLS libraries worked in extremely close coordination with the Centre, the consortium of VTLS libraries has to be considered to be a tight one, even though individual subconsortia, such as Lublin, were relatively loosely structured. Having done its job, there seemed no further *raison d'être* for the implementation team and it was therefore reasonable in early 1999 to disband it.

There are three final comments worth making. (1) One of the greatest enemies of effective technology transfer is delay, and in the Lublin case the delay was primarily due to repetitive software inadequacies. In this respect, it is probably quite sharply differentiated from the types of delays that have occurred in CASLIN, where tasks were parceled out among the libraries and where each library may have had to wait for another library to finish its appointed tasks.[25] (2) An unintended consequence of the forced technology transfer was the somewhat drastic changes in personnel at the participating libraries, although many librarians benefited greatly in terms of their enhanced human capital and greater income from extra pay received for participating in retroconversion. (3) The case of Lublin was a wonderful example of collaboration between two foundations, Mellon and OSI, and of very effective mobilization of government resources in response western philanthropy.[26]

The Kraków Library Group. As early as October 1992, when I visited Kraków once again, I was considering the possibility of creating a consortium of libraries in the city. Two of them were already users of VTLS, but there were numerous other educational institutions that could benefit from library automation. It soon became clear that AGH librarians, particularly Ewa Dobrzyńska-Lankosz, and the new Jagiellonian University librarian, Krzysztof Zamorski, as well as AGH Rector Mirosław Handke and Jagiellonian Rector Andrzej Pelczar, thought that this idea was worth exploring. After my October visit, I wrote to the rectors of the five largest institutions and reemphasized the importance of a joint approach to automation; I said, "By a 'unified and joint approach' to library automation I mean a plan for library automation that is worked out together by several institutions and that results in a single proposal, prepared by all the participating institutions and signed by the Rectors of all of them."[27]

A steering committee was established, chaired by Professor Janusz Orkisz of the Polytechnic University, possible additional participants were identified, and after intense correspondence that lasted for a year, a final proposal was prepared by December 1993.[28]

The institutions that were invited to participate in the Kraków Library Group (as it became known) were Jagiellonian University, the University (formerly Academy) of Mining and Metallurgy, the Academy of Economics, the Agricultural University, the Polytechnic University, the Pedagogical University, the Academy of Music, the Academy of Fine Arts, the State Academy of Theater Arts, the Pontifical Academy of Theology, the Medical College of Jagiellonian University, Polish Academy of Science, Institute of Nuclear Physics[29] and the University of Physical Education; although the last of these dropped out of KLG fairly soon, only to renew its membership in 2000.[30] The Rectors' Conference of Kraków institutions met frequently and worked quite harmoniously, and this created substantial hope that such a complex enterprise would become viable. It was obviously necessary to connect each of the institutions to the metropolitan area network, operated by the Academic Computing Center, CYFRONET, a supercomputer center in charge of networking, and to build local area networks in each of the institutions. This could not be done overnight and would be extremely expensive. The total cost, including software, hardware, networks, etc., was estimated at $8,820,000, of which $2.7 million had already been spent ($470,000 from the Mellon Foundation's original grant to Jagiellonian University, $1 million from KBN, and $1,230,000 from the institutions' own resources). The remainder, $6,120,000, was to be raised from the same sources in varying proportions.

Initially, there was some question as to whether the consortium would select VTLS (used by some libraries) or TINLIB (used by the Polytechnic University Library). I was particularly concerned because it had been rumored that CYFRONET or the Polytechnic University were actually distributors for IME, the vendor of TINLIB, and I had serious questions about potential conflicts of interest if Professor Orkisz of the Polytechnic University were to participate in system selection. But I was assured by the chief executive officer of IME, Kate Noerr, that these institutions were not, in fact, representing IME in any form.[31] By March 25, 1993, KLG decided to adopt VTLS as its automation system;[32] the only library that did not want to go along with this was the Polytechnic University Library, but to this day it participates in KLG. Several librarians and administrators in KLG were upset about this defection and felt that we should not fund this institution's automation efforts.

The project envisaged three different management stages and three implementation stages. The first management stage corresponded to the planning phase and had largely concluded by the time the proposal arrived. In the second, implementation, stage, the Jagiellonian University rector's office, in the person of the outstanding Vice Rector Marek Szymoński, exercised oversight over the technical project coordinator, who would periodically report to the Council of Library Directors and oversee the activities of the working committee, with subcommittees such as the library group and the hardware and network group; the working committee would also maintain close relations with the VTLS Coordinating Group, in which Jurand Czermiński always played a significant role. In the final operational phase, the rector's office oversaw the steering committee and the chairman of that committee; the functions of that committee were to formulate overall policy for KLG, make financial decisions, and appoint and oversee the project manager. The project manager chaired the board of management, assisted by various working groups and administrative services, which was the executive body and reported to the chair of the steering committee and maintained relations with CYFRONET and the VTLS Coordinating Group. It was the project manager's responsibility to undertake the implementation of the project, to set up performance schedules, and to report to the steering committee.[33] The latter, appointed by the council of rectors, consisted of representatives of the four largest universities, the director of the Academic Computing Center, and the director of the Jagiellonian Library. For purposes of implementation, the institutions were divided into three groups: group 1 consisted of Jagiellonian University and AGH, group 2 of the Academy of Economics and the Polytechnic University, and group 3 of the remaining institutions. The implementation of the project was also divided into three phases, with a set schedule of what stage of implementation was to be reached in each of the three phases by each of the three groups of institutions. This is in sharp contrast to the procedures of CASLIN, where management seemed to be much looser and gave more room for individual libraries to follow their own schedules.

The metropolitan area network (MAN) was intended to reach not only institutions of higher learning, but other organizations as well, such as city offices, vojvodship offices, Polish Railways, and Polish Telecommunications. MAN had an overlapping FDDI ring as well as ATM structure. As of September 1995, eight institutions were not yet connected to MAN, but by May 1996, this number had been reduced to four.[34] KLG had determined that it could serve the entire library community with only three HP-9000 servers of which one would serve eight institutions, another would serve the remainder, and a third server would be devoted to the union catalog for the entire city.

All these plans were entirely reasonable and the dedication of the participating personnel extraordinary. In December 1993, I could write to George Soros and wholeheartedly recommend OSI's participation in the project. In 1994, the Mellon Foundation provided a grant of $1.2 million, followed up two years later with a supplemental grant of $200,000. Together with a generous contribution of $300,000 by OSI, the KLG received almost exactly the amount of outside funds that it needed beyond its own resources and those it could reasonably expect from KBN and the Ministry of Education. What is more, the council of rectors became so enthusiastic about the project that it committed itself to covering the running expenses of the consortium in future years.

KLG represented a major agglomeration of resources and readers. The combined libraries contained approximately 12 million volumes and had about 67,000 students and 18,000 teachers in 1995–96. The achievements of KLG in implementing automation were equally impressive. Progress is well measured by the fact that as of March 1, 2000, the VTLS system in Poland as a whole held 786,312 bibliographic records, 1,508,091 item records, 982,862 authority records, 31,476 holding records, 188,108 patron records, and 162,547 circulation records.[35]

We should note in conclusion that all the VTLS groups had substantial problems with the vendor, not only Wrocław and Lublin, and communication problems and software bugs affected all VTLS consortia. Jurand Czermiński complained[36] that communications with VTLS staff on a technical level were very ineffective and that the VTLS development department was not doing a good job fixing bugs—he suspected that they were concentrating all their efforts on a new product, VIRTUA, at the expense of fixing the old one. A few weeks later, he complained that the new VTLS94 was slower than VTLS92 and pointed out particular bugs, such as the fact that copying a record from the Warsaw Central Authority Data Base to the Gdańsk library resulted in the assignment of an erroneous record number and even changed some of the Polish diacritical characters.[37] But eventually, the problems were ironed out, even if librarians' patience was wearing thin.[38]

Dynix/Horizon Libraries

On the occasion of my trip to Poland in October 1993, I also visited the city of Toruń, the seat of Copernicus University (and birthplace of Nicholas Copernicus). I met with Rector Andrzej Jamiołkowski; Vice Rector Jerzy Ludwichowski, who was acting as the rector's plenipotentiary for automation; the library director, Dr. Stefan Czaja, who did not speak English; and Maria Śliwińska, the deputy director, who fortunately did. We had already been corresponding for a while and Copernicus University was definitely leaning towards VTLS; I was rather surprised when they told me that after a careful search, which surveyed GEAC, TINLIB, VTLS, Innovative Interfaces, Aleph, Carlyle, Oracle, and Dynix (Marquis), they opted for the latter. But Toruń had asked Richard Heseltine, the library director at the University of

Hull, to be a consultant during the selection process and the search was conducted carefully and responsibly. Moreover, there was really no cogent argument for making one more library choose VTLS. To be sure, if Copernicus University had picked VTLS, it could have benefited from the work done by all the other VTLS libraries. But I had some misgivings that if too many libraries in a country selected the same vendor, that vendor would unavoidably gain an excessive amount of market power, to the disadvantage of all libraries. Under the circumstances, I had no choice but to abide by the decision.[39]

Prior to 1989, library computerization at Copernicus University was inchoate. As early as the 1970s, some effort was made to control circulation by running batch jobs on a RIAD 10 mainframe computer, but this was not satisfactory. Śliwińska appears to have been the first person to be hired (in 1989) to think seriously about automation but was given hardly any resources for it. However, as a result of a considerable amount of personal study, trips to the United States and U.K., and taking advantage of a network of librarians in Poland and the U.K. who were better informed than the Toruń staff, Śliwińska and a few colleagues started to learn about modern library management. A Ministry of Education grant in 1991 permitted the purchase of a few PCs and the staff began to convert the bibliographic descriptions of the national bibliography provided by the National Library (Przewodnik Bibliograficzny) into CDS/ISIS format. But CDS/ISIS was soon determined to be inadequate for managing a major library. After a nine-week study tour for three Toruń librarians to the university library at Hull, the Copernicus Library was ready to address the preparation of a proposal to the Mellon Foundation. That proposal arrived in the spring of 1992 and was very well done; the Foundation made a grant of $500,000 in June 1992.

While Śliwińska made substantial progress in changing the structure of the library by merging some departments, progress in automation was painfully slow. Much of the responsibility for this rests on confusion in the software vendor's organizations and consequent lack of responsiveness. Because of corporate reorganizations, Toruń actually had to deal with more than one vendor: Dynix was bought by Ameritech, the Toruń contract was negotiated with the office in the U.K., and responsibility for administering the contract was finally handed over to the German branch, which was purchased by an investor who made it *de facto* an independent company. To be sure, some problems were due to the software (for example, authority records had limited field length), but some training sessions that were promised were not delivered, Dynix staff was not as competent as could be desired, and personnel changes at the Provo, Utah office meant that promises made by one person were not fulfilled by his/her successor. The software was actually installed on February 14–16, 1995, but the library discovered that what was installed was only a beta version. Reports of software bugs were dismissed by the company as "disfunctionalities." Training sessions for the project manager that were to take place in Provo, and the scheduling of which had been repeatedly confirmed, were never provided. The library was told that it was not possible to install Solaris, Sybase, and the library system in Toruń, and the server had to be sent to Berlin for the installation. In April, the library was invoiced for partial payment and the vendor had to be reminded that according to the contract, no payment was due until acceptance testing was passed. Cataloging actually started on September 21, 1995, but was interrupted several times as new problems surfaced and the library temporarily returned to using CDS/ISIS. Progress was by fits and starts until at least November 1996, but even after that there were still some persistent bugs: authority records created in the system were not compatible with USMARC and the Polish diacritical characters were not displayed correctly. Throughout the 1995–97 period, Dynix responded to queries with unconscionable delays. By June 2000, cataloging, OPAC, serials, and circulation for the open access books were in operation, but circulation for the closed stack books and the acquisitions module were still not running. Nevertheless, Toruń did start a fruitful collaboration with the Łódż and Poznań consortia, and most importantly with Warsaw University. Thus, Toruń became the gateway for the Dynix libraries to the Warsaw University authority database, which represented a substantial advance for libraries in general, but its own progress did not match original expectations (Śliwińska).

Łódż. In January 1994, I received a letter from Rector Michał Seweriński of the University of Łódż, who was also president of the conference of rectors of the higher educational institutions in that city, in which he indicated his desire to apply for a grant to automate the libraries of all eight higher educational institutions in Łódż. This was a major breakthrough: it was the first time that a consortium had formed without any prompting from me. The rectors and librarians at the various universities had obviously observed the progress that was being made in the VTLS libraries and the formation of the Poznań Consortium, and decided to follow suit. I replied encouragingly and the next communication I had was from Jan Krysiński, rector of the Technical University of Łódż, in which he informed me that his institutions had been selected as the lead institution to manage the project and that, henceforth, my contacts would be Błażej Feret and Bogusław Święcicki. He also enclosed the proposal itself, which proposed the automation of the libraries of nine institutions.[40] The combined holdings of the libraries amounted to about 4.5 million volumes. The proposal also segmented the implementation into two phases; Phase 1 was to cost $1,813,288, and Phases 1 and 2 together $2,407,233. This was more than Mellon had granted any library project before, and it was not clear that we could provide funding on this scale. Since I also had a number of technical issues that needed clarification, I replied very promptly. My principal technical questions were about the state of the metropolitan area network, which seemed more rudimentary than I had encountered in Kraków and which was essential here, because the proposal envisaged only three servers (Sun SPARC 1000E, 2 processors, 196 Mbyte RAM) to run the Dynix software, and about the local area networks within the various institutions, the funding for which seemed dubious. I then addressed the issue of cost, and, in particular, the price for the Dynix software, which seemed to me outrageously high. I wrote,

> Normally, the Foundation does not require potential grantees to adopt one or another type of software, except for requiring that the software to be acquired (a) accomplish the tasks that need to be accomplished, (b) be obtainable at moderate cost. Dynix appears to violate the second of these criteria in two ways. First, by having a high unit price, and secondly by charging each of the eight libraries (we know that there are nine in all, but Appendix G must have been prepared before the College of Film, Television and Theater joined the Consortium) exactly the same amount. However, *five of the nine libraries are negligible in size and I cannot understand the "marketing strategy" of Dynix that rests on the apparent belief (a) that we would not notice this, (b) that we would not consider this unfair and unreasonable.*[41]

After much additional work by the consortium, the Dynix price was slashed from about $600,000 to $363,000, other economies were effected, and I was given assurances that the metropolitan area network's FDDI ring would be ready by the end of 1995 and was being financed by a 184,000 ecu European Union grant, a $350,000 grant from KBN, with a further $370,000 expected from KBN. Moreover, the rectors of the participating institutions had committed themselves to funding their local area networks out of their own general funds. My admonitions about attention to the management aspects were well heeded and the consortium committed itself to engaging three full-time persons for the management team and as many part-time technical experts as needed. I strongly felt that this was a low-risk project. The amended proposal requested $987,000; I recommended to the Foundation an outright grant of $800,000, which was awarded in June 1995, with the possibility of a follow-up grant some two years later. But once again, dealing with the vendor did not prove easy. The consortium was originally dealing with Dynix U.K., but when Ullie Kreuzhagen bought Dynix Germany in 1995, it obtained exclusive rights to distribute Horizon in Poland (and some other countries). This created substantial delays and difficulties in communication, and the consortium leadership felt that it had not been provided with relevant information about the administrative changes.[42]

In October 1996, I recommended a $30,000 grant in order to allow the Piłsudski Scientific Library to join the consortium and a further $240,000 supplemental grant to the consortium itself. In recommending that latter grant, I was able to report that by October 1995 the consortium had a revised contract with Dynix in which the consortium drove a very hard bargain.[43] By November the hardware had been selected, in December a training center had been established, and shortly thereafter, the delivery of the hardware began. The hardware included 176 PCs, 48 printers, 7 scanners, and 34 barcode readers, in addition to the three servers. By October

1996, the Technical University (the most advanced institution in the consortium) had 100,000 item records in a machine-readable database, and by the end of the year all libraries had the client software (Horizon Version 4.1.1) installed, although testing took until the middle of 1997. By the end of 1997, some of the libraries were cataloging in Horizon, and by 1998 all libraries were cataloging in the system. By December 31, 2000, cataloging and OPAC were operational at all nine institutions, acquisitions and serials control at four, and circulation at two institutions; moreover, all modules were operational at the Technical University and the Medical University.[44] Each institution decided what fraction of its total collection had to be ultimately cataloged in the new system; by the end of 2000, the Technical University had cataloged 74.2% of its target, although none of the others exceeded 24.3% of target as of that date.

This has been one of the smoothest and problem-free consortia in which progress was solid and the management effective. That, in turn, is characterized by substantial central direction, with overall supervision provided by the rector of the Technical University, who appointed the project manager (Feret) and who is assisted by a project management team in the first instance and by an interuniversity library team appointed by the conference of rectors.[45]

Poznań. In the middle of 1993, I received a letter from Artur Jazdon, university librarian of Adam Mickiewicz University (AMU) in Poznań, in which he informed me that the institutions of higher learning in that city had agreed on a common and unified approach to library automation and that a proposal was in preparation. This was welcome news, since we already a favorable impression of that institution on the basis of its continuing performance under a previous grant to establish computer networks within the university, a project that was completely finished a year later. By the end of the year, the new rector, Jerzy Fedorowski, sent me a complete proposal, but it had many loose ends and was by no means ready for funding. By June 1994, the rectors of the Poznań institutions[46] had entered into an agreement committing them to joint computerization of their libraries. By October 1994, they had a completely new proposal, which had just been finished as I was arriving in Poland and was sent by courier to my hotel in Warsaw prior to my visit to Poznań on October 27–28. There was a great deal in the proposal that was good, but I also found a great many technical issues that needed clarification. While in Warsaw, I prepared an analysis for later distribution in Poznań that contained 39 separate queries or criticisms. The team that had prepared the proposal was chaired by Artur Jazdon, and had five other members, among whom were Bogdan Maruszewski of the Technical University, who was to play a continuing role for some time, and Jacek Małuszyński, who wanted to have a continuing role but was soon to be frozen out.

The principal meeting in Poznań, at which the proposal was to be discussed, was a very large meeting attended by rectors, librarians, and computer people. I spoke first and took about two hours to go through the proposal and discussed the various problems that I saw. The major ones were these: (1) the proposal indicated that they wished to have a completely distributed system with 12 servers, and the arguments for such a solution were not at all clear; (2) they proposed to acquire the Dynix/Horizon system, even though the Dynix bid was $320,000 higher than the Ex Libris bid for Aleph; (3) there was inadequate attention paid to the training of librarians and systems people; (4) it did not address the questions of management. It turned out that some of these points had already been considered by the design team, but they had been unaware of the amount of detail I required in a proposal; they promised to include a much fuller treatment of these and other issues in the next iteration of the proposal, which they promised to get to me by December 15.

After the meeting broke up, I visited the computer center (to admire the achievements under the previous grant) and met with several people, including Jacek Małusziński, who painted a rather more pessimistic picture of the Poznań situation. First of all, it turned out that a major power struggle was going on for control of the consortium. Professor Jan Węglarż, the director of the Supercomputer Center in the city, insisted that any grant should be made to his center, which would run the software on one of his supercomputers. Węglarż was reported to know nothing about libraries and library automation and his ambitions seemed to be completely motivated by personal ambition, but he was a powerful person—he was a member of the committee in Warsaw that approved habilitation theses—and everybody seemed afraid to cross him. Since nobody seemed to think much of the idea of a supercomputer as a single server, in self-defense they invented the idea that every institution needed its own server! In any event, I communicated to Rector Fedorowski my view that putting the project under the auspices of the Supercomputer Center was a very bad idea and that I hoped they could do with less than 12 servers.

Another event occurred, of which I became aware only later and which had rather more fateful consequences. Arthur Brady, who was the Dynix/Horizon representative in Great Britain, wrote a memorandum to Małusziński, in which he recommended the creation of a separate legal entity for administering the grant and implementing the project:

> There are several different organisational structures in use, but the most common—and the one we strongly recommend here—is the formation of a separate legal organisation. I am consulting with Ameritech's legal counsel in Warsaw about what forms might be best, and I will report to you what I find out. Probably some form of non-profit corporation, comprised of a board of directors (the rectors of each institution, or perhaps the directors of each library), an executive committee empowered to sign contracts with Dynix, the hardware supplier, etc., and the project management team.[47]

The new version of the proposal duly arrived and was much improved. Professor Węglarż and the Supercomputer Center were no longer in the picture, the number of servers had been reduced to three, and while no final choice of a software vendor had been made, the consortium was now strongly leaning towards Aleph, which at this point was priced about $450,000 lower than Dynix.[48] (In fact, Dynix did lower the price drastically and its final quote was only $247,500, not much higher than the price of Aleph.)[49] Significantly, there was no discussion in the proposal of setting up a new organization to handle the project, and in my recommendation for a grant of $850,000 I assumed (erroneously, as it turned out) that Adam Mickiewicz University would be the appropriate grantee as the lead institution of the consortium. The grant was duly made to AMU on April 18, 1995, but I had been informed on March 23 by Małuszyński that the original proposal preparation team had been disbanded in January, a supervisory board had been created and that Bogdan Maruszewski, elected to chair this board, was to legally register it as the Poznań Foundation of Scientific Libraries.[50] Six days later, Maruszewski let me know that he had become the president of the new organization (which did not as yet exist legally).

But the legal registration of the new foundation turned out to be much more time-consuming than anybody had thought. While this legal process ground on, nothing was happening on the automation front. The Poznań Foundation of Scientific Libraries (PFSL) had no legal status to sign contracts or mobilize any kind of resources, and AMU was unwilling, as well as unable, to act, the inability stemming from the rectors' decision to assign all functions and powers to PFSL. Further, there seemed to be serious disagreements within the consortium about software, as well as PSFL's statutes. There were also no signs on the horizon that this period of inaction would come to an end, and on November 5, 1995, I wrote a joint letter to all the rectors of universities and the directors of the two independent libraries in which I said, among other things,

> As far as I understand, there are two principal problems. The first one is that the members of the Consortium are not unanimous in their views as to what automation system would best serve the needs of the Consortium or their own needs. The second problem is that the document entitled "Statutes of the Poznań Foundation of Scientific Libraries" provides no suitable escape clause for individual institutions desiring to leave the Consortium at some future, undetermined time. In other words, while a complete dissolution of the Consortium would presumably lead to an apportionment of the assets of the Foundation among the members, an individual institution could not take any fraction of the Poznań Foundation's assets with it, including those assets that were contributed to the Foundation by the institution itself for the purpose of promoting joint library automation. My sense is that these two factors, combined with the slow-moving legal process that has to be undertaken to legally register a foundation with the courts, have led to a year-long delay in implementation, which is extremely disappointing from the Mellon Foundation's point of view.
>
> The issue of what system to acquire (Dynix's Marquis or ALEPH or any other) may pose a difficult problem. The Mellon Foundation is completely

> determined that the Consortium members *must* use an identical system. This is deemed by us to be much more important than precisely what system you end up using, provided that the system you use (1) is satisfactory for your needs, (2) does not cost more than you can afford. In the event that the preferences of the different institutions are not identical, *no collective decision is possible that simultaneously is "best" for everybody.* Some institutions will have to accept using and living with a second-best system. Hence, whatever organizational plan you adopt . . . , the member institutions have to face the possibility that one or the other institution will be outvoted. In situations such as this, one hopes that the power and reach and overall low cost achieved through the consortium will compensate the institution that is the "loser" in the vote on what system to purchase. If that is not the case, i.e., if one or more institutions have so great a preference for one system, and other institutions for another system, so that neither side is willing to take a chance on being outvoted on the system choice, then agreement is not possible. In that event, *the Mellon Foundation will cancel and revoke the grant and demand the immediate return of all moneys, including accrued interest.*[51]

I then went on to outline some possible solutions, including modification of the statutes, liquidation of the (as yet nonexistent) Foundation and gave a deadline of April 1, 1996, for PSLF to resolve all issues, or else the grant would be forfeited.

As 1996 rolled around, progress became more visible. Fedorowski wrote to me on April 2 and indicated that (1) the legal status of PFSL had been settled, with final approval expected within a few weeks; (2) hardware and software tenders had been issued and were in their final stages; (3) a timetable for 1996 had been agreed to by the Consortium participants; (4) a letter of intent had been sent to Dynix; (5) cooperation had been established with Copernicus University and the Łódż Consortium; and (6) funds had been obtained from KBN for the purposes of automation. Since early in the year, the extremely well-organized Andrzej Nikisch had been appointed Chairman of PFSL and PFSL actually became a legal entity on May 6. Also, IBM RS/6000 servers had been selected as the hardware platform, and the cost of software, servers, and PCs fit comfortably into the Mellon grant, with about $42,000 to spare. All this was just as I had hoped and, fortunately, I did not have to carry out the threat in my letter. But AMU could not just transfer the money to PFSL, without PFSL qualifying as an acceptable foreign donee (for which it had to complete the Foreign Organization Questionnaire) (see chapter 3) and receiving the Mellon Foundation's permission to do so (which was essential, because in authorizing the grant, the trustees of the Foundation approved a grant to AMU, not to PFSL). All these steps took until September and the Mellon Foundation finally authorized the transfer of the funds to PFSL.

Alas, the problems were not over. When I arrived in Poznań in October 1996, I was informed by Maruszewski and Nikisch that the amount transferred by AMU to PSFL was not $850,000 but $805,000; the discrepancy arising out of the facts that the money had been kept in a złoty-denominated account rather than a dollar account and the złoty had depreciated rather severely in the meantime, and that the type of account the money was held in earned a very low rate of interest compared to the alternatives that were available. I assured them that I held AMU responsible for the loss and expected it to make PFSL "whole." Later I met Jerzy Fedorowski, who had just vacated the office of rector, and he apologized for the error committed by AMU and agreed that the university should replace the missing sum. My meeting with the new rector, Stefan Jurga, was by no means as amiable. He first argued that Polish universities were not entitled to dollar-denominated accounts (whether they were or were not seemed irrelevant to me, because in fact many Polish universities did have such accounts), and then blamed Maruszewski for the fiasco, on the grounds that he was responsible for the proper investment of the funds. I replied to this latter charge that until recently PFSL did not even exist legally and Maruszewski had no powers to dispose of funds that AMU held in trust; in any event, the Mellon Foundation would expect him to restore the missing money. We parted politely but not amicably by agreeing to disagree. The next day I also met with the new rector of the Academy of Economics, and he promised to support my point of view. Returning to the United States, I wrote a letter to all participating institutions' rectors in which I again expressed my views that AMU was responsible for the loss and had to make it good. Fortunately, by December 2, a compromise was reached: Rector Jurga decided to make good $38,000 of the missing $45,000 and the Mellon Foundation came to the rescue for the remaining $7,000.

Progress by March 2000 has been excellent, in spite of the rocky history of the consortium. The library system is used by about 68,000 readers and the automated system circulates about 900,000 items per year. Good progress is being made in cataloging: as of March 2000, nearly all institutions were cataloging 100% of new materials in the new system and five institutions had upwards of 95% of teaching resources cataloged.[52]

Overall, about 20% of all printed material and 75% of all teaching material in the research libraries of Poznań has been cataloged in the system, and the current rate of growth of bibliographic descriptions is about 90,000/year.[53] What is even more gratifying is that the Poznań libraries are in close cooperation with the other Dynix libraries, and with the VTLS libraries as well, through the authority file database created by the VTLS libraries. This database has been maintained at Copernicus University (a Dynix library) since 1996, and also in Łódź since November 1997, and the institutions holding the database have served as the exchange agents for the two types of libraries.[54]

The National Library: An INNOPAC User

While the Mellon Foundation had made two grants to the National Library of Poland in 1991 and 1992 ($100,000 in 1991 for book acquisitions and a computerization feasibility study and $10,000 in 1992 for miscellaneous computer equipment), my discussions with National Library staff about automation were largely disappointing until 1993. As early as 1990, the National Library's director, Dr. Stanisław Czajka, had the misguided ambition of making the National Library develop its own automation software, from which I tried to deflect him strenuously. Even so, I received several small proposals aiming at the development of such a software package, to be built around the library's IBM 4361 computer. In November 1991, when the libraries of Warsaw University, Jagiellonian University, and the University of Gdańsk were on the verge of choosing VTLS, I even tried to interest him in joining in a consortium with those libraries. Although by 1992 he had given up the plans to develop proprietary software, he must have suffered from the "sunk cost fallacy," because he kept insisting that any software to be acquired had to run on the IBM machine.

The turning point for the library came when, in the summer of 1992, Dr. Czajka had to retire for health-related reasons. The position of director was advertised, and for the longest time there appeared to be no applicants. Fortunately, one outstanding candidate did surface, namely professor Adam Manikowski, a historian at the University of Białystok, who was named director in early 1993. He introduced himself to me in a letter dated August 20, 1993, and it was quite obvious that he was extremely interested in the possibilities for automation and, in anticipation, he had already started some personnel changes in the library. An important testimonial to the changed atmosphere at the National Library was the fact that The Commission on Preservation and Access had signed a contract with the National Library, and provided $120,000 of financial support, for converting the Polish register of microform masters to machine-readable, U.S.-compatible format and to share the data with the *European Register of Microform Masters* and with the U.S. bibliographic utilities.[55] Hans Rütimann at the Commission had a very high opinion of Manikowski as well as of the staff of the National Library[56] and I was looking forward to meeting him in October. Another important event occurred during the summer, for which one must again give credit to Manikowski: the National Library formally announced its selection of USMARC as the standard cataloging format for all of Poland.

My meeting with Manikowski lived up to my expectations, and by January 1994, I had the National Library's proposal for an automated system in hand. While I had some reason to believe that the library favored VTLS, no decisions had been made and the library costed out four systems: VTLS, Aleph, TINLIB, and Voyager.[57] There did not appear to be compelling financial reasons for one over another system, although I strongly felt that TINLIB was not appropriate for a library the size of the National Library. In all respects, the proposal was careful and methodical, and the Mellon Foundation responded with a grant of $850,000 in June 1994. Final system selection was expected to occur by the end of the year.

In fact, the final selection of a system was somewhat delayed, and the contract to purchase INNOPAC by Innovative Interfaces, Inc. was signed only in March, 1996. In December 1999, the library migrated to the newer Millennium version of the software. Since 1998, cataloging has been done in INNOPAC, the old databases kept in the MAK system had been converted to INNOPAC, and work on implementing the acquisitions, serials, and circulation modules continues.[58] As of June 2000, the National Library had a machine-readable database containing 360,483 bibliographic records, 681,283 item records, 176,458 authority records, and 18,062 serials records—a considerable achievement.[59] There is no doubt that the library staff considers the choice of INNOPAC to have been an excellent one.[60]

A Polish Union Catalog?

As early as 1996, I thought that it would be a crowning achievement of library automation if all the Polish research libraries could pool their resources and efforts and create an all-Poland union catalog. The realization of this idea would represent a major step towards shared cataloging, merge the databases of the largest research libraries, provide a single computer interface for access to Polish materials, unify bibliographic standards, and speed up interlibrary loans. On the occasion of my visit to Kraków that year, I made a point of discussing this idea with Krysztof Zamorski and Ewa Dobrzyńska-Lankosz. At that stage, the idea was completely inchoate, but I asked them to think about it and discuss it with other librarians; the only condition I made was that such a union catalog would have to include not only VTLS-based libraries, but the Dynix libraries, the National Library, and possibly other libraries as well.

Sometime in the first few months of 1997, they reported to me that it was not plausible to think of forging an agreement in as diverse a group as I was contemplating. On the principle that "you can lead a horse to water but you cannot make it drink,"[61] I thought it wiser to drop the idea.

At the October 1997 Mellon conference on library automation in Warsaw,[62] I was approached, to my enormous surprise, in various coffee breaks by librarians from Dynix libraries, with suggestions that we revive the idea of a Polish union catalog. I immediately assured them that I continued to be interested in creating a union catalog, but reaffirmed my conviction that it had to rest on collaboration between VTLS and Dynix-based libraries and the National Library. I made a point of telling Joanna Pasztaleniec-Jarżyńska, the deputy national librarian,[63] that the National Library's cooperation in such an effort was an essential ingredient for Mellon support and ultimate success, and also informed several librarians from VTLS libraries that my originally stillborn idea might still have some life in it.

Indeed, discussions to define the features of a hypothetical union catalog began almost immediately; by early 1998, the VTLS libraries, the Dynix libraries, and the National Library had created a task force to sift through the many issues that had to be decided. It is fair to say that a great deal of work was done in the first eight months of 1998. But it soon became clear that a wide gulf divided the VTLS and Dynix libraries. The lesser problem was that if either VTLS or Dynix were adopted for running the union catalog, approximately one-half of the libraries would have to purchase a union catalog client for accessing it, and nobody was happy about having to use a different system than the one they were used to. But the big difference in the approach to the union catalog rested on what might be best called "cataloging philosophy." One way of characterizing this is to say that some librarians think that the advantages of "quick-and-dirty" cataloging outweigh those of "slow and careful," while others feel the exact opposite. A hint of the former position is reflected by Górny, et al., in which they argue that (1) entrusting the provision of bibliographic descriptions to the National Library is less expensive and more accurate than having individual libraries catalog their collections using a central authority file; and (2) using such a central catalog would "make it necessary to accept the introduction into local catalogs of simplified catalog descriptions."[64] The VTLS libraries insisted on impeccably clean records and on the creation of authority records *before* the creation of a corresponding bibliographic record: only thus could the consistency and accuracy of the database be assured. The Dynix/Horizon system did not permit the creation of an authority record independently of a bibliographic record,[65] and several of the Dynix libraries questioned the importance of having authority records at all. I preached the gospel of compromise, but it was not clear that a compromise was possible in these matters. The discussions were intense and, if I interpret the signs correctly, heated at times. In late June 1998, I was visited in Princeton by Błażej Feret (in the Dynix camp) and continued my missionary work. Finally, by the middle of July, Henryk Hollender (director of the Warsaw University Library) reported that all was well:

> I am just back from the "ultimate" meeting at the National Library. Everything is under control. No one has left. No one did too much shouting. All the conflicting issues have been written down in [such] a way that they are no longer conflicting, even if the philosophies behind the formulas are still a little different. To give you an example. VTLS: All the consequences of the particular libraries' management of their own bibliographic files is their own problem. Horizon: All the consequences of the particular libraries' management of their own bibliographic files is their own joy. Eventually, the mention was deleted.[66]

By October 1998, I had received a full-fledged proposal from Warsaw University (which was to be the manager of the union catalog grant). The actual project managers were Hollender, Nikisch and Pasztaleniec-Jarzyńska. The issues seemed to have been resolved, largely, it seemed to me, in favor of the meticulous approach represented by the Centre for Formats and Authority Control. No final decision on software had been made, but the preparation phase seemed to be in the home stretch. Forty-four libraries (including some using Aleph) had committed themselves to the effort and a further

30 had expressed some interest. There was no more reason to delay, and I recommended, and the Mellon Foundation approved, a grant of $705,000 to Warsaw University for creating a union catalog for Poland. The grant was paid to Warsaw University in January 1999.

A year later, I was planning another trip to Poland, this time in March, and it suddenly dawned on me that I had not heard any glorious tales of success about the union catalog and not a penny of the Foundation's grant appeared to have been spent. It was essential to get to the bottom of this, and I set up separate meetings in Warsaw with the VTLS group, the Dynix group, and the National Library. My worst fears were confirmed. What appeared earlier to be a fair amount of agreement dissolved in further argumentation and mutual accusations of unreasonableness and of not fulfilling earlier promises. The VTLS libraries would not accept a system that was not controlled by authority records and the Dynix libraries claimed that most of them did not need authority records at all. I suggested a compromise: that Dynix libraries be permitted to contribute bibliographic records to the union catalog without authority records, but their records would be flagged as "dirty." However, it was not clear that this solution was technically feasible. I decided that the state of affairs was not tenable, because the grant (the principal of which was growing daily because of interest earnings) might sit idle indefinitely.

Upon returning to the United States, I drafted a letter to be sent under William G. Bowen's signature to Warsaw University Rector Piotr Węgleński (with copies to all the principal players), which recapitulated the state of affairs and the causes of the continuing disagreement, and said,

> The Foundation is unwilling to wait indefinitely for a satisfactory solution to emerge. Accordingly, it is my regrettable duty to inform you that if we do not receive from you a signed commitment by the VTLS libraries, Dynix/Horizon libraries, and the National Library to a single, unified plan for the union catalog, together with an implementation scheme promising substantial success in the following year, we shall terminate the grant to Warsaw University and require the immediate repayment of the grant, together with accumulated interest. The deadline for receiving such a commitment is July 15, 2000.[67]

By May 22, Rector Węgleński replied to the effect that the Dynix libraries had accepted the leadership of Warsaw University and of its Centre for Formats and Authority Control, and, while not all Dynix libraries were going to participate, many of them would, as would some libraries using Aleph and Prolib. This was also confirmed by Henryk Hollender. Considerable credit must be given to all participants, in particular university rectors and prorectors, for having reached a new point of agreement. The agreement, signed by Henryk Hollender, Joanna Pasztaleniec-Jarzyńska and Andrzej Nikisch, specified that (1) the union catalog would rest on cooperative cataloging in accordance with the cataloging rules prevailing in Poland; (2) authority records would be created online in the Central Authority File prior to the bibliographic records in which they are utilized; (3) temporarily, passive partnership would be permitted for some libraries; (4) a separate catalog would be created for libraries to deposit records without creating authority records. In addition, a time-table of implementation was prepared, which anticipated the opening of the union catalog by January 2002. While this solution is not the 100% solution that I had hoped for, it was a substantially better one than would have been reached if the Dynix libraries had opted out completely. I replied to Hollender very positively and only required that, henceforth, quarterly reports on progress be sent to me. By November 2000, the various working groups had been reconvened, hardware configurations and prices had been studied, and a time-table through the year 2001 had been set up. The principal amount of the grant had grown to $790,000 and the the project management was maintaining close contacts with the State Committee for Research.[68] By the end of the year 2000, agreement appears to have been reached on VTLS's VIRTUA system as the software to be used. All this sounded promising, but only time will tell whether the union catalog consortium is too fragile to survive.

Reflections on Library Automation

The first question one needs to ask is what the overall impact of philanthropy may have been. That a large number of libraries introduced new software and hardware, and that librarians were trained in the new systems is not in question. But the interesting question asked by Manikowski is what would have happened if the western foundations had not appeared on the scene.[69] He argued persuasively that automation would have (eventually) taken place anyway, particularly because the amounts contributed by foundations were really rather modest fractions of the receiving institutions' annual budgets. But because of the severe financial stringencies throughout the region, the introduction of automation would have been postponed, and perhaps postponed by many years, thus severely delaying the educational and research recovery in these countries. In other words, the foundations interested in libraries (OSI, Pew, Mellon) appeared on the scene just in time.

One of the principal impacts of library automation was the reconsideration and drastic reformulation of standards. This obviously occurred in the Czech Republic and Slovakia, but if anything, it occurred in Poland with even more attention focused on important details. The adoption of USMARC, the recognition of the importance of authority files and the concentration of authority work in the Centre for Formats and Authority Control, the deep attention paid to creating a subject authority file called KABA,[70] and analogous milestones in the Czech Republic and Slovakia, all attest to the importance of creating unified standards. While in some instances these standards were largely borrowed from a specific foreign source, as in the case of USMARC, the Polish subject authority file KABA relied on several sources, namely on RAMEAU (Répertoire d'Autorité Matière Encyclopédique et Alphabétic Unifié), RVM (Répertoire de vedettes-matière), and Library of Congress Subject Headings (Paluszkiewicz, p. 227). KABA records are entered in USMARC format and the file, maintained at the Warsaw University Library, is linked to the name authority file and is accessible via the Internet. As of July 1, 1999, it provided over 130,000 search keys.[71] The agreement of the libraries that no subject heading may be used in a bibliographic record unless it was already present in KABA was a significant step in the direction of ensuring the complete accuracy and consistency of bibliographic records.[72] But it is precisely the argument as to which standard should be adopted that makes for friction and makes it at times difficult for libraries using different standards to work together, as in the case of VTLS and Dynix libraries—a point that has been noted earlier by Robert Hayes in reference to the choice between AACR2 and ISBD,[73] although there exist efforts to harmonize different standards.[74]

It is the adherence to standards that makes systems portable and enlarges the scope of individual libraries to have global reach and importance; the transfer of the technology of library protocols is one of the most important impacts of western library-oriented philanthropy. Library software is quintessentially designed to be completely portable; yet, as much as library protocols and standards enhance portability, concrete library software is, in a sense, less portable than could be hoped for in an ideal world, because (1) different systems use different conventions and can talk to each other only with appropriate interfaces, such as Z39.50; and (2) even the same system cannot always talk to its counterpart at another institution because of the proliferation of versions that are partly or completely incompatible. While much of the history of bringing library software online is straightforward and uneventful, in other cases the road was more than a little bumpy. In all cases, both with respect to the design of standards and the bringing of automation systems online, the librarians had to perform feats not previously attempted, and rose to the occasion splendidly; and while it would not need to be commented on in a western context, it needs to be noted, given the history of the Communist period, that they did this on their own, in a grassroots kind of way.[75]

Many important lessons were learned. Because of the absolute importance that the language in a contract between software vendor and user be of utmost clarity and free from ambiguity, it is essential that a major library have staff members who are completely fluent in English. Contract language must pay scrupulous attention to acceptance criteria. In order to secure the most advantageous financial terms, Czermiński strongly recommended that "up to the day the license agreement is signed, the vendor must not be certain that his offer is the only acceptable offer and the end user has already decided to purchase his software,"[76] a precept that is generally difficult to carry out in practice. Purchasers of library systems need to weigh ahead of time the burdens of translating (localizing) into the local language the software's manuals, error messages, and system messages and they must pay particular attention to the status of diacritical characters. Henryk Hollender sent a questionnaire to a sample of 22 libraries, in which he asked the librarians to indicate the amount of emphasis they placed on a variety of library functions in automation; the handling of diacritics was assigned the highest importance (followed by circulation, conversion of databases, and cataloging).[77] It also became clear that system bugs and unresolved developmental issues are endemic in all software, and the buyer must tread carefully between forcefully requesting assistance from the vendor and antagonizing the vendor with excessive demands. Another issue that needs to be decided is how to sequence the bringing online of an automation system; since not everything can be done at once, decisions have to be made about what will be done first, second, and so on. In Hollender's sample, the production of bibliographic records, OPAC, authority control, www availability, and circulation were the clear leaders for early implementation, with other activities such as serials control, subject access and classification further down the line (Hollender, pp. 342–43).

There is no doubt that the emphasis on consortia was important and led to much faster progress than would have been possible otherwise. While it is true that some consortia have worked less well than could be hoped for, either because of lack of university support or inadequate understanding within the consortium of the importance of subordinating individual objectives to the overall aims of the consortium, the cost savings that I anticipated were realized, both in keeping down hardware and software costs and also in the efficiency with which the localization of systems and the adoption of new standards were realized. In addition, the public impact of consortia, which was much greater than could have been achieved by the same libraries automating their systems on an individual and piecemeal basis, had much to do with the increase in the flow of state funds to the libraries. However, whether the consortia will be able to continue on the cutting-edge level remains to be seen. Funding by the governments has not increased *adequately*, and some consortia have belatedly realized that equipment becomes obsolete and needs to be replaced periodically. To the extent that state or EU funds do not step in to take up the slack created by the withdrawal of western foundations, it is distinctly possible that some consortia will lose members and will not be able to keep up with modern developments in automation.[78]

It also appears to be the case that the precise format of a consortium, whether "tight" or "loose," had a smaller impact on the ultimate outcome than the dedication and talent of the participating librarians. Consortia can be set up to be merely resource-sharing (for example, HUSLONET), or organized as a separate organization (Poznań), or working under the general supervision of a lead institution (Łódź), or as a loose alliance (CASLIN+), or an alliance with highly distributed functions (VTLS Group); there is no magic formula that makes one model work and another not work. What does seem useful is a tri-partite management structure, with a highest level council that meets infrequently and makes very general long-term policy (for example, a rectors' group), an intermediate level that meets more frequently and makes shorter term decisions (for example, a steering committee consisting of librarians), and an operations group that does day-to-day management.[79]

Nadia Caidi[80] asked the interesting question of whether one can identify factors that predispose consortia to develop as loose or tight ones. First one should note that the loose consortia described in this volume correspond roughly to my earliest conceptions of collaboration, while the formation of the tight ones tend to fall in later periods; thus representing to some extent where I was on my own learning curve. But what is also true is that the tight consortia consisted (with one exception) of institutions that were either in the same city or in countries small enough to be equivalent to one large metropolis (namely the Baltic countries), while the loose consortia were national or international in scope. The one exception to this generalization is the initial VTLS consortium which allied four institutions in three different cities; however, they were held together by strong personalities driving toward a common goal. It suggests the tentative conclusion that geographic proximity favors the type of collaboration characterized as tight.

What has also been especially impressive in all the countries of the region is how rapidly the librarians transformed themselves from a group that was not held in high regard during the Communist period and rightly regarded itself as a victim into one of self-confident professionals, who became inveterate problem solvers, because there was no other way to realize their vision.

A final question is how the East European librarians themselves have viewed the developments of the past ten years. In a pathbreaking study based on a sample survey of 49 respondents in Hungary, the Czech Republic, Poland, and Slovakia, Nadia Caidi has examined attitudes towards a national, information infrastructure as well as toward library governance, cooperation, and resource sharing.[81] Library governance was deemed to be one of the most important issues and librarians generally believed that the attitudes of government agencies and universities toward libraries needed to be improved—a point also made by Manikowski (Manikowski, p. 26). They noted that western support was flexible enough to permit them to experiment with alternative modes of organization and that this support also succeeded in mobilizing additional government moneys by drawing attention to the libraries. All respondents agreed on the value of cooperation and resource sharing, and it is gratifying that this new mode of business was adopted relatively quickly by most libraries. Whether cooperation was forced on the libraries, as suggested by Manikowski, or whether they would have coalesced into consortia even on their own is unclear; what is clear is that cooperation did take place and is here to stay.

10

Baltic Interlude

Two important shifts in the Mellon Foundation's interests in the second half of 1994 were to have a significant impact on my activities in the last five years of the decade. First, it seemed reasonable to consider whether it might not be useful and opportune to expand the Foundation's activities in Eastern Europe to the Baltic countries, which had recently achieved their independence from the Societ Union. Second, in the light of the favorable political changes in South Africa, the Foundation decided that the time had come for a major program in that country.

The first of these ideas did not represent a firm decision to start philanthropic activities in the Baltic countries; rather it indicated an open mind that suggested that it would be desirable to undertake an in-depth investigation of the pros and cons of such a program. Given my background in Eastern Europe, it was natural that I should undertake such an investigation. In contrast, the South African program already had the full commitment of the Foundation and it had been entrusted to my colleague, Thomas Nygren. However, since it was fairly obvious from the very beginning that the focus of such a program would be on higher education, which tends to imply, among others, assistance to research libraries, I was asked to assist him. However, the present chapter is devoted entirely to the Baltic countries, except for a brief allusion in the concluding section to some of my experience with a South African library consortium.

The Baltic countries are similar in size, geography, and the fundamental characteristics of their economies, and in several respects of their recent, shared history. Estonia is the smallest, with about 1.5 million inhabitants in 1995, of whom a round half-million lived in the capital city of Tallinn and another 115,000 in the important university city of Tartu. Latvia had 2.6 million inhabitants, with 920,000 living in Riga, the largest city in the Baltics, and another 100,000 in the southeasternmost city of Daugavpils, where only 13% of the population is ethnic Latvian.[1] The largest is Lithuania, with 3.7 million inhabitants; 600,000 people lived in the capital city of Vilnius and another 430,000 in the provincial city of Kaunas.[2] Prior to World War I, the three constituted the Russian Tsarist provinces of Livonia (roughly, northeast of the Daugava river) and Courland (southwest of the river). In 1918, the three countries proclaimed their independence,[3] and soon afterwards they all adopted democratic constitutions. Democracy lasted only seven years in Lithuania and was overthrown in Estonia and Latvia in 1934.[4] The Molotov-Ribbentrop Pact of August 23, 1939 assigned Estonia and Latvia to the Soviet sphere of influence and Lithuania to the German, but in September of that year a second secret protocol between Germany and the USSR assigned Lithuania to the USSR, as well.[5] In June 1940, Soviet forces occupied all three countries, and they were formally annexed by the Soviet Union a few weeks later. They were, of course, occupied by Germany during World War II, and reverted to the USSR at its end.

By the late 1980s, the spirit of perestroika and glasnost took hold in the Baltic countries. Nationalist movements were appearing and demonstrations with a nationalist agenda aiming toward greater autonomy began to take place in 1987. Andrejs Plakans discussed persuasively the subtle ways in which demonstrations to commemorate events or protest environmentally undesirable policies shaded into expressions of sovereignty and independence.[6] Although the 1989 events of Eastern Europe, the realization that the Soviet Union was not going to intervene there, and the new spirit of glasnost certainly contributed to the growth of the independence movements, Lieven noted that to western observers and even to Gorbachev, these aspirations seemed to "spring from nowhere" (Lieven, pp. 219–20), but that view betrays a certain short-sightedness on the part of both western and Soviet observers who tended to ignore Baltic independence aspirations. By October 1988, the Baltic Popular Fronts held inaugural congresses in which the ideas of free speech and free assembly were endorsed. In November, the Estonian Supreme Soviet passed a constitutional amendment that gave Estonia a veto over Soviet laws that infringed on Estonian autonomy and adopted a declaration of sovereignty,[7] which provoked a sharp reaction from the Presidium of the Supreme Soviet. The Lithuanian popular front (Sajūdis) endorsed the Estonian stance, although events unfolded somewhat more deliberately in Latvia. In August 1989, in a massive demonstration, some two million people created a human chain reaching from Vilnius in Lithuania to Tallinn in Estonia in commemoration of the August 23, 1939 Molotov-Ribbentrop pact and in protest of the resulting annexation of the countries by the Soviet Union.[8] But when independence declarations were passed in the three countries, the Soviet attitude began to harden; after the Lithuanian declaration on March 11, 1990, Soviet armed forces entered Lithuania and the USSR shut off oil supplies to the country. Some of the Soviet reaction has been attributed to Gorbachev's inability to resist pressure from the army to crack down on Lithuania.[9] The years 1990 and 1991 were characterized by an alternation between negotiations and threats and violence; a severe military crackdown took place in January 1991, and the violence of the Soviet military claimed the lives Latvian and Lithuanian citizens. But the independence movement was irreversible, as it was in Eastern Europe, and on September 6, 1991, the Soviet Union was forced to recognize Estonia, Latvia, and Lithuania as independent countries. In the same month they joined the United Nations.[10]

One of the first political problems that each country had to face, to varying degrees, was the presence of large Russian, Ukranian, and Belorussian minorities. The minorities accounted for 34% of the population in Latvia, 30% in Estonia, but less than 10% in Lithuania.[11] The latter granted automatic citizenship rights to the non-Lithuanian speaking population satisfying certain residency requirements, but the other two imposed rather onerous conditions for achieving Estonian or Latvian citizenship. Latvia, for example, reserved automatic citizenship only for those who were citizens on June 17, 1940,[12] and it was only in the second half of the 1990s that Latvia abandoned the restrictive quotas on naturalization (Iwaskiw, p. xvii). The latest step in the ongoing saga of nationalities in Latvia is the passage of a new state language law, passed on August 22, 2000, which the government has claimed to be nondiscriminatory, but which western agencies (such as the Organization for Security and Cooperation in Europe) have not yet pronounced acceptable.[13]

The three countries faced the same economic challenges as the East European ones, but to an even greater extent because of their heavy dependence on the Soviet economy. Market reforms needed to be undertaken and the liberalization of prices, begun in 1990, created immediate and drastic inflation: between the end of 1989 and June 1992, consumer prices had increased 35.5-fold in Estonia, 19.2-fold in Latvia and 16.1-fold in Lithuania. Macroeconomic stabilization was an immediate and burning objective. All three countries introduced an interim currency in 1992 to replace the ruble (the rublis in Latvia, the talonas in Lithuania), which was a modest first step towards monetary reform. In June 1992, Estonia introduced a new currency, the kroon, pegged to the German mark; in October, Latvia introduced the lats; and in August 1993, Lithuania the litas, with a floating peg to the dollar. At the same time, the initial inflationary burst of raw material prices was tapering off and the behavior of consumer prices reflected the improvement in the macroeconomic environment. Macroeconic stabilization began to be effective, and by 1995, annual price inflation was 29% in Estonia, 23% in Latvia and 35% in Lithuania.[14]

Privatization was another immediate objective, and all three countries initiated privatization programs. The process of privatization witnessed some of the same techniques as had been employed in Eastern Europe, with a relatively successful voucher scheme in Lithuania, the introduction of a fairly successful privatization agency in Estonia reminiscent of the German Treuhandanstalt (Iwaskiw, pp. 51–55), and some spontaneous privatizations throughout the region, in which management acquired the assets of companies at low prices and then sold these for large profits. Estonia, in particular, introduced legislation against such corrupt practices as early as 1992 (Lieven, p. 340). Overall, Estonia appears to have made the fastest progress in most areas, which agrees with the assessment of Alfred Stepan: "In 1991, Estonia appeared to have some of the most favorable economic and political conditions for a successful transition to democracy of any of the former Soviet republics."[15] As a result, on March 30, 1998, Estonia alone among the Baltic countries was able to begin formal European Union accession negotiations.[16]

In spite of the modest improvement in the situation in all three countries, it was clear that vast western assistance was needed in a variety of areas that had suffered under the Soviet regime, such as economic restructuring, civil society, higher education, and many others. Governmental and international agencies and private organizations responded in significant numbers. In particular, early assistance was provided by the Council of Europe, PHARE and TEMPUS, Nordic governments, USAID, USIA, and the various Soros country foundations. Some of this was extremely successful, as in the case of Estonia, and by May 1994, USAID announced that its assistance to Estonia under the SEED Act would be phased out over the next three years.[17]

Exploration in the Baltic Countries

When the Mellon Foundation decided in 1994 to investigate the desirability of starting a program in the Baltic countries, it was obvious that sooner or later I would have to go there. I knew nobody in any of the three countries and, except for a handful of institutions that I was aware of (such as Tartu University in Estonia). I knew nothing about the academic landscape. I started by drawing on resources in the United States: as early as March 1994, I contacted Jonathan Fanton, then president of the New School for Social Research, and Frederick Starr, then president of Oberlin College, both of whom I believed to have numerous contacts in the Baltic region; indeed, they both provided me with invaluable pointers. By May, I was in touch with the United States–Baltic Foundation and by June, with the Association for the Advancement of Baltic Studies (AABS), which was to play a major role in my activities during the next half dozen years. This fortunate contact came about because someone—I no longer recall who—pointed out to me that it would be senseless to try to learn about the Baltic countries without speaking with Valters Nollendorfs, professor of German at the University of Wisconsin, Madison, and academic executive director of AABS. A lively correspondence ensued and Nollendorfs has helped me over the years in more ways than I can properly acknowledge. Soon, I also met the administrative executive director of AABS, Jānis Gaigulis, and he also made a significant impact on what was to evolve later.

I had decided that it would take me until the spring of 1995 to set up a two-week trip that was sufficiently well prepared so as to promise to be informative. In principle, the Foundation could have decided to undertake projects in the Baltic countries in all of the functional areas in which it had been active in Eastern Europe, and it seemed therefore important to contact people in libraries, in universities in general, in economics and management education, and in agriculture. I started a massive campaign of writing to university officials, librarians, and others in preparation for my visit. The letters that I wrote had to be delicately phrased because, unlike the situation in Eastern Europe where the Foundation had a commitment to start a program (of some kind) from the very beginning, my explorations were to determine whether there should be a program in the Baltics at all and, if yes, of what kind. I was very much aware of treading a narrow path between the Scylla of promising more than I could deliver and the Charybdis of being so cautious as to generate no interest on the part of my correspondents at all. In any event, between November 21 and December 21, 1994, I wrote to some 38 different persons about my forthcoming visit and what I hoped to learn from it.[18] I prepared myself for the trip by trying to inform myself as much as possible about the state of higher education in the Baltics and learned much from a perceptive letter by Gundar King in which he noted, among others, that

> The academic establishment is overstaffed, underpaid, and fearful of changes. It lacks internal and public support for purposeful and energetic actions. . . . From the Baltic perspective, it is almost impossible to suggest priorities. Almost everything can be included. . . . I see academic reform not so much as a grand program, but more as separate projects making a larger whole. . . . The libraries of the Latvian Academy of Sciences, Riga Technical University, and the University of Latvia are technically backward. They also lack traditions of service to the public, and they lack periodical resources.[19]

It gradually became clear that the Soviet legacy in the Baltic countries was more damage, and probably substantially more lasting damage, to higher education than was the case in Eastern Europe. Nollendorfs commented on the Soviet period's impact on higher education in several articles:

> Perhaps the greatest burden of Communist rule is the interruption of human and social development. . . . Some of the problems inherent in education systems of formerly Communist countries [are]
> - Compartmentalization and overspecialization.
> - Academic elitism, status consciousness, and lack of social responsibility.
> - Isolation from mainstream intellectual developments in the West.
> - Fragmentation of knowledge and lack of scholarly coherence.
> - Dislocation of scientific and scholarly priorities; emphasis on natural science and technology; lack of emphasis on the humanities.
> - Lack of analytical, critical, and comparative tools in the humanities.
> - Gaps in knowledge of Western languages.
> - Huge lacunae in library holdings in the humanities and social sciences.
> - Deficits in modern information and communication technologies.
> - Absence of any coherent new educational policy; lack of curriculum specialists.[20]

If only a few years ago I was enthusiastic about the possibilities of educational reform with some reasonable external assistance for assessment

> and follow-up, I had not reckoned with the grave internal obstacles facing educational reform, quite aside from the lack of receptivity among possible partners in the U.S.A. Whatever energy there was at the beginning of independence is becoming bogged down because of bureaucracy, inertia, inability to produce a coherent educational policy, and the absence of anything close to adequate financial support. . . . What are the dangers of underfunding education in Latvia? The result will be continued stagnation and decrease in quality. To wit: 1. The best and most capable teachers will leave. . . . 2. Teachers and researchers will find second and even third jobs to eke out an existence. . . . 3. Those who will not be able to find better positions elsewhere will persist as long as they can to avoid being retired at present substandard retirement pensions. . . . 4. Even when new positions open up, new teachers and researchers will not be attracted to positions which are not competitive. . . . 5. Attempts at academic reform and improvement will be ineffective at best, impossible at worst.[21]

The words of Nollendorfs and Gaigulis represented a severe indictment of the higher educational system and suggested that there were many potentially useful actions that Mellon could undertake.

A final comment on what has happened since Nollendorfs wrote those words. In July 2000, I asked him to reflect on those words, and his reflections were published recently. He noted with regret that "seven years later so little has changed for the better, at least as far as formulation and implementation of a national educational policy is concerned."[22] He blamed the lack of a political will, which results in "bland programmatic statements" but not in action, and endorsed an OECD recommendation to create a commission at the level of the Latvian president's office for achieving consensus for "sustaining attention to overall conception and strategy." Whether this will come to pass remains to be seen.

The schedule I had set up started on March 12, 1995, in Riga; on the 16th I went by car to Tallinn, and I was accompanied on most if not all my appointments by Nollendorfs, who by that time was the director of the Riga office of AABS; he had deep knowledge of the Baltic countries and I was fortunate to be able to rely on his advice on many occasions. He was also invaluable as a translator on the Latvian part of the trip. We returned to Riga on March 19, dropped off Nollendorfs, who was replaced by his deputy, Eriks Leitis, for the Lithuanian portion of the trip. I was in Vilnius and Kaunas from March 19 to 23, when I returned to Riga for two more days before returning to the US.[23]

Since that first trip to the Baltic countries, I have made many more and every one was filled with learning new things, not the least about the music of the region. I had long been familiar with the works of Eduard Tubin and Arvo Pärt; but on these trips I gradually learned about Estonian composers Veljo Tormis and Erkki-Sven Tüür and Latvian composers Jānis Kalniņš and Pēteris Vasks, and more, which relieved the rigors and eased the tensions of the work I was doing.

Assistance Programs. As in Eastern Europe, the liberation of the countries produced a rapid response by numerous western agencies. TEMPUS and PHARE were substantially involved in the region; for example, in the academic year 1992–93, out of a total of 635 TEMPUS projects funded, 30 were in the Baltic countries;[24] a not unreasonable percentage given the proportion of the population and of academic institutions in the Baltics relative to the whole of Eastern Europe. While some of the activities of TEMPUS were aimed at upgrading the facilities at institutions of higher learning,[25] most of its activities concerned curriculum development and teaching materials. Many other organizations supported small business development, trade promotion, management assistance, privatization, assistance for foreign investment, securities markets, and, in general, economic restructuring: PHARE provided a credit line of 3 million Ecu in 1994, AB Svensk Exportkredit lent $28 million, the Export/Import Bank made long-term loans and guarantees in the amount of $40 million, the Canadian Export Development Corporation provided $10 million (Canadian), and the World Bank and the European Bank for Reconstruction and Development proposed assistance in the amounts of $20 million and $15 million, respectively.

In the educational sphere, smaller scale activities were being carried out by the European Science Foundation, which aimed at promoting European cooperation in basic research and promoting the mobility of research workers; however, its 1992 budget was only 27.3 million French francs (Quandt, p. 10). Fairly substantial contributions had been made by the Nordic countries and some of these activities were being coordinated by the Nordic Council of Ministers, with Finland being primarily responsible for Estonia, Sweden for Latvia, and Norway for Lithuania. USAID and USIA concentrated their support on developing democracy and local governments, nongovernmental organizations, the energy sector, the environment, and from my point of view most importantly, on management training and economic development. USAID had been active in promoting the privatization of large industries in Latvia, but these activities were largely failures as of 1995.[26]

Finally, the local Soros Foundations had geared up rapidly. In Estonia, activities started in 1993, in which year the Open Estonia Fund spent $788,000 in the aggregate, of which $291,000 was spent on travel grants, $300,000 on culture, $122,000 on medicine, health, and the environment and smaller sums on educational programs and civil society developments. But the Estonian program increased rapidly, and in 1994 a total of $1,958,000 was spent, of which $877,000 went for educational reforms, $180,000 for media training, $339,000 on health, and smaller amounts on culture, subvention of publications, human rights, and legal reform. In Latvia, the Soros program began in 1992 with expenditures of $97,000; by 1993, spending had reached $747,000, of which $280,000 was spent on educational reforms and $120,000 on cultural projects. Most importantly, by 1994, the Soros Foundation–Latvia had started a library program, spending $118,000 on equipment, creating Internet connections, training librarians, and the annual maintenance fee for the VTLS automation system that the National Library of Latvia had purchased with a grant from the Swedish Crown.[27] In Lithuania, the Open Society Fund-Lithuania spent $1,720,000 in 1993, with the largest single item being for education ($452,000), including curriculum development, teacher training, textbook development for secondary schools, and assistance to the presecondary level, as well. As has often been the case with country Soros Foundations, the grants that were made were numerous and small in size, attacking problems on a broad front, but by small steps.[28]

The General State of Academic Institutions. There is a substantial similarity in the structure of educational institutions in the three Baltic countries. Each one has one general university focused on the arts and sciences (University of Latvia, University of Tartu, University of Vilnius), one or two technical universities, an agricultural academy, a medical academy, assorted academies of arts and music, and an academy of sciences.[29] During the Soviet period, the pattern was the same as in much of Eastern Europe: the academies of science were the favored research institutions. However, with independence, the situation changed drastically; many academy institutes were closed or drastically dismembered. For example, the Latvian Academy of Sciences had a staff of 4,604 on January 1, 1991 and this had shrunk by December 31, 1993 to a mere 723 persons (Quandt, p. 16). Academies of Science essentially became honorary institutions and the Latvian Academy of Sciences divested itself of its institutes.[30]

Communications and access to the Internet seemed better developed in the Baltic countries in 1995 than in the Visegrad countries five years earlier, but not nearly as good as in the Visegrad countries in 1995.[31] Some 22 libraries had internet access in 1994, but most of them only the dial-up variety. Often, access was limited to 15–20 minutes per session.[32] At the time of my visit, all the major libraries had Internet connections and their window to the outside world had a bandwidth of 184Kbit/sec. The best endowed institutions already had substantial LANS: the Estonian National Library had 47 and Tartu University 20 workstations connected to their respective LANS. Estonia had generally better connections than the other countries, but backbones were usually no better than 64Kbits/second—at the present time, three out of 30 Internet service providers in Latvia have links to the outside world with bandwidths of 3 Mbit/sec, 6Mbit/sec and 34Mbits/sec, respectively.[33] Computing capabilities in universities differed substantially, with technical universities generally much better endowed, but all tended to rely on IBM-compatible Soviet mainframes, with a sprinkling of IBM 4381 computers. PCs were predominantly of the 286 and 386-chip variety and RISC machines (Sun, RS/6000, etc.) were definitely rare.

Although much thought had been devoted to academic reforms,[34] real progress seemed to exist only in Estonia, and my own experience fully confirmed the pessimistic Nollendorfs quotes on the Latvian situation earlier in this chapter. At Tartu University, I met with Prorector Jaak Aaviksoo (who subsequently became minister of education and then rector of Tartu University). Tartu was shifting from a university structure and curriculum following the German model to the Anglo-American model, with bachelor's, master's, and doctoral degrees. It had started a school of social sciences, with expatriate Rein Taagepera as its dean. The startling change was that the teaching, which used to be organized around "chairs" and research around "institutes" was being reorganized by combining these entities into departments along the Anglo-American pattern. Tartu was also introducing a mandatory retirement age of 65 (all the better to get rid of deadwood capable of teaching only scientific socialism and similar thrilling subjects), temporarily abolished tenure, and most daringly, also introduced a system by which salaries could be set to reflect merit. Tartu was probably also leading in the quality of research. Unfortunately, comparable progress was not taking place at the University of Latvia. Much power was in the hands of an old guard that seemed to have little or no interest in serious academic reforms. I noted that "The rector is thought to have no stomach for the fight that would accompany serious reform efforts," (Quandt, p. 29). In general, Latvian higher educational reforms remained stalled by inertia and bureaucracy for a long time, and while not many professors left academia (as Nollendorfs had feared), they did indeed tend to take second and third jobs to keep the wolf from the door.[35] The faculty in Latvia was aging and there was no coherent policy of attracting younger faculty members. Instead of concentrating on younger cohorts, the government created "state professorships" for senior faculty, leaving a large academic proletariat at the bottom ranks, where take-home pay was less than Ls 100 per month (about $166 in July 2000).[36] The Latvian problem was exacerbated by the fact that the individual faculties exercised enormous power, even to the extent of setting their own tuition fees (it is estimated that more than 50% of students currently pay tuition), with the result that certain desirable fields such as economics became accessible only to the rich. Teaching and research have not improved substantially and there are persistent rumors of fiscal mismanagement and improprieties.[37]

A significant structural change is the spread of private educational institutions that charge tuition. Even Tartu University and the University of Latvia are charging tuition for some programs, although the top tier of students in Latvia do not pay tuition and receive a small stipend for living expenses. Government funding for higher education and teachers' salaries have also improved in Latvia and Estonia, although academic reform movements appear not to have had much government support in Lithuania.[38]

As other countries that had been in the Soviet orbit, the Baltic countries maintained competence, and at times excellence, in natural sciences and engineering, but fell hopelessly behind the West in the humanities and social sciences. In Latvia, a major report was prepared in 1992 by the Danish Research Councils, which evaluated the quality of research at all the major Latvian institutions in the natural sciences, engineering, and, in a limited number of humanistic fields.[39] The evaluation was carried out at the request of the Latvian Council of Science and involved the detailed evaluation of the research programs and the scholarly papers produced in the various institutions.[40] The report found a great deal of unevenness, but in the sciences the assigned grades tended to range from good to excellent. The picture was much less good in the social sciences and the humanities. History suffered from the split between historians at universities and at the Academy of Science, from the two-month summer closure of the archives, and from the fact that historians generally had inadequate command of English and German. Archeologists tended to concentrate too narrowly and their ability to apply the scientific method of archeology was poor. None of the work in philosophy seemed outstanding, and in languages and literatures much work dealt with dictionary compilations of the Latvian language, Latvian literature and folklore, but ignored comparative or critical approaches. The quality of research in economics was poor, knowledge of microtheory, macrotheory, and econometrics was largely absent, and economics was not understood to be an empirical social science. In all fairness, poor library collections must bear some of the blame for this state of affairs, but as Toivo Miljan pointed out, "The isolation of Baltic academics from modern developments in the social sciences was nearly absolute under Soviet rule."[41] But the stalling of reform in academic and research programs and in academic institutions has to be largely attributed to the lack of academic entrepreneurs and to the entrenched bureaucratic administrators with "fixed mindsets and deeply rooted procedures."[42]

Economics and Management Training. Since the Foundation had done so much in Eastern Europe to improve the condition of economics teaching and research (chapter 6) and management education (chapter 11), it was logical to examine whether the experience gained there might be transferable to the Baltic countries. But it became clear rapidly that by the time I arrived in the Baltics, there was no burning need to invest additional resources in this area. In Riga, the Riga Business School[43] had come into being as a part of the postgraduate division of the Riga Technical University, with substantial assistance from the State University of New York, at Buffalo, the University of Ottawa, and the University of Gothenburg, and with funding from USIA and the Soros Foundation.[44] Unfortunately, the funding tended to be short term (although enduring), which has made it difficult to develop a Latvian faculty for the program. While the curriculum may not have been the latest, it has graduated a cadre of Latvian managers over a period of ten years that makes a notable difference for Latvian economic restructuring and development.[45]

The Swedish government provided the initiative for the Stockholm School of Economics (SSE), a distinguished institution, to start an undergraduate program in Riga under the name of Stockholm School of Economics in Riga.[46] It was started in 1994 and is managed by SSE with the cooperation of the Latvian Ministry of Education and is funded by the Swedish government for ten years. Its objective is to provide education leading to a bachelor's degree in economics and business; most of its teaching staff is Swedish, but the language of instruction is English (as it is at the new Stockholm School of Law).[47]

Several efforts improved management education in Estonia. Bentley College has had cooperative programs with several institutions of higher education for reforming business and management education. Funded by USIA, Bentley has been working since 1990 to operate faculty and curriculum development programs with the Estonian Agricultural University, the Estonian Business School, the Estonian Management Institute, Tallinn Technical University, and Tartu University.[48] The Estonian Business School was started in 1988 as a private institution and grew from 84 students in 1989 to 1,830 in 1999; it offers bachelor's degrees in business administration, international business administration, public administration, information technology management, and a Ph.D. in business administration.[49]

Finally, perhaps the most ambitious effort is EuroFaculty, created on the initiative of then German Foreign Minister Hans Dietrich Genscher and the Danish Foreign Minister Uffe Elleman-Jensen in 1993. With financing from the European Union, its objectives are to reform higher education in economics, business administration, public administration, and law by maintaining sites at the three principal universities of the region, namely the University of Latvia, Vilnius University, and Tartu University. Teaching is both by local faculty and western faculty; in the first phase, the primary objective was to upgrade the Baltic teaching staffs, with undergraduate and postgraduate teaching of students relegated to a second phase.[50] EuroFaculty represents a "long-term commitment of the Baltic universities and the donor countries to work toward substantial and permanent reform," (Miljan, pp. 3–10). But the requalification of local faculty did not work out as hoped for and professional development was refocused on young academic staff and graduate students. Instruction is in English, and full-time teaching of economics began in Riga in January 1994, with the other fields coming online a bit more slowly. By the spring of 1995, 32 semester-long courses were being offered at the three universities by 15 faculty members from the institutions that supported EuroFaculty, three faculty seconded by the Civic Education Project and two local faculty members.[51]

After the first three years of operation, a distinguished panel was asked to evaluate the operations of EuroFaculty.[52] The evaluation team found that long-term planning was difficult because the funding of EuroFaculty was annual, that not all Baltic academic staff were capable of being retrained, that the physical environment for teaching was seriously deficient, that EuroFaculty's emphasis on research (by local staff) was premature and ill-advised, and that the organizational structure of EuroFaculty seemed unnecessarily cumbersome. But the impact has nevertheless been quite dramatic, particularly in economics, and the three Baltic universities agreed to make microeconomics, macroeconomics, and econometrics compulsory subjects for all students. Progress in law has been less spectacular because the local leadership has largely not accepted the need for curriculum reform, and efforts have been less effective in public administration because of the fragmentation of the field over a number of departments and institutes. This underscores the immense influence of entrenched bureaucracies and vested interests on the success or failure of attempts to transfer technology. The kind of technology that is being transferred by programs such as this is not like, say, installing a new telephone system, where the attitude of the user is irrelevant if it can be shown that new switching apparatus is faster, more reliable, etc.; everything in reforming academic institutions and curricula depends on the cooperation of people who cannot be dislodged in the near term. In spite of all the problems, however, EuroFaculty has done a great deal to transform curricula and the culture of teaching, to train and develop staff, and to develop libraries and computer networks.

The only other related area that seemed worth investigating was agriculture, and on my first trip I visited the agricultural universities and academies in each country. In Estonia, I could only meet with the dean of the School of Forestry, and I sadly concluded that he had some trouble seeing the forest for the trees. The Latvian Academy of Agriculture was housed in a very large, magnificent but decayed palace built by Rastrelli, the architect of the Winter Palace in Saint Petersburg, and the administrators seemed sharp and enthusiastic, and they had recently signed an agreement of cooperation with the College of Agriculture of Michigan State University. The administrators of the Lithuanian Agricultural Academy made the best impression on me by their presentation of the status quo and their aspirations, and had already entered into collaboration with Purdue University. They each had needs that could be met with money, but providing such help would have involved a lot of piecemeal activities without a central organizing principle.

The training efforts by western partners in the area of economics and management, as well as in agriculture, appeared to be very substantial, particularly given the sizes of the countries, and I concluded that there was no need for the Mellon Foundation to duplicate efforts that had been ongoing in the region for several years. While the agricultural academies clearly had great needs, there did not seem to exist a coherent motif around which Foundation activities could be organized.

Libraries. Attempts to revive the library sector began fairly soon after independence was achieved.[53] Each of the three countries has three truly major libraries: a national library, an academic library, and a university library, as ordained by the Soviet system for member republics, and a number of other significant libraries belonging to higher educational institutions.[54] The approximate holdings in 1995 of the three most important libraries in each country are shown in table 10-1:

Table 10-1 Collections of major Baltic libraries in millions of items.

National Library of Estonia	4.3	University of Latvia Library	2.1
Estonian Academic Library	3.3	National Library of Lithuania	5.2
Tartu University Library	5.0	Lithuanian Academy Library	3.8
National Library of Latvia	5.6	Vilnius University Library	5.0
Latvian Academic Library	3.1		

Note:
a. Quandt, Richard E., *Report on the Baltics*, New York: The Andrew W. Mellon Foundation, 1995, p. 48.

The figures in table 10-1 are somewhat misleading because Baltic libraries tended to have many more duplicates than western libraries. Under the Soviet system, the national libraries (known as state libraries before independence) had a right to an automatic deposit of every book published within a republic, and the acquisition of Soviet periodicals was also straightforward. With independence, this system collapsed, and while new mandatory deposit laws were enacted, they were not being scrupulously observed by publishers. Foreign acquisitions were handled separately under the Soviet system and only a limited number of libraries had the right to handle their own foreign acquisitions (the nine libraries in Table 10-1 did have that right). Acquisitions budgets exhibited a large range of variation: Tallinn Technical University had an acquisitions budget of about $300,000 in 1995, but many other universities and academies had one of less than $5,000. Some material was obtained from western donations (see chapter 7), and much by exchanges; but in the Academy of Sciences Library in Vilnius I saw a pile of more than 1,000 wrapped packages of books and serials ready for mailing that the library wanted to send to its partner institutions in the West as part of an exchange program, but could not do so for lack of money to pay for the postage. Cataloging in the Soviet period was handled not by the libraries but by so-called book chambers (although libraries also maintained cataloging departments). Fortunately, this aberration was corrected after independence was achieved and the chambers were merged into the national libraries.

In each country, with the help of AABS, I managed to meet the library directors of the most important libraries at a single plenary meeting, at which I outlined the Mellon Foundation's activities in the realm of library automation and stressed the importance of unified, consortial approaches. These remarks seemed welcome in Estonia and Lithuania, but I was not at all sure that a consortial approach was going to be feasible in Latvia.

Estonian libraries were ready to cooperate and the seven largest research libraries had already committed themselves to a consortial approach. The National Library had examined a number of systems and after rejecting ALISE (see discussion below on automation in Latvia), VTLS, GEAC, Dynix, and TINLIB, had tentatively settled on the Finnish product KIRI. KIRI seemed designed for smaller libraries and the hope was that using relatively inexpensive Estonian labor, it could be extended to handle larger libraries. But in the end, the automation group at the National Library decided that KIRI was not the way to go. Tartu University had developed portions of an integrated system called INGRID, but was quite ready to abandon it if something better appeared on the scene. Lithuania, in turn, seemed not very far along in automation (although Vilnius University and the National Library already had fairly sizeable LANs), but my sense was that the librarians were ready for a common approach using identical hardware and software and have unified their library protocols by agreeing on UNIMARC, ISBD, and Lithuanian cataloging rules.

The situation was quite different in Latvia. The "mass meeting" of librarians that Nollendorfs had organized was attended by the directors of the National Library, Academy Library, University of Latvia Library, the Riga Technical University Library, the University of Agriculture Library, the Medical University Library, and the Patent and Technology Library. As noted earlier, the National Library had already acquired through the generosity of the Swedish Crown some of the modules of VTLS (although it used ALISE for maintaining the Latvian national bibliography—see below).[55] Its director, Andris Vilks, was an outstanding librarian whom I had met previously at a conference in Washington, D.C., and he seemed quite happy with VTLS. The Academy Library had acquired the French system LIBER on its own, and its director, Edvīns Karnītis, seemed reluctant to accept a common solution if that meant abandoning LIBER. Most of the other libraries were already using the home-grown Latvian software, ALISE.[56] I had asked the librarians ahead of time to prepare a position paper for me on the state of Latvian libraries, and they did so; the document made it quite clear that they intended to stick with the systems that they had already selected. It also became clear to me that a significant rift existed between the National Library and the Academy Library, and that Vilks and Karnītis cordially disliked each other.

My initial impression of LIBER and ALISE were not particularly favorable. The LIBER messages that I saw on a computer screen contained numerous misspellings, and while Eric A. Johnson had a relatively favorable view of ALISE (Johnson), searching in it seemed extremely slow. The future of the ALISE organization seemed uncertain because of its extremely small size. In any event, I made it clear that the Foundation would not undertake a project guaranteed to perpetuate the use of three different systems. I tried to explain the immense advantages of common solutions, but my arguments were clearly not persuasive, at least not at that time. They all felt that resources were better spent on improving the interfaces between the systems than investing in a new one. The Latvian librarians also had a more ambitious vision of a Baltic consortium that would unite the libraries and OPACs of all three countries; but I demurred and suggested that it might be premature to think on such a large scale, since the problems of implementation would require heroic efforts.

My sense was that library automation was a worthwhile objective in these countries and suggested that the librarians of each country work out a consortial plan for automating the principal research libraries, using identical hardware and software, and further proposed a time-table for proposal writing that would put finished proposals in my hands by the summer of 1996. I indicated that the Mellon Foundation would first have to formally decide to start such a program, and I anticipated that it might take until June 1995 to receive approval for it; thus, they would have about a year to organize themselves and prepare proposals from the time that I could tell them that we were serious about supporting the Baltic libraries. In fact, by the summer of 1995, the Mellon Foundation had decided in favor of library programs in the Baltic countries, and I so informed the national librarians.

Early in 1996, I received an invitation to present a paper at a conference organized by NORDINFO, to be held April 25–26, 1996, in Tallinn, on national information policy and cooperation issues within the library sector, relevant government agencies, and the telecommunications sector. I accepted the invitation with alacrity, if for no other reason that it would give me an excellent opportunity to find out what the Baltic librarians had been doing about consortia and library automation proposals. I made a point of setting up meetings with the relevant people and found satisfying progress on many fronts. Mihkel Reial and Asko Tamme (deputy directors of the National Library and the Academy Library, respectively) reported that the Estonian libraries had formed a consortium in September and it was, in effect, functioning, although its legal registration was still pending. The consortium had sent out some 30 requests for proposals to software vendors, received nine replies, and had narrowed the choice down to two: INNOPAC and Aleph. This seemed to me excellent progress on all counts. Vladas Bulavas, the Lithuanian National Librarian, reported in turn that a Lithuanian library consortium had been created, consisting so far of the academic libraries, but that it intended eventually to draw in public libraries, as well. The Lithuanians had narrowed the software choice down to four vendors: Aleph, Dantek, a Lithuanian system, and one other. VTLS was not among the finalists because Lithuania chose UNIMARC as its standard and VTLS support for that did not seem unambiguous. My meeting with the Latvian national librarian, Andris Vilks, revealed that the Latvians had also done a great deal of useful technical work in the past year, but it was not at all clear that any progress had been made in reducing the rivalry and healing the rift between the National Library and the Academy Library.

Many useful things were said at the conference itself. Victor Montviloff of UNESCO bemoaned the absence of a political will (in unnamed Baltic and other countries) towards creating an information policy and infrastructure, thus echoing remarks by Nollendorfs concerning the educational establishment in Latvia, and Bulavas ably summarized the automation efforts to date in the Baltic countries. In my remarks, I stressed that (1) automation should be consortium-based; (2) members of consortia must use identical software; (3) adequate intercity networks must be built and since these are public goods, it is the governments' responsibility to fund them; (4) implementation teams must be created with the power to allocate resources and adjudicate disagreements, and methods of conflict resolution and of ensuring transparent accountability must be agreed on; (5) uniform library standards must be agreed on; (6) consortium members must internalize the view that nobody can have his or her way in all things; (7) good sense must prevail in choosing systems and the trade-offs between functionality and cost must be kept in mind; (8) Baltic librarians must exert pressure on their governments to participate vigorously in funding libraries; and (9) they must exert themselves to mobilize Western European foundations to assist Baltic library efforts. Point (8) was aimed especially at the government representatives present, among whom there were the advisor to the prime minister of Latvia, the advisor to the president of Lithuania, the Latvian minister of culture, the chairman of the Estonian parliamentary commission for culture, and many others. At the end of my remarks I had an inspiration and announced that the Mellon Foundation would only make matching grants in the Baltic countries, requiring a 1:1 match by the government of the country in question. This visibly depressed some people, for example, the Estonian librarians present, who thought that the government would never provide the required match. But Helena Demakova, the Latvian prime minister's advisor, was much more upbeat and cornered me after the session and asked me to explain how matching grants would work. I explained and she thought that this would be easy: she would simply ask the prime minister to put a $750,000 appropriation into next year's budget for satisfying the match. As it turns out, the Latvian government's contribution to the eventual Latvian project turned out to be even more generous.

In the early summer of 1996, I received fully worked-out automation proposals from each of the three groups of libraries. The Estonian and Latvian proposals asked for a sturdy Volvo and were funded in due course. Alas, the Lithuanian proposal, competent though it was technically, asked for a Rolls Royce, and with considerable regret, the Foundation declined to fund it.

The Estonian Library Network (ELNET)

ELNET was created to coordinate and manage the automation of the principal Estonian research libraries. It was the logical successor of an earlier project undertaken by the Estonian Library Association, but in spite of a generous $210,000 grant from the Soros foundation, that project did not come to completion for lack of funds. However, the Soros infusion was important because it permitted the installation of basic networks.[57] The charter members of ELNET were the National Library, the Library of the Academy of Sciences, Tartu University Library, the Estonian Literary Museum and Archival Library,[58] the Tallinn Technical University Library, the Tallinn Pedagogical University Library, and the Agricultural University Library. These seven libraries contained about 80% of all the collections and accounted for about the same fraction of library patrons in Estonia; they were also the ones with the most experience in automation. The purpose of automation was seen to be the creation of a user-friendly information and communication system, the automation of routine library functions using identical software, joint collection management, the production of statistical and financial information and access to databases via the Internet. USMARC was to be the accepted standard, and the proposal not only discussed articulately the ways in which automation would improve services for users, but explicitly anticipated changes in the work patterns within the libraries, thus echoing the reorganizations that automation brought about in the East European countries. In addition, and equally importantly, the proposal came to grips with the issues of management and indicated that the statutes of ELNET specified each member institution's responsibility to provide human and financial resources to management.

The short list of software products consisted of Aleph, INNOPAC, VTLS, OLIB, Dynix, and GEAC Plus. A careful evaluation of system functionalities and reliability, the vendor, and the price led to the selection of INNOPAC, the same system that the National Library of Poland had selected. The hardware solution turned out to be particularly economical: only two servers were needed, one each in Tallinn and Tartu. The total cost of the system was projected to be $1,460,800, of which software accounted for $500,000, servers for $80,000, PCs and other hardware for $804,100, and LAN improvements for $76,700. The proposal made it clear that $600,000 would be obtained from the state budgets for 1997–98 and $280,800 from the state budget for 1999 and from other foundations, thus leaving a sum of $580,000 to be provided by the Mellon Foundation. The government match was more than adequate by my requirements, and I recommended and the Foundation awarded a grant of $580,000 in the fall of 1996.

Watching the development of ELNET in the subsequent years was sheer pleasure. I visited Estonia roughly one year after the award of the grant (in early October 1997) and found out that the hardware had been installed, a contract with III had been signed (for a new price of $390,000, instead of the originally quoted $500,000!), that they were busily at work at adapting themselves to USMARC, and that they were awaiting the arrival of the TestPac module. Most importantly, although they had never had much use for authority files before, the study of the capabilities of INNOPAC revealed to them how enormously useful authority files would be for the libraries.[59] I also visited Tartu, where I met the librarian, Peeter Olesk, who thought that automating with INNOPAC was wonderful, but that the best result of the Mellon initiative was that the libraries started to cooperate. In Tartu, I also met prorector Toivo Maimets, who did not have much use for libraries (since, *horribile dictu*, they used up as much as 2% of the university's budget!), was unsure of his facts, and seemed to be united with Olesk in a bond of mutual dislike.

Next day I met with the ELNET Board, consisting of Mihkel Reial, the chairman of the board; Toomas Liivamägi, deputy director and later director of the Tartu University library; Jüri Järs; Ann Indla; Margit Tersalu; Kersti Kikkas; and the new executive director, Asko Tamme. I learned that the preparation of the request for proposals started with each library determining its own requirements, and the board brought in specialized teams as needed. The structure was loose and nonhierarchical, and they just talked about the issues until a majority was willing to make a decision. It struck me as quite extraordinary how amiably they resolved differences and how much they were *not* bothered by one or the other of them not getting his or her way or not getting precisely the same fraction of resources as some other library—a striking contrast to the case of KOLIN (see chapter 8). I concluded that this process could not have worked without a great deal of mutual trust.

By April 1998, Reial was able to report substantial progress in training library staff and converting enough of the database to the INNOPAC format to permit serious testing of the system. Preparations had started for converting the entire card catalog. The working group established for this purpose, headed by Ülle Soosalu, determined that the Estonian catalog had been compiled over the years obeying different and frequently changing rules, so that no automatic conversion mechanism seemed feasible and they would have to resort to manual conversion.[60] In the same letter he asked whether the Foundation would entertain a follow-up grant to assist with the conversion, and soon after I replied in the affirmative, I received a proposal requesting $165,000 for converting the Estonian national bibliography. The Foundation had also just approved a grant of $75,000 to the University of Pennsylvania for intensive advanced training of ELNET librarians, and this additional request seemed quite in keeping with what we were trying to accomplish and was also approved. Total grants for the benefit of ELNET (not counting a minor grant for attending a conference) amounted to $820,000. In the meantime, the Estonian government had fulfilled its promise to provide a match for the original grant and I returned to Estonia in October 1998 for the official "opening" of the system (although the practical opening was to take place on January 2, 1999). ELNET management was thoroughly satisfied with the performance of Innovative Interfaces, Inc., which was attributed to the fact that III fixed problems exclusively remotely from California; a practice that differs from the Ex Libris practice, which frequently involves a local intermediary firm. This is quite likely the reason that I heard fewer complaints about III than about Ex Libris.[61] Another achievement of ELNET consisted of the fact that it had instituted a modest membership fee, which was enough to pay the salary of the executive director and the office expenses of ELNET: in 1998, they had collected 156,000 kroon (about $12,000). Progress had generally been so satisfying that two new institutions declared their desire to join, namely the Academy of Arts and the Academy of Music. Not all of the system was to go online at the same time: the Tallinn Technical University was going to defer starting the circulation module and Tartu the acquisitions module for various good local reasons. By March 2000, when I visited Estonia again, most of the system was operational in all the libraries, and the expectation was that by the end of the year everything would be in full operation. The remarkable thing was that the librarians were aware of a number of "nonfunctionalities" in INNOPAC, but were willing to either wait and see how big a problem this would become or actually produce fixes.[62]

In all of these matters, the librarians exhibited resourcefulness, patience, flexibility, and an unfailing spirit of cooperation. These characteristics, together with the sensible, relaxed, and nonhierarchical management style of Tamme and the board were the keys to the Estonian success.

The Latvian Library Information Network Consortium (LINC)

The Start. In contrast to the evolution of the Estonian consortium, the Latvian development was rife with conflict, the staking out of territory and with posturing for gaining imagined or real advantages. By the end of December 1995, I had received letters or e-mails from Gunārs Mangulis (University of Latvia Library) extolling the virtues of ALISE, from Andris Vilks explaining the unsuitability of ALISE, and from Edvīns Karnītis extolling the virtues of LIBER and pointing out the high cost of VTLS. The last e-mail from Karnītis was particularly amusing, because he calculated the cost of VTLS (hardware and software) for a single large library and essentially assumed that in an *n*-library consortium this figure would have to be multiplied by *n*.[63]

The idea of consortial automation was not entirely new in Latvia. As early as September 1994, a working paper on a potential Latvian library information network was made public by the Latvian Development Agency.[64] But when I met with the librarians in Riga in March 1995, I did not get a sense of either enthusiasm or unity; in fact, most librarians at that meeting were extremely passive and I recall only Vilks and Karnītis saying more than just a few words or asking a question. But after my departure, they clearly settled down to work—although my guess is that the bulk of the work was done by the staff of the National Library. I received a rather upbeat e-mail from Vilks on June 15, 1996, in which he said that the "project" (i.e., the proposal) would be mailed to me two or three days later and informed me that the project had the full support of the minister of education and science, Māris Grinblats, and that after a meeting with the librarians, the minister of the economy, Guntars Krasts, would secure the necessary cofunding by including the project in the State Investment Program for 1997. He also reported that the prime minister, Andris Šķēle, would submit this to the Latvian cabinet for its approval, which was necessary in the light of the fact that the Mellon Foundation had asked for guaranteed government cofunding.[65] I was entirely reassured when four days later I had an e-mail from Helena Demakova, the prime minister's advisor, who confirmed all these details.

The proposal of the Latvian Library Consortium was transmitted by Vilks on June 17. The participants were the National Library of Latvia, the Latvian Academic Library, the University of Latvia Library, the Riga Technical University Library, the Patent and Technology Library, the Fundamental Library of the Latvian University of Agriculture, the Medical Academy Library, and the Research Medical Library of Latvia. I was gratified that the eight largest and most important Latvian libraries had all agreed to use the same software and that the fundamental reasoning for undertaking such an automation project was on the mark.[66] Unlike the Estonian case, the final software decision had not yet been made, but the criteria that the software had to satisfy were reasonable[67] and the short list included VTLS, Aleph, and INNOPAC. Many of the libraries already had acceptable LANs and the plans for connecting these to various FDDI rings in Riga were promising. While hardware evaluations had not progressed as far as the software evaluations, it was clear that the consortium would need no more than four servers.

The total cost of the plan was in the range of $2.25–2.84 million (depending on the software and hardware ultimately selected); the consortium expected a government contribution of $1,284,000 and a Soros Foundation contribution of $129,000; thus a Mellon contribution of $840,000 would permit realization of the less expensive plans. Overall, my sense was that the proposal was even better technically or more carefully executed than the Estonian one, that the benefits of the project were immense and that my fears of disunity among the Latvians that I had observed earlier were, at this stage, groundless. Accordingly, I recommended and in the fall of 1996, the Foundation gave a grant of $840,000 to the National Library for the benefit of the consortium. It subsequently became clear that in my haste I had neglected to make certain that the proposal's assurances of unanimity were more than hyperbole.

The Crisis. After that, the project seemed to develop as expected. On February 21, 1997, the director of the department of informatics of the Ministry of Transport faxed Mellon Foundation President Bowen to the effect that the first tranche of Ls 180,000 (about $320,000) of the government's contribution to the project had been included in the Public Investment Program for 1997.[68] In the same fax he inquired whether the Foundation had any special requirements or conditions for the software and hardware tenders that were going to be sent to vendors, and I replied the same day with a list of reasonable requirements. The National Library was the lead institution in the consortium and Andris Vilks was managing the project. I received the first inkling of problems on February 28, when Karnītis inquired whether the Mellon grant could be used to upgrade the Academy Library's version of LIBER in the event that the other libraries chose a different system. I heard nothing more until I grew impatient and, on April 17, sent an e-mail to Vilks inquiring about progress. My timing could not have been better. On April 24, I received a reply from Vilks in which he recounted the following developments: (1) On April 10, he received instructions from the new minister of education, J. Celmiņš, to the effect that the government had appointed a supervisory council, that Vilks had been dismissed as project manager and Karnītis appointed in his place, and that a new bank account would be set up to hold the pooled Mellon and government moneys; (2) On April 17, Vilks received a letter from Karnītis with a request that the Mellon grant be transferred from the National Library's bank account to the bank account of the Academic Library; (3) At two meetings of the Tender Commission (on April 11 and April 18), the Mellon grant was divided (on paper) in certain proportions among the eight libraries based on the assumption that the libraries would acquire either ALISE or LIBER; (4) The Tender Commission also decided that (a) each library was free to choose any software it wanted, (b) each library would have its own server, (c) there would be no union catalog, but some browser at the Academic Library would permit limited union catalog functions, (d) each library could use any search method it wanted, (e) there would be no provision for retrospective conversion, (f) the Mellon grant could be used only for purchasing workstations and network components, (g) there would be no subproject to be financed by Soros or PHARE moneys, (h) government moneys could be used only for the costs of managing the project.

This was extremely upsetting, since it completely subverted the understandings we had concerning how the project would be implemented. Because I had been away, I did not actually receive Vilks' e-mail until four days later, but replied to him via e-mail as soon as I could:

> The information that you have provided in your e-mail which I just received (I was away for about 4 days) is EXTREMELY UPSETTING AND DISTURBING. Certain items are explicitly contrary to the promises made in the proposal and therefore are contrary to the conditions under which the Mellon Foundation made the grant. Among these are
> 1. Each library can choose its own software
> 2. Servers will be bought for each library
> 3. No union catalogue
> 4. No common search algorithms
> 5. Monetary allocations based on the cost of ALISE or LIBER
> 6. Mellon moneys to be used for LANS and WANS
> 7. No investment by the State except for management
>
> Under these circumstances:
> 1. You are instructed NOT TO TRANSFER ANY PART OF THE MELLON GRANT TO ANY OTHER LIBRARY OR AGENCY WITHOUT OUR EXPRESS PERMISSION;
> 2. You are instructed NOT TO SPEND ANY MONEY FROM THE MELLON GRANT FOR ANY PURPOSE WHATSOEVER UNTIL

SUCH TIME THAT THE FOUNDATION IS SATISFIED THAT ITS PURPOSES WILL BE ACHIEVED; thus no moneys may be spent or obligated to be spent on hardware, software, personnel, networks, or anything whatsoever;

3. You are instructed to request Mr. Celmiņš and Dr. Karnītis to explain to the Foundation their understanding of what has happened and how this may be consistent with the Latvian Proposal on the basis of which the grant was originally made;
4. It is the intention of the Foundation that there be a total freeze on the grant until further clarifications have satisfied the Foundation that the purposes of the grant will be achieved.
5. Please note further that the grant was made to the National Library of Latvia, and under the rules of the Trustees and under U.S. Law may not be transferred to any other agency without the express permission of the Foundation.[69]

I immediately informed Gaigulis and Nollendorfs of these developments and a frantic ten-day period of planning began. Between April 28 and May 9, Gaigulis, Nollendorfs, King, I and various officials in Latvia and at the Latvian Embassy in Washington, D.C. exchanged some 40 e-mail messages. As best as one could reconstruct, what had happened was that some of the ALISE-using librarians got cold feet, whereupon Karnītis presented an alternative automation plan, LatLibNet,[70] and by pretending that it was just a minor variant of the proposal submitted to the Mellon Foundation, had it accepted by the minister of education as *the* plan. Gaigulis wrote to me that the latter's dismissal of Vilks was clearly arbitrary and lacked the endorsement of the minister of culture and violated the explicit procedures specified for the Mellon project by Prime Minister Šķēle. Gaiguilis further noted in the same mail that

> Apparently an internal struggle is going on in the highest circles of the Latvian government and warrants close watching and careful evaluation. At stake is the implementation of the Mellon project in strict observance of the conditions that have been stipulated. It is very strange that the leader of the opposition (Dr. Karnītis) should suddenly have become the Project Director, on the strength of an improperly issued order.
>
> All of this points to the need for proceeding with great caution when it comes to further implementation of the Project, once this current struggle is resolved. WHAT IF there should be a new Prime Minister who could overrule earlier decisions? A new Project Director could be appointed at will, with new plans—God knows in what direction.[71]

Gaigulis, Nollendorfs, and King thoroughly agreed with my decision to freeze the funds, and Nollendorfs, who resided in Riga, made himself extremely busy tracking down information and conveying the Mellon Foundation's attitude to the various principals. By April 29, it seemed clear that I would have to go to Riga in the very near future, although Gaigulis and King thought that I should agree to go there only if I could secure an appointment to speak personally with Prime Minister Šķēle. Nollendorfs was in touch with the prime minister's office and, with the help of Gaigulis, I enlisted the assistance of the Latvian ambassador in Washington, D.C., Ojars Kalniņš, to put in a good word to Šķēle to see me. It was not clear at this early stage what I could accomplish, but in the light of Gundar King's words, I needed to convey some hard opinions to the prime minister and others; King said, among others,

> Clearly, several symptoms are apparent: Low trust within and of the Latvian government, improper use of positions of power, unwillingness to stick to agreements and contracts that are not personally or institutionally convenient, temptations to easily distribute what may be considered windfall funds, deep provincialism, and clearly a misunderstanding of what the Mellon Foundation (or for that matter, the Association for the Advancement of Baltic Studies, IREX, Soros Foundation, sponsors of EuroFaculty and other European ventures in higher education, USIA, and USAID) want to accomplish and on what terms.
>
> Because the workings of a competitive and yet cooperative environment in the West is not really understood, the damage done to the reputation of Latvia becomes a major issue. (Given the practices of secrecy in Latvia, I am sure that Latvian authorities, on any level, have no conception of the rather unrestricted discussions and consultations western organizations and their representatives have among each other.)[72]

Nollendorfs met with Celmiņš on April 30, and Celmiņš assured him that the "Mellon conditions" for the project would be adhered to; that only those portions of the LatLibNet plan would be adopted that did not conflict with the Mellon Project; that a board of trustees would be set up for the project, and that under the circumstances neither Vilks nor Karnītis would be considered eligible to become project manager. According to Nollendorfs, the implication of this was that we were making progress and I agreed, but we also agreed that it was still necessary for me to make an appearance in Latvia. On May 5, I faxed a letter to Prime Minister Šķēle in which I said,

> It has come to my attention that the standards and conditions that the Foundation has imposed on the technical implementation of the project are in serious danger of being compromised. For this reason, I respectfully request that you grant me the opportunity of meeting with you on any day next week. . . . in order for me to acquaint you with the issues and possibly to suggest remedies. I shall be happy to provide a brief (5-page) background paper on the Foundation and on the basic issues of library automation; I am sure that Dr. Andris Vilks will be able to provide a copy of the original proposal by the Consortium which formed the basis of the Foundation's grant.[73]

Nollendorfs was working on a schedule for me and by Wednesday, May 14, when I arrived in Riga, I had appointments with Rihards Piks, the minister of culture; Juris Celmiņš, the minister of education; and Andris Priekulis, the deputy minister; Andris Virtmanis, director of the department of informatics in the Ministry of Transport; Ugis Berziņš, advisor to the prime minister; of course, the prime minister, himself; and, in addition, with my *bête noire*, Karnītis himself. While I had no idea how my interview with the prime minister was going to play itself out, I had fairly strong expectations that my meeting with Karnītis was not going to be unadulterated pleasure. It had been variously reported to me that his principal arguments supporting his position and plan were (1) that he never approved the proposal that was submitted by the National Library; (2) that traditional librarianship was *passé* and only electronic libraries would matter from now on; (3) that my arguments had to be resisted because I was merely trying to "sell American software."[74]

One of my early and not very memorable meetings was with Virtmanis. He hoped that the two competing proposals could be merged, and I tried hard to disabuse him of that idea. But in any event he agreed that the search for appropriate software would not be restricted to ALISE. In the afternoon, I met at the offices of AABS with Karnītis and three other librarians, at least two of whom were very much in his camp: Agnese Buholte of the Patent and Technology Library, Ināra Lejniece of the Library of the Academy of Medicine and Raisa Kazankina of the Research Medical Library. Nollendorfs was present and, once again, did yeoman's work translating, but only for the benefit of Buholte, Lejniece, and Kazankina, because Karnītis is completely fluent in English.

I started out by telling them that the introduction of a rival plan was an outrageous betrayal of trust, that I had not been informed during the past six months of the existence of the LatLibNet proposal, which had still not been translated into English, and that under the circumstances it would not be possible for us to work with them. Karnītis said in his reply that none of the librarians had seen the project as a whole, only small pieces of it, and they nevertheless had to agree to Vilks' conditions, because it would not have been possible to continue otherwise. He described Vilks as a dictator and denied the need for a consortium as a legal entity because there already existed an Association of Academic Libraries with 15 members. He rejected an approach that would employ four servers and said that such distributed systems were no longer necessary. He also said that acquisitions were of no further importance in Latvia, because the availability of electronic materials made the acquisitions of hardcopy materials unnecessary. He insisted that libraries should be able to use whatever software they wanted (as long as they were "compatible") and ended by saying that his suggestions had been accepted by the council of ministers; in other words, checkmate for Mellon.[75]

I did not mince words in my reply. I told him that the Latvian libraries and he, in particular, broke faith with the Foundation and breached an agreement we had, since a grant is made in response to a proposal that is in effect a contract between the grantee and the Foundation. I pointed out that nearly every factual assertion he had made was wrong and that unless the derailed project were restored to its original condition, the Foundation would altogether terminate the grant. After an increasingly heated discussion, in which I at one point pounded the table with my fist, he drew himself up and said, "You should know that I was decorated by the Soviet Union for my contributions to information technology," to which I replied "I wouldn't brag about that," and curtly terminated the meeting.

In reflecting on this meeting and the fact that seven months had gone by since the project was funded without any progress, I realized that Karnītis was an information technology expert, but not a library expert. I concluded that he did not understand the role of librarianship in a modern library. I also concluded that there was such intense and mutual dislike between Vilks and Karnītis that the project might never succeed, and would certainly not succeed if either of them had a significant role in it. Finally, I did not believe that the library community in Latvia had understood yet that horizontal cooperation and the sharing of tasks is an essential ingredient for success. The librarians still seemed to think in the Soviet mold of vertical chains of command—somebody gives an order, and the thing gets done. For example, in a conversation with Minister Piks (a person entirely sympathetic towards Vilks and the National Library) I asked him what role would be given to university rectors on the supervisory board that would be created for the library consortium, and he replied "none;" in other words, the institutions that were to implement automation would have no say in it. That is not the way to world works any longer, and I believe it is one (of the many) reasons why the Soviet Union faded away, and in the following year I spent much time and effort trying to persuade Latvian academics, librarians, and government officials of the truth of my position.

That evening I was scheduled to go to a performance of Brahms' *Ein Deutsches Requiem* with Valters Nollendorfs and Dagnija Staško and I thought the plot was definitely thickening when earlier in the day, Vita Matiss, former executive director of the Soros Fund–Latvia, told me that the prime minister's advisor, Helena Demakova, wanted to meet me after the concert. She would also be attending it, but it would not do to be seen leaving the concert with her, because she would then be perceived as being partisan in the dispute, and so we would be driven separately to a restaurant for a late supper. She definitely favored the original Mellon approach to automation and mostly wanted to caution me not to condemn Karnītis too harshly when I talked with the prime minister next day, because he rather liked him. That was useful advice. When I met with Šķēle next day, he started out by asking me to review the events and problems, which I did, explaining as clearly as I could the difference between the "Mellon plan" and the "Karnītis plan," and expressed my annoyance at never having been given the opportunity to comment on the rival LatLibNet proposal. I told him that while ALISE represented a considerable achievement on the part of Latvian programmers, it was simply not a good business decision to use software that had not been vetted in scores or hundreds of international applications and to allow the consortium to depend on a very small operation with hardly any depth.[76] I ended up by saying that it was my considered opinion that the Latvians broke faith with us and that I did not see an alternative but to cancel the grant altogether. He first asked me whether I agreed with him that neither Vilks nor Karnītis should be project manager, and I said that I did. He then indicated that he himself was not sure how events had reached the point that they did, but he would like me to defer the cancellation of the grant until the middle of July, so as to give him an opportunity for an in-depth investigation. I said that since we had already lost seven months, a couple of extra months would make no difference and agreed to wait until mid-July. He promised that I would receive a letter from him indicating the outcome of his investigations.

The morning of my departure, I met with Minister Celmiņš, his Deputy Minister Priekulis and Undersecretary Vocišs; Celmiņš actually apologized for the "misunderstandings," which gave me an opportunity to tell him about zero-sum games versus cooperative games in which everybody may end up winning.

The Denouement. I returned to the United States on May 17 with the feeling that I succeeded in presenting clearly the Mellon Foundation's point of view, that we probably could not have done much more, but without any conviction that this project would revive. I immediately realized that I needed to do two things right away. First, the Mellon Foundation had scheduled for October 16–18, 1997, a major conference in Warsaw on library automation in Eastern Europe, and both Karnītis and Vilks had been invited to attend. I definitely did not want Karnītis there, and I also thought that the symbolic nature of "punishment" would do him some good; but under the circumstances it would have done more harm than good to disinvite only him and not Vilks; accordingly, I withdrew both their invitations. The second thing that I needed to do emerged from my fear that the Karnītis faction would somehow be able to generate government pressure on Vilks to yield up the $840,000 that was being held in a National Library bank account. Accordingly, on May 22, the financial vice president of the Foundation, T. Dennis Sullivan, wrote to Andris Vilks,

> As a result of recent developments regarding the Foundation's grant of $840,000 to the National Library of Latvia for use by the Latvian Library Consortium, the Foundation requires the immediate repatriation of the full amount of the grant. Please note that this action *does not constitute a cancellation of the grant.* The Foundation awaits the outcome of the review of this matter which was agreed upon by Prime Minister Šķēle and Dr. Richard Quandt. This review will be completed by July 15, 1997. If the outstanding issues are resolved successfully, the funds will be returned forthwith.[77]

Shortly thereafter, Vilks, an honorable man, returned the funds.

In the meantime, there was not much to do but wait. Nollendorfs reported that the board of the consortium had met and that there were enough people on it who wanted to salvage the project. The LatLibNet proposal was being translated into English for my benefit, and there was a possibility that a director of the consortium would be appointed in July.[78] The LatLibNet proposal was duly sent to me by Deputy Minister Priekulis, and on June 27 I sent him a devastating critique of it. Omitting the more technical discussions, I said,

> It is not clear why this proposal should be considered a consortium proposal rather than just an amalgam of eight separate library automation proposals. What is in this document that spells out the advantages of a consortial approach? Where do we gain an insight into the collaborative advantages of a consortium? In fact, is the proposal the result of work done by the members of a consortium? If yes, what working committees addressed the issues of hardware, software, LANs, WANs, and library issues (more about this item later)? My sense is that under this proposal every library is doing exactly what it wants to do; that is to say, nobody has to give up anything. But usually, to gain some advantage from collaboration, we all may have to sacrifice something. It is as if the Latvian libraries had not yet learned the true meaning of "collaboration."
>
> Deep analysis is seriously lacking in the document. Most assertions are not supported by reasoning or analysis. The document does not permit doubt and does not entertain the possibility that one solution might (or might not) be better than another; nor does it entertain the possibility that the very process of implementation might reveal inadequacies in the plan, necessitating a "mid-course correction." The document is a terse recitation of putative facts, not giving me reason to believe that any of the choices in the document are good ones. Just to give one simple illustration: for every library there is a list of hardware and software to be installed and used in various places. How were these decisions made? What reason do we have for assuming that they are the right decisions? There is not one word of explanation.
>
> The document stands in stark contrast to the proposal originally submitted to the Mellon Foundation by the National Library of Latvia. *The latter proposal* discusses in separate sections *and in detail* the requirements that software has to satisfy, the choice of hardware, the choice of software, the national union catalogue, the national bibliography, retrospective conversion, local area networks, wide area networks, management structures, and a whole host of other issues. On the occasion of my visit to Riga this past Spring, Dr. Karnītis attempted to persuade me that there was no substantial difference between the LATLIBNET Proposal and the proposal submitted by the National Library. *Nothing could be further from the truth.* The LATLIBNET document is a superficial and outright unprofessional document.[79]

The resolution to the impasse we were in arrived in July. The acting minister of culture, Aleksandrs Kiršteins wrote:

> The letter is prepared on request of Prime Minister Andris Šķēle. The content of it is approved and accepted by the Prime Minister. . . . Taking into account the estimation of the objectives and contents of both above mentioned projects (the opinions of both authors of the projects E. Karnītis and A. Vilks were considered), as well as based on your letter of June 27 addressed to Mr. A. Priekulis, the Supervisory Board has decided to support the project of the National Library, that has received the A. W. Mellon Foundation grant. No decisions contradicting this project may be allowed in the future.[80]

He went on to ask my assistance in evaluating library automation systems, including ALISE and LIBER, stated that the project will be under the supervision of the Ministry of Culture, and indicated that he would appoint a government representative to the board of the consortium, and that representatives of various branches of education, librarianship, and information (technology) would be appointed, as well. Most importantly, he announced the selection by the board and the appointment of Aldis Ābele as project manager and Ieva Vītoliņa as his deputy;[81]—two outstanding persons, as I was to learn later, who have shepherded the project from that point on, through waters that continued to be extremely turbulent until the inauguration of Aleph 500 on March 24, 2000. Nollendorfs strongly endorsed this new team and, as an informal advisor to the consortium, took steps to ensure that the team had support and would not be undermined. On July 16, 1997, Acting Minister of Culture Kiršteins and Deputy Prime Minister A. Gorbunovs signed a decree which created the Library Information Network Consortium (LINC).

It took until August to settle some additional administrative matters, such as creating a three-member board of regents for LINC (all appointed by the minister of culture) and a seven-member executive council (the meetings of which Nollendorfs would attend as an informal advisor). Since the original grant was given to the National Library, the grant could not be used by LINC directly; LINC itself first had to qualify as a grantee and the Foundation had to reauthorize the grant for the new grantee. However, since "once bitten, twice shy," when reauthorization was complete, we were not going to pay out the grant all at once, but make payments to LINC against actual invoices as necessary. The Foundation also made a $50,000 grant to AABS in order to formally buy some of Nollendorfs' time for the duration of the project to advise, supervise, and generally be the Foundation's eyes and ears in Latvia.[82] This was a good move on our part, I thought, because governmental instability continued in Latvia: during July, several ministers resigned amidst charges of corruption,[83] and by the end of July, Šķēle himself was gone. On September 17, the minister of the economy, Atis Sausnītis wrote to me at my request, confirming the Latvian government's guarantee of $1,248,000 for the project,[84] and on September 26, I received a friendly and constructive letter from the new minister of culture, Ramona Umblija, in which she commented on the resumption of constructive work by LINC and the good working relations that had developed between Nollendorfs and Ābele.[88]

Soon I had to leave for the automation conference in Warsaw, but before going to Poland, I wanted to stop for two days in Latvia. Nollendorfs' view was that Karnītis was being difficult, but perhaps precisely because of that, he seemed to be losing credibility with Deputy Minister Priekulis, as well as among the library directors who had been in his camp. My meetings with Ābele and Vītoliņa covered a myriad of technical and organizational issues and I was entirely reassured that the fate of LINC was in good hands. I also met with the board and the executive council, and their members also thought that the librarians were beginning to see through Karnītis' rhetoric and reported that six of the eight libraries had signed an undertaking to participate in the work of LINC (the two holdouts being the Academic Library and the Technical University Library—the latter because of its total commitment to ALISE). The Academic Library was prepared to participate in the work of LINC, provided it would be allowed to keep LIBER as its integrated system, and also agreed to abide by LINC's decisions concerning the network and the placement of servers, as well as those concerning the union catalog—a very substantial retreat from its position a few months earlier. But six of the other libraries were ALISE users and wanted to be assured that ALISE would be able to compete in the tenders, and I agreed, in spite of my substantial doubts about ALISE, that it should be permitted to do so.[86] Ābele and I made it clear that we would permit the Academic Library to keep LIBER as long as it prepared the necessary interfaces so that it could seamlessly communicate with whatever other system was chosen by the remaining libraries. The government had already paid the first $200,000 portion of its contribution to LINC, with another $100,000 expected before the end of the year and a further $330,000 in 1998.

Later, I met with Ramona Umblija in the company of Anita Dūdiņa, chairperson of the board, and reported that I felt more optimistic than I had been six months earlier, but I still saw problems, such as the fact that the various libraries reported to six different ministries, causing the librarians to wonder what would happen to long term financial support once the Foundation's grant ran out. I had another meeting with board and executive council members the next day, and I felt more pessimistic on that occasion, which greatly worried Dūdiņa, who saw this as a preamble to a new grant termination. But I reassured her that I was just reacting to problems as they surfaced and Ābele pointed out that the situation with the two libraries that were still holdouts was really not as bad as I thought. My last meeting was with the library directors, and I decided that instead of harping on what had happened, I would give them an inspirational seminar on cooperation, collaboration, compromise, and conflict resolution. I spoke for 45 minutes to a group that, in my judgment, looked rather glum. Finally, one librarian interrupted me and said, "Dr. Quandt, how would you define the essence of a consortium?" I replied without hesitation that the essence is the libraries themselves; that it is the living organism that the libraries create by joining one another in order to decide jointly how to solve a variety of problems. She clearly liked my answer, but then said, "Do you realize that the bylaws of the consortium that had been promulgated by the government hardly mention libraries and hardly assign any role to them in the consortium?" It suddenly became clear to me why their were dragging their feet: their view was that LINC was a structure imposed on them, with essentially no input from them, and they feared that the integrated system would be equally imposed, with no consultations with the libraries. I was frankly astonished that the hierarchical ways were so ingrained in Latvia that this state of affairs could have come about. But these ways were deeply ingrained and prevented the development of mutual trust and, hence, cooperation. As Alan Ryan put it,

> The moral that many commentators drew. . . . about the failure of communism in Russia and Eastern Europe [was that] Communist governments had failed to encourage the development of civil society. Indeed, they had systematically undermined it. They had not encouraged the growth of the voluntary organizations that teach the citizenry how to cooperate with one another without the supervision of the state[87]

I promised them that I would write to Minister Umblija and try to involve the libraries in a more essential way; I also told them that in my opinion they had to be involved from the beginning of RFP preparation all the way to system selection and implementation. I promised to try and "fix" this problem—however, without any sense of whether I had a chance of success. I said in my report to Bowen concerning the government's failure to assign an explicit role to the libraries, "This must be changed and if we cannot, then in spite of everything else that was accomplished, we should terminate the agony, because the libraries will never develop the requisite enthusiasm and the project will fail."[88] I also recommended that we ask for a formal declaration by all eight libraries that they will operate as a consortium, and that if the Academic Library insists on continuing to use LIBER, it would have to produce an interface between LIBER and whatever other system is chosen by the other libraries; moreover, that two is the maximum number of different systems that we will tolerate.

By the time the Warsaw conference started, Ābele, whom I had invited to attend, brought me a document signed by all eight library directors in which they recommended that the board of the consortium be enlarged to include all eight directors. This was something of a breakthrough and I felt encouraged. On October 21, after I received a translation of the bylaws from Ābele, I wrote to Minister Umblija,

> I have received a Letter of Intent, signed by all eight librarians of the Latvian Library Consortium, in which they express their intention to recommend to you that the Board of the Consortium be expanded to include all eight Library Directors. I am writing to you to request urgently that you accede to the Library Directors' desire to be thus represented on the Board. . . . They appear concerned that solutions (hardware, software,

training, organization of workteams, etc.) might be *imposed* on them rather than selected by them.

The fact is that the installation of a jointly adopted integrated system cannot be undertaken without the expertise of the libraries, and the choices to be made need to be decided upon by the librarians, in consultation with their various system experts. . . . *The history of every successful consortium shows that librarians and their expert advisors have been directly involved in making hardware, software, and other necessary choices.*[89]

A week later I also wrote to Ābele and imposed the strict conditions that (1) a LIBER interface was to be prepared by the Academic Library; (2) no more than two different integrated systems were permitted in LINC; and (3) the Foundation reserved the right to examine the criteria and the analysis leading to a system choice. By December 21, Ābele was able to report that (1) at a meeting in the Ministry of Culture on December 16, the board had been expanded to include the eight library directors; and (2) seven of the eight directors had signed an agreement to abide the Mellon Foundation's conditions, with the Academic Library planning to write a separate letter signifying its acceptance of the conditions,[90] which it did by February 10, 1998.

Return to Normalcy. During the first four months of 1998, vendor demonstrations were held in Riga and everything seemed back on track. The librarians were evaluating systems with the able help of two consultants who wrote a useful report.[91] There was only one more hurdle to be overcome. Ābele and I both thought that selecting ALISE would be a mistake, but ALISE staff were persuading the libraries vigorously that they ought to support the local product. The extent of ALISE lobbying was described by Nollendorfs as follows:

It has been rather surprising how aggressively ALISE has been ingratiating herself [*sic*] with the local libraries. The local library network must be 100% ALISE's. I participated in evaluation of library proposals for Soros, and was surprised by the number of proposals specifically stating ALISE as the system of choice (although I also have heard that ALISE has difficulties meeting its obligations as stated). What was even more surprising was that the chair of the committee had invited the president of ALISE to be present at the evaluation session! The big idealistic local argument, of course, is that it is a local product, that supporting it supports local business, and that it is less expensive.[92]

What was worse, it had been reported to me that some of the LINC librarians had a financial interest in ALISE; a matter that I considered to be the crassest form of conflict of interest. I immediately wrote to Ābele to the effect that if the report turned out to be true, "it would be the Foundation's view that the selection process failed to use objective criteria and could, therefore, not be expected to make the best decision for the libraries involved."[93] He replied that the two persons in the libraries who had had some connection with ALISE have had other responsibilities assigned to them that would eliminate any suggestion of conflict of interest.[94] By July, ALISE took itself out of the running, and the two short-listed products became Aleph 500 and INNOPAC. By July 17, 1998, Aleph 500 had been selected as the system for LINC: of the seven voting libraries (the Academic Library not voting because of its decision to stick with LIBER), five preferred Aleph 500 and two preferred INNOPAC. Contract negotiations with Ex Libris began, and just to be on the safe side, I asked Andrew Lass (see chapter 8) to read the contract language before signing. Ex Libris provided significant discounts, the servers turned out to be less expensive than had been anticipated, and the mood of the librarians had improved during the past six months of working together to determine which system would best suit their needs. They also seemed to begin to realize that membership in LINC had significant advantages, such as joint collection development. Ābele took on new initiatives, such as beginning contract negotiations with Springer-Verlag for electronic versions of 400 journals. He generally proved himself to be a cool and shrewd manager and negotiator, had no personal axe to grind with Karnītis, and brought LINC into the home stretch with the skill and aplomb that would do credit to a senior statesman. I made a point of going to Riga to celebrate the formal signing of the contract at the end of October.

By May 1999, when I was in Riga again, developments had assumed a routine character. Training in Aleph 500 was taking place, the wide area network was being improved, Aleph was up and running with a test database, and the National Library and the Patent and Technology Library were planning to open the cataloging modules for production by November. Karnītis had in the meantime become an information technology advisor to the prime minister, once again Andris Šķēle, and had no direct stake in LINC as a librarian; in any event, the reports I heard were that he had substantially reformed and was supportive of LINC. He asked whether some small funds could be used from the LINC grant for improvements at the Academic Library and I decided that it would be churlish and counterproductive not to reward good behavior. The librarians were reported to me as still thinking in terms of what they could get out of LINC rather than view it as a common effort, but I no longer thought that new ways of thinking could be introduced overnight.[95] I did meet with the new minister of culture, Karina Pētersone and both praised the work that had been done in the past two years and stressed the importance of her support for the library project. In March 2000, I returned to Riga for the official inauguration of Aleph 500. Not only was the system officially introduced, but I attended a meeting of librarians with Noam Kaminer, the Ex Libris project manager, and was gratified to see the librarians no longer glum, but actively participating in discussions of technical issues. While problems obviously remain and there have been some communication problems between LINC and Ex Libris, the situation appeared to have improved considerably. Money continues to be a problem, but even there, LINC now has friends in high places. While full implementation of Aleph 500 will take some time, particularly because the conversion of the various libraries' databases to Aleph format is proving to be a nontrivial task, there is substantial hope that in another year or two LINC will be fully functional.

A Comparative Evaluation

The difference between my Estonian and Latvian experiences could not be greater. The proposals to automate the research libraries in the two countries arrived on my desk within about two weeks of each other and both proposals were funded at the same time. Yet by the fall of 1998, the Latvian project was about one year behind the Estonian one, and by the summer of 2000 it was about one-and-a-half-to-two years behind, and until the last quarter of 1998 there was a serious possibility that the Latvian project would come unraveled altogether. Yet the two countries are geographically adjacent and during the past 55 years shared the same grim history of being annexed by the Soviet Union, occupied by the Germans and then reverting to Soviet socialist republics. After independence, both had substantial western assistance from governments and foundations and both attracted the support of expatriates living in the West.

What were the technologies that the Mellon grants attempted to transfer? In the first instance, the library technologies involved in automation, which also required the transfer of other ancillary technologies, such as electronic networks, servers, and PCs, many of which were being funded by other organizations, including the countries' own governments. The library technologies themselves consisted of two principal parts: the automation system that needed to be purchased and the library protocols (such as USMARC or AACR2 or authority files) that needed to be agreed upon and learned. But a second "technology" that needed to be transferred was the technology of collaboration. This the Foundation more or less forced on the librarians of the two countries, and I am quite certain that if we had permitted each library in Latvia to automate in whatever form it chose, on the basis of whatever protocols it desired, many of the problems encountered in Latvia might have been avoided, but at the cost of much greater investments in time, effort, and money and a much smaller degree of standardization and, hence, portability.

There is no doubt that certain historical accidents played a significant role in slowing down the Latvian development. Among these are the conflict between Vilks and Karnītis and the fact there existed a Latvian software, ALISE, which had powerful advocates in many consortium libraries. It is difficult to say whether these two factors would have been sufficient to bring about delay or failure; but there were two other circumstances, inexorably bound with tradition, that retarded progress. The first was a lack of understanding on the part of the relevant government agencies that a hierarchical, "top-down" type approach made no sense in forming a consortium and that such a system of management was thoroughly antiquated. As Gundar King put it,

> The management system is a parochial set of power relationships. The management processes include tributes for concessions, and rewards of apparent loyalty. The steep vertical structures permit the accommodation of various courtiers, technical experts, and even leaders with followings of their own, in their allocated territory or function.
>
> In technical matters, this system sees technical experts and technical systems as part of the implementation of orders given. They normally lack a cooperative economic dimension. Tsar Peter had his table of ranks to help sort these things out. Currently, the technical advisory functions are troubled. The actors may be perceived as technical experts or part of a loyal coterie, or both. In any case, this system is anchored more in stability than in change. Transition is very risky, and the system abhors a loser. Thus, loyalties are very important, but they can change very quickly. There is, so to speak, a serf's shrewd and complete understanding of the privilege of the powerful.[96]

The second was a lack of understanding on the part of the librarians of the whole ethos of cooperative action: that by a partial, and often very small, sacrifice, one can achieve a common goal more effectively and at lower cost, and that such an approach actually empowers libraries and leads to results that are not obtainable otherwise. Both of these failures are failures of imagination and creativity, which characterized Baltic educational systems on both the secondary and tertiary levels and retarded academic reforms.[97] Neither of these failures of imagination were operative in the Estonian case—but then the Estonians had neither a domestic product like ALISE to stiffen attitudes, nor personal conflicts to reduce everybody's expectations of cooperative successes. In a different environment, the parochialism about ALISE and the damaging effects of conflict might well have been overcome faster, and it is fortunate that LINC had a leader who had the stomach for battles on all fronts.

It might be useful, in conclusion, to compare several of the "problem installations." I shall include one South African consortium in this comparison, namely the Cape Library Cooperative (CALICO) in Cape Town, even though its details have not been discussed. Any such comparison will necessarily be impressionistic; since it is likely that every library or consortium has experienced the same sets of problems as every other, the major differences among libraries rest on the *extent* to which certain problems have been experienced. One must therefore exercise judgment about how serious a problem was in a particular case. As a practical matter, I shall attribute the difficulties that a library experienced to a particular cause only if I believe that cause to have been more than only moderately responsible.

The principal causes of problems, in rather general terms, are the following:

- Vendor failure in software, that is to say bugs in the software, promised functionalities not delivered, etc.;
- Vendor failure in responding to problems, which includes unusual delays, lack of documentation, unannounced software fixes, responses that claim to fix particular problems without actually doing so, and similar problems;
- Frictions between vendor and customer going beyond the previous reasons; that is, a relationship between vendor and customer that poisons the atmosphere and makes it difficult to correct substantive problems;
- Hardware failure or wrong hardware configuration;
- Government interference in the operations of the library(ies);
- Personal frictions or hostilities within a consortium;
- Failures of governance in the broad sense; wrong structures or lines of command or lack of leadership, including lack of support from the parent organization(s) of the library(ies);
- Inadequate training for the staff;
- Inappropriate rent-seeking; that is, behavior that pursues individual gain at the expense of the whole.

The principal installations at which serious problems had arisen are the Lublin Consortium, CASLIN, KOLIN, Copernicus University, LINC, and CALICO. My subjective evaluation of the reasons for the presence of significant problems are displayed in table 10-2.

While there are no necessary conditions among those listed in the table, there are plenty of sufficient conditions for serious problems to emerge. Vendor failures of any kind do not bode well for a consortium, nor do gover-

Table 10-2 Causes of problems.

	Lublin	CASLIN	KOLIN	Copernicus	LINC	CALICO
Vendor failure in software	Yes		Yes	Yes		Yes
Vendor failure in response	Yes	Yes	Yes	Yes	Yes	Yes
Vendor/customer Frictions		Yes	Yes	Yes		Yes
Hardware failure or configuration						Yes
Government Interference		Yes	Yes		Yes	
Failure in governance			Yes		Yes	Yes
Personal frictions		Yes			Yes	Yes
Training inadequate				Yes	Yes	Yes
Inappropriate rent-seeking			Yes		Yes	

nance failures, personal frictions or enmities, and training inadequacies. LINC and CALICO had had more than their share of these latter factors, and while LINC appears at long last to have overcome most of its problems, CALICO has just recognized the severity of its situation.[98] One must also be careful to note in assessing blame that a particular cause of dysfunctionality in a consortium might not have been present at all times. For example, the government's attitude towards CASLIN was initially obstructionist, but through the massive efforts of Andrew Lass and Vojtěch Balík, this attitude turned very positive in later years. Or, to cite another example, the rector of Šafárika University, Lev Bukovský, was extremely supportive of KOLIN, but when he was replaced by Dušan Podhradský, the atmosphere turned to icy neglect. But one obvious lesson for philanthropy and successful technology transfer, even at the cost of larger grants than a donor would normally contemplate, is to insist on substantially more training than either the vendors would normally and automatically provide or the grantees would initially request. While lack of adequate training did not cause problems everywhere, this is the only factor that predisposes projects toward difficult implementation that can be influenced by donors relatively easily, and in retrospect I wish I had realized this earlier.

Personal frictions may be difficult to anticipate, and if anticipated, may be difficult to deal with. But three aspects of potential failure are in the realm of culture (attitudes, values, and beliefs, as noted in the introduction), namely government interference, failures in governance, and inappropriate rent-seeking. All three of these were present in KOLIN and LINC, and one of these was present in CALICO,[99] and these are the consortia that easily qualify as the most problem-ridden ones. Only one other consortium suffered from government interference in a significant way (CASLIN), but the interference there was only temporary and soon turned from interference to support. It is difficult to escape the conclusion that the cultural factors had a major role to play in predisposing a consortium to problems.

11

Management Training at Universities and Colleges

It was clear from the beginning that training efforts directed toward entrepreneurs in small business enterprises would not be sufficient to create an adequate human resource base for the economic transformation that needed to take place in Eastern Europe. Those efforts, discussed in chapter 4, were and are continuing to be important for developing a small business sector, but they did not address the deficiencies in managerial knowledge in large or even medium-sized enterprises. The demands on managers in these are generally more complex than in small ones, and larger size also permits functional specialization, which in turn usually requires greater depth of knowledge. All this seemed to point toward establishing management training programs at universities that would provide in-depth knowledge in a number of specialized areas through programs lasting one to two years. Such programs are the focus of the present chapter.

Each of the Visegrad countries had some institutions devoted to university-level managerial training during the Communist period. The preeminent institution of this type in Poland was the Warsaw School of Economics (Szkoła Główna Handlowa w Warszawie[1]), established in 1906. A number of university-level "academics" of economics also provided training in economics and management in provincial cities (for example, Kraków, Poznań, Wrocław, Katowice). In the Czech Republic, economics and management training were offered by the Prague School of Economics (Vysoká škola ekonomická, now University of Economics), while in Slovakia, the corresponding institution was the Bratislava School of Economics (later promoted to university status). In Hungary, the Budapest University of Economic Sciences (formerly Karl Marx University) was created as an independent institution in 1948 and was the intellectual successor to earlier efforts to teach economics and management at the Budapest Technical University (Müegyetem).[2] In 1999, as a consequence of the policy of consolidating institutions of higher learning, it absorbed the College of Public Administration and became the Budapest University of Economic Sciences and Public Administration.[3] Its principal divisions are general social science, economics, management, and public administration, and it has a recognizably modern PhD program in economics.[4] Other Hungarian college-level institutions in management at the beginning of the 1990s were the College of Finance and Accounting (Pénzügyi és Számviteli Főiskola), the College of Foreign Trade (Külkereskedelmi Főiskola) and the College of Commerce and Tourist Industry (Kereskedelmi és Vendéglátóipari Főiskola);[5] in accordance with the move toward integration, these three have been amalgamated and have become divisions of the new Budapest School of Business (Budapesti Gazdasági Főiskola).[6]

During the pre-1989 period, these institutions predominantly did not teach the kind of economics or management that would be useful in a free-market context. In approaching the question of how to improve the level and quality of management training, foundations had to answer two questions. First, to what extent ought one to concentrate on the existing institutions dealing with management as opposed to starting brand new programs at institutions that might have had no tradition on this field? Second, since much of the demand for training was for the retraining of managers already employed in enterprises,[7] and since enterprise executives could hardly take a one-to-two year leave from their jobs to get MBAs, to what extent should one support regular MBA programs versus executive MBA programs that would perhaps be substantially shorter in duration and utilize weekends for instruction? As one might expect, there was no precise answer to these questions and both types of programs were supported.

The management programs differed from other types of programs (say, library programs) in two significant respects: (1) the initiative to implement them frequently came from western institutions such as business schools or schools of agriculture; and (2) the East European partner in such ventures frequently maintained several western alliances, with each devoted to some particular worthwhile project, without too much attention to whether duplication was wasteful or not. Thus, for example, the Warsaw School of Economics maintained a close partnership with the Carlson School of Management of the University of Minnesota in MBA and Executive MBA programs, while also maintaining relations with the University of Calgary and the University of Quebec in Montreal for similar projects.[8] Western professors of business made frequent trips to Eastern Europe throughout the 1990s and made many wonderful suggestions to East European rectors about streamlined MBA programs, faculty development, etc.; it did not seem to be in the nature of East European rectors to refuse these overtures, and I had the feeling that declarations of intent and protocols of collaboration were readily signed, without too much thought about what would ultimately have to be delivered in terms of physical and faculty resources and what kind of wasteful duplication might ensue from excess zeal.

In judging the effectiveness of the various East–West collaborative efforts, at least two considerations have be be kept in mind. First, did a collaboration succeed in establishing and maintaining quality education? Second, did a collaboration succeed in creating a long-term partnership between two or more institutions, or did the partnership fizzle after an initial period of enthusiasm? The case studies discussed below provide examples of both.

MBA and MBA-Equivalent Programs

Introduction. The demand for management and business training that manifested itself in Eastern Europe after 1989 was immense. The growth of privately owned companies and the need to enhance the competitiveness of industry in the East European countries, which was important partly for replacing the exports lost to the COMECON countries and partly for conforming to EU accession standards, substantially contributed to creating a demand for business school education. While there do not appear to be detailed analyses of the demand for manpower, the insightful study of Leslie Grayson concluded that the demand was very substantial: in one of the smaller countries, Slovakia, the Association of Slovak Entrepreneurs had about 200,000 active members in 1993, of whom some 5% were estimated to have the potential of benefiting from a one-year course in management and some 15% from a two-to-six week course in management.[9] Two western-inspired and supported institutions that achieved early distinction were the International Management Center (IMC) in Budapest[10] and the Czechoslovak Management Center (CMC) in Čelakovice.[11] IMC was favorably reviewed as late as 1996, and both the quality of the permanent professors and the academic administration were praised.[12] Both institutions succeeded in becoming high-quality and high-profile institutions fairly rapidly; they had been closely associated with the University of Pittsburgh which took a leading role in setting them up, with IMC relying on substantial Soros support and CMC on support from USAID.[13] Both operated on a nonnegligible scale and have made a substantial contribution to the stock of well-trained managers: IMC has graduated over 450 persons to date and CMC some 200. But both have been experiencing financial and other academic difficulties of late. In 1996, only slightly less than 70% of IMC's revenues came from commercial sources (Portwood), and, in 1995, it lost three important philanthropic funders: the Hungarian-American Enterprise Fund, the Rockefeller Brothers Fund and the São Paolo Bank.[14] Pittsburgh severed its connections with IMC, which then affiliated itself with the Weatherhead School of Management at Case Western Reserve University in 1996 and gained the right to grant CWRU MBA degrees; but that connection was terminated by CWRU in the summer of 2000 on the grounds that there was no longer any "strategic synergy between the two schools."[15] IMC clearly suffered from not having a chief executive officer (dean) for some 18 months and there were complaints about the quality of the part-time faculty, about cheating on examinations, and about a lack of strategic planning.[16] Both institutions experienced difficulties in finding continuing sources of funds as of the autumn of 2000. CMC started to experience financial hardships when USAID funding ended in 1997 and the Katz School of Business at the University of Pittsburgh also withdrew; eventually the Katz

School did invest some funds in an Executive MBA program at CMC, but it is reported to be doing less than well; declining student enrollments as well as faculty retention were some of the problems. At the present time, the future of the two institutions is uncertain. Existing older institutions, such as universities and academies of economics, were eager to get on the bandwagon and provide the types of training that appeared to be in demand, and new institutions sprang up like weeds in an untended garden.[17] The Czech educational system had no provision for MBA degrees, and so any MBA-granting institution had to be affiliated with a foreign institution. The Czech Ministry of Education currently lists some 17 programs of this type; all of these are affiliated with British, French, Spanish, German, or American institutions.[18] Quality assessments of these institutions are difficult to come by. In Poland, the largest country with the largest number of programs, MBA or MBA-equivalent programs have been instituted at a variety of distinguished institutions.[19] Another source lists 77 independent schools in Poland with names like "Salezian College of Economics and Marketing in Łódź," "School of Banking and Finance in Katowice," "Economic College in Warsaw," etc.[20] Grayson reported that in 1992–93, "there were 300 programs teaching business in Poland (80–100 in Warsaw alone), about 150 in Hungary, 100 in the Czech Republic, and more than 50 in Slovakia," (Grayson, p. 10). An excellent review of the more important programs in existence in 1993 is provided by Grayson, who notes that it was too early to "pick winners or national champions" among the many management education efforts under way (Grayson, pp. 41–88).

But it seemed clear that, particularly in the early years, both funding and the maintenance of respectable quality were serious problems. The former problem suggested that much foundation support would be needed; the latter, that the continued involvement of western partners for some period of time was essential. As indicated in chapter 5, accreditation was almost nonexistent, and even today, only one institution, the Leon Koźmiński Academy of Entrepreneurship and Management in Poland, has been accredited by EQUIS, the accreditation agency of the European Foundation for Management Development.[21]

While librarians may have been deprived during the Communist period of access to the right hardware and software, they were largely in tune with modern developments in the western world. Unlike them, professors of economics and/or business had a completely different outlook, ideologically, pedagogically, as well technologically, and some of the common approaches to teaching business administration, such as the case study method, were as good as unknown in the region. As a result of inadequate prior preparation in the field of management, two significant efforts by East European organizations to obtain Mellon funding had to be declined.

The first of these efforts, which might be illustrative of a number of similar initiatives, was mounted by the Mirosław Dzielski Institute of Industry and Commerce (MDI), an organization created by the Kraków Industrial Society. The Mellon Foundation had indirectly supported it with two small grants to the Polish Society for Arts and Culture (Chicago), which ran summer seminars for businessmen in 1990 and 1991 in Kraków jointly with MDI. MDI's first proposal was lush and expensive and requested an annual budget of $452,000 for (1) summer programs with the Polish Society for Arts and Culture; (2) a Polish-American business school envisioned as a middle-executive non-degree-granting training program; (3) a junior business college; (4) a business development center; (5) miscellaneous business seminars. MDI seemed to have no sources of funds other than the hoped-for Mellon grant, faced grave difficulties in finding qualified Polish teachers, did not have an American partner who could give advice about the curriculum and textbooks, and in general, could not articulate the "big picture." The proposal was fragmented and ignored the drop from 95 to 60 in summer enrollments in the Mellon-funded courses, not to mention the fact that the Mellon Foundation was considering major grants to the University of Hartford for establishing business training programs at the neighboring Jagiellonian University.

The second abortive effort to obtain Mellon funding was by the Kraków International School of Management (KISM), an organization created by the not-for-profit Progress and Business Foundation in Kraków. KISM claimed to have the enthusiastic endorsement of the rectors of Jagiellonian University, the Kraków Academy of Economics, and the Academy of Mining and Metallurgy, and was intended to "build a top-quality, accredited, international MBA program in Kraków, that delivers management education adapted to the needs of the rapidly changing East Central European economies."[22] The president of KISM, Maciej Zebrowski, a serious and committed person, and I discussed the objectives and proposed programs of KISM, but the position papers never got beyond vague generalities.[23] The actual proposal, dated December 23, 1992, purported to assemble faculty in economics, accounting, finance, marketing, organizational behavior and human resource management, but there was no operational plan of any sort, it was not explained why such a business school was needed at a time when the Mellon Foundation was already funding the development of a management program at Jagiellonian University and the Academy of Mining and Metallurgy, and I could not understand how such a vague plan deserved a grant of $2.7 million. There was really no choice but to decline to participate in this.

Both the president of the Mirosław Dzielski Institute, Marian Kania, and Maciej Zebrowski were serious, decent, and committed people. How could they have hoped to get funding for projects that seemed so vague and ill thought-out? I think the answer is threefold. First, neither of them was familiar with the detailed project planning that is customary in the West. In fact, many observers of the East European transition have noted that the term "planning" became anathema in Eastern Europe, because it was invariably associated with the economic planning of the Communist regime; an attitude that I encountered in many other instances. Second, the previous regime did not provide any training in how to design a project. Projects were typically designed and managed on the level of ministries and one should not be surprised that many grant applicants were inexperienced in project design. And finally, in the early 1990s, money seemed easy to get. It became abundantly clear that certain catchwords or phrases, if used in judicious combination, magically unlocked many coffers: words such as markets, privatization, economic development, small enterprises, civil society, international links, foreign direct investment, and global business environment, tended to have quite magical effects.

But there were also other reasons for not acting favorably on certain proposals. The U.S. Business School in Prague was established in 1989 under the auspices of the Rochester Institute of Technology and it solicited funds both from USAID and the nongovernmental sector, such as foundations. By October 1990, the idea of operating such a school had received the strong endorsement of Václav Klaus:

> I want to emphasize as strongly as possible that I do *not* consider this effort premature in any way; on the contrary, no new business and corporate culture will emerge here before there is a concentrated intellectual effort to put it in place. It is a *conditio sine qua non* for the removal of all that marxist-leninist rubbish we still encounter in many subtle or not-so-subtle manifestations.[24]

The proposal was detailed and professional in its approach to creating an institution with a long-term horizon (although initially only a five-year one), and the proposed faculty members, who would spend three weeks at a time in Prague to deliver courses consisting of 40 teaching hours, included such illustrious names as Roman Weil, Michael Mussa, Douglass C. North, and others. The problem I had with the proposal was the price tag. USAID was being asked for $2.5 million for the first two years, which in itself need not have been objectionable. But the total price was the result of detailed budget items that I did object to. Faculty members were budgeted at $25,000 per (40-hour) course, which seemed excessive.[25] In any event, the Mellon Foundation did not contribute to this project. But the project flourished and is very much in business.[26] It offers a full-time one-year program, a part-time executive program taken over two to four years, and certificate programs in finance and accounting, marketing, and human resources. Its faculty comes predominantly from prestigious institutions such as Carnegie-Mellon University, the London Business School, Northwestern University, Rochester Institute of Technology, UCLA, the University of Chicago, the University of Rochester, Yale, and others. Tuition is currently $16,000 for the full-time program, although scholarships of up to $9,000 are available for Czech and Slovak citizens, and up to $8,000 for other East European citizens. While the tuition level is high, it is not unusual for MBA programs: the tuition for the Executive MBA program at the Warsaw School of Economics, created with the assistance of the University of Minnesota and funded by Mellon, is a comparable $15,750.

As I indicated in earlier chapters, I became convinced rapidly that there is a substantial difference between the efforts required to create a Ph.D. program and a business school program. First, the former requires the prolonged presence of faculty with distinguished research accomplishments; doctoral dissertation advising cannot be done effectively if faculty advisors are on location for two or three weeks and then gone for the rest of the year, and it cannot be done well by academics who have not had substantial research experience. In the latter, a dissertation is not required and the course material can be easily divided into two or three- or four-week segments. Second, business schools are not primarily for creating the ability to do cutting-edge scholarship but for educating practitioners; hence, it is not as important that the teachers be faculty members from the most prestigious business schools. I was firmly convinced that if the price tag was too high, using faculty from somewhat less prestigious institutions would represent a cost-effective trade-off.

A Summary of Mellon Efforts. Management programs tend to exhibit smaller variability with respect to content and plausible implementation sequence than other types of programs, such as library automation; they are also generally less controversial or contentious. For these reasons, analyses of Mellon programs in this area will be somewhat less detailed than those in other areas. In general, the objectives of grants were to develop faculty capabilities, create business management programs, and upgrade institutional infrastructure. All grants aiming at university-based management programs were made between 1991 and 1996 and totaled $6.9 million, distributed in 31 individual grants. A fairly standard pattern was for a certain number of American professors spend some time in Eastern Europe, teaching students as well as assisting the local faculty with course design, curriculum development, and teaching methods, and for selected East European faculty members to spend periods of varying duration at the American partner university. The courses offered in a newly created MBA program typically included managerial statistics and quantitative analysis, organizational behavior, managerial economics, macroeconomics, microeconomics, finance, financial accounting, marketing, production, global and international management, legal and social issues, and business policy and strategy. Courses would typically be taught in modules of three to four weeks with daily sessions of three to four hours, and the length of the MBA programs as a whole tended to range from two to four semesters.[27] An important outcome of some of these efforts was the creation of East European case studies, an important tool in teaching business in the UNITED STATES.[28] Most importantly, in certain instances the American partner institution was willing to award its own American MBA degree at the conclusion of the program, which did a great deal to enhance the East European partner institution's prestige. In some cases, the Mellon efforts were not at all directed toward creating new instructional programs but toward shoring up the infrastructure of institutions; with minor qualifications this was the case at the three Hungarian business colleges; these institutions seemed less appealing to potential American partners and they were in dire need of improvements in their infrastructure. Much of the Mellon efforts in those three institutions was devoted to upgrading library and computing facilities.[29]

Warsaw University and University of Illinois. In early 1991, John Hogan, at the time dean of the School of Business Administration of the University of Illinois at Urbana-Champaign, conceived the plan to start an MBA program under the auspices of the School of Management of Warsaw University. The plan seemed sound and Mellon funded it in two segments, but I was concerned that continuity would be lost when later in the year Hogan decided to move to Georgia State University. However, at that point the administration of the project was placed in the capable hands of Case Sprenkle and my anxieties dissolved. Rector Andrzej Wróblewski provided suitable renovated space for the program and the highly capable Krzysztof Obłój was appointed as its director.

The program was a 16-month (three-semester) Executive MBA program offering 15 core subjects in 44-hour courses and two 28-hour electives each semester; courses meet twice a month on Thursdays, Fridays and Saturdays. Each year, four faculty members from Warsaw spent the year at Illinois and anywhere from two to four Illinois faculty members taught in the program in Warsaw. The first three cohorts received University of Illinois MBAs in addition to their Warsaw degrees; after the initial period the graduates only received certificates from the University of Illinois, because the number of courses taught by Illinois faculty was not large enough to meet the degree requirement. The average age of the students was 31 and the program, which has continued to function after the University of Illinois participation ceased, has graduated, as of October 2000, 311 alumni, and the ninth enrolled cohort has 53 students.[30] Tuition has grown from $3,500 in the first year to $9,000 at present, which is substantially less than what some other programs are charging, but this still makes the operation financially viable. Some 58% of the graduates hold top managerial positions, many of them in foreign companies, and the two partner institutions can be justifiably proud of their achievements.

John Hogan, who conceived the Warsaw University program, successfully transplanted that model, with a modest Mellon grant, to the Poznań Academy of Economics. It much resembled the Warsaw program and succeeded in creating an MBA program that is highly rated in Poland today. The success of the project was due, among others, to the unfailing courtesy and respect with which Hogan treated the Poznań faculty. Astonishingly, the project was carried out with what turned out to be the smallest grant made by Mellon for a program of this type, namely $200,000.

Jagiellonian University and University of Hartford. In mid-1990, a correspondence sprang up between me and Humphrey Tonkin, the president of the University of Hartford, who was interested in establishing a variety of management-oriented training programs in Eastern Europe. The suggestion that I found most appealing was that a business curriculum, and eventually perhaps a full business school, be set up at Jagiellonian University, a distinguished institution with no track record in this area. Intense negotiations and planning activities by Hartford, Jagiellonian and myself led to a proposal by February 22, 1991, by which a consortium of the University of Hartford, the University of Massachusetts, Boston College, and Columbia University would assemble a faculty team to work with Jagiellonian University in creating a business school curriculum and a faculty development and exchange programs. Robert S. Hager of the University of Hartford was appointed coordinator of the program (succeeded in 1993 by Demetrios Giannaros), and a phased approach to training Polish faculty and implementing the program in Poland was adopted. The Mellon Foundation made a total of six grants for this project, covering a total of 10 formal phases; the first five of the grants were to the University of Hartford and the last one directly to Jagiellonian University when it became clear that the program was here to stay.

The first grant covered the training of eight Jagiellonian faculty members in the United States, an intensive three-week training program in Kraków in January 1992, and the launching of the program with 49 carefully selected students. The second grant was to carry these activities further through 1992 and start a Regional Small Business Center for providing consulting services. A second cohort of students was admitted, the training activities had been broadened, and by the spring of 1993 there were 85 students in the program.[31] By Phase 5 of the program, envisioned for 1993, Hartford was assisting the development of computer facilities, providing textbooks, and anticipating a close collaboration with the Kraków Academy of Economics. By 1994, the project planned to establish a Center for Faculty Development in Business so that training activities could continue in the indefinite future, long after the American presence was ended. The quality of the effort was recognized by USIA, which had made grants totaling $425,000 to the project within the first three years of its existence. The Jagiellonian faculty were largely satisfied with the program but wanted to become more involved in determining the content of the curriculum and complained initially that the program was too difficult, and certain midcourse corrections became necessary from time to time. There was definite progress, but it was uneven. The Hartford team had to deal at times with Polish faculty members whose English was quite inadequate.[32] The Hartford team attributed many problems to the lack of careful planning and management at the Polish end of the project.[33] It did not help matters that since the inception of the program, it had already had five different Polish directors,[34] some of whom had neither an interest in it, nor the ability to cope with standard administrative problems.[35] One American professor believed that cultural differences also played an important role.[36] Most serious was the criticism leveled at Jagiellonian Rector Andrzej Pelczar, who was perceived to have a less than enthusiastic attitude towards the project, and many project participants welcomed the reelection of Aleksander Koj as the rector of the university. The arrival of

Koj represented a crucial step in the development of the project, because on May 25, 1994, the Jagiellonian University Senate decided to establish the School of Management (now referred to as the Jagiellonian Business School) as an independent unit of the university in the expectation that it would become a separate "faculty" of the university, and Tadeusz Borkowski became the director of the new unit.[37] Without a doubt, this was a major shift for a university concentrating on traditional subjects.

By the spring of 1998 it had become reasonable for the U.S. Consortium to phase out its activities. It had made key contributions to the establishment of the business school and may be credited with training over 200 faculty members from over 40 institutions, with special emphasis given to the faculty at Jagiellonian University. Its workshops in 1996 and 1997 had led to the establishment of the Association for Business and Education Development, which was created for promoting faculty and curriculum development, faculty exchanges and collaboration among business schools and business firms.[38] In the fall of 2000, Jagiellonian Business School had about 200 undergraduate students, 440 master's level students, 73 postgraduate students and about 80 Ph.D. students.[39] A recent article in a Polish daily ranked the program 7th in Poland.[40]

Warsaw School of Economics and University of Minnesota. In May 1992, I began a dialogue with Professor Mahmood Zaidi of the Carlson School of Management at the University of Minnesota. Minnesota already had a three-year project funded by USAID for a "train the trainers" kind of program at the Warsaw School of Economics and the Olsztyn University of Agriculture and Technology.[41] Zaidi was now thinking of deepening the activities at WSE and creating a program leading to the MBA degree. By March 1993, a full-fledged proposal had been submitted, which described the usual course offerings that would constitute the backbone of the MBA degree, provided for 12 WSE faculty members to spend three months (one quarter) at the University of Minnesota during the next two years and for 12 Minnesota faculty members to spend two weeks each in Warsaw in curriculum design and assistance to WSE faculty. The program promised to provide some competition for the Warsaw University—University of Illinois project, which I thought was healthy, the design seemed solid, and the Mellon Foundation made its first grant for the project in June 1993. Ultimately, a day MBA, was created as well as an Executive MBA program (WEMBA), although only the latter has survived. The design of the programs aimed at a balance between quantitative and behavioral skills and recognized that whatever program was put in place, it could not be just a copy of the Minnesota MBA program.[42]

Little did I know that by the time the grant had been extended, serious frictions were developing between WSE and the Minnesota project management. I found out about all this in August 1993 from a letter by (outgoing) rector Aleksander Müller, in which he told me, in rather aggrieved tones, that the portion of the budget designated for WSE and approved by Mellon was not satisfactory and that there seemed to be a substantial disagreement between his institution and Minnesota concerning both budgets and curriculum.[43] Checking with Zaidi revealed that (1) WSE appointed a new dean for the program, Aleksander Sulejewicz, who wanted to do things his own way; (2) Sulejewicz, apparently without consultation, changed the curriculum fairly drastically; (3) WSE insisted that $60,000 of the grant be transferred to WSE for administrative expenses;[44] and (4) Sulejewicz demanded that the number of Polish professors trained at Minnesota be increased by three to eight people. While Müller did write me a second letter on his last day in office, in which he declared that his earlier letter was no longer relevant and that peace and harmony reigned again, it seemed fairly clear that some renegotiating needed to be done and that I had to make sure that the two parties did in fact agree with one another.

The most important thing seemed to be to keep the two parties talking to one another and to find a middle ground between their rather extreme perceptions of each other. On the occasion of my visit to Warsaw in October 1993, WSE administrators complained that University of Minnesota administrators had not consulted them properly, while at the same time, the latter said that WSE administrators had seen and "signed off" on every part of the original project plan and should not seek to alter it. But both parties wanted the project to succeed and approached the problems with good will, and by the fall of 1994, peace had truly been reestablished, and the spirit of cooperation that pervaded the effort bore rich fruit. WSE seemed determined to become the top management program in Poland, and so it has (Pacioreck). In fact, by October 1994, the top WSE administrators (Rector Janina Józwiak, Dean Sulejewicz, and Prorector Krzysztof Rutkowski) already believed that WSE was the best place in Poland for management training: my suggestions that they participate in a Polish effort to establish business school accreditation was rejected by them on the grounds that they did not want to be associated in a joint effort with marginal programs. They were quite adamant that most programs, with perhaps a handful of exceptions, were worthless—mere façades created by the presence of a few Americans.

Whether this was true in 1994 is difficult to know, but WSE, with the help of the University of Minnesota, continued to become stronger. The number of Polish professors trained in Minneapolis was increased and, by late 1994, some 40 of them had received training there. At that time, WSE was also beginning to develop plans for an Executive MBA program. In October 1997, I was fortunate to be able to attend the first formal graduation ceremony that was held for three cohorts (two regular day-MBA cohorts and one Executive MBA cohort; 91 students in all) with great pomp and circumstance and the graduates received WSE, as well as University of Minnesota, MBAS. Unfortunately, the day-MBA program was terminated at that point for lack of funds: under the Polish constitution, no tuition may be charged for regular students. But the Executive MBA program has continued and is very successful; it is a two-year program that has classroom meetings only on weekends, for which a tuition of $15,750 is charged.[45] The entering class in the fall of 2000 numbered 46, over 160 students have graduated from WEMBA, placement from the program has been excellent, and the program is clearly self-supporting and here to stay (Sulejewicz et al., pp. 50–51).

Comenius University and Cornell University. During the early summer of 1992, I met with David Lipsky and Linda Gasser of the School of Industrial and Labor Relations at Cornell University. Their approach to management training in the Czech Republic and Slovakia had the novel element of concentrating on human resource development. I thought that this was an outstanding idea, since incentives, compensation schemes, arbitration, and conflict resolution were topics altogether missing from management education during the previous regime, and previous U.S.-funded efforts were notably thin in the area of human resource management.[46] The Cornell team wanted to establish a presence at Charles University, but was less certain about how to proceed in Slovakia; I suggested in early July that they approach the Bratislava Economics University (BEU), and they wrote to the rector almost immediately. But they received no reply for a long time and it began to dawn on me, which I realized more fully only later, that BEU was a very weak institution, with a leadership that was lacking in initiative, as well as know-how. I then steered the Cornell team to Comenius University, where they received a much more hospitable reception.

The Cornell plan was designed to provide education in human resource development, train faculty members in this area, provide enduring institutions that would be able to continue such training in the future, contribute to resolving labor and employment problems, and develop a clearinghouse for information on employment issues. The concrete plan envisaged the teaching of four courses each year (two at Charles University and two at Comenius) in which a junior faculty member from the host institution would be paired with an experienced Cornell professor. The four junior professors would then spend some time at Cornell in a specially designed orientation program that would permit them to interact with unions, firms, government agencies, and professional associations. Cornell professors would also participate in developing bibliographies and the curriculum and would participate in the host country in seminars with faculty and representatives of unions and management. All this seemed extremely promising and the Mellon Foundation made its first grant for the project, which started in earnest in the autumn of 1993. But it was obvious that further grants would be needed to complete the project and, by the spring of 1994, I was in frequent correspondence with Linda Gasser at Cornell, who reported good progress in the initial phase. I wrote in February,

> There is only one thing that gives me mild concern, which is that the number of local faculty who have gotten involved is rather small. What worries me about this is that we might not be creating sufficient depth in Prague and Bratislava. Faculty are underpaid in those countries and brain drain is a significant factor—what will happen to HRM [human resource management] if one or two of these people gets seduced by industry?[47]

During the spring we also discussed an important midcourse correction. Up to that point, the pattern was for Cornell faculty members to commute weekly between Prague and Bratislava; an exhausting schedule that wasted a lot of time. Henceforth, some Cornell professors would be stationed exclusively in Prague and others exclusively in Bratislava. In June 1994, I visited both the Prague and Bratislava operations. The program in Prague was housed at the Institute of Economic Studies of the Faculty of Social Sciences and the people I met there had high praise for the Cornell team; but the most laudatory thing they could say about the project was that it diversified the Faculty's program and that it did wonders for the students' ability to speak and understand English. Overall, I thought that the Charles University faculty sounded lukewarm at best about human resource management. I got the distinct impression that the traditional arts and sciences orientation of Charles University made them turn up their noses at lowly subjects such as management. By the time Cornell and I were discussing a substantial follow-up grant, I was convinced that we should drop the Charles University part of the project and concentrate on Comenius University, where the faculty was bubbling with excitement. In fact, I determined on the occasion of my June 1994 visit to Bratislava that over 100 students had taken the Cornell-run courses there, whereas the enrollment at Charles University was a mere 15.[48] While it seemed reasonable to accede to the Cornell request not to drop Charles University abruptly, only negligible efforts were invested in continuing the program there.

More grants followed and the program officially continued through 1997. Cornell did yeoman's work in developing the Comenius University faculty through courses in human resource management, organizational behavior, strategic management, international human resource management, compensation, cross-cultural management, and conflict resolution. In the later phases of the project, considerable attention was devoted to improving the library of the School of Management at Comenius, in which librarian exchanges were highly instrumental. The Slovak faculty members had outstanding academic experiences at Cornell, enriched by site visits to many companies, and as many as seven Slovak teaching assistants ultimately returned to Cornell to earn Master's of Professional Studies degrees. Cornell assisted the development of top Comenius University administrators and helped create a center called Management Development Center.

By a number of criteria, the project was a success. The teaching of human resource management has become a permanent feature at Comenius. Seven core courses are offered annually and two electives; while enrollment in the two core courses is required of all students and is correspondingly high (about 200), enrollment in the other courses tends to be limited to those specializing in human resource management, about 20–25 per year, which may not be as high as in the trendier subjects of finance and marketing, but is a very respectable showing.[49] The library is much better than before. Important and durable connections have been made between Comenius and Cornell professors. The Faculty of Management has established a Slovak management journal, *The Journal of* HRM. Research and publications in human resource management have created an impact that simply did not exist before. But as noted in an insightful paper by Gasser,[50] there were frustrations, and the overall impact may well have been less than initially hoped for. The principal reasons for lack of complete success were the following.

1. Cross-cultural differences may have impeded the effectiveness of interactions between the faculties:

   > For the CEHREI [Central Europe Human Resource Education Initiative] project partners, some of the differences encountered stemmed from different ways of thinking and operating, different perspectives on teaching styles and faculty roles, organizational hierarchy/power distance needs, communication patterns, language idiosyncrasies, historical influences, family, work, and other value differences, focus on the collective instead of the individual, value and use of funds, personal responsibility and accountability, and perceptions of openness to change, as well as simple logistical influences like time zones. (Gasser, p. 7)

2. Infrastructure limitations included uneven access to telephones and computers, restricted library access (by American standards), difficulties in making logistical arrangements, and others.
3. Problems arising from the fact that Slovakia was a country in transition: this state was characterized by the influx of substantial funds from other sources (USAID, European Union) for related purposes that often diverted the faculty's attention from the project at hand. A case in point may have been the USAID-funded project implemented by the University of Pittsburgh at the Faculty of Management and may illustrate the point made before that in their eagerness to cash in on what the world at large was offering East European universities, rectors often failed to make sure that their resources were commensurate with the tasks undertaken.
4. Problems arising from brain drain: the fear I had expressed in the spring of 1993 became a reality when talented individuals left for greener pastures. While they almost certainly continued to work in the human resource management field in their new occupations, their departure impoverished the program at hand.

Gasser's paper makes a number of sound practical recommendations for how projects of this kind should be implemented. It is clearly very important to understand the local environment: the political and economic situation, and most importantly, the cultural values of the country. It is important to know what commitments mean, what the sense of time-urgency is among local partners, and what the concepts of commitment and accountability mean. And perhaps most importantly, commitments, motivations for, and expectations from projects must be honestly discussed and clearly understood from the beginning, lest "unenlightened self-interest" interfere with the management of the project. On balance, in spite of all caveats and qualifications, the project succeeded in creating a valuable, new, and permanent program.

Loyola Marymount University and TQM Training. In September 1994, I was contacted by Professor Zbigniew Przasnyski of Loyola Marymount University in Los Angeles. He reported that earlier in the year he and his LMU colleague, Bohdan Oppenheim, had attempted and failed to get funding from the Stefan Batory Foundation (the Soros "country-foundation" in Poland) for a project to create a technology-oriented management program in Poland, and wanted to know whether Mellon might have an interest in such a project. We had already assisted the creation by Case Western Reserve University of a technology-oriented MBA program at the University of Veszprém in Hungary, and that program seemed to produce good results. The Przasnyski suggestion therefore struck me as another case promising classical technology transfer, and I encouraged him to submit a formal proposal.

The justification for such a program was the fact that in modern, technology-oriented enterprises, successful management must rest on a combination of traditional management tools and engineering subjects such as statistical process control, design and process optimization methods, design of experiments, benchmarking, and concurrent engineering.[51] Traditionally trained managers tend to know little about the engineering aspects of the business, and traditionally trained engineers know correspondingly little about finance, marketing, and other business subjects. Przasnyski and Oppenheim were very familiar with the nature of management training in Poland and felt that relatively little had been accomplished for the benefit of larger companies. The quality of Polish products tended to be low, and Poland was being urged by other European countries to adopt the ISO 9001 standard as a minimal implementation of a Total Quality Management (TQM) standard.[52]

The plan was simple and straightforward:

1. Two technical universities, the Technical University of Gdańsk and the Technical University of Szczecin would establish and operate, together with LMU, the Center for Management of Technology in Poland. Przasnyski and Oppenheim had explored the feasibility of this on several previous visits to Poland and had secured the enthusiastic participation of the two universities' rectors. The Center would have a site director at each campus: Ireneusz Durlik in Gdańsk and Czesław Sułkowski in Szczecin.
2. During the summers of 1996 and 1997, Przasnyski and Oppenheim would each give two courses in TQM and technology management (TM), with a target enrollment of 50 in each; hence, assuming a 50% certification rate, by the end of the second summer, 100 persons would have been certified. Courses would meet in 90 contact hours and be equivalent to two 3-credit courses at LMU. Successful completion of the course would earn a completion certificate. Each of the certified trainees would be expected to train within six months another 10 persons at his or her home institution; so, by the end of the project over 1,000 additional persons would have been trained. Tuition would be

charged for the summer courses; the tuition income would be sufficient to pay the directors' salaries and other office expenses, and the Center would thus become self-sufficient.

3. Two persons from each Polish institution would take courses at LMU for a whole academic year; upon their return they would be qualified to take over the teaching in the Center.
4. The American team would provide consulting services to Polish firms on an ad hoc and pro bono basis.

All this seemed extremely promising and the grant request was modest; Mellon awarded $163,000 for the project in June 1995.

During the summer of 1996, in Gdańsk, 56 individuals were trained; of these 25 were from industry, 21 from academic institutions and the rest from consulting firms.[53] But by the summer of 1997, only 40% of the trainees had actually delivered on their promise to train additional people and the academics had the worst showing in this respect. Consequently, a larger percentage of industry-based trainees was selected for the 1997 summer course in Szczecin: 31 persons out of a total of 55 were industry-based. In both summer programs the trainees described themselves well satisfied with the courses and with what they had learned. Tuition had been set at $225 for the first year and at $300 for the second; this produced enough income to pay the expenses of the Center as anticipated. One associate and three assistant professors had been selected to spend a year at LMU; all four turned in outstanding performances and earned the MS in Engineering and Production Management degree. In addition to training more than 100 individuals in the formal part of the program, the American team also exposed nearly 2,500 others to TQM through various consulting activities during the subsequent years, and the number of secondary trainees was appreciable, although it did not reach the theoretical maximum of 1,000 persons.

As far as these tangible results are concerned, one would have to be satisfied with the results. However, all was not well with the program. Neither site director was competent in TQM, nor especially interested in it. The Gdańsk part of the Center got involved in a financial dispute with LMU, particularly concerning the remuneration of the director, and attempted to renegotiate the original contract.[54] He was replaced in 1997 by one of the faculty fellows who earned an LMU degree, Zbigniew Celmerowski, which represented an definite improvement. The Szczecin director refused to attend any lectures on TQM on the grounds that it would be embarrassing for him as a senior faculty member to do so.[55] The situation at Szczecin went from bad to worse: not only was the director there not interested in TQM, but he prevented the first faculty fellow returning from a year at LMU from organizing future courses at Szczecin; also, he did not provide any information on secondary trainees as was required by the original agreement with LMU and was generally unresponsive and lacked initiative. The academic year 1996–97 was largely wasted there, and the American team decided to close down that operation.[56] Oppenheim and Przasnyski wrote to Rector Stefan Berczyński:

> In accordance with paragraph 1.5 of the Agreement between LMU and Politechnika Szczecińska, we hereby inform you that, the Politechnika Szczecińska branch of the Loyola Marymount Center for Total Quality Management and Technology Management (hereafter referred to as "the LMU Center in Szczecin") is now closed. Although this decision is made with regret, it is irrevocable and effective immediately.
>
> The decision is due to the noncompliance of the LMU Center in Szczecin with the formal agreements signed by the Rektor, i.e., the Project Proposal to the Andrew W. Mellon Foundation, the Agreement between LMU and Politechnika Szczecińska; and the agreement with the Faculty Fellows.[57]

The Gdańsk site continued to be active until 1998, when the new director, Celmerowski, was promoted to dean and could no longer supervise the activities of the Center. Only eight people had signed up for the 1999 summer program, which had to be canceled, and no course was offered in 2000 either. There is some hope that when Celmerowski resumes normal faculty duties, he will revive the Center's operations.

Much blame attaches to the rectors of the two institutions who had selected initial site directors with no interest in the program and did not mobilize adequate human resources for it. And they clearly did not understand or care about the long-term impact that TQM might have in Polish industry. Two of the four fellows trained at LMU defected to private industry, in contravention of their original agreement with LMU: the usual, financially induced brain drain. The faculty fellows who remained were too junior in academic rank to dare rocking the boat, and it is plausible that they were nominated for such training precisely because their junior status prevented them from becoming a threat to the site directors. The rectors and site directors were happy to have an American presence, because that conveyed prestige, but they were not willing to do the work that would have made it a long-term success. Professors are motivated by the need to earn more, but are not willing to make the effort that learning the subject matter requires.

The pessimistic outlook of Oppenheim and Przasnyski is based on a single example and many rectors, professors, library directors in Poland as well as in the other East European countries are committed to raising the intellectual and professional standards of their institutions. But examples of venal or indolent behavior occur often enough so that donors and project leaders should pay particular attention to the recommendations of Oppenheim and Przasnyski:[58] (1) select the key collaborators in the East European institution with considerable care; (2) document all requirements in full detail, specifying what kinds of activities are required and what kinds not permitted under the terms of the project; (3) obtain the endorsement of the presiding officers of the partner institution, being mindful of the fact that the desire for prestige may make these officers sign letters of intent even when they do not have the intention or ability to mobilize resources for the project; and (4) do not be misled by extreme forms of hospitality. The transfer of technology cannot be accomplished without committed people who understand what long-term goals are at stake. In a sense, people's hearts and minds must be open to commitment, hard work, accountability, and decency; and these are part of the social foundations that Dahrendorf thinks may take 60 or more years to achieve (see note 72, chapter 1).

An Assessment. One motif that emerges from the case studies discussed above is that cultural differences are enormous, that they affect perceptions of the meaning of "commitment" and "accountability," and that American technology transfer teams did not always anticipate the range of East European responses to new opportunities. Project failures were sometimes caused by the brain drain created by opportunities for lucrative alternative employment, in spite of commitments to stay with the project. The grant to the College of Finance and Accounting in Budapest had a subproject in which faculty members would be commissioned to write case studies, for which they would receive handsome honoraria; yet only a few of these were ever written, because the faculty had attractive alternative opportunities, which *dominated* the in-house project financially, if not intellectually. Failure was also caused by a lack of interest in the outcomes and by the apprehension of older faculty that the new knowledge acquired by the younger faculty members would diminish their status.

Another problem, which I mercifully experienced only infrequently, was the lack of transparency in financial matters. One crass example of this was the case of the Bratislava Economics University. City University of Bellevue, WA, was running outstanding programs in Slovakia, teaching undergraduate management, English, and other subjects, which Mellon had funded since 1994; it was an amazing operation that by 1998 had 1,500 students at its various campuses in Slovakia such as Bratislava, Trencin, Poprad, and had an active program at Slovnaft, the large oil company. CU also had some cooperative programs with BEU. In 1996, the University of Pittsburgh, funded by USAID, was running a weekend Executive MBA program (WEMBA) at BEU, and City University, the University of Pittsburgh, and BEU were in some ways cooperating toward a common objective. It therefore seemed reasonable in 1996 to make a $126,000 grant, to be managed by City University, for a triple purpose: $33,333 to be used by City University for its own programs, $33,333 to be used by the University of Pittsburgh for library acquisitions for its program at BEU, and $59,334 to be used by BEU for library acquisitions of its own and for creating an effective Internet node. But in October 1996 I received an extremely disturbing report from Kevin Sontheimer, the University of Pittsburgh project director:[59]

1. BEU was repeatedly advised to maintain a separate bank account into which WEMBA tuition revenues would be deposited and from which disbursements for allowable expenses would be made; in spite of this, BEU had commingled these funds with other university funds, which made the tracking of disbursements and the attribution of interest earnings very difficult;
2. BEU had made excessive use of WEMBA revenues for paying administrative and other expenses that were not related to WEMBA activities;

3. BEU had been unwilling or unable to provide coherent and consistent reporting of its management of WEMBA-related funds;
4. BEU had contradicted its own commitments concerning the correction of excessive charges and the level of net income for BEU faculty development efforts.

I immediately notified City University to put a moratorium on any and all payments to BEU out of Mellon funds and wrote to the president of the Mellon Foundation,

> If I could convince myself that this is a one-time aberration. . . . I might recommend that we ignore the lapse and trust that there would be no repetitions in the future. Unfortunately, the unresponsive manner in which Sontheimer's demands for clarification and correction have been handled by the Kvestor (chief financial officer of the university), the Rector, a Vice Rector, and others suggests that this is an endemic problem; that the Foundation's moneys made available to BEU through the grant to City University of Bellevue, WA, could also end up commingled with other funds and inappropriately spent.[60]

We ultimately rescinded the moratorium on condition that City University audit every disbursement proposed by BEU, including verification that invoices for equipment were matched by actual deliveries. But Kevin Sontheimer shifted the site of the WEMBA program to the Faculty of Management at Comenius University, which was a reasonable decision. What was also distressing was that when I visited Rector Juraj Stern, unquestionably an honorable man, on May 27, 1997, he was unable to understand that the financial and accounting practices of his university were unacceptable and that the university officer who was in charge of BEU's side of the WEMBA program had made incomplete, inconsistent, and inaccurate statements. All he could see was that "his honor had been impugned." This, of course, is just an illustration of the cultural divide: what is accepted as standard practice differs widely from one culture to another, and suggests that success cannot be assured unless cultural norms are discussed and clarified before projects start.

How successful have been the efforts to create management programs in the region? Have winners emerged as yet? This may be a less controversial question for the three smaller countries, namely the Czech Republic, Slovakia, and Hungary. In the first of these, the U.S. Business School in Prague is an undoubted success and in Slovakia, Comenius University has no effective competition. Both of these institutions are examples of successful technology transfer. In Hungary, the Budapest University of Economic Sciences and Public Administration (BUESPA) and the International Management Center have dominated the scene (although the future of the latter may now be in question). The former is, of course, a Central European powerhouse in economics, as well. The other institutions that have economics programs of some note are the Universities of Miskolc, Debrecen, and Pécs, but they do not match the level of BUESPA.

The situation is more complicated in Poland, because the institutions offering management training are much more numerous. A recent article in a prominent Warsaw daily provided a ranking of the top 45 programs (Paciorek). The rankings were obtained from questionnaires sent to 115 professors of management and marketing at 38 institutions, who were asked questions about the quality of teaching, the market value of the graduates, and the research achievements of the faculty. The rankings were not easy to arrive at because of attribution problems: of the 597 tenured professors at 40 universities, 59 had positions at more than one institution. The article reported that the minister of education was aware of one professor who had 17 simultaneous positions at different institutions. In any event, the top 10 institutions according to the survey are, in descending order, (1) the Warsaw School of Economics;(2) the Academy of Economics in Kraków; (3) Warsaw University; (4) the Leon Koźmiński Academy of Entrepreneurship and Management; (5) the Academy of Economics in Wrocław; (6) the Academy of Economics in Poznań; (7) Jagiellonian University; (8) the Academy of Economics in Katowice; (9) the University of Gdańsk; and (10) the University of Łódż.

All the institutions that have management programs supported by Mellon are in the top ten except for the Technical University of Gdańsk, which ranked fourteenth. The decision of the Warsaw School of Economics to renegotiate the terms of its collaboration with the University of Minnesota was very reasonable, even creative, in retrospect: the attractiveness of the program at WSE seems to rest in part on the breadth of its course offerings, for which Sulejewicz must get credit. Overall, Poland has done extremely well. In spite of the occasional frictions between local and western teams, the differences in cultural norms and perceptions, the inadequacies of the infrastructure, and the episodic appearance of unenlightened self-interest, modern management methods have been permanently implanted in all the countries.

Some Agricultural Programs

By the middle of 1991, it was reasonably clear that useful work could be done to improve the state of agriculture in the various countries of the region. Agriculture has been a significant sector in the economies of all four Visegrad countries, but particularly in Hungary and in Poland. In the former, in 1991, agriculture accounted for 16.1% of the workforce and 15% of the Gross Domestic Product, while the corresponding figures for Poland were 27.3% and 15% respectively.[61] As usual, the recognition that there was room for work in this area was followed by a search for potential projects.

While I was keenly aware of my own limitations in this field, it seemed likely that the possible interventions in agriculture could resemble the more classical forms of technology transfer to a greater extent than in general management training, in the sense that in agriculture one could come to grips in a more direct way with specific agricultural processes and would not have to rely to the same extent on just classroom learning. Indeed, some of the early projects that I recommended for funding were of a rather practical sort. It is probably due to my lack of experience in this area that some of these were not particularly successful.

An early project of this type was proposed by the International Executive Service Corps (IESC), an admirable organization that provides "volunteer industry experts [who] provide assistance . . . to improve productivity or efficiency, to develop new products, train workers, or assist in the development of new systems and organizational structures."[62] In 1991, IESC had been very active in Hungary, Poland, Czechoslovakia, and Yugoslavia, and was giving serious thought to expanding its activities to the countries of the former Soviet Union. IESC proposed a number of alternative projects to Mellon, and we ultimately narrowed the discussion down to a research project that would investigate the hypothesis that in Poland one might profitably substitute rye feed in raising poultry for wheat and the other more usual feed grains. Rye had the enormous advantage of being much cheaper than alternative grains, and this appeared to be important, because Polish poultry could not effectively compete with imported West European poultry, which had a price advantage of 30%. The problem with using rye feed was that rye-fed poultry weighs only 69% of what corn-fed poultry does at the same age. The specific hypothesis to be investigated was whether combining rye with enzymes and antibiotics could compensate for this weight disadvantage. The project had the support of the Polish Ministry of Agriculture, and the research was to be carried out by a scientist at the Polish Institute of Animal Physiology; the role of IESC was to negotiate with U.S. enzyme manufacturers and persuade them to participate in the project and possibly assist in the creation of an enzyme manufacturing facility in Poland. But the initial hypothesis was not supported by the research data; some further work was done to explore the relationship in feed between viscous polysaccharides, enzyme supplements and fats, but no recommendations for drastic changes emerged from the work.

Another practical project was proposed by Heifer Projects International (HPI); an organization that has worldwide programs in sustainable agriculture and agricultural technology transfer.[63] The project, proposed in early 1993, was simplicity itself. Five thousand rabbits would be purchased and given to 1,000 Polish farm families in the Kielce-Radom regions; each would receive four does and one buck and undergo technical training in rabbit farming and elementary business training in accounting, marketing, and distribution. The project would have the assistance of the Polish Extension Service and of TechnoServe, another U.S.-based organization engaged in agricultural technology transfer. At the end of one year, each family would be expected to return one animal for each animal it had received, which would then be used to seed additional families with rabbits. Rabbit producer clubs would be started in each village where the project was operating in order to monitor progress and trouble-shoot problems. The project was expected to give farm families the experience of running a small business, provide low-cost, high-protein meat for farm families, improve Poland's export earnings by supplying rabbits to Western European countries, and provide high-quality, low-cost fertilizer for the farms. Alas, it did not quite work out as expected.

Enthusiasm for the project was uniformly high, and farmers generally expected that the experiment would lead to viable commercial activity. They adopted the rabbit farming techniques introduced by HPI quite easily and the producer clubs were effective. The administration of the project left nothing to be desired. But three serious problems drastically reduced the effectiveness of the program: (1) a new rabbit virus and a protozoa infestation caused high mortality; both were preventable, but only with high cost vaccinations and water treatment; (2) low income farmers used an unexpectedly high proportion of the breeding stock for home consumption and were unable to make a sufficiently high investment in cages, which had been shown to reduce the spread of infection among the rabbits; (3) the demand side of the market had not been properly investigated and the relevant cost–benefit analysis had not been carried out.[64] In fact, HPI was buying rabbits from the farmers at a price 50% higher than the market price and was also covering the cost of vaccinations, raising questions as to what would happen once HPI withdrew from the project. Vaccination, in particular, was expensive, because the timing of vaccinations required trained personnel to visit each farm at least once every six months. Overall, the project was not particularly successful.

A third, and much more successful project, was implemented by TechnoServe, which created the Center for the Promotion and Support of Agricultural Enterprise as a "development resource for farmers, entrepreneurs, unemployed factory workers, students, government agencies and other non-governmental organizations in rural, southeastern Poland."[65] It provided business advisory services, helped farmers with specific development projects and assisted communities with strategic planning; it was entirely successful in these ventures and by 1999, TechnoServe expected that it would be financialy independent within two years.

All three projects attempted to change agricultural processes directly. The fact that two of them were not particularly successful does not alter the fact that much could be gained from changing agricultural processes by direct intervention. It was also the case that each of these project had only a single facet, and the lack of diversification within the projects made them relatively high-risk. It was likely that university-based projects with a broader reach might be less risky.

Purdue University in Poland. My initial contacts with Purdue University were through its department of agricultural economics, which was much interested in improving the state of agribusiness in Eastern Europe. The initial venture was a short but intensive workshop in agricultural marketing for some 40 persons, which was held in Budapest, at the Budapest University of Economic Sciences. This initial contact led to a more enduring relationship with Purdue's International Programs in Agriculture and with Richard E. Gelzleichter, who managed a series of projects funded by Mellon. I was particularly interested in improving the state of agricultural extension services, because I felt that extension services in Poland were not working as effectively as they could. In the United States, the federal, state, and local governments collaborate with the land grant universities, and there is also close collaboration between university specialists and county-level extension educators on the one hand and producers, processors, and consumers on the other. Extension work is a two-way street in which problems flow from producers to experts, and solutions flow in the reverse direction.[66] In contrast, extension work in Poland had been the responsibility of the Ministry of Agriculture through its Agricultural Advisory Center (Ośrodek Doradztwa Rolniczego, ODR), which carried out its functions through 48 (now 15) administrative regions of Poland. But universities did not get involved in extension work in more than an episodic manner and there had never been adequate funding for university extension work (Gelzleichter, p. 8). Purdue seemed singularly well suited for improving the conditions of agricultural extension work, because Gelzleichter and the dean of agriculture at Purdue, Robert Thompson, began as early as 1990 to work with Polish agricultural experts, and, in particular, with Joseph Kania of the Kraków Academy of Agriculture. Trips between Poland and the United States, partly financed by the U.S. Department of Agriculture, further paved the way, and Polish and American faculty members, some ODR advisors and the newly elected rector of the Kraków Academy of Agriculture, Kazimierz Kosiniak, started in 1993 to explore the development of the Malopolska Association for Agricultural Extension.

I had suggested to Gelzleichter that it might be useful to broaden the reach of Purdue's involvement with Polish agriculture by also including in any potential project the Agricultural University of Warsaw and the Poznań Academy of Agriculture, whose rectors I had met before. Purdue developed a many-faceted program, which the Foundation had funded with $178,000 in early 1994, with an additional grant of $458,000 in late 1994, a third grant of $280,000 in 1996, and a final grant of $15,500 in 1998.

While minor details of the work changed over the years, the substance remained well focused on certain key aspects of agricultural technology. The principal areas in which know-how and technology were to be transferred were as follows:

1. Food science, stressing technology exchange and industry collaboration. The project stressed outreach to the food industry, created collaborative research opportunities, and offered technical seminars; it also established a food in-

dustry advisory group at each university. Training was to concentrate on preservation technology, the state of processing, packaging, food handling, quality standards, etc.

2. Water quality, since it is significantly affected by agriculture. Purdue proposed to provide technology transfer leading to the preservation and improvement of water quality.
3. University resource development, an area badly lacking in expertise throughout the region. The objectives were to improve the universities' communications techniques and media relations.[67]
4. Leadership development, based on the recognition that in a market environment, rural development cannot occur without the kind of personal leadership that was discouraged under the Communist system.
5. Rural development, including the training of substantial number of experts in this area. Emphasis was to be placed on how rural communities can resolve local problems.
6. University/ODR farm advisor extension training proper, covering business practices, agricultural businesses, extension services, government organizations, etc.
7. Integrated pest management, concentrating on pest and beneficial organism identification, monitoring techniques, economic management decision techniques, and the application of control tactics.
8. A poultry and pork research project, with a view towards improving meat quality.
9. Forage production and improvement of management practices in this area.
10. Entrepreneurship training and cooperative business internships.[68]

It is clear that the project had many relatively independent parts and needed to coordinate not only the parts, but the activities at the three Polish institutions. Not everything was done in all places; thus the water quality project concentrated on the Agricultural University in Warsaw and the poultry and pork project on the Agricultural Academy of Poznań. But most of the effort was expended in Kraków. Gelzleichter organized and managed this very substantial effort, which involved many Purdue professors in Poland and many Polish faculty members and ODR advisors at Purdue, with ease and finesse. Most importantly, Gelzleichter had a very important insight at the time of the second grant application, namely that a sum of $20,000 be reserved for each Polish university, to be awarded on a competitive basis for creative research proposals, a provision that was repeated at the level of $10,000 each in the third grant. This provision of the grant did much to accustom Polish scholars to the system of competitive grant applications.

The project transferred enormous amounts of technology to Polish agriculture in a number of areas. It contributed to the institutional development of Polish universities and placed extension and outreach on a solid footing, and created collaborative links between university scientists, regional farm advisors, government agencies, educational institutions, and business enterprises. The individual aspirations of the Polish institutions were respected, and while the Polish professors often traveled to Purdue as a group, they had latitude in cultivating their own interests. It is fair to say that the extension philosophy prevailing in the United States had been transplanted into the Malopolska region of Poland, solving the problem that the technology discovered in universities cannot be used easily or adapted quickly for the benefit of producers and consumers if universities are not involved in extension services (Gelzleichter). The links that had been forged between the academic institutions in Poland, ODR, and Purdue have become permanent.

The project transferred massive amounts of technology, and it is reasonable to ask what the key elements of success were. On the "mechanical" side, the obvious contributors to success were the excellent project management provided by Gelzleichter and the Purdue team, the solid administrative support from the participating Polish institutions, and the effectiveness of the faculty and student exchanges that characterized the project from the beginning. But even more important was the sheer enthusiasm for the project on both sides, the conviction of the Purdue faculty members that something important could be accomplished, and the receptivity to change of the Polish faculty members and institutions. Mutual trust developed rapidly, and the Polish participants were able to overcome the skepticism that was inbred during the Communist era and recognize opportunities (serendipities) when they presented themselves. Without these factors, in another place or at another time, the same project might not have succeeded. The combination of down-to-earth, practical technology with an idealistic approach was unbeatable. As Gelzleichter said, "To reach for the stars while our feet are still on the ground is what stirs the passion to search for more" (Gelzleichter, p. 18).

Cornell University and the University of Limerick in Hungary. Numerous West European universities had joint programs with East European ones within the framework of TEMPUS (see chapter 2). The University of Limerick (Ireland), created in the 1970s with assistance from the World Bank to stimulate economic development in Ireland, was one of these. Limerick had significant capabilities in business administration and in agricultural economics and management, and had participated in a wide variety of technology transfer projects in various parts of the world. It happened to have a joint project with the Agricultural University of Gödöllő (GATE), and when Barra O' Cinneide of Limerick discovered that Mellon had already made a grant to GATE, he began a dialogue with me and with faculty members of GATE and of the College of Agriculture and Life Sciences at Cornell University, with a view toward an ambitious knowledge transfer project to be implemented at GATE.[69] On the Irish side, Limerick also secured the participation of Teagasc, an Irish government agency that performs agricultural and food research, engages in education and training for the agricultural sector, and provides agricultural extension services. The dialogue culminated in a grant of $787,000 to Limerick in June 1994.

The extensive experience that Limerick staff had already accumulated at GATE convinced the project manager, O'Cinneide, that the rural economy of Hungary bore many resemblances to the Irish situation in the 1950s and early 1960s, and one could hope that the adoption of solutions that had worked in Ireland might also hasten agricultural developments in Hungary.[70] The proposal focused on five principal areas: (1) training in agricultural economics and management; (2) curriculum development; (3) technical assistance, (4) extension services; and (5) rural development.

The design of the project seemed tailor-made for Hungarian circumstances. The former state farms and cooperatives were being privatized, and the law on restitution placed some farm land in the hands of individuals not qualified to run a farm business. Even those units that chose to remain in a cooperative setting faced the problems of having inadequate technical and management skills. The quality of food products was generally believed to be lower than necessary for successful export to the EU, and efforts to provide training in all the relevant areas was likely to have a high payoff. The training was to take place both in Ireland and in the United States, where participants would be exposed to ten-week courses and workshops. The principal subjects were microeconomics, macroeconomics, agricultural engineering and food management, information systems, agricultural marketing, agricultural policy, rural development and land reform, and entrepreneurship. Faculty members from several agricultural universities, managers and technical staff from agricultural businesses, and state agencies were the primary audience for the training. Curriculum reform was to be undertaken at the Faculty of Agricultural Economics of GATE and the technical assistance was for improving GATE's holdings of books and teaching materials, providing computing, other equipment, software and various online services. The initiative to improve extension services was a collaborative effort with GATE staff as well as Ministry of Agriculture staff, aiming at a detailed assessment of the state of extension services and recommendations for improvement. The study was to provide a blueprint for the long-term development of advisory and training activities at the farm level by describing current practices, assessing the needs of producers and agribusiness managers, designing the extension system itself, and describing an appropriate funding system. Finally, the rural development component aimed at providing a pilot series of rural development projects with particular emphasis on agri-tourism or village tourism.[71]

The project shared many of the characteristics of the Purdue project and was comparably successful. Some 53 Hungarians were trained in Ireland or the United States for periods ranging from three weeks to four months. GATE passed the Hungarian accreditation commission's examination with ease. The project produced five new texts in economics, marketing, and management, and a new curriculum was produced. Several trainees obtained important positions in the Ministry of Agriculture and a joint MBA program was in operation between 1996 and 1998. Most importantly, close and durable research links developed among the participating institutions. The culmination of all these efforts was the "Vision 2000 Symposium" on June 3, 1998, which was attended by staff from universities and colleges, several government ministries, county agricultural offices, producers' lobbies, research institutes, food industry executives, and farmers. The symposium was so successful that planning began almost immediately for a "Vision 2001 Symposium," in the hope that it would become a regular event.

Summary Comments

It is quite clear from a comparison of the various management-oriented programs discussed in the present chapter with the CERGE-EI project (see chapter 6), that the creation of a doctoral program is substantially more difficult and costly than the creation of an MBA or MBA-equivalent program. To be sure, CERGE-EI was created *ex nihilo*, while the various management programs were mostly spliced onto existing institutional frameworks. But doctoral programs require both a breadth and a depth of analysis and knowledge that is not matched by the requirements of the MBA degree; nor does the latter involve students in the arduous research that is the hallmark of the doctoral dissertation. As a rough and perhaps superficial comparison, one may note that it takes a minimum of five years to provide a person with the economics education required to teach economics on the college/university level, while a year of study may well be enough to retrain a person with some management knowledge so that he or she can teach a particular course in a business school.

But that raises the question of whether the MBA projects in Eastern Europe provide the kind of training that western business schools aspire to. Criticisms of American practices at various times in the second half of the twentieth century argued that core business subjects should be taught in full-year courses, that rigor should be at a high level, that the MBA degree should be a general management degree rather than a highly specialized one, and that managerial problem-solving should be emphasized. In recent decades, sensitivity to the social and political environment has also become an increasingly important aspect of business education, and courses on business ethics are nowadays regarded as normal components of the curriculum.[72] There is no comprehensive analysis as yet of the extent to which the East European programs have a satisfactory coverage of core subjects and a reasonable set of undergraduate prerequisites. And it is not clear yet to what extent the East European management training programs are going to serve the above objectives.[73] Technique is easier to teach and to learn than insight and judgment, and it remains to be seen whether the broader approach to management training will have been successfully transplanted. But it is a good sign that some management training efforts, for example, at the Warsaw School of Economics, have recognized that behavioral skills and creative activities are important components of such programs.

A second major issue is the financial viability of educational programs, and this question pertains as much to programs that are not management oriented as to those that are. It is certainly the case that programs that are embedded in a functioning institution have a greater a priori potential for survival than those that are created out of nothing. It is invariably the case that new programs entail new costs, and an existing institution has at least the potential of diverting some of its regular revenues to fund new efforts. But it is not always the case that the institution perceives the program as meriting such bail-outs after the western support ceases. Substantial support did materialize in the case of the Kraków library consortium (from the Kraków institutions of higher learning), in the case of CERGE-EI (from Charles University) and others, but Šafárika University has not yet committed adequate resources for its own library and thus for KOLIN, nor has the Technical University at Gdańsk for the TQM program. But even CERGE-EI has found it difficult to survive in the long run without an endowment, which it is now in the process of raising, making it a unique institution in this respect. CEU is likely to survive in the long run, because recently George Soros endowed that institution with a gift of $250 million. Management schools have the rare (for Eastern Europe) opportunity to charge tuition, at least for executive MBA programs, which can result in long-run self-sufficiency. But institutions that are not embedded in a larger organization or do not receive adequate support from the parent organization and have only modest tuition-raising potential clearly face serious long-term problems. Cases in point are Academia Istropolitana Nova and, possibly, the International Management Center and the Czechoslovak Management Center. In retrospect, western expectations about how certain institutions will be able to "take off" on their own after an initial period of laying foundations and consolidating were far too sanguine, and it is not even clear that in the heady days of the early 1990s anybody seriously expected that western support would substantially diminish by the end of the decade.[74] It seems that while everybody agreed that institution building was an important objective in the region, we failed to realize that indigenous support for many of these institutions would be slow in coming.

Success or failure is hardly ever decided by technical matters as such. On the part of the recipients of the knowledge transfer, there are several important factors predisposing a project to failure, or at least to implementation difficulties. First among these is the recipients' inability to communicate in English; a problem that has tended to recede over time. The second is a legacy of the Communist period, namely a lack of initiative and motivation, an indolent attitude toward the benefits that students and professors can derive from improving one's own institution, and a lack of imagination with respect to what effort, innovation, and cooperation can accomplish. The third factor is the tendency of East European rectors and other institutional leaders to undertake commitments in excessive numbers, perhaps for reasons of prestige, without fully understanding the claim on resources that a project may entail. And finally, rent-seeking behavior on the part of participants may be a serious hindrance to projects. Some of the pursuit of self-interest is understandable in the light of the extremely poor pay that academic staff receives and as a result of the newly internalized free-market message that "the sky is the limit" as far as personal gain is concerned. But there is little doubt that cases of unenlightened self-interest have occurred.

Other problems may arise from inadequacies of the infrastructure at the recipient organization; not because such inadequacies may directly prevent successful completion of a project, but because the knowledge providers' perception of deficiencies may cause irritation and frictions. A related issue is the thinness of the management infrastructure at many East European institutions, which has the consequence that the unexpected departure of a key person is likely to have much more unfavorable effects than in a western context. Another broad category of problems emerges from inadequate attention to cultural differences; these may pertain to teaching methods, work habits, financial and accounting practices, and a whole host of other details. Lack of understanding of these may also cause frictions and retard implementation. Many of these cultural factors are legacies of the Communist period and it is not clear that western donors or service delivery organizations could hope to do much about this in the short run. Still another factor is the respect accorded to the members of the recipient organization as reflected by the extent to which the deliverers of technology seek or fail to seek sufficient input from them; this also means that in dividing up the budget between the two institutions (and mindful of the fact that American honoraria are vastly greater than East European ones) the bulk tends to be spent by westerners on westerners, leaving the members of the recipient organization with the feeling that they have been left out of the process and short-changed. A final factor is the degree of enthusiasm exhibited by the providers; this is partly a matter of demonstrated personal attitude, but is also partly a matter of organization: providers who fly in for three days, collect huge honoraria, and are never seen again create an entirely different impression from those that come regularly and stay for longer periods of time.

All of these factors stress the human side of the equation. It is easy to write equations on the blackboard; it is much harder to induce people with a 45-year history under Communism to "buy into" the process of enthusiastic innovation.

12

Economy and Society

The previous chapters have dealt with a varied menu of activities focused overwhelmingly on management training and economics education, on improving the state of national and university computing capabilities, and on modernizing research libraries. These activities had substantial effects on the development of free-market economies, on the modernization and westernization of educational systems, and on the integration of the East European countries into a mainstream western-style educational and research environment. But the same objectives could also be pursued by somewhat different means; that is, one could also think of reforming particular institutions of the economy, the political system, or society at large by more direct and practical actions. Thus, for example, it seemed entirely reasonable to help streamline the privatization process in the several countries.[1]

There is, of course, an endless list of potential direct interventions, but it made good sense to place some special emphasis on areas that I knew something about. This involved first and foremost the development of particular business enterprises or of regional industrial centers and the improvement of the financial institutions in the several countries. Closely related were projects to improve the quality of government operations. Most importantly, many of the useful activities that one could undertake were to recognize the importance of the shift from a strongly centralized, pre-1990 system to one that was becoming increasingly decentralized in several important respects.

Improving Business Practices

The East European countries were just emerging in 1990 from 45 years of centralized control, and the bad habits that people in universities had learnt during the socialist period characterized the business sector as well. Initiative was lacking, accountability and transparency were not well-understood concepts, even elementary methods for measuring performance were largely absent, and appropriate compensation and incentive structures had not been put in place. Adding to the difficulties of operating a competitive business was the confusion created by the multiplicity of new laws that had been enacted: Laws were frequently changed or simply added to prior laws without regard to contradictions or burdens on the private sector. Corruption within both the public and private sectors blossomed, to circumvent irrational regulations and gain wealth and power rapidly in what became a vacuum of both authority and responsibility.[2]

The business firms that were hardest to reform were the large state-owned enterprises, which had long been part of the socialist bureaucracy, and most projects designed to improve business practices found it more productive to concentrate on small to medium-sized enterprises, many of which had been created more recently. Indeed, the business advisory activities carried out by organizations such as the SEED Foundation and PASBAF (see chapter 4) did just that. But even small businesses had to operate in a stultifying legal environment. An example is provided by The River Company, founded in 1988 by three partners in Kraków. Their plan was to sell pure bottled still (that is not sparkling) water; but to do so, they first needed an appropriate government permit. They were told that pure water was not on the official list of permitted business products, and, hence, they could not sell this product; but if they wanted to, they could sell orange juice or sparkling water, or whatever. They then decided to sell computers; but at that time only individuals could import computers. Hence, they had to import computers from abroad as individuals, and then had to resell them to their business![3]

Small Projects. Efforts to improve the operations of concrete business entities ranged from the modest to the huge. Two projects of the former type were initiated in early 1992 by the University of Iowa and by Georgetown University. They were very similar to each other, and the two universities received seven small grants between them over the next four years.

Professor Henry Madden of the College of Business Administration at the University of Iowa created the Emerging Free Market Economy Program, which would send (initially) six MBA students under the supervision of Madden to Czechoslovakia for the three-month period of the summer holidays to provide hands-on consulting assistance to small and medium-sized businesses, although after 1994 the project concentrated all its efforts on the Czech Republic. Iowa's Czechoslovak partner was the Prague School of Economics, which identified the businesses at which the consultants would work. The project was to begin with a few days of seminars on business development, financial control, and general management issues, but the bulk of the summer was to be devoted to consulting activities, that is, to activities that would actually "fix" some troublesome aspect of a business entity. The first grant was for $20,000, which covered airfares and cost-of-living; no honoraria were paid to any member of the team. The project was designed with care and, before departure, the consultants received intensive language training and training in what they might expect with respect to the business practices in place. Project accomplishments ranged from identifying product diversification strategies for selling products in western markets to establishing communications between a Czech business and a U.S. manufacturer and to concluding successful negotiations between a Slovak automobile dealership and Volkswagen.[4] Because of the low cost of the project, it seemed extremely cost-effective and three more grants of comparable size were extended in 1993, 1994, and 1995. A total of about 64 person-months of consulting activities were carried out.

The Georgetown University project was largely MBA student initiated and operated within the framework of an organization called Students for East European Development, created in October 1990 under the auspices of the School of Business Administration. The project actually began in that year and student volunteers spent the summer months drafting business plans, revising production schedules and revamping corporate structures in the host countries. The 1992 project covered Czechoslovakia, Poland, Hungary, and Germany; 25 students and a supervisor were assigned to these countries (although only one to Germany). There was much variation among the types and sizes of firms to which students were assigned. They included a Slovak accounting firm, a large manufacturer of jet-trainer aircraft fuselages, a privatization consulting firm, the largest Czech producer of chemical equipment, and some six others. Most firms received a single student consultant, but a few of the firms were assigned several.[5] The initial grant was $32,600, again spent on airfares and cost-of-living, without any honoraria for any participant. The project's preparatory phase included the provision of training materials for the host firms, the prior analysis of software to be installed, market research and the identification of potential U.S. partners for the host firms, language training, and so on. Among the accomplishments of the consultants were the development of a functional organization strategy for a manufacturer, an evaluation of the financial and technical feasibility of producing pagers and other electronic products for the U.S. market, the financial analysis of a proposed methanol plant project, and the evaluation of the operations of a commercial bank and the development of a strategic business plan.[6] A total of about 110 person-months of consulting were provided for a Foundation outlay of $82,600, making this also an extremely cost-effective project.

To illustrate the difference between project conceptions, it is worth noting that in 1990, I received another proposal from the School of Business Administration of Georgetown University. This one envisaged holding two 15-week courses for 20–25 Czechoslovak managers at a cost of $675,000;[7] it should not be surprising that Mellon declined to fund this project.

The MBA Enterprise Corps. In a class by itself was the MBA Enterprise Corps (MBAEC). It was a project developed at the Kenan Institute of Private Enterprise at the Graduate School of Business of the University of North Carolina at Chapel Hill (UNC), and it, too, was planning to send young American MBAS as interns or consultants to East European business firms for rather more extended periods. I received a proposal in October 1990, and I initially declined it because it seemed as much intended to benefit the

American interns/consultants as it did the East European host companies.[8] It seemed expensive and it involved large-scale "people-movements" without a clear-cut training objective. But I changed my mind some months later because I realized that I had been wrong. I became convinced that the benefit of overseas consulting assignments to American interns was coincidental and that the bulk of the benefits would accrue to the East European companies; I also realized, as I should have from the beginning, that this was not a haphazard training program but one that would apply crucial remedies to particular carefully selected companies. In sum, the project seemed well worth supporting and a 1991 grant of $200,000 was followed by equal grants in 1993 and 1994, by a $150,000 grant in 1995, and a $75,000 grant in 1996.[9]

The not-for-profit Consortium of the MBA Enterprise Corps, Inc. was formed and its board was composed of faculty members of the 16 business schools (out of 18 invited) that had decided to join MBAEC. Volunteers for the Corps were sought out at the top 20, later the top 50, business schools, and they were asked to commit themselves to serve in the Corps for a 15-month period (of which three months were spent on training and orientation).

Extremely important for the success of MBAEC was the painstakingly careful preparation of the candidates prior to assuming their positions in some foreign country. It is fair to say that the criticisms one occasionally hears that American trainers or consultants were not sensitive to the nuances of the local environment were emphatically not applicable here. The training and orientation were carried out at UNC during the early years and in the host countries in later years. In each country, a country director was put in charge who saw to it that Corps members were familiarized with the environment and developed some facility with the local language. The training included segments on the responsibilities of Corps members, the role of the country directors, the process for switching assignments, relations with USAID, and a host of others. The position of country director was challenging, since they often had to supervise, make arrangements for, and facilitate the activities of the members, including the finding of living quarters. Each Corps member was required to sign an agreement which spelled out the terms of his or her engagement. In this agreement, they had to accept responsibility for the success of the program, commit themselves to the tasks assigned to them, exercise good judgment, and abide by the laws and customs of the host country. It was understood that the Corps members were in-service volunteers and not employees of the consortium, nor of his of her graduate school, nor even of the host company. Members were expected to adhere to a demanding code of conduct which included items such as the following:

> 1. I will be an optimistic and confident force for change in my assignment while maintaining a sense of humility, patience and tolerance, especially when facing real and/or perceived differences between me and local management.

> 2. I will always keep the interests of the host company primary and treat all information that I obtain as confidential. . . .
> 3. I will act as an adviser rather than a line manager when possible, even when a task could be completed better and more quickly by me; I will consider how I can offer value in the long-term and construct my behavior around this plan.
> 4. I will observe, or at least show appreciation for local customs except when in violation of the local law or personal ethics. . . .
> 8. I am committed to serving at my assigned host company for 12 months. . . .
> 11. I will accept no increase in local support/per diems (either in cash, equity, goods or deferred incentives) above the country maximum during my assignment. . . .
> 12. I will strive to conduct myself in a manner that reflects positively on the activities and the image of the Corps, and if I act unprofessionally so as to embarrass the Corps, I will accept the penalties specified by the Board of the Consortium. (Behrman, p. 72)

The standards were high and the financial compensation modest. Corps members' cost of living was paid by the host companies; in addition to this they received approximately $100/month from MBAEC as spending money and accrued for each month of service the sum of $333.33, which was paid to the members upon separation from the Corps, as a way of easing reentry into the American environment.

In the initial years, there were many more applicants than could be engaged with the available funds, and MBAEC was active in many countries. A total of 346 Corps members were employed in the Czech Republic, Estonia, Hungary, Poland, and Slovakia between the pilot phase in 1990 and 1996, with Poland accounting for the lion's share with 161 members.[10] In addition to these core countries, substantial numbers of Corps members worked in Bulgaria (54), Lithuania (26), Russia (27), and Ukraine (33), and smaller numbers in Brazil, China, Croatia, Ghana, Hong Kong, Indonesia, Kazakhstan, Kyrgyzstan, Moldova, Tanzania, Thailand, and Uganda. Total Corps membership in all countries was 525.

The scale of this operation was huge. The aggregate 11-year cost of the project (1990–2000) was $19,984,310, of which USAID provided (in round numbers) $11 million, host enterprises $3.6 million, the Kenan Institute and other consortium members $2.4 million (in kind), and foundations, as well as corporations, provided the rest. Among notable private contributors were the Czech and Slovak American Enterprise Fund, the Mellon Foundation, Sara Lee Corporation, the William R. Kenan, Jr. Fund, the Bulgarian-American Enterprise Fund, Aon Corporation, the Pew Charitable Trusts, and many others.

It is difficult to be precise about what MBAEC accomplished, partly because of the size of the project, but most importantly because of the highly varied nature of the Corps members activities. It is clear that they developed financial plans for privatization, created new marketing strategies, analyzed alternative product and market opportunities, developed feasibility studies, assisted in negotiations, facilitated reorganizations, redesigned retail store layouts, improved product quality, developed western contacts, and generally assisted the companies' managements. The formal positions that they held included those of account officer, advisor to the president, advisor to the general manager, marketing and financial consultant, investment officer, office products director, new business development manager, director of strategic planning, and many others. One Corps member reported to MBAEC as follows:

> To fulfill my responsibilities, I have done the following things during my first few months: (1) hired a marketing director for one of the old companies, (2) helped negotiate a contract dispute between two of our old companies, (3) designed a basic cost accounting system . . . , (4) made a valuation and cash flow projection for one of our clients to assist in a merger. . . . (5) networked with various US government support groups . . . , (6) designed a patent policy and license for a discovery, and (7) began designing a course for the Spring semester at the Warsaw Polytechnic Business School.[11]

This is from just one of literally hundreds of quarterly reports that acknowledge, without any immodesty, that the Corps members' contributions to the host companies were often of vital importance. Host company satisfaction with the program was extremely high and was regularly checked by MBAEC with the aid of questionnaires that host companies had to complete and submit. In my own travels I encountered quite a few persons who had high praise for MBAEC. But perhaps the most important output of the entire project was to change the mind-set of the host company managers. Initially, managers were often not even aware that improvements in company operations were needed, and even if they were vaguely aware, they often did not know what had to be changed; they had to be persuaded, which frequently proved to be a difficult task, because host company "managers thought anyone under 30 could not be experienced" (Berhrman, p. 18). One manager said, "We do not know what we do not know." Persuading managers to recognize the true problems and accept advice from consultants as young as they were was often a major human resource management problem. But a year was typically long enough to allow mind-sets to change. In the final analysis, the quantity of knowledge transfer that was affected by all three of these projects was enormous; but perhaps the change of mind-sets was their most important contribution. The fact that this was accomplished at an average cash outlay by U.S. agencies and foundations of approximately $24,500 per person-year is nothing short of remarkable.[12]

Regional and Sectoral Development

It is well known that under the Communist system central control of all aspects of society and of the economy was paramount, and regional or sectoral autonomy was essentially nonexistent. Autonomy by any subunits of society was perceived to be as dangerous to the state, as were alliances among similarly situated entities. The fact that alliances were frowned upon and not truly feasible is one of the reasons why collaboration among universities was rarer or weaker than in western countries. This fact is also one of the principal reasons why voluntary associations, such as foundations and other nongovernmental organizations, did not find fertile soil for their development prior to 1989.[13]

After 1989, the East European countries undertook, to varying degrees, a substantial decentralization of their governments. The drive toward decentralization was considerably aided by the insistence of the European Union that economically backward regions have credible development plans, which the EU supported with Structural Funds; hence, regional development became a means of advancing the cause of EU accession for the East European countries. George Peterson notes that the most pertinent EU principle is that of subsidiarity, which "holds that governance in different functional areas should be carried out by the lowest level of government which fully incorporates the external benefits associated with the function and can be administratively efficient."[14] Several local government laws were passed, for example, the Local Governments Act of 1990 in Hungary. The power to provide certain classes of service was devolved upon local authorities. New financing arrangements were introduced, which permitted the local governments to share in the central government's revenues and also permitted them to establish their own tax base.[15] It therefore seemed to make good sense to undertake programs for improving the functioning of the newly autonomous or semi-autonomous regional units.

Regional Programs in the Czech Republic. In January 1992, Bern Ewert, a principal in the firm of Ewert and Boone, visited the city of Plzeň (Czechoslovakia) and met with a large cast of characters, which included Rector Jiří Holenda of the University of West Bohemia;[16] the mayor of the city, Zdeněk Mraček; the members of the City Council, the minister of finance, and other high government officials.[17] Ewert was the former city manager of Roanoke, Virginia, and as I subsequently learned from a number of sources, an extremely successful one. His interest in rebuilding the city's administrative structure and enhancing its economic viability coincided with Rector Holenda's agenda, who wrote to Prime Minister Petr Pithart,

> With your permission and support, Plzeň would like to enter into a great experiment to launch into the mainstream of the fully developed world. Through the Department of Development at the University of West Bohemia, we propose retaining the services of an American consulting firm, whose general partners are a former city manager and a businessman with experience in owning and operating small businesses and banking. This firm would advise the City Council, coordinate with the University of West Bohemia, and assist in the establishment and operation of small and medium size businesses.[18]

I was approached by both Ewert and Holenda for funding such a project and it was fortunate that they found each other congenial, because it would not have been possible for the Mellon Foundation to fund a for-profit enterprise such as Ewert and Boone; however, it was straightforward to make a grant to the University of West Bohemia, which could then easily engage any consultant to the project it wanted. The basic thrust of the project was to enhance management skills in West Bohemia, particularly the management skills in local businesses and in the city government of Plzeň, and to revitalize the functioning of the economy in an around the city. The detailed agenda for the project consisted of (1) the creation of a master's program in public administration at the university; (2) the establishment of a management and development program that could be replicated in other locations; (3) the introduction of modern management techniques in the city government; (4) the provision of assistance to small and medium businesses; (5) the provision of other services to local businesses, such as matching Czech business capacity to western demands, and the enhancement of the region's appeal for tourism; and, finally, (6) the establishment of a development (fundraising) office at the university. While the first and the last subprojects were specifically directed toward improving the university, the others corresponded broadly to the agenda described in Holenda's letter to the prime minister. Little did I suspect at that point that items (2), (3), and (5) contained the seeds of a much larger and more demanding project. By February 6, I received a formal proposal from Holenda, which stated that the total project cost was $420,000; however, funding from the Ministry of Education and some other sources left only $243,000 to be covered by Mellon.

The grant was made and the project began. Much progress occurred during the next two years, but not as much as had been hoped and not without some significant frictions. The proposed two-year master's program of public administration had to be scaled down to a one-year program; otherwise it would not have been compatible with the Czech educational system. The university found it very difficult to attract competent teachers in the field and decided that it had to "grow" its own talent; accordingly we permitted some of the grant funds to be used for Czech students to study public administration in the United States, an activity not anticipated in the proposal. Much of Ewert's efforts was devoted to streamlining the Plzeň city budget. By June 1993, substantial progress had occurred and I noted that,

> While the process of entirely revamping budgeting procedures is not yet complete, enormous progress has been made and the budget, which used to be a 2-page document, is now beginning to look like the City of Roanoke budget, from which one can identify sources and uses of funds in complete detail, with ample budget narrative, and with a clear distinction between a regular and supplementary budget.[19]

But frictions did develop. Ewert had a temporary falling out with the deputy mayor (subsequently mayor) of Plzeň, Zdeněk Prosek. Ewert also got into an unpleasant altercation with an American Civic Education Project teacher who was teaching in the public administration program; I suggested to Holenda that he help them to work out their disagreements. But nothing came of this attempt at peacemaking.

Later in 1993, another grant ($50,000) had been made for the project, but this time to George Mason University (GMU), which had indicated a willingness to appoint Ewert as an adjunct professor for the continuation of the original University of West Bohemia project, as well as for a large project that was under discussion. During the course of the grants, a great deal of knowledge was transferred through explicit training and several visits by Czech officials to Roanoke and GMU. The most important accomplishment was to open the Czech officials' eyes to the functions that a budget can and should perform: (1) it is a document that expresses budget policies; (2) it serves as a financial plan by listing revenues and expenses, incorporates an accounting system and measures performance; (3) it is a guide to operations (personnel, timing, specific work performance); and (4) it has to serve as a communications device.[20] The existing software that had been used for budgeting consisted of several independent packages and was extremely inefficient; the effort to introduce a single, integrated financial system did a great deal to enhance local understanding of how a budget system should function. On the whole, we may conclude that the Plzeň officials learned that budget preparation should reveal which services are necessary, what minimum and maximum levels of services can be provided, what resources are necessary for providing particular levels of service and how resources can be used most effectively (Ewert et al., p. 12). By February 1995, I could report to the Mellon Foundation that the recent municipal elections in Plzeň did not, contrary to expectations, disrupt the budget reform process, that about 100 Plzeň officials in 45 city departments had been trained, and that the new director of the city budget had submitted his first budget based on the newly defined objectives and activities.[21] But in other areas the project did not live up to expectations: in particular, the reorganization of the entire city government did not go beyond an outline of a plan, and the model of administrative and budget reform had not spread from Plzeň to other cities, as had been hoped.

A new project proposal was submitted by GMU on October 8, 1993. In part, it described the steps proposed for continuing the work that was already under way in Plzeň. But the most important part of the proposal dealt with the establishment of a National Economic Development Certification Program. The proposal pointed out that the Czech Republic was widely regarded as "an ideal industrial location by many European, Asian, and North American companies. Unfortunately, many of these companies cannot find industrial *land* which is zoned and for sale on which to build their factories and are, therefore, unable to begin production and employment in the new market economy."[22]

In fact, an informal submission by Ewert in February 1993 had indicated that in the preceding four months, the city of Plzeň had received 12 requests for land from foreign companies with an employment potential of 1,400 persons, and the city had only two hectares (less than six acres) of land available. The certification program was intended to overcome the unpreparedness of cities for attracting foreign companies by establishing a comprehensive industrial development program in each of seven cities (regions) and was to rest on a partnership between the national and local governments. The program, to be endorsed by the Ministry of the Economy and the Ministry of Trade and Industry (headed, respectively, by two old friends, Karel Dyba and Vladimír Dlouhý) was intended to become a standard for all Czech communities. The key feature of the plan was that the program would be modeled after the National Industrial Certification program of the Commonwealth of Virginia.

In Virginia, a community must meet certain standards in order to become "certified." In order to achieve certification, a community must

- have a local economic development organization primarily responsible for industrial development;
- meet community information requirements, such as have a local marketing book and promotional brochure, a detailed community profile ready for distribution, community maps, audio-visual presentations, etc.;
- meet quality-of-life requirements, including a community assessment, a plan for dealing with deficiencies, and a leisure services program;
- have a local contact team for meeting with industrial prospects and one that is ready to provide a community presentation in a local briefing room;
- meet finance requirements by demonstrating the existence of an active industrial development authority, an industrial development corporation, local finance resources, and have available a local person well versed in local government finance;
- have an existing local industry program, consisting of an industry visitation team and an assistance program that facilitates access to local and state governments and that can provide local technical training.

In addition, there are site and building requirements; a community of the size of Plzeň (at the time about 200,000 inhabitants) needed to provide one hundred acres in four to ten sites, complete site and building profiles, and had to have detailed information about local construction capabilities.[23]

It is difficult to exaggerate the complexity of the tasks envisaged. The project was to select seven Czech cities; establish an economic development authority in each; train 15–20 economic developers; acclimatize Czech officials to the importance of providing timetables of events, accurate data, regular reviews, and updates; undertake community assessments and improvement programs; and a host of other tasks. In other words, the project had to undertake many activities in which, in the pre-1989 period, city officials had no experience nor an incentive to gain experience. It was hoped that a partnership would spring up between the relevant ministries in Prague and the various cities, and that local restructuring and economic development would have a beneficial feedback on each other. In my recommendation to the Mellon Foundation for a $559,000 grant to GMU, I said,

> For some time, I have felt that [the] funding [of] advice and reform on the central government level, and particularly in macro policy making, is no longer as cost-effective as measures taken to improve economic performance on the micro or at least more disaggregated level. Improving the performance of city governments strikes me as having very high priority. Particularly because more powers are devolving on cities than used to be the case under the Communist system, the lack of rational management on that level runs the risk of much misdirected effort, when the intrinsic potential of city economies is quite high. This view is clearly shared by Karel Dyba, the Czech Minister of the Economy, who told me in June that he had the highest opinion of the efforts that we had funded in Plzeň.[24]

The principal GMU faculty involved were Brack Brown and Louise White of the GMU Department of Public and International Affairs; Brown was budgeted to spend one-eighth of his academic year time plus two summer months on the two-year project; White would spend only one summer month on it. Ewert was to spend 100% of his time for two years. Various other persons were to be engaged as consultants: for example, two Czech trainers for a total of 260 days; a training coordinator and an economic development advisor, both for 260 days; a finance officer for 50 days; and a program coordinator in the Czech Republic, Michael Kunz, for 65 days; all of which adds up to a fairly significant commitment of resources.

News of the project spread rapidly and, by December 1993, the president of the Foundation for a Civil Society (FCS), Wendy Luers, had expressed an interest in having her foundation become a partner in the venture. She had a very impressive track record in the Czech Republic and Slovakia and had available to her an unmatched network of contacts. She made a number of useful suggestions for how the project could be improved, and an additional argument for FCS's participation was the expectation that Michael Kunz would agree to head up her foundation's Prague office. But in the end, no agreement was reached between Luers, Brown, and Ewert; nothing came of this and Kunz did not join FCS's Prague office. Some interest was also exhibited by USAID,[25] and in the end PHARE became an enthusiastic participant in the project.

The project clearly advanced the Czech government's interests in regional development policy, since EU accession criteria required the existence of such a policy, and so the Ministry of Economy became the project's sponsor, with the minister's personal assistant, Zdeněk Degen, representing the minister and chairing the steering committee. It was agreed that cities that earned certification would be awarded the certificate by the minister of economy on the advice of an advisory committee. Early in the game, the ministry requested CzechInvest to assist cities with training and advice on meeting program standards.[26] CzechInvest members were on the steering committee, as were two EU PHARE advisors, Howard Moody and Andrew Thorburn, and, for a brief time, the mayor of Plzeň.[27] PHARE had already allocated funds for city-based training and PHARE trainers were especially interested in the development and marketing of industrial sites by cities; they gladly participated in the training activities conducted by the project. Finally, the promise by the ministry and CzechInvest to steer foreign direct investments to the cities provided the right incentives for the cities to participate in the project and to abide by the requirements set up by the ministry, thus creating a desirable national–local partnership (Brown, p. 6).

A lot of progress occurred in the first year of the project. Nine cities were selected, since the original group of seven was augmented by two more at the minister's request. The nine cities were Brno, Hradec Královė, Louny, Nový Jičin, Olomouc, Pardubice, Plzeň, Zlin and Znojmo.[28] City officials were beginning to be trained in all the details of the certification procedure, including how to start assembling the relevant databases. Throughout the period of the project, it was quite evident that PHARE had a substantial interest in seeing the project become successful. In fact, by the end, seven cities had been certified and there is no doubt that the project as a whole was successful. But there were many turns in the road toward success at which the project could have gotten derailed, and many circumstances made it less successful than it could have been.

There is a saying that "the devil is in the details," and it is instructive briefly to recount that problems that were caused by personalities, bureaucracy, political instability, and obtuseness. First of all, all was not well within the project team. Brown and White (the "home team") were mostly in the United States (although Brown took frequent trips to Prague), and Ewert and Kunz (the "Prague team") were mostly in the Czech Republic. It was essential that the Prague team, plus whatever consultants they hired, be in regular contact with the home team. By July 1994, Brown felt that Ewert was not keeping him in the picture and suggested that I require him to report regularly, say every two months, both to him and to me, so that we would always be *au courant*.[29] I wrote to Brown and Ewert and suggested that a 5–10 page report would be desirable every four months, but I continued to receive very scanty and infrequent information from Ewert. Frictions between the home team and the Prague team became more frequent and vocal; Ewert made unilateral decisions and, perhaps not being familiar with the ways of universities, made salary promises that could not be delivered. Unpleasant arguments ensued about this, and much energy was uselessly expended on this issue. By November 1994, I wrote to Brown as follows:

> Let's first take on the issues of Bern's [Ewert's] lack of adequate prior consultation on various things, including commitments to new activities. I think it should be a matter of general policy that none of the senior project people make commitments without adequate advance consultation with the others. . . .

> Now as a practical matter, I am prepared to meet with all of you together if necessary, but I would prefer not to have to, because it would be an implicit admission that the senior investigators cannot work with one another without my presence. For this reason I think that there is real value (still) in trying to resolve all this among yourselves. But if it does not work, I will attend a meeting held at a suitable time. . . . Until this becomes necessary, my ground rule would be that all senior staff conduct themselves in such a manner with respect to the project and the commitments that are made, that AT ANY RANDOM TIME I CAN ASK ANY SENIOR PROJECT MEMBER TO TELL ME WHAT IS GOING ON AND THEY WILL BE EQUALLY IN THE PICTURE TO DO SO.[30]

None of this made much difference; while Ewert and Kunz were both technically competent and did good work, they just did not get along with the home team. Disagreements arose even with respect to who said what. Bern Ewert had made a suggestion at a meeting that was recorded in the minutes,[31] but he subsequently roundly denied ever having made such a suggestion.[32] Other administrative problems beyond the dynamics of the home and Prague teams arose, as well. The Prague team had office space at CzechInvest, which was irritated by its telephone bills; in fact, pressure was applied by the Czech side to make the Prague team pay for the office space it used in the Ministry of Economy This prompted me to write to Minister Dyba to ask him that some minimal office space be made available *gratis*, as a Czech contribution to the project.[33] Fortunately, my letter had the intended effect.[34] But the Prague team complained about the irritability and irrationality of some of the Czech participants and about the fact that the conflict between the Ministries of Economy and Industry and Trade forced Kunz to spend his time "putting out fires at CzechInvest."[35] It was fairly clear that in spite of its technical know-how, the Prague team did not only not get along with the home team, but had frequent troubles getting along with its Czech hosts as well. But *mirabile dictu*, in spite of all these frictions, by December 1994, the teams were working on a proposal for a supplementary grant. In February 1995, the supplementary grant request for $88,000 arrived and Brack Brown could report that

> The nation-wide Czech Certification program has earned a positive reputation in the Czech Republic and has reached a fairly advanced stage of development. Cities in all regions of the country are participating and external funding (U.S.A.I.D) has been secured to meet some of the U.S. training module expenses. Other partnering funds have been secured for in-country training from E.U. Phare.[36]

And six weeks later, the mayor of Plzeň, Zdeněk Prosek wrote,

> Our national government and the city participants can expect benefits from your program. Plzeň now has a system which it can operate with its own staff and without assistance of foreign advisors. The benefit of in-

dependence was demonstrated last summer when we developed and presented a proposal to Mercedes Benz for the sale of Plzeň's industrial site for their use to manufacture Mercedes Mini-Vans.[37]

All this was very good, but not as good as it should have been. In May 1995, the Advisory Committee unilaterally postponed the date by which the first cities were ready to be certified from December 1995 (the official end of the project) to May 1996, for largely specious reasons;[38] unfortunately Brown did not learn this until September (Brown, p. 10). The lack of adequate communication persuaded me that in spite of the good work done earlier by Ewert and Kunz, there was no point in reemploying them after the expiration of their contracts, and alternative arrangements were being looked into for the final Phase II of the project, to take place in the spring of 1996. But Degen had great admiration for both of them,[39] and since no city was to be certified before the end of 1995, we finally did reemploy them on a consulting basis for a month in 1996, so as to complete the various training modules and provide technical advice.[40] But even more delays and troubles lay around the corner. While the first few cities had been properly certified by March 1996, Degen left the Ministry of Economy for a new position. At that point, the Czech side of the project started to unravel:

- Oversight of the project was assigned to Jaroslav Novotný who promptly left for a month.
- During the month before a proper successor was named, the program received no guidance from the Czech side.
- In April, Minister Dyba appointed Vik Stepanek to succeed Degen; however, Stepanek was not enthusiastic about the project.
- In spite of initially agreeing to launch the Phase II training, Stepanek reversed himself and ordered all training to stop on the basis of a report by a junior official, Oldřich Praus, who knew next to nothing about the project.
- On the basis of this official's report, Stepanek ordered an extensive review of the whole project, delaying it even more.
- While some progress was made at a May 1996 meeting attended by Brown, Havelka, Stepanek, Thorburn, and Praus, and myself, the situation did not improve a great deal.
- Dyba was defeated in the elections and Stepanek himself left for greener pastures.

By this time, I was anxious about the level and quality of the Czech support for the project and about the future of the project as a whole. On November 22, 1996, I wrote to Prime Minister Václav Klaus as follows:

> I would like to call your attention to a project we have supported in the Czech Republic through George Mason University—the Economic Development Certification Program. . . . Its aim has been to prepare a first group of nine pilot cities to adapt some American methods of forming economic development partnerships between localities and the state. . . .

> To our satisfaction, five of the original nine pilot cities have received from the Ministry of Economy hard-won certificates which attest to their careful development of new organizational arrangements, properties, and informational materials. These efforts seem to have substantially improved their prospects for attracting new investors. . . .
>
> The present situation appears to be one where the organization and commitment necessary to sustain or expand the certification program are disappearing. We know that CzechInvest, with some help from PHARE along with a modest additional input from ourselves, is willing to try and advance the program. . . . We see no reason why it should be abandoned before another dozen or so cities are helped. . . . Our impression is that certification objectives are completely compatible with your objectives of encouraging more local initiatives and entrepreneurship throughout the country. . . . What seems to be missing is a strong signal from the Government that this is a highly valued and workable program.[41]

He replied in a terse letter to the effect that the project had the full support of the ministry. But not much more happened (only two additional cities were certified) (Brown). In October 2000, I had a useful lunch conversation with Havelka, in which he assured me that the project still had enormous value because, even though certification was no longer an ongoing activity, the principles of certification and of attracting foreign investors had permeated local governments on a broader scale. In fact, the new all-Czech phase of the project changed the thrust of the certification program and eliminated several of the certification requirements. What ensued was the creation of a few commendable regional development programs, but without the widescale spread to numerous individual cities empowered to seek out foreign direct investment, as had been hoped.[42]

In a sense, it was a miracle that many good results were obtained. The relationships within the American team left a great deal to be desired. Brown blamed himself to some extent for not defining the ground rules for Ewert more clearly, but I do not believe that this would have made a difference; Ewert was marching to the beat of his own drummer. It was fortunate that there was a great deal of competence and expertise within the team and that Brown suffered the peccadilloes of the Prague team with relative equanimity; otherwise the whole project would have been a disaster. The Czech bureaucracy was substantially dysfunctional: the Ministry of Economy had had three different ministers during the term of the project, new officials assigned to it had no familiarity with it and pursued their own questionable agendas, information flows within the American team as well as among the Czech participants were inadequate, and Czech officials often exhibited indifference, indolence, or obtuseness toward the project. Brown noted that "many department heads . . . were ex-communists or simply old leaders with set habits who could not adapt to fast paced change," and "Czech organization culture . . . can be said to be characterized by damaging levels of interpersonal as well as interdepartmental rivalries and jealousies."[43] Once again, the recipient group did not understand how much a project could accomplish if personal ambitions were swept away; it may be possible for a dedicated donor team to overcome the legacies of the previous regime, but it is very difficult to do so if the team itself is divided.

A Program in Logistics. In April 1992, Anne Morris of the Center for Logistics and Transportation of the City University of New York and Alain Kornhauser of Princeton approached me with a novel idea. The two of them had a long history of running a distinguished series of symposia on upgrading the management of global logistics operations and they thought that they could design a very valuable project to raise consciousness about market-oriented logistics in Czechoslovakia. While in my earlier years, I had done some research in transportation economics, I had never thought of logistics management as an area for Eastern Europe.

Their proposal noted that the former Communist countries were characterized by low levels of service in distribution and transportation; an obvious reason for this was the perpetual state of excess demand, as a result of which suppliers could easily ignore the needs of the end users. While the East European experts in transportation had excellent skills in technical subjects, they had very little conception of what it took to operate efficiently in a market environment and had not introduced an integrated mode of conducting transportation activities, which in the West leads to a continuous flow in the supply chain.[44] The proposal stated that the Czechoslovak rail and motor carrier enterprises were extremely inefficient: they employed 250,000 and 450,000 workers, respectively, in a rail and road network that would require five to ten times fewer workers in a U.S. or West European setting. It was not sufficient for efficiency that trains arrive on schedule, but the entire transportation problem would have to be embedded in an optimizing framework that would lead to cost and inventory control and provide service to customers. It seemed clear that in the new market economy, at least as far as road carriers were concerned, many new companies would spring up, and that there would be a crying need for managers well versed in market-oriented logistics. The core of the proposal was to "train the trainers"; that is, to train a cadre of persons in academia, as well as in the industry, for the sake of maximum technology diffusion. Designed originally as a three-week (later reduced to a two-week) training course, the program would concentrate on a business curriculum, a logistics management curriculum, and on case studies presented by senior logistics staff of western companies. The program would (1) teach functional excellence in managing a total logistics cycle; (2) emphasize evaluation of options and measurement of productivity and quality; (3) teach an understanding of the concept of integrated product development; and (4) attempt to establish an association of logistics professionals.

This was a simple but effective conception, and upon checking with Czech transportation sources, I found that the demand for such a program was considerable and that nothing along these lines had been offered by other western organizations. But the simplicity of the program should not obscure the immense efforts that Morris undertook to organize the project, line up participating institutions,[45] hold preworkshop planning conferences, and recruit senior executives, as well as academics, for the program. In particular, she was extremely conscious of the difficulty of getting senior executives to take three weeks off from their jobs and conscientiously to stick with a training program; hence, she made detailed arrangements with each participant's employer and received commitments that not only would they be released from their jobs for the workshop, but would additionally be protected by the employer from sudden and distracting work assignments. The Mellon Foundation made a grant of $175,000 for the project, and the workshop took place from June 7 to June 21, 1993. Forty-three persons registered for the workshop and by all standards it was very successful. Critiques by the participants were favorable and the small group sessions that were held regularly, as well as the case studies, became extremely popular. Morris had invited an impressive list of western industry experts to address the workshop, including Timothy Mellon, a trustee of the Mellon Foundation. The Czechs were so enthusiastic about the impact of the workshop that they decided on their own to run a logistics workshop in November 1993 and January 1994. What was particularly important from the point of view of technology diffusion was the varied background of the participants, ensuring that the knowledge transferred would be spread widely within the industry, and that the Czechs participants actually succeeded in setting up a professional logistics organization.

It was a great success, and by May 1994, Morris was considering a second workshop, modeled after the successful one in Prague, to be held in Budapest. In her usual fashion, she methodically explored the avenues of organizing the workshop, starting with getting commitments of participation and organizational help from the Budapest University of Economic Sciences. The proposal was not materially different, and Mellon made another grant, this time for $137,000.

The evolution of this project could not have been more different. Preparations started in December 1993 and the workshop was held in June 1995. The intervening 18 months were a period of broken promises and indolent neglect of commitments by professors and staff of BUES. Professor József Berács was to be the liaison person between BUES and the American team and became one of the bottlenecks, for he seemed unable to carry out agreed upon tasks in a timely fashion.[46] Relevant people were excluded from the steering committee. Efforts to address key issues (such as recruiting interpreters) were unsuccessful for a long time. Professor Attila Chikán, a distinguished economist and now rector of the university, declared that the project was unnecessary because BUES had taught integrated logistics for 30 years, and that, contrary to the previous agreement, the participants could not be expected to attend all sessions (Morris, p. 18). Periodic questions about money surfaced: BUES staff would have liked City University of New York to transfer funds to BUES, so that BUES could then pay local providers of services (such as interpreters), not understanding that the City University financial system required it to pay these service providers directly. The purpose of the preparatory meetings was to develop an administrative system for the workshop, but this was not achieved. Morris blamed the lack of common goals for the numerous failures, and she was right. But there was a deeper set of problems as well. First, BUES staff tended to think of BUES as the apex of the economics/management pyramid in Eastern Europe and showed signs of resenting the masterminding of the workshop by westerners. Second, the original agreement to participate in and perform certain housekeeping services for the workshop may well have been the result of the endemic desire to participate in prestige-enhancing activities, without either realizing the magnitude of the resource commitment that such a step may entail or feeling obligated to contribute to the success of the project. In the spring of 1996, after it was all over and I had received a final project report from Morris, I felt forced to write to Rudolf Andorka, rector of BUES and a personal friend. After describing the nature of the workshop and the general problems of organizing such events, I continued as follows:

> The general conclusion is the she [Anne Morris] encountered unremitting noncooperativeness and obstructionism, primarily from Professor József Berács, Professor Chikán, and Dean Kerekes. She clearly ended the experience with a bad taste in her mouth, and I must say that what she recounts is a major disappointment for me. I should hasten to say that the reception she received is hardly her fault, irrespective of what Berács, Chikán, or Kerekes may say: Professor Morris organized a similar workshop in Prague two years earlier and had nothing but praise for the cooperation she received from the Czech local organizers.
>
> The kind of conduct that she describes casts a very unfavorable light on Hungarian academics and on BUES in particular. I am, regretfully, calling these matters to your attention, in the hope that in the long run they will not be beyond remedy.[47]

I received a gracious reply from Rector Andorka as well as from Vice Rector József Temesi. While Temesi felt that Morris may have overreacted in the face of some problems, both of them acknowledged that mistakes occurred and both of them correctly pointed out that cultural differences accounted for part of the problem: that Americans are used to streamlined, efficient organization and that Hungary was not on that same level yet. Andorka, in particular, noted somewhat sadly in reference to my report of the project's success in the Czech Republic that "unfortunately Czech culture is more western than ours, nota bene has been more western for centuries."[48]

In spite of the difficulties, the workshop turned out to be a success. The participants were prepared to engage the issues, attendance was excellent, enthusiasm for the subject was high, and substantial technology transfer took place. What was most promising for the future was that Morris had met some very enthusiastic faculty members at the Technical University of Budapest (TUB) and felt that she could, with a one-week additional workshop to be held at TUB, cover the remaining tasks that the previous workshop at BUES failed to accomplish, namely deepen Hungarian faculty members' understanding of the usefulness of the case method and create an executive program in market logistics. Her request for an additional $32,000 was granted and the last workshop took place in April 1997, with considerable success. The TUB organizer, Katalin Tánczos, was enthusiastic and highly capable and the Hungarian episode thus ended on a positive note.

What is one to infer from contrasting the national industrial certification program in the Czech Republic with the logistics program's record in both the Czech Republic and Hungary? It is tempting to accept Andorka's pessimistic judgment that Hungary is not as "western" as the Czech Republic and that the failures in setting up the logistics program in Hungary were due to Hungarian inefficiency. But the manifest inefficiency on the Czech side in the certification program was equally pronounced. At the same time, we have seen numerous examples in both countries in which projects were implemented efficiently and with high degrees of competence—to wit, various business training programs, CERGE-EI, and numerous library and computer projects. My sense is that what almost derailed the two projects at hand was not primarily a question of efficiency, but rather of commitment and of alternative agendas in the recipient countries. In the midst of the drastic political changes occasioned by elections and ministerial appointments, the last thing that the Czech leadership cared about was keeping the certification project on an even keel. For an institution of the prominence and size of BUES, the logistics project was a mere speck, with no long-term consequences from pleasing or disappointing the visiting team. Fortunately, the certification team persisted (in spite of internal divisions), as did the logistics team, and hence the results were satisfactory. But both the Czechs and the Hungarians had erred in not being able to say no to a suggestion that they participate in a project when, in fact, they did not have either the resources or the commitment for wholehearted participation.

The Development of Banking and Finance

All sectors of the East European economies needed to adapt to western types of institutions before they could truly function as market-based economies, but perhaps none more than the banking and finance sector. There were, with sporadic exceptions toward the end of the socialist era, no credit or equity markets. One minor exception was in Hungary, where a rudimentary bond exchange functioned during the late 1980s. I recall attending a session of this exchange in 1988. The traders sat at one end of a large room and visitors such as myself at the other end. At one point, the presiding officer intoned lugubriously, "The Budapest Exchange is now open for trading. Do I hear any bids or offers?" Then there was silence. After a few minutes, one trader announced that he wanted to buy X bonds at a price of Y. Then there was silence again for a few minutes. Then somebody offered to sell some other bonds. Again silence. I observed the proceedings for some 40 minutes and in that period of time two trades were made! There is also an apocryphal story—and if it is not true, it certainly could have happened—about George Soros, who needed to provide immediate funds for some worthy philanthropic undertaking in Poland; but when he wrote out a check for $500,000 on the spot, no money could be put into the account of the donee, because the one and only bank officer capable of doing so was on holiday. It was therefore quite obvious that asset markets would have to be created as the economies made a transition to a system based on privately owned enterprises, and that in an economic system in which the state was to have a drastically reduced role, modern banking institutions were essential for mobilizing savings, extending credit, and providing a payments mechanism.

Developing Banking Institutions. I assigned high priority to projects that promised to train banking officials, as did all the countries in Eastern Europe. I knew from an early November 1990 meeting with Stefan Kawalec, director general in the Polish Ministry of Finance, that officials responsible for the transformation of the Polish economy considered training in banking the single highest priority. Much training was accomplished by the banks themselves. In Poland, for example, 23 banks trained some 90,000 of their employees in 1993 and about 100,000 in 1994, which accounted for 50% of the total employment in the banking sector.[49] But much was needed that could not be accomplished without assistance. One of the earliest, relatively small-scale, Mellon-funded projects of this type was initiated by K. J. Walraven, affiliated with Georgetown University. In the spring of 1990, he requested a small grant for undertaking an exploratory trip to Lublin and made good connections at Catholic University Lublin. The Lublin faculty was enthusiastic about a banking workshop, and Walraven's proposal envisaged two separate two-week training workshops, with each having two parallel sessions, in which Georgetown professors and some consultants would teach bank accounting, risk analysis and credit evaluation, commercial bank management, and commercial banking and financial services. Mellon funded the project and the workshops were very useful.

Another small-scale project, funded in 1992, was a three-week training workshop to be held at the University of Illinois (Urbana) for bank legal officers from Czechoslovakia, Hungary, and Poland. The rationale for the project was that the participation of legal officers is essential for the safe and efficient functioning of banks in a modern environment. Legal officers ensure compliance with regulatory standards, participate in risk management, promote transactional efficiency by supervising finance and credit operations, and participate in across-border transactions. The workshop was held in the United States. It included seminars on corporate and organizational structure, bank regulatory environment, payment systems, credit transactions, problem loans, and foreign bank operations in the United States. In addition, the 18 workshop members received practical training at a number of Chicago banks and law firms. Finally, a third small project was a series of workshops held under the auspices of the International Law Institute; it was addressed to high-level banking and trade officials in several countries on topics such as international loan negotiations and debt management.

A much larger and enduring project was one initiated by the Institute for EastWest Studies (IEWS). IEWS was formerly known as the Institute for East–West Security Studies, the principal activity of which was to conduct research and disseminate research papers on security issues such as arms reduction; most recently it has been renamed EastWest Institute.[50] The hope had been that the communication channels created by the Institute would reduce Cold War tensions.[51] But after 1989, it made sense to rename it and redirect its efforts toward all the problems of economic, social, and political transition in the former Communist block countries. In fact, as early as 1989, IEWS placed international economics and restructuring at the top of its research agenda.[52] By 1990, IEWS had established a permanent presence in Czechoslovakia by creating the European Studies Center at Štiřín Castle near Prague.[53] IEWS's president, John Mroz, clearly shared my views that banking was a matter of the highest priority and IEWS held a meeting on banking and monetary reform at Štiřín in November 1990.[54] Most importantly from my point of view, it created the Banking Finance Assistance Center (BFAC) and placed it in Budapest.[55]

Sometime in the spring of 1992, I was approached by Douglas Kruse, a young man with substantial experience in banking and finance, who wanted my advice about becoming the director of a Budapest center for advising banks. At the time I had no idea that this had anything to do with IEWS, but in any event, I suggested that for a person of his background and experience, such a position would be of considerable interest and importance. In the autumn, I met with John Mroz and board member Anthony Solomon,[56] whose idea BFAC was, and we discussed the possibility of Mellon support for this undertaking. The design of BFAC and of its mission also had the benefit of advice by Kálmán Mizsei, a distinguished Hungarian economist who had written important papers on bankruptcy. BFAC planned to address itself primarily to senior bank officials, senior government officials dealing with financial and banking matters, key politicians, local professional training institutions, and the media. It thus differentiated itself from certain other local organizations such as the International Bankers' Training Center in Budapest, which appealed primarily to middle-level management and taught more basic banking skills. IEWS sensibly decided to conduct such training in Eastern Europe, because it did not seem practical to bring the relevant senior people to the United States for training. I was also pleased that the IEWS proposal was extremely sensitive to the fact that there had already been a surfeit of seminars in Eastern Europe (on many subjects) and that diminishing returns were beginning to be noticeable; BFAC was to respond to these problems by selecting for workshops only those topics that were in the greatest demand at the moment. Being in the center of the relevant region meant that BFAC could respond extremely quickly to a perceived need. Workshop were planned on (1) the banks' role in advising companies in which they had an interest; (2) techniques for working out bad loans; (3) building an effective internal audit function; (4) managing the credit process in a branch banking system; and (5) selecting the right hardware and software for the payments system. Among policy-oriented topics, the most immediate needs were for workshops on (1) managing bank privatization; (2) the role of deposit insurance; (3) the formulation of appropriate regulations for the thrift sector; and (4) the role of the government in addressing the bad loan problem.[57] BFAC had excellent connections with western financial institutions (for example, Chemical Bank, JP Morgan, Credit Lyonnais, Price Waterhouse, and the European Bank for Reconstruction and Development) and was going to obtain for the workshops the *pro bono* services of senior officials from these and similar institutions. BFAC seemed tailor-made for the East European situation, and the Mellon Foundation provided a total of six grants amounting to $422,500.

BFAC started to run well-attended seminars on the various topics, and was fortunate to have outstanding senior bankers from JP Morgan and Credit Lyonnais as lecturers. The seminars were usually one to two days in length and were attended by 15–20 people, representing a half dozen banks. Activities focused on Hungary, Slovakia, and Poland and were generally targeted to the nationals of the country in which the workshop was held, because mixing the audience might have created difficulties as a result of the various countries' different regulatory environments. In 1994, Gene Spiro took over as director of BFAC and provided it with outstanding leadership until 1999. A great deal of knowledge was transferred, and Hungary and Poland in particular developed into mature banking environments, aided in part by the influx of foreign banks. The major banks in Hungary were privatized (Hungarian Foreign Trade Bank[58] and National Savings Bank[59]) or sold to foreign interests (Hungarian Credit Bank[60] and Commercial and Credit Bank[61]). Since this circumstance made it increasingly difficult for foreign banks to enter the banking market, the perceived benefits to the western banks of contributing their top executives' time on a *pro bono* basis diminished, and it became harder to obtain the services of these experts for BFAC. The Hungarian, and to some extent the Polish banking sectors, became the most sophisticated in the region. Spiro was entirely right in guiding the activities of BFAC in later years both toward more general types of economic reforms[62] and toward other countries, some located in the NIS. Albania, Bulgaria, Romania, Kazakhstan became the next targets, and Spiro strongly encouraged "East–East" assistance, which relied on Visegrad country experts to teach banking. Also, as the banking communities in the original countries became more familiar with the nuts-and-bolts problems, the emphasis in banking itself shifted to the policy area and topics such as bank privatization, banking supervision, deposit insurance, and the like became more important in BFAC's menu of offerings.

In the period 1995–98, IEWS gradually returned to its original mission of dealing with security issues, and in 1999 it turned its attention to Russian and Bulgarian reconstruction. It made sense for BFAC to seek a new home, and on June 1, 1999, it merged with the East–West Management Institute, a not-for-profit organization established by Soros in 1989. It continues to function under these new auspices. BFAC has had a major impact on renewing the banking sector. As Spiro summed it up:

> In the context of providing assistance to the formerly centrally planned economies, supporting countries and societies that are turning toward transparent, democratic systems of governance, the mission was and continues to be critical. The empowerment of the middle classes, the ability and opportunity for individuals to earn, save, invest, build and derive profit (or loss) are preconditions to democracy and civil society taking root. Thus BFAC's agenda was, and is, on target. . . .
>
> Deploying experienced practitioners from mature markets in the United States and Great Britain is of significant value. . . . The fact that Central European countries are, in their turn, quickly becoming experienced with the west and, indeed, are becoming globalized, plays into the hands of assistance providers who recognize the ongoing need for their work in new and more distant region. (Spiro, pp. 10–11)

One last organization that had significant support from Mellon in this functional area was the Financial Services Volunteer Corps (FSVC). Over the years, it received three grants totaling $350,000. It was created in April 1990, in order to provide *pro bono* technical assistance from the U.S. private sector to the countries of Eastern Europe.[63] FSVC's initial, as well as much subsequent, funding derived form USAID, and in many ways its activities paralleled those of BFAC. Two significant differences were noted by Spiro between FSVC's and BFAC's modes of operation: the former used almost exclusively U.S. personnel and its training activities were mostly "one on one," meaning that a U.S. advisor would work closely with a particular bank or government official for a few weeks.

FSVC's activities covered many countries. Initially, it concentrated on Hungary, Poland, and Yugoslavia, but by the fourth quarter of 1998, it was operating simultaneously in 11 countries, namely Albania, Croatia, Hungary, Macedonia, Poland, Romania, Kazakhstan, Kyrgyz Republic, Moldova, Russia, and Ukraine. In Albania, FSVC dealt with the privatization and rehabilitation of the commercial banking sector, the development of the central bank's regulatory capacity, and with legal and technical assistance to promote development of capital market institutions; in Croatia it promoted the transparency, efficiency, and liquidity of capital markets and privatization through training; in Hungary, it improved banking and capital markets supervision, and the national payments system; in Poland, it introduced pension reform and promoted sound insurance practices; in Kazakhstan, it improved the financial payments system and central bank auditing; in Russia, it strengthened the financial sector infrastructure.[64] The program in Russia was started in 1992 and since that time has encompassed 221 project activities.[65]

The total number of live projects in the fourth quarter of 1998 was 54, the total number of institutions served 165, and the value of donated services was estimated at $2,070,000. There is no doubt that FSVC's activities had a massive impact. In the spring of 1999, John C. Whitehead, cochairman of FSVC's board, wrote,

> Mellon's support has enabled FSVC to address a number of critical issues related to financial sector development in Russia, the Central Asian states of Kazakhstan and the Kyrgyz Republic, and, to a lesser extent, the Caucasus region. In Russia, the Mellon funding has helped underwrite our work with reform-minded officials of the Central Bank of Russia and the Duma in support of their efforts to restructure the banking system; FSVC is undertaking this work in close cooperation with the World Bank and the IMF. Mellon's support has also helped to underwrite numerous FSVC seminars and workshops for Russian central and commercial bankers on important payments system issues, including the development of a nationwide real-time gross settlement system, regulation of payment cards, and prevention of money laundering. While economic and financial conditions in Russia remain dire, FSVC is playing an important role in channeling Western expertise into Russia to address complex financial problems, and in promoting an ongoing dialogue between Russian and American financial practitioners at a time when many Western institutions are turning away from Russia.[66]

The breadth of FSVC's activities is unsurpassed and there is no doubt that it has had a lasting effect on the region.

Municipal Credit. Decentralization of governmental functions was taking place throughout the region, partly as a reaction to the highly centralized system in place during the socialist era, and partly as a result of a growing recognition that decentralization empowers people by giving them some control over their fate and, thus, enhances civil society. As early as the Round Table Conference in Poland in the spring of 1989, the Solidarity representatives demanded that local councils be released from dependence on the central government and that municipal governments be reestablished as legal entities with property rights.[67] Hungary was very much in the forefront of decentralizing the responsibility for providing government services at the local level. One of the first acts of local finance reform dates to 1986, which gave local governments some small responsibility for investment decisions.[68] But on the whole, during the Communist era, the central government was responsible for capital investments of all types on the regional and local levels. In 1988, the personal income tax and the Value Added Tax were introduced in Hungary, followed in 1990 by the Local Government Act, which devolved decision-making about most public services upon the local governments. The local governments had four sources of finance after 1990: shared taxes, normative grants from the central government, their own tax base and other resources (for example, sales of their assets), and borrowing. Allocations from the central government have been generally declining: thus, for example, local governments were allocated 50% of the personal income tax in 1992, but only 40% in 1999 (Pigey, p. 22) It is generally believed that a lack of political will and insufficient tax administration capabilities have caused the local governments to fail to exploit adequately their taxing powers; moreover, they have sold so many of their assets in the 1990s that this source of funds will also not be able to provide future revenues at satisfactory levels. An additional problem is the very large number of municipalities (about 3,200 in 1999), a large fraction of which is extremely small, with the result that they are unable to provide certain services efficiently. These considerations have called for the possible creation of an intermediate tier of government which would be able to provide these services more efficiently.[69] Lending sources were extremely limited in the early 1990s, and while some municipal governments had actually issued bonds, the issues were very small and illiquid, and no genuine bond market existed. Substantial portions of investment budgets had to rely on cash on hand or the sale of municipal assets.

It is against this backdrop that I was contacted during the summer of 1993 by Thomas H. Cochran, executive vice president of the Resources Development Foundation (RDF), a not-for-profit organization specializing in the transfer of technology that promotes the sound use of natural resources. RDF already had a Pew Charitable Trusts-financed program in Poland, where the Local Self-Government Act of 1990 had turned over to the local authorities a large set of new obligations.[70] The RDF resident advisor, Anthony Levitas, was assisting the Polish government with the development of local infrastructure financing mechanisms. The discussions between Cochran and myself led to a proposal by November 1993. RDF proposed the creation of a Municipal Credit Program (MCP) in Hungary that would be able to manage a complex loan structure and possibly issue its own debt as well. The MCP would foster a professional approach to project management, increase economies of scale, help mobilize foreign sources of funds, subject projects to the discipline of the market, and allow infrastructure projects to be more demand driven. While creating such mechanisms in the framework of an MCP might have seemed too optimistic, its key feature was to place an expert advisor in the appropriate Hungarian government agencies for a minimum of one year to (1) assist in evaluating the current status of the municipal market mechanism; (2) evaluate and recommend improvements for credit review mechanisms; (3) improve the nature of legal debt instruments and advise on the development of application procedures for MCP loans; (4) assist in the development of foreign sources of credit; and (5) train Hungarian officials. Cochran, who had gained great experience in the field as senior vice president in the public finance division of Lehman Brothers, was to spend 25% of this time on the project, as well.

Mellon made an initial grant of $240,000 to RDF in the spring of 1994. Soon thereafter, the Ministry of the Interior, the Ministry of Finance, and RDF agreed to establish an advisory position in municipal capital market development, and the talented and experienced Hungarian-American

Charles Jókay was hired for the position. By June, he was working with ministry officials on the enforceability of contracts in the municipal sector, bankruptcy provisions for municipal governments, the pledging of "nonfundamental assets," debt service reserve funds, "trip wires" to warn of possible debt service payment problems about to occur, the nature of prospectuses, the municipalities' cost of capital, and other related issues.[71] Cochran was enthusiastic about the process that had been initiated because of the excellent relations that had been established with Hungarian officials and because the discussions revealed some fundamental problems that needed to be solved, such as the fact that regulations required a municipality experiencing financial difficulties to continue to pay for water service, police, etc., but none required that debt service continue to be paid. Another even more serious problem was that there was at that time no "nonimpairment" clause in the Hungarian constitution, that is, one that prohibited the Parliament from enacting legislation that would impair pre-existing private contracts.[72] By December, it was clear that more work had to be done: Hungarian officials needed more training and more help in setting up credit review mechanisms and guidelines for appropriate interest rate differentials, and further assistance was needed for the design and implementation of externally financed pooled lending and relending mechanisms. Another grant of $220,000 followed in the spring of 1995, which would carry the project into 1997.

What did the project accomplish? First, it transferred a broad range of technical knowledge to the responsible Hungarian officials. But equally important was the concrete embodiment of the work by RDF in two acts of parliament: the Municipal Debt Adjustment Act of 1996 and the Securities Act of 1997. The former contains provisions "that are quite sophisticated and impose definite financial and moral costs on local governments who default on debt or other payments" (Pigey, p. 35). The law aims to (1) prevent municipal defaults; (2) provide clear administrative procedures; (3) provide for reorganization and workout; (4) maintain public services; and (5) allow for an increase in borrowing provided that local taxes are increased.[73] It also specifies the mandatory services that have to be maintained during debt adjustment. In fact, 11 cases have been adjusted under the Act up to the autumn of 2000; seven of these ended with a workout agreement, two with liquidation, and two were still in process. The Securities Act deals with the issuance and trading of local government bonds (Pigey, p. 35); the first bond under the Act was placed in 2000.

Hungary has clearly joined Poland and the Czech Republic in the forefront of municipal finance reform in the East European countries. It is obvious that there was a strong commitment on the part of Hungarian officials to streamline municipal finance, and it is equally obvious that the personalities on the two sides meshed well. But Cochran notes three very special points. First, when the advisor was selected for the position in the ministries, the Hungarians were asked whether they preferred a generalist, who knew the country, or a specialist, who probably would not. They strongly preferred the former, which is what Jókay was at the time of his appointment. Secondly, Cochran believes that working with a Mellon grant (rather than, say, a USAID grant) provided much flexibility in designing the project and in dealing with Hungarian officials. He recounted that

> I was comparing notes with a senior USAID official in Warsaw who was setting up analogous initiatives . . . I mentioned that my initial entrée with the Polish government had been through the nation's powerful budget director—she expressed surprise (and some indignation) because State Department and USAID protocol restrictions had made it impossible for her to even approach the Budget Director (and numerous other officials).[74]

Finally, being funded by a private foundation persuaded the Hungarians that the American team had no vested interest in any one particular solution; that the advice given "would come without any governmentally imposed limitations, biases or agendas." This last point was certainly true for all three, largely complementary, major programs in the banking and finance area described in this chapter, and they were all successful in transferring important technology.

Government Efficiency and Civil Society

Attempts to improve the efficiency of government operations was an entirely natural accompaniment of democratization and of the initiatives to put business enterprises on a more solid footing. Government officials were no better trained than enterprise managers, and projects to remedy this lack were fairly numerous, although—in my experience—not as numerous as those attempting to transfer technology to business enterprises.

One of the earliest projects that came to my attention was directed by Professor Tomasz Mroczkowski of The American University. He correctly recognized the trend toward decentralization and his project focused on local government officials in Poland. The political changes in that country had placed considerable burdens on local officials who were unfamiliar with the implications of restructuring and who were to play increasing roles in privatization, employment transfers, training, local government finance, and the provision of services now devolved upon the local governments. It was deemed essential that such officials be capable of making informed decisions about the relationship between central and local governments, the legal and policy aspects of private business development, the promotion of entrepreneurship and small businesses, public relations, and local budgeting and taxation. The program he proposed, and which was funded by a $134,000 grant had two parts: first, various experts from American University traveled to Poland and conducted seminars in several locations, and, in the second part of the project, 24 Polish local government officials visited the United States and participated in discussions with business and government officials in Washington, D.C., Chicago, and Springfield, IL.

Another early project, proposed by the Executive Council on Foreign Diplomacy, planned to arrange one-year internships for Hungarian, Polish, and Czechoslovak diplomats at large U.S. corporations, where they were to be trained in commercial and investment banking, financial management, industrial production, business–government relations, strategic business planning, and related subjects. Only small numbers of individuals were targeted to come each year, and the $150,000 grant from Mellon was to pay only a fraction of the total cost: the interns' home ministries and the host companies were to provide the bulk of the needed funds. In retrospect, I thought that this was too little to make a substantial difference and it was probably not as successful as I had hoped.

A third project was truly off the beaten path. I was contacted in the fall of 1991, by Zsolt Szekeres of Information for Investment Decisions (IID) in Washington, D.C., who explained that IID was engaged in a project to modernize the office of the Hungarian prime minister. IID had undertaken a preliminary assessment and determined that much of the hardware and many of the procedures used in the prime minister's office were antiquated and needed drastic overhaul. The proposed overhaul dealt with the schedule of the prime minister, the preparation of government sessions, the information flow and the preparation of decisions, work procedures including a computerized document control system, security measures, and the like. The project was funded with a grant of $205,000, but the State University of New York at Albany was asked to be the administrator of the grant, because IID was not a 501(c)3 organization. While I had no doubt that much useful work was done in this project, I declined to entertain the thought of a follow-up grant.

These were straightforward projects; but there was a class of related projects that not only aimed at improving the functioning of some level of government, but also shaded more or less explicitly into creating and strengthening civil society. Kevin Quigley notes that probably not more than 1% of the $108 billion dollars of assistance provided by the international community to the former socialist countries between 1989 and 1996 was spent on civil society projects.[75] Moreover, the evaluation of NGOs in the region is not unambiguous. Erik C. Johnson views one class of these, namely public policy institutes or "think-tanks," as having been quite successful in capturing the attention of policymakers.[76] But Quigley's analysis of the effectiveness of U.S. efforts, mostly undertaken by USAID, is somewhat pessimistic. Americans tend to have an abiding faith that NGOs, functioning in a realm not adequately covered by the government or market forces, create or encourage citizen participation in their own fate, which is an essential characteristic of civil society. Quigley's reasons for pessimism include the substantial difference between the American and East European mind-sets, the cultural difference that I have alluded to before, the focus of USAID efforts on public policy–oriented NGOs rather than those with a service orientation, which made them easy targets for becoming politicized, and the bureaucratic struggles between USAID and other organizations.[77] He concludes that NGOs, civil society, and democracy are only loosely related and that donors had only a modest impact on civil society or democracy (Quigley, pp. 191–215).

While the Czech Republic and Slovakia probably had the weakest nonprofit sector, the growth of this sector in Hungary and Poland was rapid and, as might be expected, gave rise to numerous early abuses, such as the creation of nonprofit foundations by for-profit enterprises in order to evade taxes (Siegel and Yancey, pp. 29–32). But in spite of some tightening of the applicable laws, growth was rapid, and in Hungary the number of NGOs grew from 800 in 1989 to over 8,000 in 1992 (Siegel and Yancey, p. 19) and to about 30,000 in 1994 (Johnson). The failures in (mostly) USAID-guided efforts to foster civil society through NGOs are balanced by the fact that precisely in the area of service delivery, which accounted for an even more vanishing fraction of total assistance given, NGOs have made some important contributions.

One of the most memorable and important of these was the Foundation in Support of Local Democracy (FSLD). Founded in 1989 by a group of Polish senators and deputies, such as Jerzy Regulski, Aleksander Paszyński and others, it initially aimed at training local government officials through regional training centers and schools of local and public administration. Throughout the 1980s, Regulski and his colleagues had been developing strategies for decentralization reforms.[78] FSLD cooperated closely with Local Democracy in Poland (LDP), a project headed by Professor Joanna Regulska, Regulski's daughter at Rutgers University.

Its general aims were "the creation of new institutions that could facilitate the spread and ideals of local democracy, and could also provide mechanisms for making local government accountable and transparent." Motivated by the importance of "accountable and transparent local government run by skilled professionals, and citizens' understanding of and ability to act within a local self-government system,"[79] early specific projects dealt with preparing local election candidates, developing a local government training network, building local democratic institutions in Poland, promoting local self-government, training for Polish mayors, and so on.[80] It was obvious that the ambitious agenda required the training of vast numbers of people. The proposal, jointly submitted to Mellon by FSLD and Rutgers and funded in 1991 in the amount of $298,000, was for the creation of a network of libraries at 15 regional training centers that served 2,500 communities and provided short-course training in economics, environmental problems, administration, management, law, office technology, and foreign languages for local officials, as well as the public at large.[81] The overall mission of FSLD was and still is[82] to undertake technical support for local government, to support the building of regional levels of government, and to support the development of civil society and improve cooperation between civic groups and government.

While training activities had to rely heavily at first on foreign experts, they were designed so as to place over time increasing reliance on local talent, until it could stand entirely on its own. By 1998, FSLD operated schools of public administration in Łódź, Szczecin and Białystok, with a total enrollment of 2,824 and 15 training centers, at which 2,172 "training events" had taken place in the year with a total of 50,433 participants.[83] FSLD held more than 30 conferences in 1998, created clubs and forums functioning on the gmina level, engaged in consulting activities, provided information technology assistance to municipal offices, helped local governments with various planning activities, engaged in civic education projects, and trained local mass media personnel. From the beginning to the present, more than 350,000 persons had participated in its training activities and over 70,000 copies of its publications had been distributed in Poland (Regulska). Most recently, FSLD has engaged in East–East assistance by exporting its know-how to Belarus, Ukraine, Slovakia, Hungary, the Baltic countries, Albania, Serbia, and Macedonia. The total budget for 1998 year was approximately $4.8 million; equally important is the fact that FSLD has become entirely self-supporting, with 60% of its revenues from fees for educational and training activities and the balance from consulting. As a result of Regulski's vision and the organization's understanding of and timely response to what was needed, it has become a profoundly influential and important organization throughout Poland and the rest of Eastern Europe, and is one of the great successes in creating effective civil society organizations.

In the Czech Republic and Slovakia, a vital role was played by the Foundation for a Civil Society (FCS; formerly Charter 77 Foundation) under the leadership of Wendy Luers.[84] The activities of FCS reached nearly every facet of training and civil society rebuilding and played a role that was far beyond the immediate impact of particular projects by working as a catalyst in initiating and developing projects, acting as a facilitator by virtue of its network of contacts, and acting as a partner in many collaborative projects.[85] Some of its early projects included a conference for Czechoslovak leaders on constitutional issues, various forms of assistance to the independent media, a civic education project which sent American professors to Czechoslovakia for teaching courses in social sciences and the Masaryk Fellowship Program, which brought 95 Americans to Czechoslovakia in its first year for teaching English. By 1994, over 3,200 Czech and Slovak students had benefited from the program.[86] FCS collaborated with the Soros Foundations, the Institute for EastWest Studies, the East–West Management Center, and others. In 1991, it started a major project entitled "Justice in Times of Transition"[87] which was ultimately spun off in 1999 to Harvard University. In March 1992, it established an "Expert Advisors Program," generously funded by the Ford Foundation and the Pew Charitable Trusts, which functioned for a number of years and placed high-level advisors in key positions at various Czech and Slovak government agencies.[88] The advisors' work was much appreciated by the host agencies, which often requested extensions of their terms of appointment. There is a strange collective cognitive dissonance concerning western technical assistance. There are those who believe that Central and East European officials have considered western technical assistance inflexible, as well as costly, that western experts have often failed to take into account the special circumstances and needs of those they were assisting or advising.[89] And there are those, myself included, who have found that much western technical assistance was based on sensitive explorations of the needs of the recipients. It is, of course, possible that both views are true for disjoint subsets of cases, but if that view is correct, it appears more difficult to generalize than is commonly believed by some of the critics of western efforts.

By 1998, the need for FCS programs was diminishing, as was the willingness of funders to support programs in the Czech Republic and Slovakia. As of July 31, 1999, FCS restructured itself: it became a much leaner organization and was able to spin off its offices in Prague and Bratislava, which had succeeded in becoming fully independent. Luers and FCS have made major contributions to the transition in the Czech Republic and Slovakia.

Finally, I want to comment on a small and lean organization, the Central Europe Institute (CEI). It started its activities in 1991, and its initial mission was to assist small business clients in drafting business plans and improving their operations, and these activities might well have been described earlier (see chapter 4). It established small offices, called Entrepreneurship Centers, and in its most active period operated four, in Bratislava, Bánska Bystrica, Košice and Prague. CEI's program was competent, needed, but unremarkable; by 1995 it had served some 500 business clients and provided a variety of training activities. It had good luck in raising funds from the Open Society Fund, the Czech and Slovak Enterprise Fund, the Bradley Foundation, the Whitehead Foundation, USAID, and the Mellon Foundation ($300,000 through 1993). But by 1994, CEI was beginning to reposition itself and started a "Business in the Community Program." This repositioning was partly based on the insight that small business advising was no longer as important as in the early years of the Velvet Revolution, but most importantly on the new insight that something else was:

> The spirit that animates this program is that people who are involved in making change will support that change despite hardships they might have to endure. The converse, as we can plainly see, is equally true: those on the sidelines cannot visualize the future but only discourage at the difficult present.
>
> To this end, the Central Europe Institute seeks to encourage in the Czech and Slovak Republics, where it has a well-established base, the founding and development of local, voluntary chambers of commerce. This effort, knows as the Institute's Business in the Community Program, is not primarily about economic development. It focuses on the business owners as a sub-set of society because, with their resources and natural leadership abilities, they are in a unique position to contribute to their communities.[90]

CEI's president, Peter McFadden, had selected five communities in the Czech Republic and Slovakia for this effort and secured the assistance of a number of chambers of commerce in Wisconsin. One of the chosen towns was Žamberk, a town of 6,000 residents in the Czech Republic. McFadden recounted how, after a speech he gave there about forming a voluntary chamber of commerce, person after person got up and said that a chamber of commerce would encourage people to work with one another again and to begin to trust each other, or how it would restore a sense of community and improve community life. The Žamberk Chamber of Commerce was in fact founded and embarked on an ambitious agenda of creating an information and referral center, establishing purchasing cooperatives among entrepreneurs, and facilitating communications between the business community and the government.[91] But McFadden would not rest still. Convinced that Slovakia needed municipal revival more than the Czech Republic and that the need for ordinary business consulting was diminishing, he decided to close or spin off CEI's Entrepreneurship Centers and to concentrate on the city of Košice, where he founded the Košice Business Forum. He had an infectious enthusiasm and persuaded a broad segment of the business community in Košice to support this organization. The Business Forum was a direct outgrowth of the Žamberk experience and in 1995 and 1996 Mellon gave CEI two final grants of $100,000 each. Its membership included journalists, business executives, educators, environmentalists, architects, musicians, city government officials, and others. The Forum organized monthly meetings and special events such as a massive pasta party on the eve of the Košice International Peace Marathon (the oldest in Europe). It mobilized the city's resources to create a wintertime skating rink in the middle of Hlavna ulica, the main pedestrian square of the city and arguably one of the great public spaces of the region: over 100 high school students volunteered to participate in its construction and several businesses donated the necessary materials.[92] McFadden believed that people could be mobilized for collective action and such action would simultaneously strengthen the NGO sector and create opportunities for urban development and renewal. The people in Košice had felt alienated and abandoned by the central government and were thrilled by the Forum's call to action. For the first time, the citizens were beginning to act together on a voluntary basis for the good of the community as a whole. But, in a sense, the crowning achievement of community action was the great macarena party in 1997. McFadden conceived the idea of mobilizing the city for a macarena dance party that would be so large that it would get into the *Guinness Book of World Records*. He determined that to qualify for the biggest dance party ever, it would have to be attended by over 60,000 persons. Everybody told him that this was utopian; the entire population of Košice was only 250,000 and it was difficult to imagine that more than a quarter of the population could be mobilized. But he went ahead and urged the inhabitants to attend and started to persuade the surrounding communities to send delegations. One community seemed particularly receptive, but its representatives said that attendance was not feasible, because the last evening train from Košice did not stop at their town and the people attending would have no way of getting home that same evening. McFadden suggested that they ask Slovak Railways to make an exception and arrange for the train to stop anyway. They thought that this was laughable; it would never happen. But McFadden persisted, and they finally contacted Slovak Railways, obtained a commitment to stop the train, and the inhabitants did attend the party (and the train did stop at their town that evening).

The fact that the dance missed by a hair from getting into the *Guinness Book* did not matter. What the charismatic McFadden conveyed to the members of the Košice community was that citizens acting together can accomplish wonders. A real sense of community pervaded the city and a sense of self-respect and accomplishment prevailed. This, too, is a part of civil society that was inconceivable during the previous regime.

Why did this experiment succeed? Certainly because of McFadden's drive and personality. But also because of the recognition on the part of large segments of the population that they were getting something by their own efforts that they had never had before. The sad part of the tale is that it did not last. Meeting the expenses of the Business Forum became increasingly difficult as U.S. funders turned to other types of activities and after a while McFadden himself needed to return to the United States. Three years later, in the autumn of 2000, the Košice Business Forum is only a memory, perhaps proving that it is harder to create permanent institutions than short-term enthusiasms.

Final Thoughts

The kind of "technology" that was diffused as a result of the activities described in this chapter was highly variable, ranging from short-term fixes for business firms or government agencies to permanent institutions embodied in new financial legislation, and from institutionalizing the training of local government officials and the public for more creative interactions to demonstrating the effectiveness of grass-roots movements. The new laws about financial institutions and local governments will stay on the books, large nationwide organizations like FSLD will continue to discharge their functions, and much of the business and banking lore conveyed to high-level executives will grow permanent roots in their circles. Where failures or near failures occurred, it was predominantly because of alternative agendas, weak commitments, and bureaucratic and personal rigidities. We may also wonder whether the business firms or the government agencies that have had interns assigned to them will be permanently altered for the better. And will those with new skills be able to transfer the knowledge they have gained in one assignment to another? The jury is still out on that question. The most successful civil society organization was clearly FSLD. Perhaps the most interesting and challenging case is that of Košice. In retrospect, it was easy to generate popular grass-roots support for community action in that city; but as quickly as the fire took hold, as quickly it was extinguished for lack of flammable materials. I suspect that there are citizens in Košice who wistfully observe today that all could be as it was before if only they had another McFadden among them, not realizing that each and every one could be a potential McFadden.

13

Lessons Learned

Our wanderings through the East European landscape were accompanied by many observations on what worked and what did not, but they were not systematic because they were always in the context of some particular project or activity. That fact does not diminish their usefulness, but it does make it somewhat more difficult to summarize the lessons we have learned.

Much has been accomplished in Eastern Europe by western foundations and governments, but there have also been some failures, which have been roundly criticized. However, just as the critics may have overstated the extent of the failures, so those of us who have been selecting and funding projects may well have exaggerated the importance of the successes. Democratic forms of government have been introduced successfully, and I think permanently, in a number of countries of the region. Market economies have taken hold, new market-oriented institutions have been created, and businesses have been streamlined. Higher education has been partially reformed and libraries have been modernized. The NGO sector has grown and civil society organizations are making an impact. But higher educational reform in some countries is badly lagging, discrimination against minorities has barely been dented, the hearts and minds of many in responsible positions show no sign of change, and advocates of authoritarianism are lurking in some corners. What has characteristically worked was the transfer of knowledge and technology; the failures have typically occurred in transferring values. While it is neither unexpected nor unforgivable for western donors to scale down their activities in the region, it is premature to declare victory.

It may be useful to pull the practical lessons of my narrative together, in the form of an admonition or exhortation to a hypothetical philanthropic organization contemplating a major program in some unfamiliar part of the world.

Dear Foundation President: the first thing I would like to suggest to you is that ten years is about the maximum length of time that you should allow a program officer to head up the efforts of your foundation in the chosen region. He or she will need a year or two to learn the ropes, and then have about eight years of work at maximum effectiveness. But after that, no matter how conscientious your program officer is, when problems arise, he or she will be tempted to think "I have been there, and I have done this, and I know how to do it, and I do not have to think about it any longer." And it is true, your program officer will have learned a lot about the region, and being "on autopilot" can be very useful and save time; but this attitude is not conducive to finding new and imaginative solutions to problems that may only be disguised as old problems, but are not. Most importantly, your program officer should remain flexible and not be too adamantly wedded to preconceptions: projects hardly ever work out exactly as originally conceived and midcourse corrections are not only reasonable, but frequently unavoidable.

At the beginning, you and your program officer will be eager to get some projects going, and your trustees are entitled to expect some positive results in the not-too-distant future; but do not be in too much of a hurry. The choice before you is similar to that faced by a young scholar who knows that he or she needs to "break into print" soon, in order to merit advancement in salary and rank. Lightweight scholarship can be turned out easily and published in second-rate journals fairly quickly, while more serious work may take years to complete. But a young scholar is poorly advised if he or she concentrates too much on the "quick and dirty." I experienced the same pressure in 1990, when I was beginning to line up projects and was very fortunate that many serious projects came my way soon, so that I did not have to squander too many resources on ephemeral ones.

Do not be tempted to stick too closely to the mainstream. The projects that most of your colleagues at other foundations or government agencies are supporting may be solid and useful, and some areas may indeed require massive interventions by many agencies to make a permanent difference. But in repetition and replication there lurks the danger of a lack of imagination and a lack of boldness. Imagination and boldness are attributes that carry high risk, but just as we have learned in financial analysis that in order to achieve above normal expected returns in our investments we must be willing to shoulder above average risks, so here too, you must be ready to face risk in order to achieve much.

There are always two, and sometimes three parties to philanthropy. There is always a donor and a donee, and the success or failure of your projects will always depend on the charateristics of both. But often the transfer of technology involves a third party, an experienced organization that can assist in the delivery of new knowledge, and the characteristics of such technology deliverers are equally crucial. If you are transferring knowledge or technology, there are, of course, always some technical aspects: some hardware or software, something mechanical or mathematical or otherwise technical. But failures are hardly ever of the technical sort, and even if they occur, they are easily corrected. If a computer fails, or the design was flawed and you underconfigured some component in a technical solution, you can always replace the offending equipment or reconfigure the design, and no permanent harm will have been done. But when projects really fail, they almost always fail because of people, and when they succeed, they do so because of the competence, imagination, enthusiasm, and mutual courtesy of people.

If failures occur, they can originate from any one of the three possible parties, and you will find that most of the problems are rooted, broadly speaking, in cultural differences that are not adequately appreciated. The greatest of the problems is a misunderstanding of what your target region really needs. If you have an academic bent, do not assume that workshops are endlessly useful and avoid the temptation to listen to the pleas of knowledge deliverers and technology providers who like to visit capital cities but are bored by the provinces or inconvenienced by their inferior hotel arrangements. When the trainers you engage design their curriculum, be sure that they do not put on their teaching agenda subjects that are fashionable in the West and can be taught without preparation or effort, because they represent the trainers' own field or discipline; they need to be sensitive to the region's needs as perceived by the trainees. You must make certain that they follow a carefully charted course between listening to signals from the trainees and, at the same time, not letting them steer you into erroneous or antiquated directions. Make sure that the trainers get adequate training themselves in local customs and practices, and at all costs make sure that they do not react to the inconveniences of life in the region with indignation or irritation. Unless your foundation has no budget constraint (a most unlikely scenario!), look carefully at the cost-effectiveness of projects, and do not hesitate to scale back the remuneration of western technology providers if you think that it is excessive. And perhaps most important of all: respect the donees and cut them some slack; let them make some of the decisions and divide the budget between western trainers and donees in such a manner that the latter do not feel that they are being treated like irresponsible children.

Your program officer must be patient and not expect perfectly crafted proposals and professional looking annual and final reports from the donees. In the region in which your foundation is active, there may be no tradition of proposal writing and your prospective donees may have no idea at all about what you expect to find in a proposal. At the same time, your program officer must insist that the proposals you receive have adequate detail about the objectives of the grant that is being sought, the alternative ways the objective could be satisfied, the reasons for the particular method chosen, and the plans for the implementation of the project and for its management. He or she should act like a loan officer in a bank who is considering extending a loan to a business, and require the same types of supporting evidence as a loan officer would. Once the grant is made, the program officer needs to remain in close contact with the project, and if problems arise, particularly conflicts between the technology providers and the donees, he or she should play an active role in problem-solving and conflict resolution. And in deciding how much money is reasonable for project implementation, do not be shortsighted by looking only to the current phase of the project, especially if you are intending to build permanent institutions in the region. Excessive optimism about the availability of future funding levels from other sources, particularly from the governments or the businesses of the grant recipients' country, has prematurely killed many a good project, and you should always be thinking, "if the current phase of the project works out all right, and I would be inclined to fund it again, how much of an additional investment would the foundation have to make in order to assure the long-term survival of the project?" And if you decide that it would take too much of your foundation's resources to ensure long-term survival, are there ways in which you can interest some of your sister foundations in joining forces with you?

But some of the failures will be due to the attitudes of your donees, particularly if the region you are dealing with has just emerged from an authoritarian system organized in a strictly hierarchical manner. Some of the institutional leaders will lack initiative and imagination because their previous political system did not reward these qualities, and will not have a passionate identification with the institution they are supposed to serve or a deep commitment to the project you are willing to finance. They will often be unable to conceive of new ways of doing things and may lack an understanding of the power of collaboration and cooperation. They will receive you in a friendly manner, but do not let the charm of their hospitality (even if it involves such quaint customs as drinking brandy toasts at 10:30 in the morning) mislead you into thinking that there is a will to back up promises and a commitment to mobilize resources. Since prestige can be derived in some regions of the world from having western contacts, particularly moneyed contacts, promises of collaboration will be given easily, but without any serious intention of marshalling resources for the project, and, indeed, often without any understanding of what resources may need to be made available. Many of these institutional leaders will betray themselves by bragging about the large number of western collaborators they have signed up with for a plethora of uncoordinated projects. In a worst-case scenario, you may even find officials who are motivated primarily by personal gain—unenlightened self-interest—or those who simply cannot be bothered to work with you; your program officer needs a sharp eye to identify these, but when you do, run the other way. Remember the old adage, "you can lead a horse to water, but you cannot make it drink;" there is no point in forcing projects on recipients who do not want them. Finally, do not forget that government interference in projects is likelier than in the environment that you are used to, and innocent-looking political changes and relatively minor upheavals can wreck a project.

But be grateful when you encounter smart, flexible, and enthusiastically committed institutional leaders, of whom you will meet many. They will frequently reveal themselves to you by speaking your language and by volunteering their own ideas; treasure them and build lasting relations with them. Persuade the best of them to work with each other; fund them generously, but get them used to the importance of matching grants, in which they either secure additional moneys from other donors or contribute to the project from their own resources. And when one or two years have passed after the completion of a project, ask your donees to provide a brief but analytical report on what has happened after the grant period itself has ended. Both you and they will find this a most enlightening and rewarding experience and it will help cement their experience for the future.

Not everything that you try will be successful, but many projects will succeed, and you will find satisfaction in knowing that you have improved the functioning of some aspect of your target region. But the greatest satisfaction will come from seeing how the transfer of knowledge and technology has empowered your target audience and put it on the cutting edge.

Notes

Foreword

1. Alvin Kernan, a distinguished Professor Emeritus of English at Princeton, served in an analogous role as the leader of Foundation programs in the humanities, including a seminar series designed to help Ph.D. candidates navigate the sometimes treacherous waters created by theoretical and doctrinal disputes. Gilbert Whitaker, now dean of the Jones School of Business at Rice University and formerly provost at the University of Michigan, is a third example; Dean Whitaker continues to serve as the Foundation's senior advisor for a program designed to test out cost-effective uses of technology in teaching. A final example is provided by Stuart Saunders, former vice chancellor of the University of Cape Town, who provides vital assistance to the Foundation in its South African program.

Introduction

1. A more or less random list of Ph.D. economists would include George Schultz, former secretary of state; Michael Blumenthal and Lawrence Summers, former secretaries of the treasury; Laura Tyson, Michael Boskin, and Joseph Stiglitz, former chairs of the Council of Economic Advisors; Burton G. Malkiel, former member of the Council of Economic Advisers; Robert Reich, former secretary of labor; Alan Blinder, former vice chairman of the Board of Governors of the Federal Reserve System; Arminio Fraga, president of the Central Bank of Brazil.

2. Ruth, Stephen and Robert Schware, "Can this Connectivity Project Succeed? Predicting and Assessing Success in Internet Implementations in Developing Countries," http://www.isoc.org/inet96/proceedings/g5/g5_4.thm.

3. It is not even clear how we should measure research productivity: any self-respecting tenure committee in an American university will recoil at the thought of measuring it by the quantity of the output, yet the notion of the quality of research is certainly more elusive, although citation frequencies and impact factors may point in the right direction. An important study of the productivity effects on teaching and learning of using digital teaching materials is provided by Bennett, Scott, "Information Based Productivity," in Ekman, Richard and Richard E. Quandt (eds.), *Technology and Scholarly Communication*, Berkeley: The University of California Press, 1999, pp. 73–94.

4. Harrison, Lawrence E. and Samuel P. Huntington (eds.), *Culture Matters: How Values Shape Human Progress*, New York: Basic Books, 2000, and particularly Landes, David, "Culture Makes Almost All the Difference," in Harrison, Lawrence E. and Samuel P. Huntington, op. cit., pp. 2–13 and Porter, Michael E., "Attitudes, Values, Beliefs, and the Microeconomics of Prosperity," in Harrison, Lawrence E. and Samuel P. Huntington, op. cit., pp. 14–28.

Chapter 1

1. While this is technically correct, the Foundation did support research in or about the region and made grants to numerous universities and research organizations. It is also appropriate to contemplate for a second the terminology "Central Europe" and "Eastern Europe." As any Hungarian or Czech or Pole will tell you, they live in Central Europe, because if you look at a map of Europe, which stretches from Portugal in the west to the Urals in the east, Hungary, Poland, etc., are in the center of that region. To them, Eastern Europe denotes Ukraine and (the European part of) Russia. But whether it is accurate or not, the term "Eastern Europe" has come to denote in the past 10 years the entire European region that used to be in the Soviet orbit, and this is the convention that I shall also follow.

2. The Ministries of Industry, of Agriculture and Food, of Finance, the National Organization of Small Industry, the National Council of Industrial Cooperatives, the National Committee of Technological Development, the Karl Marx University of Economics (soon to be renamed the Budapest University of Economic Sciences), the International Management Center, the National Organization of Commerce, the Hungarian Chamber of Commerce, the National Banking Association, the State Insurance Inspectorate, and the Technical University of Budapest. All of these organizations are strictly Hungarian, with the exception of the International Management Center, which had a cooperative program with the University of Pittsburgh and had funding from the Ford Foundation and from the Rockefeller Brothers' Fund (and, much later, from the Mellon Foundation as well). See also chapter 11.

3. In Eastern Europe, the chief officer of a university is typically called "rector" and not "president."

4. In southwest Hungary; subsequently known as Janus Pannonius University.

5. A notable exception to this is the Catholic University in Lublin (KUL), Poland.

6. Even the Central European University, created and funded by George Soros, avoided these disciplines.

7. After Mathias Corvinus, a great Renaissance king of Hungary in the fifteenth century, who founded an important royal library, the surviving books from which are referred to as "Corvinas." Mathias also founded a university, Academia Istropolitana, in Bratislava (Pozsony in Hungarian), Slovakia. See chapter 6 for details.

8. Fábry, G., *Higher Education and Research in Hungary during the Period of Social Transformation: 1990–1992*, Vienna: Institut für die Wissenschaften vom Menschen, 1993.

9. One need think only of the subsequent hostilities between Azeris and Armenians, Croatians and Serbs, Serbs and Bosnians, Serbs and Kosovars, Russians and Chechnyans, the economic collapse in Russia, and of the evolution of dictatorship in Belarus to see the wisdom of this restriction. See, for example, Kaplan, R. D., *Balkan Ghosts*, New York: St. Martin's Press, 1993. But while it remained true that the southern tier of countries was relatively neglected, other foundations did undertake activities in countries other than the three in which Mellon was to be active.

10. See, for example, Kornai, J., "Resource Constrained vs. Demand Constrained," *Econometrica*, 47 (1979), 801–20, *The Economics of Shortage*, Amsterdam: North Holland, 1980, *Growth, Shortage and Efficiency*, Oxford: Blackwell, 1982, "The Soft Budget Constraint," *Kyklos*, 39 (1986), 3–30.

11. Goldfeld, S. M. and Quandt, R. E., "Budget Constraints, Bailouts, and the Firm Under Central Planning," *Journal of Comparative Economics*, 12 (1988), 502–20, "Output Targets, the Soft Budget Constraint, and the Firm Under Central Planning, *Journal of Economic Behavior and Optimization*, 14 (1990), 205–22, "Rationing, Defective Inputs and Bayesian Updates Under Central Planning," *Economics of Planning*, 23 (1990), 161–74, "Effects of Bailouts, Taxes and Risk Aversion on the Enterprise," *Journal of Comparative Economics*, 16 (1992), 150–167, "Uncertainty, Bailouts, and the Kornai Effect," *Economics Letters*, 41 (1993), 113–19, and Quandt, Richard E., The Kornai Effect Revisited," in Eric Maskin and András Simonovits (eds.), *Planning, Shortage and Transformation: Essays in Honor of János Kornai*, Cambridge, MA: The MIT Press, 2000, pp. 209–29.

12. Portes, R. and Winter, D., "Disequilibrium Estimates for Consumption Goods Markets in Centrally Planned Economies," *The Review of Economic Studies*, 47 (1980), 137–59.

13. Portes, Richard, Richard E. Quandt, David Winter, Stephen Yeo, "Macroeconomic Planning and Disequilibrium: Estimates for Poland, 1955–1980," *Econometrica*, 55 (1987), 19–41; Portes, Richard, Richard E. Quandt, David Winter, and Stephen Yeo, "Estimation de la taille des erreurs de planification," *Annales de l'Insee*, No. 55/56, (1984), 245–55; Portes, Richard, Richard E. Quandt and Stephen Yeo, "Tests of the Chronic Shortage Hypothesis: The Case of Poland," *Review of Economics and Statistics*, LXX, No. 2, (1988), 288–95.

14. Charemza, W. and Quandt, R. E., "Models and Estimation of Disequilibrium in Centrally Planned Economies," *Review of Economic Studies*, 64 (1982), 109–16,

15. Charemza, Wojciech, Mirosław Gronicki, Richard E. Quandt, "Modelling Parallel Markets in Centrally Planned Economics: The Case of the Automobile Market in Poland," *European Economic Review*, 32 (1988), 861–83.

16. Tříška, Dušan and Quandt, Richard E. (eds.), *Optimal Decisions in Markets and Planned Economies*, Boulder, CO: Westview Press, 1990.

17. Quandt, Richard E., *The Econometrics of Disequilibrium*, Oxford: Blackwell, 1988.

18. An obvious consequence was that management training in Eastern Europe, such as it was, never had to concern itself with the field of marketing.

19. Hare, Paul G., "Hungary: In Transition to a Market Economy," *The Journal of Economic Perspectives*, 5 (1991), 195–201.

20. *The Nation*, Vol. 248, No.7, February 20, 1989, p. 220.

21. Blazyca, G., "Industrial Structure and the Economic Problems of Industry in a Centrally Planned Economy: The Polish Case," *The Journal of Industrial Economics*, 28 (1980), 313–26.

22. Council for Mutual Economic Assistance, founded in 1949, with an initial membership consisting of the Soviet Union, Hungary, Poland, Czechoslovakia, Bulgaria, and Romania.

23. Marer, Paul, "Foreign Economic Liberalization in Hungary and Poland," *American Economic Review, Papers and Proceedings*, 81 (May, 1991), 329–33.

24. See, for example, Portes, Richard, "Economic Reforms in Hungary," *American Economic Review, Papers and Proceedings*, 60 (May, 1970), 307–13.

25. Crane, Keith, "Institutional Legacies and the Economic, Social, and Political Environment for Transition in Hungary and Poland," *American Economic Review, Papers and Proceedings*, 81 (May, 1991), 318–22.

26. Kornai, János, "The Hungarian Reform Process: Visions, Hopes, and Reality," *Journal of Economic Literature*, 24 (1986), 1687–1737.

27. Much of the economic analysis in this section is based on Quandt, Richard E., "Recent Economic Developments in Eastern Europe," Princeton University, Financial Research Center Memorandum No. 125, May 1991.

28. *PlanEcon*, May 11, 1990.

29. I recall that in 1985, when I gave a paper at Professor Welfe's conference in Poland, I was paid a small honorarium in Polish złotys, which could not be converted into any other currency, nor taken out of the country. Since I had to spend them, I decided to have the best possible dinner on the eve of my departure from Warsaw. I studied the menu at the Victoria Intercontinental, and determined that caviar was the most expensive dish. When I ordered it, I was told that they did not have any. Smoked salmon was the second most expensive, but they did not have any of that either. I was finally reduced to having a medium-priced meal and was left with some unusable złotys.

30. Lipton, David and Jeffrey Sachs, Creating a Market Economy in Eastern Europe: The Case of Poland," *Brookings Papers on Economic Activity*, 1, 1990, pp. 75–147.

31. Kornai, J., *From Socialism to Capitalism: What is Meant by the 'Change of the System,'* London: The Social Market Foundation, 1998, p. 35.

32. *The Economist*, February 18, 1989, p. 51.

33. http://www.wdi.bus.umich.edu/transiti.htm

34. Ash, Timothy Garton, "Refolution: The Springtime of Two Nations," *The New York Review of Books*, June 15, 1989. The word "refolution" is Ash's wordplay on "reform" and "revolution." Ash has written a large number of thoughtful and provocative essays on Eastern Europe between 1983 and 1989, republished as *The Uses of Adversity*, New York: Random House, 1989.

35. *The Economist*, January 21, 1989.

36. *The Nation*, October 9, 1989, p. 376.

37. On the occasion of which Hungary's arguably greatest poet, Sándor Petöfi, declaimed "talpra magyar, hí a haza, itt az idő, most vagy soha" ("arise Hungarians, the country calls, the time has come, now or never").

38. *The New York Times*, Section 4, p. 3, col. 1, July 2, 1989.

39. http://www.armonkinstitute.org/eiseltext.htm.

40. Laber, Jeri, "Fighting Back in Prague," *The New York Review of Books*, April 27, 1989.

41. Ash, Timothy Garton, "The Revolution of the Magic Lantern," *The New*

York Review of Books, January 18, 1990. See also Ash, Timothy Garton, *The Magic Lantern*, New York, Random House, 1990.

42. Whipple, Tim D. (ed), *After the Velvet Revolution*, New York: Freedom House, 1991, p. 15.

43. A good case in point is that of "public goods." It is well known among economists that public goods, that is goods the enjoyment of which by one person does not diminish the amount available for another person (national defense is an obvious example), require substantial government intervention in their provision, because the free market is unable to provide these goods in the requisite amount. I recall a moderately heated argument I had with Klaus about whether public goods existed at all and whether the government should provide for them.

44. Professor of Economics at Columbia University, former Chairman of the Council of the President's Economic Advisers and former Vice President of the World Bank.

45. Klvačová, Eva, "Nečekejme na Godota, nepřijde" (Don't wait for Godot, he is not coming), Interview with Václav Klaus, *Ekonom*, 24/1999, 12–14.

46. Czechoslovakia, the most repressive country during the Communist period, was the only one with a tradition of democracy during the interwar period.

47. See the outstanding article by Hankiss, Elemér, "European Paradigms: East and West, 1945–1944," *Dædalus*, 123, No. 3, Summer 1994, pp. 115–26.

48. Dahl, Robert, "Transitions to Democracy," Paper presented at the US—Hungarian Political Scientists First Roundtable, The International Research & Exchanges Board, San Francisco, CA, August 28–29, 1990. But note the discriminatory treatment accorded to Roma (Gypsy) people in all countries, or of the Russian speaking population in Latvia, etc.

49. On the former, see Coricelli, F., *Macroeconomic Policies and the Development of Markets in Transition Economies*, Budapest: CEU Press, 1998; on the latter, see Frydman, R., Rapaczynski, A., and Earle, J. S., *The Privatization Process in Central Europe*, Budapest: CEU Press, 1993.

50. Kemme, David M. "Economic Transition in Eastern Europe and the Soviet Union: Issues and Strategies," Institute for East-West Security Studies, New York, November 1990.

51. Brada, Joseph C. and Arthur E. King, "Sequencing Measures for the Transformation of Socialist Economies to Capitalism: Is There a J-Curve for Economic Reforms?" Conference on the East European Transformation, Princeton University, May 3–4, 1991.

52. See Giannaros, Demetrios, "Did the 'Shock Therapy' Approach Work in the Economic Restructuring of Eastern Europe? Some Evidence from Poland and Russia: A Brief Review," *Global Business & Economics Review*, 2/1(2000), 53–66.

53. Johnson, Simon and Gary W. Loveman, *Starting Over in Eastern Europe*, Cambridge, MA: Harvard Business School Press, 1995, pp. 28–32. See also Lipton and Sachs, op.cit.

54. CEPR *Bulletin,* No. 37, 1990.

55. Chapman, Bruce, "A Recipe for a Free-Market Hungary," *The Wall Street Journal*, April 6, 1990.

56. George Soros has taken a rather critical view of the effectiveness of the Blue Ribbon Commission. See Soros, G., *Underwriting Democracy*, New York: The Free Press, 1990, 1991, p. 118.

57. Exceptions might be the Budapest University of Economic Sciences (then Karl Marx University), the University of Gdańsk, Moscow State University, Novosibirsk State University, and CEMI (Central Economics and Mathematics Institute) in Moscow.

58. For an account of the economic transition in the Czech Republic, see Svejnar, Jan (ed.), *The Czech Republic and Economic Transition in Eastern Europe*, New York: Academic Press, 1995.

59. For example, Czechoslovak exports in 1990 to other East European countries and to the USSR declined by 33.7 and 20.3% respectively and Hungarian exports by 15.4 and 16.8% respectively. See Quandt, op. cit., p. 9. Hungary was also more successful than the other countries in replacing exports to the COMECON countries with exports to Western Europe. See Soros, G., *Underwriting Democracy*, New York: The Free Press, 1990, 1991, p. 120.

60. For a theoretical treatment of privatization and privatization in Great Britain, see Vickers, John and George Yarrow, *Privatization: An Economic Analysis*, Cambridge, MA: The MIT Press, 1988.

61. See Kotrba, Josef, "Privatization Process in the Czech Republic: Players and Winners," in *The Czech Republic and Economic Transition in Eastern Europe*, ed. by Jan Svejnar, New York: Academic Press, 1995.

62. Sulc, Zdislaw, *Strucne dejiny ekonomickych reforem*, Brno, Doplnek, 1998, pp. 86–87.

63. Bernasek, Anna, "Hungarian Rhapsody," *Fortune*, Nov. 8, 1999, pp. 46–48. For a brief account of the role of Viktor Kozeny and his Harvard Capital and Consulting, see Stroehlein, Andrew, Jan Culik, Steven Saxonberg, Kazi Stastna, "The Czech Republic 1992 to 1999," *Central Europe Review*, 1, No. 12, September 13, 1999; http://www.ce-review.org/99/12/stroehlein12.html.

64. Friedrich, Alex, "Bad Market Ends Well, Kozeny Laughs," *The Prague Post Online*, January 7, 1998; http://www.praguepost.cz/archive/busi10798a.html.

65. Orbán, Viktor, "The Case Against Restitution," *Uncaptive Minds*, IV(2), 1991, 33–38.

66. Demszky, Gábor, "Building a Market Economy in Hungary, *Uncaptive Minds*, IV(2), 1991, pp. 44–52.

67. Rondinelli, Dennis A. and Martin R. Fellenz, "Privatization and Private Enterprise Development in Hungary: An Assessment of Market-Reform Policies," *Business & The Contemporary World*, 5 (Autumn, 1993), 75–88; Rondinelly, Dennis A. and Jay Yurkiewicz, "Privatization and Economic Restructuring in Poland: An Assessment of Transition Policies," *American Journal of Economics and Sociology*, 55 (April 1996), 145–60.

68. Mokhtari, M. and Caner, S., "The Problem of Arrears in the Russian Federation," United States Government Technical Assistance, Working paper, June 1999.

69. Slightly different figures are given in *SEED Act Implementation Report Fiscal Year 1998*, Washington, D.C.: Department of State, March, 1999.

70. There is some disagreement between the GNP per capita figures for 1990 reported by *The World Factbook* for 1990 and the December 28, 1990 issue *PlanEcon*. The former gives GNP per capita in 1990 for Czechoslovakia as $7,700, for Hungary as $5,800, and for Poland as $4,200.

71. Flemming, John and John Micklewright, *Income Distribution, Economic*

Systems and Transition, Florence, Italy: UNICEF, Innocenti Occasional Papers, Economic and Social Policy Series No. 70, May 1999.

72. Dahrendorf, Ralf, *Reflections on the Revolution in Europe*, London: Chatto & Windus Ltd., 1990.

73. *New York Times*, December 4, 1999, p. A3.

Chapter 2

1. Gordon, Robert Aaron and James Edwin Howell, *Higher Education for Business*, New York: Columbia University Press, 1959.

2. Cooke, Ian and Paul Mayes, *Introduction to Innovation and Technology Transfer*, Boston and London: Artech House, Inc., 1996.

3. Williams, Frederick and David V. Gibson, *Technology Transfer: A Communication Perspective*, Newbury Park, CA: Sage Publications, 1990, p. 45.

4. Reddy, Allan C., *A Macro Perspective of Technology Transfer*, Westport, CT: Quorum Books, 1996, p. 8.

5. Navaretti, Giorgio Barba and Andrea Bigano, "R&D Inter-firm Agreements in Developing Countries. Where? Why? How?," in *Creation and Transfer of Knowledge*, ed. by G. Barba Navaretti, P. Dasgupta, K.-G. Mäler, D. Siniscalco; Berlin: Springer, 1998, p. 33.

6. Blomström, Magnus and Ari Kokko, "Foreign Investment as a Vehicle for International Technology Transfer," in *Creation and Transfer of Knowledge*, ed. by G. Barba Navaretti, P. Dasgupta, K.-G. Mäler, D. Siniscalco; Berlin: Springer, 1998, p. 285.

7. For a first-rate review and assessment of democracy assistance programs between 1989 and 1994, see Quigley, Kevin F. F., *For Democracy's Sake: Foundations and Democracy Assistance in Central Europe*, Washington, D.C.: The Woodrow Wilson Center Press, 1997.

8. Quigley, Kevin F. F., "Lofty Goals, Modest Results: Assisting Civil Society in Eastern Europe," in *Funding Virtue: Civil Society Aid and Democracy Promotion*, ed. by Marina Ottaway and Thomas Carothers, Washington, D.C." Carnegie Endowment for International Peace, 2000, pp. 191–215.

9. Kirk, Mary E., *Where Walls Once Stood: U.S. Responses to New Opportunities for Academic Cooperation with East Central Europe*, New York: Institute of International Education, 1992.

10. Indeed, I myself was the beneficiary of IREX travel grants, which directly promoted my research in the 1980s with Wojciech Charemza.

11. Quigley, Kevin F. F., *Conversations on Democracy Assistance*, Washington, D.C.: East European Studies, The Woodrow Wilson International Center for Scholars, 1996.

12. Soros, G., *Soros on Soros*, New York: Wiley, p. 112.

13. *Building Open Societies: Soros Foundations 1995 Report*, New York: Open Society Institute, 1996, p. 166.

14. *Building Open Societies: Soros Foundations Network 1998 Report*, New York: Open Society Institute, 1999, pp. 174–75.

15. Quigley, Kevin F. F., "For Democracy's Sake: Background Paper," Paper delivered at the Bratislava Workshop, November 2–3, 1995.

16. *Mott Foundation Civil Society Grants in Central Eastern Europe/Russia*

and Central Eastern Europe/FSU, Flint, MI: The Charles H. Mott Foundation, 1999.

17. *Programs Supporting Reform in Central and Eastern Europe*, The German Marshall Fund of the United States, Washington, D.C.: 1995.

18. *The German Marshall Fund of the United States 1995 Annual Report*, Washington, D.C.: 1995, p. 21. ELTE is the acronym for Eötvös Lóránd Tudományegyetem; the Eötvös Lóránd University in Budapest.

19. *The German Marshall Fund of the United States 1996 Annual Report*, Washington, D.C.: 1996, p. 28. See also Chapter 12.

20. *The German Marshall Fund of the United States 1997 Annual Report*, Washington, D.C.: 1997, p. 26.

21. *The German Marshall Fund of the United States 1998 Annual Report*, Washington, D.C.: 1998, p. 17.

22. Szanton, P. "DIALOG," *TransAtlantic Perspectives*, Autumn 1994, No. 30, pp. 6–9.

23. Jordan, P. and Przbylska, B., "BORIS," *TransAtlantic Perspectives*, Autumn 1994, No. 30, pp. 13–15.

24. *Program Guidelines*, Philadelphia: The Pew Charitable Trusts, 1994, p. 24.

25. *Pew Charitable Trusts Grants List*, Philadelphia: The Pew Charitable Trusts, 1999.

26. *Rockefeller Brothers Fund Annual Report 1990*, New York: 1991, p. 30.

27. *Rockefeller Brothers Fund Annual Report 1991*, New York: 1992, p. 43.

28. *Rockefeller Brothers Fund Annual Report 1992*, New York: 1993, p. 34.

29. *Rockefeller Brothers Fund Annual Report 1994*, New York: 1995, p. 41. See also chapter 11.

30. Ibid.

31. Armenia, Azerbeijan, Belarus, Estonia, Georgia, Kazakhstan, Kyrgyzstan, Latvia, Lithuania, Moldova, Russia, Tajikistan, Turkmenistan, Ukraine, and Uzbekistan. See http://www.macfdn.org/programs/gss/printing/fsu_guides.htm.

32. See http://www.carnegie.org/guidelines.htm.

33. There is no guarantee that the figures in the tables are complete.

34. Quigley, Kevin F. F., "For Democracy's Sake: How Funders Fail—and Succeed," *World Policy Journal*, Spring 1996, pp. 109–18 and Quigley, Kevin F. F., "Lofty Goals, Modest Results: Assisting Civil Society in Eastern Europe," in *Funding Virtue: Civil Society Aid and Democracy Promotion*, ed.by Marina Ottaway and Thomas Carothers, Washington, D.C.: Carnegie Endowment for International Peace, 2000, pp. 191–215.

35. Quigley, Kevin F. F., "For Democracy's Sake: How Funders Fail—and Succeed," *World Policy Journal*, Spring 1996, p. 116.

36. *SEED Act Implementation Report Fiscal Year 1998*, Washington, D.C.: Department of State, March 1999, p. 1.

37. Poland, Hungary, Czech Republic, Slovakia, Bulgaria, Romania, Yugoslavia, Albania, Estonia, Latvia, Lithuania, Bosnia, Croatia, Macedonia, and Slovania.

38. Quigley, Kevin F. F., "For Democracy's Sake: Background Paper," p. 43.

39. *East European Initiative, Summary of USIA Programs*, Washington, D.C., United States Information Agency, October 1991.

40. Newberg, Paula R. and Thomas Carothers, "Aiding—and Defining—Democracy," *World Policy Journal*, Spring 1996, 97–108.

41. Active in Hungary, Poland, Bulgaria, the Baltics, Romania, and Albania.

42. A notoriously polluted region due to the metallurgical plants in Nova Huta.

43. See http://www.info.usaid.gov/regions/eni/. Evaluative reports can be seen at http://www.dec.org/usaid_eval/.

44. European Bank for Reconstruction and Development, *Annual Report*, London: The Bank, 1991, 1992, 1993, 1994, 1995, 1996, 1997, 1998.

45. http://www.rsl.ru/tacis/webdraff.htm.

46. Bulgaria, Czech Republic, Hungary, Latvia, Lithuania, Poland, Romania, Slovakia, Slovenia.

47. Albania, Bosnia and Herzegovina, Macedonia.

48. See http://europa.eu.int/comm/dg1a/PHARE/programme_types/index.htm. See also Vodichev, Evgeny, "The EU TACIS Program as a Tool for Technical Assistance to Russia," Appendix D.3 in *The Humanities and Social Sciences in the Former Soviet Union: An Assessment of Need*, The Kennan Institute, Woodrow Wilson Center, Washington, D.C.: August 1, 1999.

49. http://europa.eu.int/comm/enlargement/intro/criteria.htm.

50. See http://europa.eu.int/comm/dg1a/PHARE/implementation/guidelines/98–99.htm.

51. Ibid.

52. http://europa.eu.int/en/comm/dg22/tempus/intro.html.

53. The European Community's program for cooperation in the field of education is *Socrates*, and its arm for higher education is *Erasmus*. See http://europa.eu.int/en/comm/dg22/socrates/bref-en.html. Both are administered by Directorate General XXII.

54. *ERASMUS and LINGUA (Action* II*) Guidelines for Applicants*, Brussels: Commission of the European Communities, 1991.

55. *TEMPUS Financial Support for Cooperation and Mobility in Higher Education between Central/Eastern Europe and the European Community*, Brussels: Commission of the European Communities, 1991.

56. http://www.britcoun.org/who/bcwork.htm.

57. *The Know How Fund for Eastern Europe*, KHF-1, Joint Assistance Unit, Eastern European Department, Foreign and Commonwealth Office, London, November 1, 1990

58. http://www.britcoun.org/poland/governance/index.html, http://www.britcoun.org/estonia/english/work/jicap.htm, http://www.britcoun.org/slovakia/work/jicap.htm, http://www.britcoun.org/czechrepublic/work/jicam.htm.

59. http://www.volkswagen-stiftung.de/english/merkblat/merkost.htm.

60. Quigley, *For Democracy's Sake: Foundations and Democracy Assistance in Central Europe*. See in particular, pp. 122–23 for an excellent country breakdown of foundation assistance between 1989 and 1994.

61. Quigley, "For Democracy's Sake: Background Paper," p. 52.

62. Quigley, "For Democracy's Sake: How Funders Fail—and Succeed." The importance of being able to act quickly is also stressed by Czermiński, Jurand, "A Coordinator's View of the Impact of Policy on the Financing and the Managing of Library Automation," Chapter 3 in Lass, Andrew and Richard E. Quandt (eds.) *Library Automation in Transitional Societies: Lessons from Eastern Europe*, New York: Oxford University Press, 2000.

63. Czermiński, Jurand., op. cit.

64. Cochran, Thomas H., "Reflections on the International Public Finance Program, 1993–1996," unpublished paper, New York: October 2000.

Chapter 3

1. Beschel, Robert P., "Western Proposals for Assistance to Poland and Hungary: An Evolving Catalog," Carnegie Corporation of New York, December 1989.

2. Letter From Richard E. Quandt to William G. Bowen, January 11, 1991.

3. The warning against concentrating on capital cities was echoed later in *Making Change Last: Lessons Learned in Providing Management Training and Economics Education to Central and Eastern Europe*, Highlights of a Conference, Washington, D.C.: Business-Higher Education Forum, December 1–2, 1992.

4. I want to stress that there were important exceptions, and not only among the outstanding expatriates such as Stanisław Wellisz at Columbia, Stanisław Gomulka at the London School of Economics, or Paul Marer at Indiana University. I have already mentioned Kornai and Charemza; see also a very early success by East Europeans in publishing in a western scientific journal: Szakolczai, György and Stahl, J., "Increasing or Decreasing Returns to Scale in the Constant Elasticity of Substitution Production Function," *Review of Economics and Statistics*, 51 (1969), 84–90.

5. Letter by G. Todd Jagerson to Richard E. Quandt, dated October 25, 1990. Underlining in the original.

6. Palmer, Tom G., *Philanthropy in Central and Eastern Europe: A Resource Book for Foundations, Corporations, and Individuals*, Fairfax, VA: The Institute for Humane Studies, George Mason University, 1990, 1991, pp. 22–23.

7. http://www.ce-review.org/99/12/stroehlein12.html.

8. *New York Times*, November 25, 1999, p. A9. See also Tomášková, Petra, "Position and Protection of National Minorities," in *Reports on the State of Human Rights in the Czech Republic 1999*, ed. by Libuše Šilhanová; Prague: Czech Helsinki Committee, 2000, pp. 11–22.

9. *After the Fall: The Human Impact of Ten Years of Transition*, Florence, Italy: UNICEF, The MONEE Project, November, 1999.

10. CAROLINA, an e-mail news service (carolina@mbox.fsv.cuni.cz), Prague, Dec. 10, 1999, reports that Norway had abolished visa requirements for Slovak citizens on October 6, 1999, but re-instituted them a month later because of the numerous Slovak Roma requests for asylum. While Foreign Minister Eduard Kukan denied that the exodus was politically or racially motivated, the Government Commissioner for Roma Issues, Vincent Danihel, definitely included prejudice among the reasons for emigration (in addition, of course, to economic reasons). See *Slovak News Update*, New York: The Foundation for a Civil Society, December 1999.

11. Brown, Brack, "Unbalanced Sectoral Development in the Czech Republic: Democracy and the Civic, Market, & State Sectors," Fairfax, VA: George Mason University, 1994.

12. Schöpflin George, "Postcommunism: The Problems of Democratic Construction," *Dœdalus*, 123, No. 3, Summer 1994, pp. 127–41.

13. Mirsky, Arlene Elgart, "The New Global Community," *The Metropolitan Corporate Counsel*, 8/7–8 (December 2000/January 2001).

14. Schöpflin, op.cit., p. 132. See also the discussion of "programmatic corruption" in chapter 8.

15. *The Economist*, Vol. 56, No. 7 (July, 1989), p. 221.

16. *Establishing an Environmental Assistance Program for Central and Eastern Europe*, The German Marshall Fund of the United States, Washington, D.C., August 15, 1990.

17. See Kosztolányi, Gusztáv, "Acquatic Chernobyl–Requiem for the Tisza and Szamos: Part I," *Central Europe Review*, Vol. 2, No. 7, February 21, 2000, http://www.ce-review.org/00/7/csardas7.html and Lovatt, Catherine, "Gold and Cyanide," *Central Europe Review*, Vol. 2, No. 7, February 21, 2000, http://www.ce-review.org/00/7/lovatt7.html. While the Romanian Government has taken an accommodating position and appears to be willing to pay compensation, the Australian firm's cynical protestation that the fish died because of "cold water" is distressing.

18. http://www.centraleurope.com/news.php3?id=142431. For maps showing the extent of the polluted area see http://www.personal.ceu.hu/departs/envsci/.

19. Palmer, Tom G., *Philanthropy in Central and Eastern Europe: A Resource Book for Foundations, Corporations, and Individuals*, Fairfax, VA: The Institute for Humane Studies, George Mason University, 1991, p. 28.

20. According to the conventional wisdom, East European universities were much more up-to-date in the sciences and engineering than in the humanities. But they were often not on the cutting edge in those subjects either. I received a proposal from an East European university for the development of a graduate course in nuclear engineering. I asked a distinguished physicist in the Princeton Physics Department to comment on the quality of the proposal. He said, "It is really very good; it reminds me very much of a similar course I took in graduate school some 25 years ago."

21. Beschel, Robert P., *Western Proposals for Assistance to Poland and Hungary: An Evolving Catalog*, New York, Carnegie Corporation, December, 1989.

22. Some of my targets were Csaba Csáki, Rector of Karl Marx University, Mária Ormos, Rector of Janus Pannonius University in Pécs, Zoltán Daróczy, Rector of Kossuth Lajos University in Debrecen, Béla Csákány, Rector of József Attila University in Szeged. Note that in Hungarian, the last name comes first and the first name comes second; in the text I will use the American convention but in the names of institutions I shall stick to the Hungarian order; hence Kossuth Lajos, rather than the American "Lajos Kossuth." Note that Janus Pannonius was a medieval chronicler, Lajos Kossuth the Hungarian leader during the 1848–89 war of liberation against the Hapsburgs, and Attila József a great twentieth century poet.

23. Hungarian is a Finno-Ugric language (with a not unappreciable importation of words from Slavic languages, Turkish, German, and others), and its only relatives in Europe are Finnish and Estonian.

24. If the grant was to be for the benefit of a university library, then the university's financial statements had to be submitted, not the library's.

25. This should not surprise anybody. Statistical systems were generally bad, and they probably became worse as the transition to market economies began,

because "with loosened controls, enterprises have been less inclined to provide complete data to statistical authorities." See Belkindas, Misha, Dinc Mustafa and Olga Ivanova, "Statistical Systems Need Overhaul in Transition Economies," *Transition*, 10(4), August 1999, 22–24. In general, they find that there is a moderately positive relationship between a country's GNP per capita and the quality of its statistical data.

26. These dinners were not only models of gracious entertainment, but permitted me to meet important intellectuals and politicians from countries in which the Mellon Foundation had no programs; such as Dimitrina Petrova from Bulgaria, Alin Teodorescu, Sorin Vieru and Silviu Brucan from Romania, Sonia Licht from Yugoslavia, and many others.

27. BITNET permitted e-mail but not remote login.

28. http://www.cuni.cz/cuni/history/historie.html.

29. Since 1990, several law faculties have reopened, in particular those at Masaryk University in Brno and Palacký University in Olomouc.

30. The original institution created in 1632 was called Academia Dorpatensis, since Dorpat is the German name of Tartu. http://www.ut.ee/dokumendid/english/history.html.

31. Darvas, Peter, "Governmental Reforms in Hungarian Higher Education: *Historical Traditions and New Actors*," in *Reform and Change in Higher Education*, ed. by Mauch, James E. and Paula L. W. Sabloff, New York: Garland Publishing, 1995, pp. 245–86. See also http://www.elte.hu and Szögi, László, *A Short History of Lóránd Eötvös University of Budapest, 1635–1985*, Budapest: Lóránd Eötvös University, 1985.

32. Founded in 1735; see Szörényi, László, *Saecula Hungariae 1701–1796*, Budapest: Széchenyi Arts Center, 1985. Selmecbánya is called Banská Štiavnica in Slovak.

33. Darvas, Peter, op. cit., pp. 249–52. In particular, Darvas points out the considerable impact of Hungarian Minister of Education, Ágoston Trefort, whose higher educational proposals supported the ideals of competition, secularization, and liberty.

34. Tucker, Aviezer, "Reproducing Incompetence: The Constitution of Czech Higher Education," *East European Constitutional Review*, 9/3, Summer 2000, 94–99.

35. See Pelczar, Andrzej, "Challenges Facing Universities in Central and Eastern Europe: A Polish Perspective," Paper delivered at the 41st Biennial CRE Conference, Maynooth, Dublin, May 5–8, 1993; also *in* CRE-*Action*, No. 102, 1993, 41–53.

36. It is no longer so narrowly specialized, since it has tried to add social sciences and humanities in recent years. The apocryphal story of its creation was that in the early years after World War II, the general secretary of the Communist Party, Mátyás Rákosi, was making a speech in which he was announcing the creation of a new university; unfortunately he mis-spoke himself and said that two new universities were going to be created. Since he could not, by definition, be wrong, officials scrambled to find a new site and objective for this unplanned institution and hence the University of Veszprém was born.

37. See, for example, McCrudden, Christopher, Jackson Newell, Karen Schilling, Karen Spear, "CDC Consultants' Report on Hungarian Higher Education," Citizens Democracy Corps, Washington, D.C., July 1993. The creation of the

Citizens Democracy Corps was announced by President Bush on May 12, 1990, and its mission was "to establish a center and a clearinghouse for American private sector assistance and volunteer activities in Eastern Europe." See *Current Policy*, Washington, D.C.: US Department of State, No. 1277.

38. Zalai, Ernő, "The First Wave of the Recent Reform of Training Management. Economics Specialists at the Budapest University of Economic Sciences," Budapest: Budapest University of Economic Sciences, May 1991, p. 18.

39. Amsterdamski, S., *Perceptions of Dilemmas: Summary of a Qualitative Study*, Vienna: Institut für die Wissenschaften vom Menschen, 1993, p. 17.

40. Wróblewski, Andrzej K., "The Future of Universities in Poland," unpublished paper, Warsaw Univerity, 1991.

41. Fábry, G., *Higher Education and Research in Hungary during the Period of Social Transformation*, Vienna: Institut für die Wissenschaften vom Menschen, 1993, p. 9.

42. Koucký, J. and Hendrichová, J., *Higher Education and Research in the Czech Republic: Major Changes since 1989*, Vienna: Institut für die Wissenschaften vom Menschen, 1993, pp. 31–32.

43. Personal communication from Wojciech Charemza, July 30, 1999.

44. Szapiro, Tomasz, "Barriers to Transforming Higher Education and How to Overcome Them," in *Education for Transition to Market Economy in Countries of Central and Eastern Europe*, ed. by Cichocki, Krzysztof and Paul Marer, Warsaw: Polish—U.S. Fulbright Commission, 1996, pp. 37–41.

45. Darvas, Peter, "Institutional Innovations in Higher Education in Central Europe. Agenda for Research and Cooperation," Vienna: Institute for Human Sciences, 1996.

46. *Conference of Rectors of Academic Schools in Poland Directory*, Kraków: Nowa Galicja, 1998, p. 45.

47. Pénzügyi és Számviteli Főiskola and Képzőművészeti Főiskola.

48. *Education for Transition*, Civic Education Project and Institut für die Wissenschaften vom Menschen, 1997; Part I. *International Cooperation in Social Science Higher Education in Central and Eastern Europe;* Part II. *Social Science Teaching at Central and East European Universities* (prepared by Daniel Calingaert); Part III. *Higher Education Policy in Central and Eastern Europe.*

49. Fábry, G., op. cit., p. 13. Italics in the original.

50. *Education for Transition*, Part II, p. 68.

51. Amsterdamski, S., p. 8. Italics in the original.

52. See, for example, "The Question of Imagination; An Interview with Professor Andrzej Kajetan Wróblewski, Rector of Warsaw University," *Polityka*, Oct. 21, 1992.

53. Pelczar uses the term "scientific employees," from which it is not clear whether he uses the term with its East European meaning, namely "scholars" or "academics" or whether he means it in the stricter (American) sense of "employees in the natural sciences and mathematics."

54. Hryniewicz, Janusz, Bohdan Jałowiecki, Agnieszka Mync, *The Brain Drain in Poland*, Warsaw: University of Warsaw, European Institute for Regional and Local Development, September 1992, p. 51.

55. *Education for Transition*, Part II, p. 27. See also chapter 11.

56. See also Bok, D., "Report to the Parliamentary Subcommittee Concerning Draft Law on Higher Education," unpublished paper resulting from a visit

to Hungary, October 4–9, 1992, by Derek Bok, William F. Massey and Robert Zemsky under the auspices of the Citizens Democracy Corps, in response to an invitation by Árpád Göncz, President of Hungary.

57. In various conversations with academic administrators, this always seemed to be a particularly sticky point. "If the State provides the funding," they would ask, "how can the State grant them autonomy? Doesn't funding imply control?"

58. Many academics argued that neither was feasible, because theirs was a small country in which everybody knew everybody and those being evaluated would always know who the evaluators were—an argument that struck me disingenuous and represented a basic fear of being found to be less than worthy.

59. Higher Education Act (1990); see Amsterdamski, S. and Jablecka, J., *Higher Education and Research in Poland: The Inherited Situation and Reforms*, Vienna: Institut für die Wissenschaften vom Menschen, 1993.

60. Amsterdamski, S., p. 11, Derényi, András, "Hungary," in *Education for Transition*, Part III, Mária Hrabinská, "Slovakia," in *Education for Transition*, Part III, Davrowa-Szefler, Małgorzata and Maria Wójcicka, "Poland," in *Education for Transition*, Part III.

61. Czermiński, J., "A Coordinator's View of the Impact of Policy on the Financing and Managing of Library Automation," in Lass, A. and Quandt, R. E. (eds), *Library Automation in Transitional Societies: Lessons from Eastern Europe*, New York: Oxford University Press, 2000.

62. Glenday, Christine, "An Overview of Science and Technology Reform in Central and Eastern Europe 1990–1993," NSF/Europe, http://www.nsf.gov/home/int/europe/reports/67.htm.

63. Koucký, J. and Hendrichová, J., op. cit., p. 18. A slightly different figure is reported in 1991 in Bollag, Burton, "Science Academies in Eastern Europe Start to Lose Grip on Research With the Emergence of Competitive Grants," *The Chronicle of Higher Education*, August 14, 1991, which reports that one third of 13,000 employees had been dismissed in the Czechoslovak Academy of Sciences. However, there is reason to believe that these figures refer to total staff, because the number of scientific staff members was reported by the Czech Academy of Sciences to have declined from 5,665 in 1985 to 3,529 in 1993. See Kadlecová, Ivana and Antonín Vítek, "Publikační Aktivita v Akademii Věd České Republiky" (Publication activities at the Czech Academy of Science), Prague: Czech Academy of Science, 1993.

64. Sarmir, E. and Zajac, S., *Higher Education and Research in the Slovak Republic: Major Changes since 1989*, Vienna: Institut für die Wissenshcaften from Menschen, 1993.

65. Hungary introducéd a token tuition payment in the late 1990s, but the newly elected government dominated by FIDESZ (Association of Young Democrats) promptly abolished tuition.

Chapter 4

1. George Soros reported that in 1987 he had difficulty receiving a favorable ruling from Washington for donating 200 personal computers to the Institute for Personal Computers in the USSR until he asked for help from Deputy Sec-

retary of State John C. Whitehead. See Soros, G., *Opening the Soviet System*, London: Weidenfeld and Nicolson, 1990, p. 21.

2. Sending one "average" faculty member to Eastern Europe for one year from the US might have cost in 1990 about $100,000 (including salary, benefits, travel, etc.) and might have resulted in his/her teaching as many as a hundred students. Bringing 100 students to Princeton for one year would have cost $2.5 million.

3. The late Fizz Harbison of the Princeton Economics Department used to joke to the effect that in order to qualify as an expert on a developing country one must at least have had the experience of flying over it in an airplane. I wonder how many East European experts were forged in the 1990s by this standard.

4. Letter from Jonathan F. Fanton to William G. Bowen, dated January 15, 1990.

5. The initial list of research centers consisted of the Warburg Institute, the Institute for Advanced Studies in the Humanities in Edinburgh, the Maison des sciences de l'homme, the Wissenschaftskolleg zu Berlin, the Herzog August Bibliothek in Wolfenbüttel, the Netherlands Institute for Advanced Study, the American Academy in Rome, the American School of Classical Studies at Athens, and Villa I Tatti. In later years we added the Institut für die Wissenschaften vom Menschen, the W. F. Albright Institute of Archeological Research, the American Research Institute in Turkey, the Center for Advanced Studies in Norway, and the Council for Scientific Research in Spain.

6. Letter from Mary Ellen Lane to Richard E. Quandt, dated February 15, 2001.

7. See, for example, Żuchowski, Tadeusz J., *Pałac papieski na Watykanie od końca V do poczatku* XVI *wieku* (The papal palace in the Vatican from the end of the fifth to the beginning of the sixteenth centuries), Poznań: Adam Mickiewicz University, History of Art Series No. 27, 1999; Lewartowski, Kazimierz, *Late Helladic Simple Graves: A Study of Mycenaean Burial Customs*, as reported in *Newsletter*, American School of Classical Studies at Athens, No. 43, Summer 1999, p. 3; or the e-mail report of February 7, 2000, by Andrzej Pieńos to Olga Spilar: "La parution de mon livre 'La naissance de la culture du regard. Le paysage européen vers 1800 entre la nature a l'histoire' est prévue à Varsovie pour l'année 2001. Il est le fruit le plus important et le plus concret de ma bourse."

8. In the western world, economics is typically taught in economics departments and management in business schools, although in the past 10–15 years economics departments have taught business subjects (e.g., finance) and business schools have taught subjects originally relegated to economics departments (microtheory, macrotheory, game theory).

9. See also http://www.hungary.com/seed/.

10. Personal communication from Robert Hisrich, September 2, 1999.

11. The charter members included the National Organization of Small Industry; the National Council of Industrial Cooperatives; the Ministries of Industry, of Agriculture, of Finance; the International Management Center; Karl Marx University; the Budapest Technical University; the Hungarian Chamber of Commerce; the National Banking Association; and the State Insurance Inspectorate.

12. Lackó, Zsuzsa and Anikó Soltész, "A SEED Alapitvány múltja, jelene,

jövője" (The SEED Foundation: Its Past, Present, Future), Working paper, The SEED Foundation, January, 2000.

13. Footnoting and citation standards in scholarly works were no better. Not infrequently, the full citation to a printed work might be as follows: "See Marx, p. 217." In the fall of 1997, I ran a large conference on library automation in Warsaw, with the intention that the papers (mostly by East European authors) would be published in a volume, which they were (see Lass, Andrew and Richard E. Quandt (eds), *Library Automation in Transitional Societies: Lessons from Eastern Europe*, New York: Oxford University Press, 2000). But since I knew that we would have to do a lot of actual text editing anyway, I wanted to minimize the work of fixing up citations. Accordingly, I sent every author a copy of Kate L. Turabian's *A Manual for Writers of Term Papers, Theses, and Dissertations*, 6th rev. ed; Chicago: Chicago University Press, 1996, with the very strong plea that everybody follow Turabian's style. While the content of the contributed papers was excellent, to my amazement and disappointment, hardly anybody followed the recommended formats, and nearly everybody used his or her own eccentric and idiosyncratic citation and footnoting method, *in spite of the fact that nearly all the contributors were professional librarians*. Were they defying an "ugly American" who was once again telling them how to do it? I do not think so. Rather, I believe, they did not have the time to make the effort I requested of them; nor did they probably see the value of it.

14. Personal communication from Anikó Soltész, October 26, 2000.

15. *Management Training and Economics Education Report*, AID Grant # EUR-G-00-91-00061–00; Second Quarterly Report (Seventh Year), Period January 1, 1998 through March 31, 1998 and Third Quarterly Report (Seventh Year) Period April 1, 1998 through June 30, 1998, State University of New York–Albany.

16. *Management and Economics Education Project 180–0029 (State University of New York) Final Evaluation*, Budapest University of Economic Sciences, November 1998.

17. At the University of Miskolc, Ybl Miklós Polytechnic School in Debrecen, the College of Food Industry in Szeged, the Pollack Mihály Technical College in Pécs, the Széchényi István College in Győr, the University of Veszprém and the University of Forestry and Wood in Sopron.

18. *Management and Economics Education Project 180–0029 (State University of New York) Final Evaluation*, Budapest University of Economic Sciences, November 1998, Section 2.1.1. The ISO 9000 (9001, 9002, 9003) standards specify the steps that a company must undertake to be eligible for certification. ISO 9000 was formulated in 1965 and is based in Geneva, Switzerland; it is a worldwide federation of national standards bodies. ISO 9000 basically establishes minimum requirements for manufacturing products and services. See http://www.iso9001compliance.com. The reader may also wish to compare these efforts to another Mellon initiative in the Czech Republic based on the National Industrial Certification program originating in the Commonwealth of Virginia, see chapter 12.

19. For the basics of these concepts see Tenner, Arthur R and Irving J. DeToro, *Total Quality Management*, Reading, MA: Addison-Wesley, 1992, Camp, Robert C., *Benchmarking*, Milwaukee, WI: ASQC Quality Press, 1989, Tenner, Arthur R. and Irving J. DeToro, *Process Redesign*, Reading. MA: Addison-Wesley, 1996.

20. Letter (undated) from Ágnes Tibor to Pier Benci, SPEDD, Inc.

21. See PASBAF Proposals to The Andrew W. Mellon Foundation, dated March 3, 1994, May 30, 1995, and July 3, 1996. It should be noted that figures of this type cannot be taken entirely seriously. In a personal communication on March 26, 2000, Dr. Mieczysław Bąk of PASBAF reported that of the 2 million plus small companies registered only about 1.3 million actively conducted operations.

22. *Central European Small Business Enterprise Development Commission, Annual Report 1992*, Washington, D.C., 1993.

23. In Hungary, centers were set up in Pécs and Debrecen and in the Czech Republic in Brno. See Hoy, Frank, and Przemysław Kulawczuk, "Outcomes Assessment of Small Business Assistance in Poland," http://www.sbaer.uca.edu/docs/proceedings/96ICSS101.txt.

24. See http://www.fairplay.pl.

25. Later called the Talent Information and Commercial Exchange Foundation.

26. Letter to Richard E. Quandt, dated October 14, 1999, by Louis Vidic, CEO of SPEDD, Inc.

27. Brynjolfsson, Erik and Lorin M. Hitt, "Beyond Computation: Information Technology, Organizational Transformation and Business Performance," *The Journal of Economic Perspectives*, 14 (Fall, 2000), 23–48.

28. Ruth, Stephen, "Measuring Long Term Effects of Technology Transfer in Developing Nations: The Case of Internet Training at the Romanian Academy of Sciences," *Information Technology for Development*, 9 (2000), 105–21. The situation is slightly better in South America than in Eastern Europe, but substantially worse in the Asian-Pacific area and in the Near East and Africa.

29. Ruth, Stephen R. and Raul Gouet, "Must Invisible Colleges be Invisible? An Approach to Examining Large Communities of Network Users," *Internet Research*, 3/1 (1993), 36–53.

30. Ruth, Stephen, "Measuring Long Term Effects of Technology Transfer in Developing Nations: The Case of Internet Training at the Romanian Academy of Sciences," *Information Technology for Development*, 9 (2000), 105–21.

31. Ruth, Stephen, "Using Academic Networks in Developing Nations: A Value-Added Services Perspective," in *International Information Systems*, San Diego, CA: HBJ Professional Publishing, July 1992, pp. 87–99.

32. E-mail from LISTSERV "Discussion of Eastern Europe Telecommunications EUEARN-L@UBVM.BITNET" dated November 25, 1992.

33. Ruth, Stephen, "Internet-related Grants in Central Europe: Insights and Lessons Learned," Fairfax, VA: George Mason University, December, 1999.

34. Ruth, Stephen and Robert Schware, "Can This Connectivity project Succeed? Predicting and Assessing Success in Internet Implementations in Developing Countries," http://www.isoc.org/inet96/proceedings/g5/g5_4.htm.

35. http://aol.pcwebopedia.com/TERM/T/TCP_IP.html.

36. http://www.ini.cmu.edu/COURSEWARE/X.25.html.

37. Olenick, Peter, "A Primer for Wide Area Computer Networks," Princeton University, CIT, August, 1992.

38. Letter from William E. Clements to James B. Conklin, dated January 8, 1990.

39. A short time thereafter, on October 30, 1990, an agreement was signed by IREX and the Academy of Sciences of the Soviet Union which effected its entry

into the BITNET/EARN network. See *IREX News in Brief*, 1, No. 1, February, 1991, p. 1.

40. For a general discussion of Internet diffusion see Goodman, S. E., L. I. Press, S. R. Ruth and A. M. Rutkowski, "The Global Diffusion of the Internet: Patterns and Problems, *Communications of the* ACM, 37/8, (1994), 27–31.

41. Barnes, Harry G., Memorandum of March 7, 1990.

42. It needed some 32 modems of various descriptions (two 9.6 Kbit/sec modems, the rest slower), software, port selector, IBM 7171 protocol converter, and extension of the disk storage capacity of the mainframe, 23 PCs, 20 3270 emulator cards, and so on.

43. Letter from Richard E. Quandt to William G. Bowen, dated October 24, 1990. Underlining in the original.

44. Kučera, Vladimír, "Project EARN/Bitnet Connection & Computer Services in UTIA, Final Report," Prague: Institute for Information Theory and Automation, June, 1992.

45. In June, 1991, for example, I discovered that he had spent some time at the Third School of Medicine of Charles University, as well as at the Center for Theoretical Study of Charles University, headed by Ivan Havel.

46. Quoted in Quandt, R. E., "Report on Grants in Eastern Europe," The Andrew W. Mellon Foundation, October 8, 1991.

47. Eunet is operated by the European association of Unix system users.

48. Gruntorád, Jan, "Research and Academic Networking in the Czech and Slovak Federal Republic," Prague: Czech Technical University, 1992.

49. http://www.sanet.sk/E/factsheet.html.

50. See http://www.eenet.ee/EENet/vorgukaart.html.

51. Letter from František Janouch to Richard E. Quandt, dated February 5, 1990.

52. *Application to The Andrew W. Mellon Foundation for a Grant to Establish the Advanced Scientific Computer Center (ASCOC)*, undated.

53. Named after Jan Amos Komensky (1592–1670), a bishop of the Moravian Church, who pioneered modern educational methods and believed that education continued throughout one's life. There is some evidence that he was asked to become the first president of Harvard College. See http://www.moravian.org/comenius.htm.

54. *Application to The Andrew W. Mellon Foundation for a Support to Building of the Comenius University Bratislava's Computer Network* [*sic*], Bratislava: Comenius University, July 26, 1991.

55. DEC gifts in kind left something to be desired: the Microvax delivered to CERGE was defective and was not accompanied by manuals. When I met the DEC manager for Czechoslovakia in Bratislava in June 1992, I told him that this just would not do.

56. Hladký, Rostislav, "Towards the 21st Century: A University in Transition," Olomouc: Palacký University paper, February 2, 2000.

57. Ruth, Stephen, "Using Academic Networks in Developing Nations: A Value-Added Services Approach," *International Information Systems*, San Diego: HBJ Professional Publishing, July, 1992.

58. Shortly thereafter, we also received a proposal from the Faculty of the Humanities for book and journal acquisition in a variety of fields in the amount of $138,000; a total grant of $274,000 was awarded by the foundation in June.

59. *Információ Technológia*, Vol. 1, No. 3, 1992, pp. 2–3.

60. Letter from Ádám Kiss to Richard E. Quandt, dated April 26, 1991.

61. He pointed out to me several times during the next few years that the Mellon grant was just like having a £1,000,000 bank note. See Twain, Mark, "The £1,000,000 Bank-Note," in *The Complete Short Stories of Mark Twain*, New York: Bantam Books, 1957.

62. The actual grant was larger, because it also included some noncomputer related items.

63. Manikowski, Adam, "Library Automation—Fortunes and Miseries in Poland," chapter 1 in Lass, Andrew and Richard E. Quandt (eds.), *Library Automation in Transitional Societies: Lessons from Eastern Europe*, New York: Oxford University Press, 2000, p. 23.

64. Ruth, Stephen, "Internet-related Grants in Central Europe: Insights and Lessons Learned," Fairfax, VA: George Mason University, December 1999.

Chapter 5

1. The members were Lord Dahrendorf, Frans A. M. Alting von Geusau, Colin G. Campbell, Umberto Colombo, Jonathan F. Fanton, Elemér Hankiss, Josef Jařab, Krzysztof Michalski, Hans-Ludwig Schreiber, Juraj Švec, and Jerzy Szacki; see Campbell, Colin G. and Ralf Dahrendorf (eds.), *Changes in Central Europe: Challenges and Perspectives for Higher Education and Research*, Vienna: Institut für die Wissenschaften vom Menschen, 1994, pp. 69–70.

2. Bollag, Burton, "World Bank Promoting a Western Approach to Higher Education in Eastern Europe," *The Chronicle of Higher Education*, July 31, 1001. But in 2001, the government of Hungary canceled the loan agreement on the grounds that better terms were available from sources other than the World Bank. See Varga, Dóra, "Felsőoktatási beruházások világbanki hitel nélkül," (Investment in higher education without World Bank loans), *Magyar Hirlap,* June 29, 2001.

3. Rupnik, Jacques, "Higher Education and the Reform Process in Central and Eastern Europe," *European Journal of Education*, 27, No. 1–2, 1992, 45–51.

4. Fogel, Daniel S. and James E. Mauch, "Restructuring Higher Education in the Czech and Slovak Republics: An Institutional Perspective," in *Reform and Change in Higher Education*, ed. by Mauch, James E. and Paula L. W. Sabloff, New York: Garland Publishing, 1995, p. 218.

5. *Concept for Higher Education Development in Hungary* (ed. István Bakos), Budapest: Co-ordination Office for Higher Education, Ministry of Education and Culture, 1991. This document assigned priority to increasing the number of students and indicated that this was to be achieved without a significant expansion of the higher educational network; see p. 12.

6. Hryniewicz, Janusz, Bohdan Jałowiecki, Agnieszka Mync, *The Brain Drain in Poland*, Warsaw: University of Warsaw, European Institute for Regional and Local Development, September 1992, p. 23.

7. Wróblewski, Andrzej K., "The Future of Universities in Poland," unpublished paper, Warsaw University, 1991.

8. Freeman, Kassie, "Equality in Higher Education in Post-Communist Hun-

gary and Poland: "Challenges and Prospects," in *Reform and Change in Higher Education*, ed. by Mauch, James E. and Paula L. W. Sabloff, New York: Garland Publishing, 1995, pp. 183–207.

9. Jařab, Josef, "Higher Education and Research in the Czech Republic," *Higher Education Management*, 5, November 1993, 309–16.

10. "A Magyar Felsőoktatás Fejlesztése 2000-ig" (Development of Hungarian Higher Education until 2000), Budapest: Secretariat of the Hungarian Rectors' Conference, January, 1992.

11. Fogel, Daniel S. and James E. Mauch, op. cit., p. 225.

12. *Statistical Yearbooks of the Republic of Poland 1990–1998*, Warsaw: Central Statistical Office. The total number of students consists of "studia dzienne," full-time students on a Monday–Friday daytime schedule, "studia wieczorowe," full-time students on a Monday–Friday evening schedule, and "studia zaoczne," part-time students on a weekend schedule. The first two categories account of about 60–70% of the totals. Personal communication from Tadeusz Poplonkowski, Ministry of Education, December 8, 1999.

13. Detailed annual data broken down by type of institution are available from the author by request.

14. Rocki, Marek, "Basic Issues Concerning the Costs of Higher Education," *Zarządzanie Współczesną Szkołą Wyższą* (Promoting Effectiveness of Higher Education: Issues of Institutional Cooperation and Development), Polish-American Seminar, Kraków, March 10–11, 1994.

15. *Statistical Yearbooks of Hungary, 1990–98*, Budapest: Hungarian Central Statistical Office.

16. Dahrendorf, Ralf, "Main Problems of the Reform of Higher Education and Research in Former Eastern Europe," in Campbell, Colin and Ralf Dahrendorf, op. cit., p. 11.

17. Bollag, Burton, "Private Colleges Reshape Higher Education in Eastern Europe and Former Soviet States," *The Chronicle of Higher Education*, June 11, 1999; http://chronicle.com/weekly/v45/i40/40a04301.htm.

18. *The Prague Tribune*, October 1999.

19. Personal communication from Professor Ádám Kiss, deputy secretary of state for education.

20. http://www.czso.cz/eng/figures/puble-c/e01/yearbks/1998/22.htm.

21. Personal communication from Ladislav Kabat, former rector of the Agricultural University in Nitra, May 8, 2000.

22. Bollag, Burton, "Enrollments Rise 50% at Russian Universities," *The Chronicle of Higher Education*, February 4, 2000; http://chronicle.com/daily/2000/02/2000020402n.htm.

23. Lajos, Tamás, "Perspectives, Hopes and Disappointments: Higher Education Reform in Hungary," *European Journal of Education*, 28, No. 4, 1993, pp. 403–11.

24. Tucker, Aviezer, "Reducing Incompetence: The Constitution of Czech Higher Education," *East European Constitutional Review*, 9/3, Summer 2000, 94–99.

25. MacWilliams, Bryon, "Conflicts over Language Impede Ukraine's Higher-Education System," *The Chronicle of Higher Education*, February 2, 2001, pp. A42–A44.

26. Bollag, Burton, "Romania Suspends University Rectors Over Allegations of Selling Diplomas," *The Chronicle of Higher Education*, October 2, 2000,

http://chronicle.com/daily/2000/10/2000100205n.htm. Another example of unethical behavior is reported by Mooney, Paul, "Students at Chinese University Reportedly Pay Large Bribes to Switch Majors," *The Chronicle of Higher Education*, November 22, 2000; see http://chronicle.com/cgi2-bin/printable.cgi.

27. Balcerowicz, Leszek, "Clamping Down on Corruption," *The Wall Street Journal Europe*, October 20–21, 2000, p. 11.

28. Personal communication from Lev Bukovský on October 24, 2000.

29. Malova, Darina and Erik Lastic, "Higher Education in Slovakia: A Complicated Restoration of Liberal Rules," *East European Constitutional Review*, 9/2, Summer 2000, 100–4.

30. Kocsis, Károly, "Business Education and Restructuring the Higher Education System in Hungary," Paper given at the Business Education's Response to Eastern Europe Conference, Columbia University, May 1, 1991.

31. Balczerowicz, Leszek, "Thoughts on Changes in Education in Post-socialist Countries," in *Education for Transition to Market Economy in Countries of Central and Eastern Europe*, ed. by Cichocki, Krzysztof and Paul Marer, Warsaw: Polish—U.S. Fulbright Commission, 1996, p. 4.

32. This clearly was taking place, as reported by Jařab, Josef, op.cit., when he notes that in the Czech Republic, rectors are elected by the university senate, academic freedom is guaranteed, deans are elected by the senates of the individual schools, and thus the number of appointed officials has drastically declined.

33. Note the coincidence of this with one of the principal aims of the TEMPUS Program discussed in chapter 2.

34. "Az Országos Atheneum Bizottság beszámolója az MTA 1991. májusi közgyűlése számára az egyetemekkel történt eddigi megállapodásokról," (Report of the National Atheneum Commission to the May 1991 meeting concerning the agreement with the universities), Budapest, April 25, 1991.

35. ELP *Newsletter*, 1, No. 1, Autumn 1999, p. 3.

36. Bollag, Burton, "Euopean Ministers Affirm their Commitment to a Unified Higher Education System," *The Chronicle of Higher Education*, May 23, 2001, http://chronicle.com/cgi2-bin/printable.cgi.

37. Handke, Mirosław, "Financing Higher Education: Its Limits and Possibilities in Poland," "*Zarządzanie Współczesną Szkołą Wyśzą* (Promoting Effectiveness of Higher Education: Issues of Institutional Cooperation and Development), Polish-American Seminar, Kraków, March 10–11, 1994.

38. Koj, Aleksander, "Jagiellonian University in the Period of Transition from State-owned to State-supported Institution of Higher Education," *Zarządzanie Współczesną Szkołą Wyśzą* (Promoting Effectiveness of Higher Education: Issues of Institutional Cooperation and Development), Polish-American Seminar, Kraków, March 10–11, 1994. Handke, Mirosław, op. cit., the rector of the Academy of Mining and Metalurgy, disagrees with the split and gives it as 65% from the state budget and 19% from KBN.

39. Personal communication from Richard Spies, Vice President for Finance and Administration of Princeton University.

40. Wheeler, David L., "Falling Behind: U.S. Trails 3 European Nations in Rate of College Graduation," *The Chronicle of Higher Education*, May 26, 2000, p. A63.

41. McCrudden, Christopher, Jackson Newell, Karen Schilling, Karen Spear, "CDC Consultants' Report on Higher Education: Higher Education in Szeged," Washington, D.C.: Citizens Democracy Corps, July 1993. A good analysis of

information systems appears in Galbally, James F., Margaret A. Miller, Anthony M. Morgan, Timothy R. Warner, "Report on Hungarian Higher Education and the Technical University of Budapest," Washington, D.C.: Citizens Democracy Corps, July 1993.

42. "*Forint Votes" for Civil Society Organizations Studies*, Series ed. László Harsányi; Budapest: Research Project on Nonprofit Organizations, 2000, "1996. évi CXXVI. törvény a személyi jövedelemadó meghatározott részének az adózó rendelkezése szerinti közcélú felhasználásárol" (Law CXXVI/1996 on the Use of Some Part of the Personal Income Tax in Accordance with the Disposition of the Taxpayer), p. 12.

43. Vajda, Ágnes and Éva Kuti, "Citizens' Votes for Nonprofit Activities in Hungary," in *"Forint Votes" for Civil Society Organizations Studies*, series ed. László Harsányi; Budapest: Research Project on Nonprofit Organizations, 2000, pp. 176–77.

44. *Népszabadság*, October 28, 2000.

45. It should be noted that, in the Czech Republic, Václav Klaus has not regarded NGOs favorably and would have certainly opposed such legislation on the ground that NGOs arrogate to themselves programmatic decisions that are "properly" within the government's sphere.

46. See Healey, Robin, "Teaching Morality," *Central Europe Review*, 1, No. 20, November 8, 1999; http://www.ce-review.org/99/20/healey20.html.

47. Also strongly underscored by Wróblewski, Andrzej K., op. cit. He also argues that the narrowly specialized training students receive must be replaced.

48. Karwacki, Andrzej, "New Tasks and the Need for Professional Administration in Polish Higher Education Institutions, *Zarządzanie Współczesną Szkołą Wyższą* (Promoting Effectiveness of Higher Education: Issues of Institutional Cooperation and Development) Polish-American Seminar II, Kraków, May 26–27, 1994.

49. Bollag, Burton, "Reforms in Higher Education Disappoint Eastern Europeans," *The Chronicle of Higher Education*, December 3, 1999; http://chronicle.com/weekly/v46/i15/15a05501.htm.

50. Handke, M., op. cit. Voting with their feet is, in effect, what has seemingly happened in higher education in South Africa, where numerous historically disadvantages institutions (HDIs), such as the University of Fort Hare and the University of the Western Cape, have experienced drastic declines in enrollments because of inadequate academic programs and poor management: enrollments at Fort Hare, for example, have reportedly dropped from 5,000 to 900 in two years.

51. But note that in the late 1970s, in Hungary, the Minister of Education, Imre Pozsgay, attempted to bring about a less autocratic, centralist system of managing higher education, by which the "Ministry could be a representative of the higher education sphere rather than a bureaucratic executor of government decisions." See Darvas, Peter, op. cit., p. 271.

52. Muraközy, László, "Szubjektiv emlékek a debreceni egyetemi integrációról," (Subjective memories of the Debrecen integration), unpublished working paper, Debrecen: Debrecen University, October 2000.

53. See http://www.gau.hu.

54. Hrubos, Ildikó, "Transformation of Hungarian Higher Education," *International Higher Education*, No. 20, Summer 2000, pp. 8–10 and Morgan,

Anthony, "Reform in Hungarian Higher Education," *International Higher Education*, No. 19, Spring 2000, pp. 23–24.

55. See *Concept for Higher Education Development in Hungary*, op.cit.

56. Kozma, Tamás, "Higher Education in Hungary: Facing the Political Transition," *European Journal of Education*, 25, No. 4, 1990, 379–90.

57. Bok, Derek, *Universities in Hungary*, Washington, D.C.: Citizens Democracy Corps, 1992.

58. Lévay, Irén, "Library Integration of Debrecen University," unpublished paper, Debrecen: Kossuth Lajos University, 2000.

59. Kabat, Ladislav, "The Czech and Slovak Higher Education System in the Period of Transition," unpublished paper, Nitra: University of Nitra, May 2000.

60. Chapman, David W., "Hungarian Higher Education Management Seminar Series: A Report to the Mellon Foundation," Albany, NY: SUNY–Albany, July 1992.

61. Pelczar, Andrzej, "Autonomy, Competition and Cooperation in Higher Education," *Zarządzanie Współczesną Szkołą Wyższą* (Promoting Effectiveness of Higher Education: Issues of Institutional Cooperation and Development), Polish-American Seminar, Kraków, March 10–11, 1994.

62. Huber, Robert, "The Reorganization of Research in Central and Eastern Europe," NCEEER *Quarterly Newsletter*, Issue Five, April 2000.

63. Bok, Derek, *Universities in Hungary*, Washington, D.C.: Citizens Democracy Corps, 1992.

64. The principal experts participating in this were Christopher McCrudden, associate treasurer of Princeton University; Anthony Morgan, vice president for budget and planning at the University of Utah; and Timothy Warner, associate vice president for university budgets at Stanford. See also note 41 and surrounding discussion.

65. McCrudden, Christopher, Anthony Morgan, Timothy Warner, "Report on Follow-up Conferences to the Site Visits by American Volunteers," Budapest: Hungarian Higher Education Project, August, 1994.

66. Róna-Tas, András, "Accreditation in Hungary and the Hungarian Accreditation Committee," *Accreditation and Background Information*, Budapest: Hungarian Accreditation Committee, 1993.

67. Law on Higher Education, §4(2).

68. Law on Higher Education, §6(1)

69. The extent to which Meciar was a "democrat" is illustrated by his insistence that the state must guarantee that the press tells the truth. See *The New York Times*, Oct. 11, 1992.

70. Personal communication from Ferdinand Devinsky, rector of Comenius University, July 30, 1999.

71. Chwirot, Stanisław, "Accreditation at Polish Universities," Paper presented at the Quality Assurance at Polish Universities Conference, Kraków, March 27, 1999; see also http://www.amu.edu.pl/~ects/uka/referat2.html. See also http://www.amu.edu.pl/~ects/uka/uka-eng.html and e-mail from Stanisław Chwirot to Richard E. Quandt, dated August 17, 2001. I am also indebted to the secretary of the UAC, Maciej Kozierowski, for much useful information.

72. For a more detailed discussion, see Morgan, Anthony and Amy Aldous Bergerson, "Importing Organizational Reform: The Case of Lay Boards in Hungary," Paper presented at The Association for the Study of Higher Education

International Forum, San Antonio, TX, November 18, 1999; final version published in *Higher Education*, 40(2000), 423–48.

73. Bowen, William G., *Inside the Boardroom*, New York: John Wiley and Sons, 1994, pp. 1–2. Italics in the original. Bowen develops 20 "presumptive norms" for boards of trustees, which could well serve as a model in the East European context.

74. Veblen, Thorstein, *The Higher Learning in America*, New York: B. W. Huebsch, 1918, pp. 65–66, where he says "So far as regards its pecuniary affairs and their due administration, the typical modern university is in a position, without loss or detriment, to dispense with the services of any board of trustees, regents, curators, or what not. Except for the insuperable difficulty of getting a hearing for such an extraordinary proposal, it should be no difficult matter to show that these governing boards of businessmen commonly are quite useless to the university for any businesslike purpose. Indeed, except for a stubborn prejudice to the contrary, the fact should readily be seen that the boards are of no material use in any connection; their sole function being to interfere with the academic management in matters that are not of the nature of business, and that lie outside their competence and outside the range of their habitual interest."

75. See http://www.kul.lublin.pl.

76. http://www.kul.lublin.pl/uk/phenomenon.html.

77. http://www.ppke.hu.

78. http://www.ppke.hu/mission_eng.html.

79. http:/www.ceu.hu.

80. The events that culminated with the removal of CEU from Prague were described as follows: "in the end Soros was forced to move the university to Budapest, while the government of Klaus pushed Soros out of the public eye, not only ungraciously, but stupidly, in ideological arrogance." See Kantůrková, Eva, "The Archives of One Foundation," in Paton, Derek and Jitka Uhdeová (eds.), *The Charta 77 Foundation [Twenty Years]*, Brno: Atlantis, 1998, p. 19.

81. http://www.ceu.hu/facts/3.htm.

82. The master's programs in the humanities and social sciences are in Central European history, economics, economy and society, gender studies, human rights, international relations and European studies, medieval studies, nationalism studies, philosophy, political science, society and politics and Southeast European studies, although the last of these has recently been discontinued. See http://www.ceu.hu/facts/4.htm.

83. It is worth noting that CEU houses the Open Society Archives, the main holdings of which are the records of the Research Institute of Radio Free Europe/Radio Liberty. See http://www.ceu.hu/facts/11.htm.

84. Agovino, Theresa, "Controversial Rector Divides New University," *The Chronicle of Higher Education*, November 24, 2000.

85. Darvas, Peter, "Institutional Innovations in Higher Education in Central Europe. Agenda for Research and Cooperation," Vienna: Institute for Human Sciences, 1996.

Chapter 6

1. See http://www.cerge.cuni.cz/.
2. "A Proposal for Establishing A Center for Economic Research and Grad-

uate Education (CERGE) at Charles University, Prague," prepared by Jan Svejnar and Josef Zieleniec, July 2, 1990, pp. 3–4.

3. Test of English as a Foreign Language and Graduate Record Examination, respectively.

4. Advanced theory, econometrics, economic history, public choice and public finance, money and banking, industrial organization, labor economics, international trade and finance, economic development, urban and regional economics, environmental economics, and transition economics.

5. This was an outstanding group of people who were not afraid to make the right decision even if it was fraught with peril.

6. Some of the acquisitions were the result of purchases and some came from donations. For example, the outstanding and large professional book collection of Stephen M. Goldfeld, a distinguished economist, was donated to CERGE by his wife after his death in 1995.

7. This was extremely unusual in Eastern Europe. I had met with many student groups at various universities, and most of the time they were deathly afraid of opening their mouths. Discussing issues with the professor was simply not part of the local ethos.

8. Quandt, Richard E., *Report on Trip to the Czech Republic, Slovakia and Hungary, May 20–June 2, 1995*, New York: The Andrew W. Mellon Foundation, 1995.

9. See later in this chapter about the role that the Economics Institute of the Czech Academy of Sciences was to play.

10. Quandt, Richard E., *Report on Projects in Eastern Europe, June 1994*, New York: The Andrew W. Mellon Foundation, 1994.

11. Letter from Deputy Prime Minister Vladimír Dlouhý to Richard E. Quandt, dated June 7, 1990: "Dear Richard: The purpose of this letter is to assure you of the Czechoslovak Government's approval of the plans for the establishment of a Center for Economic Research and Graduate Education (CERGE) at Charles University in Prague. In particular, the Czechoslovak Federal Government guarantees that a suitable building for CERGE will be provided before the end of 1990. Sincerely, Vladimír Dlouhý"

12. Memorandum dated May 24, 1991 from Jan Svejnar to Radim Palouš, Miloslav Petrusek, and Deputy Minister Jan Koucký.

13. Letter from Václav Klaus, Minister of Finance to Radim Palouš, dated May 30, 1990.

14. Letter from Richard E. Quandt to Radim Palouš, dated March 26, 1993.

15. Letter from Richard E. Quandt to Rector Karel Malý, dated January 25, 1999.

16. Final Report to The Andrew W. Mellon Foundation from CERGE-EI, Ann-Arbor, MI: CERGE-EI Foundation, November 1999.

17. See Ashenfelter, Orley, Erik Berglof, Konrad Stahl, *Evaluation of CERGE-EI*, Prague: September 3, 1999.

18. Carter, Anne P., Michael Grossman, Michael S. McPherson, Sherwin Rosen, Michael Rothschild, "Site Visit Report," Report to the New York State Education Department, Bureau of Doctoral Program Review, January 27–28, 2000. Carter, Grossman and (late) Rosen are professors of economics at Brandeis, City University of New York Graduate Center and University of Chicago, respectively; McPherson is president of Macalaster College, and Rothschild is

dean of the Woodrow Wilson School of Public and International Affairs at Princeton.

19. Other outstanding institutions, such as the New Economics School in Moscow, CEU, or the Kiev-Mohyla Academy have only had MA programs. The Central European University plans to start a Ph.D. program in the fall of 2000; see http://www.ceu.hu/econ/intro.htm. For the most current information, see http://www.cerge-ei.cz.

20. See http://www.ainova.sk.

21. http://www.ce-review.org/99/12/stroehlein12.html.

22. Pozsony played an important role in Hungarian history: it served as the Hungarian capital from 1541 to 1784 and the Hungarian Diet met there until 1848. See http://www.bartleby.com/65/br/Bratisla.html.

23. http://lib.elte.hu/ebooks/memoria/eng.html. The two previous attempts to found a university were in the city of Pécs (1367) and Óbuda (Old Buda) in 1395. Neither lasted more than a few years.

24. Such as the Deutscher Akademischer Austauschdients, the Austrian Chancellor's office, the Friedrich Ebert Stiftung, the Volkswagenstiftung, the Austrian Ministry of Science and Research, and most importantly in certain ways, the Institute for Advanced Study and the Technical University in Vienna, both of which were going to provide significant teaching support for the program. See letter from Kevin Sontheimer to Richard E. Quandt, dated April 18, 1994.

25. AI had remarkably broad appeal among donors and raised substantial funds from TEMPUS, NATO, the Council of Europe, the Austrian and Dutch Governments, the Pew Charitable Trusts and many others.

26. Which was probably lucky for me because I had no business promising a grant to USAID.

27. The audit was carried out in a legally highly questionable manner. I quote from Alena Brunovska's formal written objection to the audit: "The inspection was performed by Jozef Bača and Dr. Ladislav Lutter on March 6–9, 1995 according to the commission of the Slovak Ministry of Education No. 9/1995–13 dated March 2, 1995. I became acquainted with the audit record on March 14, 1995 without having enough time to study it and advance my opinion. The inspectors were loath to deal with my objections that had arisen as my immediate reaction to the record and I was not allowed to input any of my remarks to the written protocol despite the fact that they should have done it [sic] (according to §21 sect. 1 Law No. 418/1991 from the Law Collection of the Slovak National Council). . . . The inspection . . . repeatedly manipulates the meaning of various quotes so as to imply wrongdoing when in fact this is not the case." Brunovska, Alena, "Standpoint Regarding the Inspection Record Carried out at Academia Istropolitana," Bratislava: Academia Istropolitana, 1995.

28. Letter from Richard E. Quandt to Ambassador Branislav Lichardus, dated April 13, 1995.

29. Letter by Andreas Wörgötter to Mikulas Luptačik, chairman of the board of trustees, dated February 15, 1995.

30. Letter to William G. Bowen from Richard E. Quandt, dated November 13, 1996.

31. Cohen had been the director of a Mellon project at the College of Finance and Accounting (Pénzügyi és Számviteli Főiskola) in Budapest. But as of January

2001, there is renewed hope that he would take on the job of organizing and directing the finance program. E-mail from Katarina Vajdova to Richard E. Quandt, dated January 4, 2001.

32. Sontheimer, Kevin, "A Proposal Submitted to Katarina Vajdova, Executive Director and the Board of Trustees of Academia Istropolitana Nova," University of Pittsburgh, Economic Policy Institute, December 1, 1999.

33. Personal communication from Katarina Vajdova, Feb. 4, 2000.

34. http://journal.ainova.sk.

35. Graduates have obtained positions in the Slovak Ministries of Culture, Finance, Foreign Affairs, in public relations firms, the media, large industrial companies, etc.

36. Letter by Deborah Duff Milenkovitch to Richard E. Quandt, dated November 7, 1991.

37. Letter by Stanisław Wellisz to Richard E. Quandt, dated January 23, 1992.

38. The program admitted about 10% of all economics undergraduates at the university (about 20 persons).

39. *Proposal for the Establishment of a Central and East European Economic Research Center*, Columbia University and Warsaw University, February 18, 1994.

40. On the Soros side, numerous individuals were involved—George Soros himself, Alfred Stepan, then the rector of CEU, and William Newton-Smith, the chair of the Higher Education Support Program of the Soros Foundation.

41. Micgiel, John and Stanisław Wellisz, *Final Report on the Warsaw University-Columbia University Cooperative Program in Economics*, New York: Columbia University, March 29, 1999.

42. Some topics were the effect of variable customs levies on food prices in Poland, the impact of the victory of the left-wing coalition on the performance of the Warsaw Stock Exchange, the influence of protectionism on the sugar industry, sex discrimination in Poland, and the effect of the wage tax on Polish small and medium enterprises. Other papers were more theoretical and dealt with finding computer solutions for Bayesian equilibria, studies of options premiums using the Black-Scholes model, and others.

43. Letter from Stanisław Wellisz to Richard E. Quandt, dated June 23, 1997.

44. *Students' Final Research Papers*, Nitra: Institute of Economics Studies, University of Agriculture, 1996.

45. One proposal would have outlawed the use of textbooks not written in the Slovak language, another one would have permitted the summary firing of university rectors.

46. Tauer, Loren, "Third Annual Report for the Continued Funding of the Institute for Economic Studies," Ithaca, NY: Cornell University, June 1999.

47. Tauer, Loren and Zdenko Stefanides, "The Empirical Impact of Bovine Somatotropin on a Group of New York Dairy Farms," *American Journal of Agricultural Economics*, 81 (1999), 95–102; Tauer, Loren, and Zdenko Stefanides, "Success in Maximizing Profits and Reasons for Profit Deviation on Dairy Farms," *Applied Economics*, 30 (1998), 151–56.

48. Hranaiova, Jana, *Delivery Options in Futures Contracts and Basis Behavior*, Dissertation in Partial Fulfillment of the Requirements for the Degree of Doctor of Philosophy, Ithaca, Cornell University, August 2000.

49. Tauer, Loren W and Ĺubica Bartová, *Funding for the Institute of Economic Studies, Final Report*, Ithaca, NY: Cornell University, July 2000.

50. Bartová, Ĺubica, personal communication on March 13, 2000.

51. E-mail from Loren Tauer to Richard E. Quandt, dated March 13, 2000.

52. It is sufficiently important for social scientists to be familiar with research methods and strategies, as well as with the management of research projects, that Mellon funded two short-term workshops on these topics in 1991 and 1992. They were organized by Zdeněk Slouka of Lehigh University and held under the auspices of the Czechoslovak Society of Arts and Sciences.

53. None of these programs are likely to draw Hungarians, because the Budapest University of Economic Sciences already does a respectable job of training in economics and appreciable numbers of its graduates get admitted on their own to first-class economics departments in western countries, including the United States.

Chapter 7

1. Pasztaleniec-Jarzyńska, Joanna and Halina Tchórzewska-Kabata, *The National Library in Warsaw*, Warsaw: The National Library Publishing House, 1998.

2. http://www.nlc-bnc.ca/pubs/nl-news/1999/mar99e/3103-02e.htm.

3. *Alma Mater Jagiellonica*, Jagiellonian University, 1999. For the *Behem Codex*, see http://gallery.euroweb.hu/html/zgothic/miniatur/1501-550/02p_1500.html.

4. Another valuable collection in Kraków is the Czartoryski Library which holds the *Codex Aureus Pultoviensis* from the eleventh and twelfth centuries.

5. Other works owned by the library include a French engraving by N. Dupin, a mezzotint by V. Green after J. Trumbull's painting, a French engraving by N. Lemire after J. B. Paon's painting, another mezzotint by V. Green, and another French engraving by J. Chevillet, after M. H. Bounieu's drawing. Personal communication from Jolanta Talbierska at the Warsaw University Library, February 21, 2000.

6. "Wrocław University Library," unpublished paper, Wrocław, 1992.

7. Padziński, Andrzej, "The Special Collections of the Hieronim Łopaciński Voivodship Public Library in Lublin," Lublin: Łopaciński Library, 2000.

8. http://portico.bl.uk/gabriel/en/countries/hungary.html.

9. For an analysis and description of medieval resources in Hungary, see Hunyadi, Zsolt, "Archival and Manuscript Resources in Hungary," paper delivered at the 33rd International Congress of Medieval Studies, Western Michigan University, Kalamazoo, MI; http://www.asu.edu/clas/acmrs/archivalresources.htm.

10. *A Magyar Tudományos Akadémia Könyvtára*, Budapest: Hungarian Academy of Sciences, 1988 and *Oriental Collection of the Library of the Hungarian Academy of Sciences*, Budapest: Hungarian Academy of Sciences, 1991.

11. http://www.drk.hu/csokonai.htm and http://www.drl.hu/arany.htm.

12. For an ecclesiastic collection owned by the Roman Catholic Church, see the Esztergom Cathedral Library; http://gallery.euroweb.hu/html/zgothic/miniatur/1501–550/.

13. *A Debreceni Református Kollégium Könyvtáráról*, Debrecen: undated.

14. See http://www.oclc.org/oclc/new/n214/n214intn.htm and http://www.nkp.cz/start/English/PublicRel;/characteristics.htm.

15. For example, Knoll, Alfred and Stanislav Psohlavec (Project mgrs.), *Chronicon Concilii Constantiniensis*, Memoria Mundi Series Bohemica 2, Prague: Národni knihovna and Albertina icome Praha, 1995; Knoll, Alfred and Stanislav Psohlavec (Project mgrs.), *Antiphonarium Sedlecense*, Memoria Mundi Series Bohemica 1, Prague: Národni knihovna and Albertina icome Praha, 1995; Knoll, Alfred and Stanislav Psohlavec (Project mgrs.), *Digitization of Rare Library Materials, Storage of and Access to Data: The Solution for Compound Documents, Manuscripts and Old Books*, Memoria Mundi Series Bohemica, Prague: Národni knihovna and Albertina icome Praha, 1997; Knoll, Alfred and Stanislav Psohlavec (Project mgrs.), *Digitization of Rare Library Materials, Storage and Access to Data*, Memoria Mundi Series Bohemica, Prague: Národni knihovna and Albertina icome Praha, 1999; Babouf, Charif and Jitka Charvátová, *Catalogue of Arabic Manuscripts of the National Library of the Czech Republic*, Prague: Albertina icome Praha, 1998; Stoklasová, Bohdana and Jaroslava Jeřábková, *Czech Books Published in the 20th Century*, Prague: Národni knihovna, 2000.

16. http://www.ulib.sk/english/history.htm.

17. It should be noted that steps are being taken to improve the situation. The Estonian National Library building and the Tartu University Library building are relatively new. Warsaw University has just opened a brand new library building. The Polish National Library and the Jagiellonian University Library have just built large modern additions to the library buildings. There is serious talk of a new building for the Latvian National Library.

18. Lass, Andrew, "CSFR/US National Library Project: Priorities and Strategic Planning," South Hadley, MA: Mount Holyoke College Working paper, October 15, 1991.

19. If the novel had been *The Magic Mountain* (which it was not) and the other book a text on tuberculosis, I might have seen some logic in the arrangement.

20. *Universitas Carolina Pragensis*, Prague: undated publication by Charles University.

21. Shim, Wonsik and Paul B. Kantor, "A Novel Approach to the Evaluation of Research Libraries," Final Report, Alexandria Project Laboratory, SCILS, Rutgers University. Shim and Kantor employ data envelopment analysis; see Charnes, Abraham, William W. Cooper, Arie Y. Lewin, Lawrence M. Seiford, *Data Envelopment Analysis: Theory, Methodology, and Application*, Boston: Kluwer, 1994.

22. Dudzińska, Elżbieta, "Import of Serials at a Time of Economic Change," *Polish Libraries Today*, 3 (1995), 49–71. Throughout the period from 1972 to 1993, the bulk of the imports was accounted for by the U.S. (29% in 1993), the U.K. (20%) and Germany (19%).

23. Kamiński, Stanisław, "Contemporary Problems of Acquisitions in Polish Libraries (Based on the Example of Warsaw University Library BUW)," *Polish Libraries Today*, 3 (1995), 35–48.

24. http://fisher.lib.virginia.edu/cgi-local/newarlbin/listyear.pl.

25. Particular thanks are due to, in Hungary, Irén Lévay and Klára Koltay of the Kossuth Lajos University library in Debrecen, Béla Mader of the József Attila University library in Szeged, Éva Zalai Kovács of the University of Hor-

ticulture and Food Industry (Kertészeti és Élelmiszeripari Egyetem); in Poland, Błażej Feret of the Technical University of Łódż, Ewa Chrzan of the University of Gdańsk library, and Andrzej Nikisch of the Poznań Foundation of Scientific Libraries, who supplied data for Adam Mickiewicz University in Poznań; in Estonia, to Asko Tamme of the Estonian Library Network (ELNET).

26. There are other examples of such dual arrangements: until recently, South Africa had two "national" libraries: the South African Library in Cape Town and the National Library in Pretoria. In October, 1999, the two became a single entity called the South African National Library.

27. http://fisher.lib.virginia.edu/cgi-local/newarlbin/listyear.pl.

28. This is a tendency that has been often noted and bemoaned; see Cummings, M. Anthony, Marcia L. Witte, William G. Bowen, Laura O. Lazarus, Richard H. Ekman, *University Libraries and Scholarly Communication*, Association of Research Libraries, 1992, p. 53.

29. Hayes, Robert M., "Staffing Patterns for Academic Libraries of Central and Eastern Europe, Russia, and CIS Countries," in Lass, Andrew and Richard E. Quandt (eds.), *Library Automation in Transitional Societies: Lessons from Eastern Europe*, New York: Oxford University Press, 2000, pp. 374–98.

30. Śliwińska, Maria, "Staff Preparation for Automation," in Lass, Andrew and Richard E. Quandt (eds.), *Library Automation in Transitional Societies: Lessons from Eastern Europe*, New York: Oxford University Press, 2000, pp. 399–406.

31. Steinerová, Jela, "Libraries of Central and Eastern Europe: Basic Dilemmas," in Lass, Andrew and Richard E. Quandt (eds.), *Library Automation in Transitional Societies: Lessons from Eastern Europe*, New York: Oxford University Press, 2000, pp. 59–73.

32. Krysiak, Ewa, "Interactions Between Library Automation and Staff and Patron Training," in Lass, Andrew and Richard E. Quandt (eds.), *Library Automation in Transitional Societies: Lessons from Eastern Europe*, New York: Oxford University Press, 2000, pp. 407–19.

33. Virgos, Márta, "The Human Aspects of Library Automation," in Lass, Andrew and Richard E. Quandt (eds.), *Library Automation in Transitional Societies: Lessons from Eastern Europe*, New York: Oxford University Press, 2000, pp. 426–42.

34. Feret, Błażej and Marzena Marcinek, "The Future of the Academic Library and the Academic Librarian: A Delphi Study," *Library Career Development*, 7/10 (1999), 91–107.

35. Many East European librarians complained that well-meaning donations often dumped useless material on the libraries. See, for example, Verrev, Viivi, "Managing the Estonian National Library with Depleting Resources," and Šmite, Sarma, "Situation in the Latvian Libraries," both in Ian R. M. Mowat and Maria Śliwińska (eds.), *Library Management: East-West Relation*, Proceedings of the seminar held in Toruń, November 6–12, 1994, Toruń: Nicholas Copernicus University, 1995, p. 145–50.

36. Greenberg, Janet, ACLS *Manual for International Book and Journal Donations*, New York: American Council of Learned Societies, 1993.

37. Roberts, Rachel, *Library Development in Central and Eastern Europe from Assistance to Cooperation: An Investment in the Future*, Proceedings of a Workshop held in Strasbourg, February 3–4, 1994, Luxembourg: European Commission, 1994.

38. For example, Eötvös Lóránd University wished to acquire such classics as James M. McPherson's *Abraham Lincoln and the Second American Revolution*; Richard M. Fried's *Nightmare in Red: The McCarthy Era in Perspective*; Carl N. Degler's *In Search of Human Nature: The Fall and Revival of Darwinism in American Social Thought*, Robert A. Dahl's *Democracy and Its Critics*; Bernard Baylin's, *Voyagers to the West: A Passage in the Peopling of America on the Eve of the Revolution*; Jagdish Bhagwati's *Protectionism*; Robert H. Wiebe's, *The Segmented Society: An Introduction to the Meaning of America*; and Sean Wilentz's, *Chants Democratic: New York City and the Rise of the American Working Class.*

39. Pénzügyi és Számviteli Főiskola.

40. Külkereskedelmi Főiskola.

41. Letter from William A. Hunt to Richard E. Quandt, dated March 2, 1990.

42. I was surprised to find on the list *Recursive Methods in Economics Dynamics* by Nancy L. Stokey and Robert E. Lucas—certainly not an elementary book.

43. Letter by William A. Hunt to Richard E. Quandt, dated March 17, 1992.

44. A meeting held at the Mellon Foundation and attended by the directors of the four arguably most important Czechoslovak libraries, U.S. library experts and U.S. foundation representatives. See the discussion of CASLIN in chapter 8.

45. Letter from Richard E. Quandt to William G. Bowen, dated October 19, 1991.

46. Letter from Richard E. Quandt to William G. Bowen, dated December 26, 1992.

47. Letter from the Chief Executive Officer of IME, Kate Noerr, to Richard E. Quandt, undated, but from internal evidence clearly written during the second half of March 1993.

48. Letter from Richard E. Quandt to William G. Bowen, dated December 17, 1993.

49. *The East & Central Europe Journal Donation Project, Proposal to the Carnegie Corporation of New York*, New York: New School for Social Research, March 2000.

50. The same people tended to surface time and again; thus in Hungary, the agent was Miklós Fogarassy, project manager for several Mellon projects, in Latvia, Aldis Ābele, later executive director of the Latvian Library Information Network Consortium, etc.

51. Letter from Arien Mack to Richard E. Quandt, dated January 13, 1993.

52. Desruisseaux, Paul, "Scholar's Remark Leads to Journal Donations for Eastern Europe," *The Chronicle of Higher Education*, 39(42), June 23, 1993, p. A28.

53. *The East & Central Europe Journal Donation Project, Proposal to the Carnegie Corporation of New York*, New York: New School for Social Research, March 2000.

54. Quoted in *The East & Central European Journal Donation Project Proposal to the Carnegie Corporation of New York*, New York: The New School for Social Research, March 2000.

55. Bejger, Peter, "American Books for Eastern Europe," *Publishers Weekly*, 237(36), September 7, 1990, p. 22. But the history of the Sabre Foundation goes back much earlier than that: it was registered as a private operating foundation

in 1969. In the first 17 years of its existence, Sabre examined the philosophy of property ownership, sought to improve the quality of political discourse and began a program for examining the basic concepts of free institutions. See http: //www.sabre.org/HISTORY.html.

56. *Annual Report 1989*, Somerville, MA: Sabre Foundation, Inc., 1990.

57. Letter from Josiah Lee Auspitz to Richard E. Quandt, dated September 13, 1990.

58. Letter from Josiah Lee Auspitz to Richard E. Quandt, dated September 30, 1990. But, as noted earlier, telephone service was atrocious in the early 1990s and by 1991–92, Sabre started to send the lists by e-mail. See e-mail from Josiah Lee Auspitz to Richard E. Quandt, dated January 2001.

59. The four countries received rather similar shares: Czech Republic 39,208 volumes, Slovakia 37,160, Hungary 32,676, and Poland 38,115. The largest share in this was accounted for by business and economics (41,046 volumes), science and technology (28,322), social science (18,830), arts and literature (18,753), and medicine (18,153). See Deehy, Suzanne M., *Interim Grant Report to The Andrew W. Mellon Foundation*, Cambridge, MA: Sabre Foundation, Inc., June 27, 1994.

60. *Civic Education Project*, New Haven, CT: CEP Pamphlet, 1993.

61. Letter from Emily Lehrman to Richard E. Quandt, dated March 25, 2000. See also *Annual Report, July 1, 1998–June 30, 1999*, New Haven, CT: Civic Education Project, 2000.

62. Becker, Jonathan A. (ed.), *Assessing the Impact of Book & Journal Donations to Central & Eastern Europe*, Prague and New Haven, CT: Civic Education Project, 1994.

63. While the Journal Donation Program, the Solidarity project, and the American Czech and Slovak Education Fund have all collaborated with Sabre, it is not clear to what extent their supply lists have affected the distribution plans of each other.

64. A remedy for this is provided by JSTOR; see later in this chapter.

65. Auspitz, Josiah Lee, "Memorandum," Sabre Foundation, Inc. October 20, 1995.

66. Recommendations to donors: (1) that cultural and educational institutions still lacked the financial means to acquire library materials on their own; (2) that materials from Western Europe were needed in addition to U.S. materials; (3) that textbooks and reference books represented the highest priority; (4) that materials other than printed ones (video, audio, CD-ROM) were badly needed; and (5) that access to other materials through discounted prices were also needed. Commitments of the East European partners: (1) to accepting responsibility to provide services to a variety of users; (2) to securing the widest possible access; (3) to seeking local financial and material support; (4) to engage in regular consultations with one another; and (5) to cooperate with donors in alleviating obstacles such as customs duties. *Dialogue of East European Partners (*DEEP*)*, Budapest: National Széchényi Library, 1995, pp. 6–8.

67. Trojanowska-Bitka, Anna, "Poland," *Dialogue of East European Partners (*DEEP*)*, Budapest: National Széchényi Library, 1995, pp. 28–34.

68. Letter by Kerry S. McNamara to Josiah Lee Auspitz, dated January 4, 1996.

69. Letter from Josiah Lee Auspitz to Kerry S. McNamara, dated April 8, 1996.

70. Halm, Johan van, "Sabre Foundation, Inc. Scientific Assistance Project—Outside Evaluation," Amersfoort, Netherlands: Johan van Halm Information Consultancy, February 1994.

71. *The Chronicle of Higher Education*, December 15, 2000; http://chronicle.com/free/v47/i16/16a04702.htm.

72. JSTOR provides over the Internet by subscription complete digital copies of 117 journals from their very inception until about 3–5 years before the present and includes a total of about 5 million page images. Subscribers have unlimited browsing, downloading, and printing privileges. See http://www.jstor.org.

73. See, for example, Chressanthis, G. A. and J. D. Chressanthis, "The Determinants of Library Subscription Prices of the Top-ranked Economics Journals: An Econometric Analysis," *Journal of Economic Education*, 25 (1994), 367–82; Noll, Roger and W. E. Steinmueller, "An Economic Analysis of Scientific Journal Prices: A Preliminary Analysis," *Serials Review*, 18 (1992), 32–37; Cummings, Anthony M., M. L. Witte, William G. Bowen, L. O. Lazarus, Richard H. Ekman, *University Libraries and Scholarly Communication*, Washington, D.C.: Association of Research Libraries, 1992; Quandt, Richard E., "A Simulation Model for Journal Subscription by Libraries," *Journal of the American Society for Information Science*, 47/8 (1996), 610–17.

74. From http://stats.jstor.org/staff; only accessible to JSTOR staff.

75. E-mail from Irén Lévay to Richard E. Quandt, dated June 18, 2000.

76. Fogarassy, Miklós, "Special Role of the Hungarian National Library in the Formation and Management of Consortia," in Lass, A. and Quandt, R. E. (eds.), *Library Automation in Transitional Societies: Lessons from Eastern Europe*, New York: Oxford University Press, 2000, pp. 420–26.

77. E-mail from Márta Virágos to Richard E. Quandt, dated July 18, 2000.

78. http://www.lib.klte.hu/huslonet/index.english.html.

79. E-mail from Irén Lévay to Richard E. Quandt, dated June 18, 2000.

80. E-mail from Márta Virágos to Richard E. Quandt, dated July 18, 2000.

81. E-mail from Darina Kožuchová to Richard E Quandt, dated June 18, 2000.

82. Burmeister, Erzsébet, "Számitástechnika a miskolci egyetem könyvtárában az 1974. és 1994 között eltelt 20 évben," (Computer technology in the library of the University of Miskolc during the 20 years between 1974 and 1994) Miskolc: University of Miskolc, January 1995; http://www.lib.uni-miskolc.hu/publ/erzsi/szamtech.htm.

Chapter 8

1. By the term "integrated" I mean a system that performs all library functions (such as cataloging, circulation, serials control, acquisitions, open public access catalog [OPAC], and statistical reporting) using the same database.

2. Bridge, Frank R., "Automated System Marketplace 1992," *Library Journal*, April 1, 1992.

3. Sold by Innovative Interfaces, Inc.

4. Czermiński, Jurand., "A Coordinator's View of the Impact of Policy on the Financing and Managing of Library Automation," in Lass, A. and Quandt, R. E. (eds.), *Library Automation in Transitional Societies: Lessons from Eastern Europe*, New York: Oxford University Press, 2000, p. 60.

5. Mader, Béla, "Automation and Academic Libraries in Hungary: Theory and Practice in the Period of New Challenges," in Lass, Andrew and Richard E. Quandt (eds.), *Library Automation in Transitional Societies: Lessons from Eastern Europe*, New York: Oxford University Press, 2000), pp. 93–103.

6. Borgman, Christine L., "Library Automation in Central and Eastern Europe: Progress and Prospects," in *International Conference on Library Automation, Conference Proceedings, Budapest, April 10–13, 1996*, ed. by Monika Segbert, Katarina Steinwachs, Peter Burnett; Luxembourg: European Commission, DG XIII/E-4, 1997, pp. 13–22.

7. For the Czech and Slovak Union Catalog see http://www.caslin.cz:7777/caslin/ENG/Welcome_eng.html.

8. Füredi, Mihály, " 'Szubjektiv' gondolatok a hazai könyvtárgésitépesről," ('Subjective' thoughts about domestic library automation), *Könyvtári Figyelő*, 38 (1992/4), 636–44.

9. Murányi, Lajos, "Das ungarische Informations-und Dokumentationswesen nach 1990," (The state of Hungarian information and documentation technology after 1990), *Newsletter*, Sozialwissenschaften in Osteuropa, November 1994, http://www.berlin.iz-soz.de/publications/newsletter/socsci-eastern-europe/nl94–95/nl94-4-5.htm.

10. Tolnai, György, "A Könyvtárgépesités Helyzete Magyarországon," (Library automation in Hungary), unpublished paper, May 1992, and Tolnai, György, "A hazai könyvtárgépesités a megváltozott világban" (Hungarian library automation in a changed world), *Tudományos és Műszaki Tájékoztatás*, 39/7–8, 1992, 303–9. This localized version was referred to as NEKTÁR.

11. Sadowska, Jadwiga, "Automation in the National Library in Warsaw: From an In-House System to INNOPAC," in Lass, Andrew and Richard E. Quandt (eds.), *Library Automation in Transitional Societies: Lessons from Eastern Europe*, New York: Oxford University Press, 2000), pp. 104–19.

12. Svoboda, Martin, "Cooperation, Consortia, Compatibility, Connectivity: The Case of CASLIN," in Lass, Andrew and Richard E. Quandt (eds.), *Library Automation in Transitional Societies: Lessons from Eastern Europe*, New York: Oxford University Press, 2000), pp. 134–52.

13. Svoboda, Martin, "Automation in Czech Libraries—How to Proceed?" Paper presented at LIBER Workshop, Budapest, June 1992.

14. Prokop, Igor, Milan Rákus, and Daniela Slížová, "Realization of the CASLIN Project in the Slovak National Library and Its Impact on the Automation of Libraries in Slovakia," in Lass, Andrew and Richard E. Quandt (eds.), *Library Automation in Transitional Societies: Lessons from Eastern Europe*, New York: Oxford University Press, 2000), pp. 153–79.

15. http://www.unesco.org/webworld/isis/isis.htm.

16. Borgman, op. cit., p. 16. Some libraries (7.1%) had converted some of their catalog records, but not by year and 10% did not respond to the questionnaire.

17. Řepišová, Zuzana, "The Implementation of an Automated System at the University Library in Bratislava, Slovakia," Lass, Andrew and Richard E. Quandt (eds.), *Library Automation in Transitional Societies: Lessons from Eastern Europe*, New York: Oxford University Press, 2000, pp. 204–18.

18. http://www.ifla.org/VI/3/p1996–1/unimarc.htm. See also http://www.tlcdelivers.com/tlc/crs/gen0001.htm for various USMARC formats and the CD-

ROM "USMARC Made Easy" by Learning Curve Pty Ltd and DocMatrix Pty Ltd, 1996.

19. http://www.tlcdelivers.com/tlc/crs/gen0001.htm.

20. According to Borgman, op. cit., at the end of 1994, 15.7% used UNIMARC, 14.3% used USMARC, 24.3% some "national" MARC, and the remainder a variety of other formats.

21. 12.% used Standard ASCII, 65.7% used either ASCII Latin 2 or IBM Codepage 852 (versions of the same format), 11.4% used a national format, and 15.7% used some other format. See Borgman, op. cit., p. 19.

22. Könyvtártudományi és Módszertani Központ.

23. Letter from Richard E. Quandt to William G. Bowen, dated July 31, 1990.

24. The participants in the consortium and the amounts requested by them were the National Széchényi Library ($60,312), the József Attila University Library ($13,800), the National Educational Library and Museum ($40,360), the Library of the Parliament ($30,000), the National Institute for Medical Education Library ($27,365), the Semmelweiss University of Medicine Library ($20,000), the Central Library of the Medical University of Debrecen ($20,885), the Central Library of the Medical University of Pécs ($20,000), and the Central Library of the Szent-Györgyi Medical University of Szeged ($20,000).

25. The group included the University of Forestry Library ($30,000), the Széchényi College of Technology Library ($30,000), the Pannon University of Agriculture Library ($30,000), the University of Veszprém Library ($30,000), the Politechnic of Dunaujváros Library ($10,000), the University of Horticulture and Food Industry Library ($30,000), the University of Veterinary Sciences Library ($28,500), the Technical University of Budapest Library ($29,500), the University of Miskolc Library ($27,000), the Agricultural University of Debrecen Library ($30,000), the Szabó Erwin Library ($10,000), the National Széchényi Library ($10,000).

26. See Bakonyi, Géza and Károly Kokas, "Könyvtári integrált rendszerek és hazai alkalmazásuk," (Integrated library systems and their domestic use), October 1996, http://dis.gau.hu/tinlib/tinbgkk.html and Tapolczai, Ágnes and Péter Popovics, "Novell rendszerben működő integrált könyvtári rendszer (Tinlib) megjelenése WWW felületen," (The appearance of a Novell-based integrated library system [Tinlib] in the realm of WWW), November 1997, http://dis.gau.hu/tinlib/tintawww.html.

27. E-mail from Miroslav Bartošek to Andrew Lass, dated May 10, 2000.

28. Gomba, Olga, "A Retrospective Conversion: The Case of Debrecen," Lass, Andrew and Richard E. Quandt (eds.), *Library Automation in Transitional Societies: Lessons from Eastern Europe*, New York: Oxford University Press, 2000), pp. 183–95.

29. Koltay, Klára, "Why, and How to, Translate a Subject Heading System: Authority Control Provides the Solution," Lass, Andrew and Richard E. Quandt (eds.), *Library Automation in Transitional Societies: Lessons from Eastern Europe*, New York: Oxford University Press, 2000), pp. 267–83.

30. Borgman, Christine L., "Könyvtári automatizálás és hálózatok, Részlet a felsőoktatási világbanki 1,3 moduljával kapcsolatos jelentésből," (Library automation and networks, fragment from the report on the World Bank higher

education module No. 1.3), *Tudományos és Műszaki Tájékoztatás*, 43(1996), 95–104.

31. Bakonyi, Géza, "VOCAL—a Corvina könyvtárak osztott katalogizálási rendszere,"(VOCAL—the distributed cataloging system of the Corvina libraries) *Könyvtári Figyelő*, 45, 1999/2; http://www.oszk.hu/kiadvany/kf/1999/2/bakonyi_h.html.

32. Gyüre, Péter, "Osztott katalogizálás a gyakorlatban—a "VOCAL" rendszer, (Shared cataloging in practice—the VOCAL system) http://www.iif.hu/rendezvenyek/networkshop/98/eloadas/html/d/pgyure/pgyure.htm.

33. "VOCAL," http://www.lib.klte.hu/rendszer.html.

34. Vajda, Erik, "Közös (osztott) katalogizálás—közös (központi) katalógus," (Shared (distributed) cataloging—shared (centralized) cataloging), *Könyv, Könyvtár, Könyvtáros*, 2000/3. I am also indebted to Erik Vajda for a most illuminating conversation.

35. E-mail from Béla Mader to Richard E. Quandt, dated July 3, 2000.

36. Personal communication from Andrew Lass.

37. Letter by Andrew Lass to Richard E. Quandt, dated June 6, 1990.

38. Letter from Elizabeth T. Kennan to Richard E. Quandt, dated March 15, 1991. In a brilliant article, Lass described his impressions on a guided tour of the Clementinum: "The guided tour of a sacred site turned into a horror picture show of crates filled with books placed in the aisles between the overflowing stacks; of poorly ventilated attics with leaking roofs filled with more rows of crates covered with plastic sheets to protect the catalogued but never shelved volumes from dust, rain, and pigeon droppings; of rotting manuscripts and old books . . ." See Lass, Andrew, "Portable Worlds: On the Limits of Replication in the Czech and Slovak Republics," in Buraway, Michael and Verdery, Katherine, *Uncertain Transition*, Lanham: Rowman & Littlefield Publishers, 1999, pp. 273–300.

39. The smaller of these grants also included a site visit to the National Library of Poland, but Mount Holyoke did not pursue that avenue further.

40. Charter 77 Foundation had provided a grant for a visit to the National Library by Norman Ross, of Norman Ross Publishing, Inc., Richard Kneeley, doctoral student at Columbia University, and Edward Kasinec, chief of the Slavic and Baltic Division of the New York Public Library. See letter from Norman Ross to Wendy Luers, dated May 10, 1991.

41. Unrelated to the immediate interest in the National Library, an important conference, "East European Bibliographies and Electronic Networks," dealing with some related issues and funded by the Rockefeller Foundation, was held on February 24–28, 1992, at the Bellagio Study and Conference Center at the Villa Serbelloni, Bellagio, Italy.

42. Klimová, Milena, "Main Problems of the Slavonic Library at the National Library in Prague," unpublished paper, Prague: Slovanská knihovna, 1991.

43. See http://www.senka.sk/martin2/ms-e.html and *Law 169 of the Slovak National Council about Matica Slovenská*, §1. See also http://portico.bl.uk/gabriel/en/reports/slovakia-99–00.html.

44. E-mail from Igor Prokop to Richard E. Quandt, dated July 21, 2000. See *On libraries, on the supplements of Act 27:1987 (Digest) of the Slovak National Council on the state care of historical monuments and on the amendments and*

supplements of Act No. 68:1997 (Coll,) on Matica slovenská, Act No, 183, Bratislava, May 12, 2000.

45. I had invited Donald Koepp, director of the Princeton University Library; Lee Varian from the Princeton University Computer Center; Norman Ross, of Norman Ross Publishing, Inc.; Hans Rütimann of the Commission on Preservation and Access; Stanley Kalkus of the Department of the Navy Library; and Declan Murphy of the Library of Congress. Of course, Elizabeth Kennan and Andrew Lass of Mount Holyoke College were there, and the foundation world was represented, in addition to myself and Kellum Smith, by Marilyn Perry, president of the Samuel H. Kress Foundation; Kevin Quigley of the Pew Charitable Trusts, Mary Albon of Charter 77 Foundation, Jillian Poole from the Fund for the Arts and Culture of Central and Eastern Europe, and Deborah Marrow of the Getty Grants Program. Zuzana Nagy, a Slavic librarian at Harvard, acted as translator.

46. This was amply illustrated later by Andrew Lass, who had taken numerous photographs inside the National Library.

47. Willard, Kristen, "Report of October 18, 1991, Meeting Regarding Czech and Slovak Libraries," New York: The Andrew W. Mellon Foundation, October 18, 1991.

48. An amusing commentary of the speed with which information technology changes is the remark by one western expert, commenting on the need to do retrospective conversion of catalogs once an automated system is installed: he felt that microfilming card catalogs was the best approach because electronic conversion of catalogs was not feasible. Yet, before the middle of the decade, one could very effectively scan card catalog cards and create computer records for them!

49. Letter by Vojtěch Balík, Miroslav Bielik, Jaromír Kubíček, Emil Vontorčík to Richard E. Quandt, dated November 11, 1991.

50. Vojtěch Balík, Miroslav Bielik, Jaromír Kubíček, Emil Vontorčík, "Letter of Intent," November 11, 1991. I did not know at the time what a pivotal role Lass had played in the drafting of the letter of intent. Later that fall, when he was in Prague, Balík asked his advice about what the next step should be. Lass strongly felt that something should be put down on paper, and he and the deputy director of the National Library, Martin Svoboda, then wrote the first draft of the letter. See letter from Andrew Lass to Richard E. Quandt, dated January 22, 2001.

51. Letter by Esko Häkli to Hans Rötimann, dated April 29, 1992.

52. That is not to say that from that point on everything was smooth and unproblematical. My own sense is that Balík was the librarian with the clearest vision of what a cooperative approach could produce. He was the principal architect of this agreement, while some others, such as Vontorčík, would ultimately be more interested in what CASLIN could do to advance their standing than in what CASLIN could do for the aggregate of libraries and users.

53. Letter by Richard E. Quandt to Eva Kantůrková, dated February 22, 1992.

54. E-mail from Richard Kneeley to Richard E. Quandt, dated March 17, 1992.

55. Letter from Vojtěch Balík to Richard E. Quandt, dated April 13, 1992.

56. Letter from Eva Kantůrková to Richard E. Quandt, dated April 28, 1992:

"Parlamentní výbor se za účasti ministra kultury několikrát výsledky kontroly v Národní knihovně zabýval a usnesl, aby konrétni šetření provedla i ministerstva financí a kontroly, nadřazená v této věci ministerstvu kultury" and later "Jsme Vám osobně, pane profesore, a Mellonově nadaci velice vděčni, že míníte pomoci tak významnému ústavu, jakým Národní knihovna bezesporu je."

57. In the interest of full disclosure, I should add that in July 1992 I received an 18-page letter, half of it handwritten, from somebody who accused Balík and others in and out of the National Library of unethical behavior, wild prejudice, and total ignorance, and claimed to be persecuted by them. A close reading of the letter convinced me that the writer was paranoid and could safely be ignored. But this was not the only complaint. In August 1993, I received a letter from one John Kafka from the Bratislava University Library, who complained that old Communists were still holding responsible positions in the library, that books from the fourteenth to eighteenth centuries were being sold illegally to western countries, and said that he had become "rather suspicious of the Mellon Foundation." Subsequent correspondence suggested that because Mellon was "associated" with this untrustworthy library, Mellon had to be guilty of something. I concluded that he was a crank and ignored him as well.

58. Letter from Miroslav Bielik to Richard E. Quandt, dated July 21, 1992.

59. The number of e-mail messages and letters sent and received by me concerning CASLIN in 1992 amounted to 82.

60. Letter from Andrew Lass to Richard E. Quandt, dated August 22, 1992.

61. Quandt, Richard E., *Report on Trip to the Czech Republic, Slovakia and Hungary, May 20–June 2*, New York: The Andrew W. Mellon Foundation, 1995.

62. For reprographic equipment, each center library would receive a camera for photographing documents, an automatic film developer, a film duplicator, and an enlarger.

63. The grant was intended to pay for only part of the costs associated with installing an automated library system. For example, the renovation of the antiquated telephone system in the Czech National Library, estimated to cost $250,000, was to be borne by the government.

64. Quandt, Richard E., *Trip Report on Projects in Eastern Europe: June 1994*, New York: The Andrew W. Mellon Foundation, 1994.

65. Řepišová, Zuzana, "The Implementation of an Automated System at the University Library in Bratislava, Slovakia," in Lass, Andrew and Richard E. Quandt (eds.), *Library Automation in Transitional Societies: Lessons from Eastern Europe*, New York: Oxford University Press, 2000, pp. 204–18. See also, for example, Lyman, Randall, "Czech Libraries to Enter 20th Century," *Prognosis*, June 11–24, 1993.

66. See http://www.caslin.cz:7777/caslin/ENG/caslin-e.html. See also letter from Andrew Lass to Richard E. Quandt, dated January 22, 2001.

67. Košice (Kassa in Hungarian) is a historic city with a great Gothic cathedral, and is the burial place of Prince Ferenc Rákóczy II, the leader of the Hungarian revolt against the Hapsburgs in the early eighteenth century. Before 1918, Košice was part of Hungary and still has a nonnegligible Hungarian-speaking population.

68. Letter from Darina Kožuchová and Ľubomira Šoltésová to Ex Libris, dated May 17, 1996; letter from Azriel Morag to KOLIN, dated May 20, 1996; letter from Andrew Lass to Azriel Morag, dated May 20, 1996.

69. Fortunately, Gerboc did not last long and was replaced in 1998 with the vastly more collaborative and capable Dr. Tibor Trgiňa.

70. Prior to 2000, I had met Androvič only once and can only guess at the reasons he might have had for wanting to replace Aleph. In Košice, Mižák's deputy argued that they needed to abandon Aleph because "its circulation module does not work;" a blatantly incorrect statement.

71. But on July 23, 1998, I received a letter from Dr. Mižák in which he denied that the Regional Library wanted to leave KOLIN.

72. Postma, Renée, "Slovakia not out of Trouble yet after Defeat Meciar," October 1, 1998, http://www.rnw.nl/budapest/html/slovakia0110.html.

73. Letter from Miloš Somora to Richard E. Quandt, dated August 24, 1998. I replied somewhat mollified, and this was followed up in September by another letter by him confirming our understanding. Similar letters were also sent by the other two rectors.

74. This point is in contrast to the Sabre Foundation position that it may be sensible to have books in professors' offices, because they are then safe from pilferage. See chapter 7.

75. Letter from Richard E. Quandt to Rector Dušan Podhradský, dated November 5, 2000.

76. Why not distribute a grant in proportion to the number of books in the respective libraries, or in proportion to the number of readers, or professors, or departmental libraries, or whatnot?

77. For the Academy Library, see http://www.lib.cas.cz/.

78. Kadlecová, Ivana, "Current State of Automation," unpublished paper, Prague: Czech Academy of Sciences, 2000.

79. It recently reverted to its pre–World War II name of Moravian Regional Library.

80. E-mail from Miroslav Bartošek to Andrew Lass, dated May 10, 2000. For searches, see http://publib.upol.cz/atpar.

81. An illustration of the complexity of even this simple question, in the case of KOLIN, is contained in an excerpt from a letter from Andrew Lass to Azriel Morag (CEO of Ex Libris), dated June 20, 1996: "You insist [that the language of instruction] should be in English (though you would 'try to train in Czech/Slovak whenever possible.') Please correct me if I am wrong to understand that whenever possible you would use CASLIN people previously trained by Ex Libris but otherwise training would be in English? If that is so, does this mean that in the case of using someone from another library, Ex Libris would reimburse them but otherwise if a translator is needed, then KOLIN has to find/pay for a translator?"

82. Přibramská, Iva, The Aleph Implementation at the National Library of the Czech Republic," in Lass, Andrew and Richard E. Quandt (eds.), *Library Automation in Transitional Societies: Lessons from Eastern Europe*, New York: Oxford University Press, 2000, pp. 168–79.

83. "e) Acceptance Report(s). The ALEPH system as installed will be tested and checked by the USER against the LIBRIS [Ex Libris] specifications and functionality marked STANDARD in the LIBRIS response to the RfP . . . USER will carry out the acceptance tests and produce the first Acceptance report within six (6) weeks of LIBRIS's notification that ALEPH is ready for such testing . . . If USER fails to carry out this procedure within the agreed period or if USER turns to

Production, then System Acceptance will have been achieved for the purposes of the third and final payment." See *Software License Agreement between Ex Libris* LTD *and Bibliotéku Informacijās Tīklu Konsorcijs* (Library Information Network Consortium of Latvia), October 24, 1998.

84. As indicated before, KOLIN, as well as CASLIN proper more recently, are served by a firm located in Budapest. Travel time from Budapest to Košice is over four hours, and while I have no evidence that this played a role, given the ethnic conflicts between Slovaks and Hungarians, I would be reluctant to assign a Slovak user to a Hungarian service company or conversely.

85. See, for example, announcement of the 7th international seminar, *CASLIN 2000: Education for Librarians;* http://www.ics.muni.cz/caslin2000/index-e.html.

86. *Souborné katalogy: organizace a služby* (Union catalogues: their organization and services), Prague: Národni knihovna ČR, 2000.

87. Deák, Nóra, "Current Library Management Issues in Hungarian Libraries or How Do Hungarian Libraries Cope with the Changes?" in Ian R. M. Mowat and Maria Śliwińska (eds.), *Library Management: East–West Relation,* Proceedings of the seminar held in Toruń, November 6–12, 1994, Toruń: Nicholas Copernicus University, 1995, pp. 137–41.

88. In rare instances, the library itself may be able to carry out the conversion.

89. Once the decision was made to ask Mount Holyoke to manage CASLIN, it made sense to place it in a supervisory position with respect to *all* library projects affecting the CASLIN libraries.

90. Stoklasová, Bohdana and Jaroslava Jeřábková, *Czech Books Published in the 20th Century*, Prague: Národni knihovna, 2000.

91. Stoklasová, Bohdana, "The Retrospective Conversion in Czech Libraries," in Lass, Andrew and Richard E. Quandt (eds.), *Library Automation in Transitional Societies: Lessons from Eastern Europe*, New York: Oxford University Press, 2000, pp. 196–203.

92. Medical School, Agricultural University, Reformed Theological College and Institute for Nuclear Research.

93. Only 60% of these figures are gross wage, with the remainder representing the employer's compulsory contribution to social insurance.

94. Paluszkiewicz, Ewa, "The Retrospective Conversion of the Catalog in the Library of the American Studies Center, Warsaw University," in Lass, Andrew and Richard E. Quandt (eds.), *Library Automation in Transitional Societies: Lessons from Eastern Europe*, New York: Oxford University Press, 2000, pp. 113–19.

95. See also Burnett, Peter, "Retrospective Conversion: Prerequisites and Problems," in *International Conference on Library Automation, Conference Proceedings, Budapest, April 10–13, 1996*, ed. by Monika Segbert, Katarina Steinwachs, Peter Burnett; Luxembourg: European Commission, DG XIII/E-4, 1997, pp. 177–86.

96. Stoklasová, Bohdana, "Cataloging Standards in Czech Libraries," in Lass, Andrew and Richard E. Quandt (eds.), *Library Automation in Transitional Societies: Lessons from Eastern Europe*, New York: Oxford University Press, 2000), pp. 219–23.

97. See, for example, "The Logical Structure of the Anglo-American Cataloguing Rules—Part 1," Joint Steering Committee for the Revision of Anglo-

American Cataloguing Rules, August 1998, http://www.nlc-bnc.ca/jsc/aacrdel.htm, and Gorman, Michael, "International Conference on the Principles and Future Development of AACR2 (1997: Toronto)," http://www.ala.org/alonline/news/aacrs.html.

98. Letter from Andrew Lass to Richard E. Quandt, dated November 17, 1993.

99. In Poland, the regulations were, if anything, even less clear than in Czechoslovakia and there appears to have been much randomness in what imported items were charged customs duty and/or VAT; the outcome often seemed to depend on the skill of the dealer. See Czermiński, Jurand, "A Coordinator's View of the Impact of Policy on the Financing and the Managing of Library Automation," *Library Automation in Transitional Societies: Lessons from Eastern Europe*, New York: Oxford University Press, 2000, pp. 45–58.

100. Quandt, Richard E., *Report on Trip to the Czech Republic, Slovakia and Hungary, May 18, 1996 through June 1, 1996*, New York: The Andrew W. Mellon Foundation, 1996.

101. In spite of the fact that one of the four original CASLIN members is called the Bratislava University Library.

102. E-mail from Andrew Lass to Richard E. Quandt, dated February 15, 1995.

103. Personal communication from Vojtěch Balík, May 22, 1995.

104. Balík, Vojtěch, "CASLIN, Priorities of the Czech National Library, and Government Policy on Information and Libraries," in Lass, Andrew and Richard E. Quandt (eds.), *Library Automation in Transitional Societies: Lessons from Eastern Europe*, New York: Oxford University Press, 2000, p. 125.

105. Lass, Andrew, "Portable Worlds: On the Limits of Replication in the Czech and Slovak Republics," in Buraway, Michael and Verdery, Katherine, *Uncertain Transition*, Lanham: Rowman & Littlefield Publishers, 1999, p. 284.

106. "Pre-requisites for the implementation of the state information policy," Office of the Czech Republic, http://www.vlada.cz/eng/vrk/rady/sip/dokumenty/sipcesta/priloha2.eng.html.

Chapter 9

1. Quandt, Richard E., "On the Optimum Number of Library Software Licenses," *Journal of Economic Behavior and Organization*, 38(1999), 349–56.

2. VTLS, Inc. of Blacksburg, Virginia. See http://www.vtls.com/.

3. Ameritech Library Services provided the systems Dynix, Horizon, and Notis. See http://www.ameritech.com.

4. See http://www.iii.com/.

5. Over 125,000 pre-1801 imprints, drawings by Dürer, Rembrandt, Piranesi, Rubens, Vasari, and many others.

6. Letter from Richard E. Quandt to William G. Bowen, dated November 4, 1990.

7. Ultimately, the grant (extended in the spring of 1991) was for $288,000, because it also included a $47,500 portion for personal computers to be used by the Warsaw University Business School. (See chapter 11). The Foundation also made a separate $40,000 grant to Warsaw University for intensive training of a librarian at Indiana University.

8. http://www.magma.ca/~pfeiffer/poland/kra_history.htm.

9. Of this sum, $466,000 was for library automation, $20,000 for a fancy color copier for the university museum, $50,000 for journal subscriptions for the American Studies Center, $20,000 for a study/training trip for some librarians to the United States, and $4,000 for a PC for the English Teacher's College.

10. Somewhat of a misnomer, since the Academy (now called University) is really an engineering school and might be considered the "MIT of Poland."

11. *Warsaw University Library's Report for The Andrew W. Mellon Foundation*, Warsaw: Warsaw University Library, October 19, 1992.

12. Letter from Józef Ziółkowski to Richard E. Quandt, dated August 28, 1992.

13. Only Wrocław had the more recent 1994 (Unix-based) version of the software; the other libraries all used VTLS92.

14. Letter from Andrzej Ładomirski to Richard E. Quandt, dated February 2, 1995.

15. The complaints about not receiving timely delivery of the proper (Polish, Czech, Estonian, etc.) character sets and when delivered, the characters not being correctly displayed, were so pervasive, that I waggishly suggested at a library automation conference organized by the Mellon Foundation in Warsaw, in October 1997, that the easy solution to the problem would be for all languages henceforth to eliminate all diacritical characters.

16. Unicode essentially solves the problem of encoding the characters in either of two ways: with a single code for, say, š, or with separate codes for the s and for the hacek. But handling this in the software was not as straightforward in the early 1990s as it might be today. See Brickell, Anthony, "Unicode/ISO 10646 and the Chase Project," in *International Conference on Library Automation, Conference Proceedings, Budapest, April 10–13, 1996*, ed. by Monika Segbert, Katarina Steinwachs, Peter Burnett; Luxembourg: European Commission, DG XIII/E-4, 1997, pp. 119–25.

17. Akademia Rolnicza w Lublinie; a school of agriculture functioning as an independent institution (as is customary in Eastern Europe).

18. See Lass, Andrew and Richard E. Quandt, "Introduction," in Lass, Andrew and Richard E. Quandt (eds.), *Library Automation in Transitional Societies: Lessons from Eastern Europe*, New York: Oxford University Press, 2000.

19. I wondered why the Medical Academy (a separate medical school) did not join the consortium and was told that there were two reasons: first, it had a very small library, and second, it was under the authority of the Ministry of Health (rather than of Education), which made consortial relations with other types of universities fraught with bureaucratic difficulties.

20. Nowak, Grzegorz, "The History of Library Automation Project for Lublin Consortium Libraries," unpublished paper, Lublin: Marie Curie Skłodowska University, 1999.

21. When an unauthorized user issued the process command, he or she would quite properly be "caught." But if the process command was mistyped in a particular way (e.g., typed as "processq") the command was actually carried out even for an unauthorized user!

22. E-mail from Grzegorz Nowak to Richard E. Quandt, dated August 7, 2000.

23. See http://hieronim.wbp.lublin.pl/ang/lopatek.htm.

24. Paluszkiewicz, Anna, "Evolution of Authority File Work in Poland," in

Lass, Andrew and Richard E. Quandt (eds.), *Library Automation in Transitional Societies: Lessons from Eastern Europe*, New York: Oxford University Press, 2000, pp. 224–34.

25. See the perceptive discussion of delays in Lass, Andrew, "Managing Delays: The Micropolitics of Time in the Czech and Slovak Automation Project," in Lass, Andrew and Richard E. Quandt (eds.), *Library Automation in Transitional Societies: Lessons from Eastern Europe*, New York: Oxford University Press, 2000, pp. 345–59.

26. For the libraries of the Lublin Consortium, see http://priam.umcs.lublin.pl.

27. Letter from Richard E. Quandt to Jan Janowski, Jerzy Mikułowski-Pomorski, Józef Nizioł, Andrzej Pelczar, Barbara Skucińska, dated November 14, 1992.

28. I wrote in my *Report on Trip to Poland, October 21–31, 1993*: New York: The Andrew W. Mellon Foundation, 1993: "I think they are thinking about all of this in a much more focused way than a year ago and the very hard time I had been giving them during the last twelve months has had some payoff."

29. Not yet an active contributor to KLG as of June 1, 2000. See e-mail from Henryk Hollender to Richard E. Quandt, dated June 1, 2000.

30. E-mail from Ewa Dobrzyńska-Lankosz to Richard E. Quandt, dated June 1, 2000.

31. Letter from Kate Noerr to Richard E. Quandt, undated, but from internal evidence clearly written during the second half of March 1993.

32. Letter from Andrzej Pelczar to Richard E. Quandt, dated March 25, 1993.

33. Dobrzyńska-Lankosz, Ewa, "The Kraków Library Group as an Example of a Local Consortium," in *International Conference on Library Automation, Conference Proceedings, Budapest, April 10–13, 1996*, ed. by Monika Segbert, Katarina Steinwachs, Peter Burnett; Luxembourg: European Commission, DG XIII/E-4, 1997, pp. 289–95; Dobrzyńska-Lankosz, Ewa, "The Polish VTLS Consortium: Six Years of Cooperation," in Jadwiga Woźniak and Robert C. Miller (eds.), *Research Libraries: Cooperation in Automation*, Warsaw: Center for Formats and Authority Control, Warsaw University, 1999, pp. 23–29; and Dobrzyńska-Lankosz, Ewa, "The Kraków Project: Problems of Management," in Lass, Andrew and Richard E. Quandt (eds.), *Library Automation in Transitional Societies: Lessons from Eastern Europe*, New York: Oxford University Press, 2000, pp. 360–73.

34. *Academic Computer Center CYFRONET-Kraków: Guide 1996*, Kraków: CYFRONET, 1996. For the current state of the network, see http://www.cyfronet.krakow.pl/grafika/obrazki/ack/ack_schemat.jpg.

35. E-mail from Henryk Hollender to Richard E. Quandt, dated March 28, 2000.

36. E-mail from Jurand Czermiński to Richard E. Quandt, dated December 23, 1995.

37. E-mail from Jurand Czermiński to Richard E. Quandt, dated January 20, 1996.

38. See http://victoria.uci.agh.edu.pl and http://www.bj.uj.edu.pl.

39. Śliwińska, Maria, "Nicholas Copernicus University Library Computerization," unpublished paper, Toruń: Copernicus University, 2000.

40. The University of Łódź, the Technical University of Łódź, the Medical

University, the Military Medical Academy, the Academy of Fine Arts and Design, the Academy of Music, the Theological Seminary, the Center of Molecular and Macromolecular Studies of the Polish Academy of Sciences, and the Academy of Film, Television, and Theater.

41. Letter from Richard E. Quandt to Błażej Feret and Bogusław Święcicki, dated December 6, 1994. Italics in the original.

42. E-mail from Bogusław Święcicki to Richard E. Quandt, dated March 11, 1996.

43. By October 1996, the consortium had paid Dynix almost nothing. The contract stipulated partial payments over time as certain milestones were reached, with no exceptions for contigencies; thus, for example, one of the servers was stolen at Berlin airport, which caused a substantial delay; this was at the expense of Dynix. The consortium even succeeded in inserting a penalty clause into the contract: for every day's delay beyond a certain deadline, Dynix was to pay the consortium 1% of the software purchase price. See letter from Richard E. Quandt to William G. Bowen, dated October 23, 1996.

44. *Final Report Academic Libraries Automation Project*, Łódź: Technical University of Łódź, December 31, 2000.

45. See http://bg.p.lodz.pl/astr_2.htm.

46. Adam Mickiewicz University, the Technical University (now called University of Technology), the Agricultural University, the Academy of Medicine, the Academy of Economics, the University of Physical Education, the Academy of Music, the Academy of Fine Arts, the Pontifical College of Theology (now part of Adam Mickiewicz University), the Poznań branch of the Polish Academy of Sciences Library, the Edward Raczyński City Public Library, and the Poznań Society of Friends of Science Library.

47. Memorandum from Arthur Brady to Jacek Małuszyński, dated September 2, 1994.

48. In my recommendation for a Mellon grant, I said, "Note that there is no guarantee that if and when the Consortium receives a grant, the pro-Dynix forces might not try to reassert their point of view. However, I do not believe that they could pose a serious threat, because the Consortium is unlikely to be able to afford Dynix. If Dynix were to come down sufficiently in price to be affordable, I would have no objection to their going with Dynix after all." Letter from Richard E. Quandt to William G. Bowen, dated December 26, 1994.

49. Letter from Bohdan Maruszewski to Richard E. Quandt, dated January 25, 1996.

50. E-mail from Jacek Małuszyński to Richard E. Quandt, dated March 23, 1995. He also indicated that he was no longer involved in the project, but might apply for the position of director of the Foundation, a position that he was not to fill.

51. Letter from Richard E. Quandt to Jerzy Fedorowski, Eugeniusz Mitkowski, Ryszard Ganowicz, Bohdan Gruchman, Janusz Gadzinowski, Wiesław Osiński, Mieczysław Koczorowski, Wojciech Müller, Bogdan Częsz, Jan Stankowski, Jacek Łukomski, Antoni Gąsiorowski, dated November 5, 1995. Emphasis in the original.

52. *Final Report to The Andrew W. Mellon Foundation*, Poznań: Poznań Foundation of Scientific Libraries, March 2000.

53. See http://www.pfsl.poznan.pl.

54. Burchard, Maria, "Union Authority File: 1993–1998," in Jadwiga Woź-

niak and Robert C. Miller (eds.), *Research Libraries: Cooperation in Automation*, Warsaw: Center for Formats and Authority Control, Warsaw University, 1999, pp. 37–43.

55. See letter from Patricia Battin to Adam Manikowski, dated August 18, 1993.

56. Letter from Hans Rütimann to Richard E. Quandt, dated October 5, 1993.

57. *Request for Grant Addressed to The Andrew W. Mellon Foundation*, Warsaw, National Library of Poland, January 20, 1994.

58. *1999 Annual Report to the Conference of European National Librarians*, Warsaw: Biblioteka Narodowa w Warszawie, 2000.

59. E-mail from Ewa Krysiak to Richard E. Quandt, dated June 5, 2000.

60. See http://alpha.bn.org.pl/screens/opacmenu.html.

61. The president of the Mellon Foundation, William G. Bowen, expresses the same idea by saying that we do not deal with "reluctant dragons."

62. Lass, Andrew and Richard E. Quandt (eds.), *Library Automation in Transitional Societies: Lessons from Eastern Europe*, New York: Oxford University Press, 2000.

63. Adam Manikowski had become by that time the victim of Machiavellian governmental machinations, and I hardly knew his successor, Professor Jakub Lichański. He did not last long and was replaced on June 1, 1998, by Michał Jagiełło.

64. Górny, Mirosław, Bogdan Maruszewski, Andrzej Nikisch, "In Search of an Optimum Model of Development," in Lass, Andrew and Richard E. Quandt (eds.), *Library Automation in Transitional Societies: Lessons from Eastern Europe*, New York: Oxford University Press, 2000, pp. 83–90.

65. E-mail from Ewa Dobrzyńska-Lankosz to Richard E. Quandt, dated June 10, 1998.

66. E-mail from Henryk Hollender to Richard E. Quandt, dated July 22, 1998.

67. Letter from William G. Bowen to Piotr Węgleński, dated April 4, 2000.

68. Hollender, Henryk, "National Union Catalogue: Report #1 for the Mellon Foundation," Warsaw: Warsaw University, November 14, 2000.

69. Manikowski, Adam, "Library Automation—Fortunes and Miseries in Poland," in Lass, Andrew and Richard E. Quandt (eds.), *Library Automation in Transitional Societies: Lessons from Eastern Europe*, New York: Oxford University Press, 2000, pp. 23–27; see quotation in chapter 4.

70. Katalogi Automatyczne Bibliotek Akademickich.

71. Głowacka, Teresa and Jadwiga Woźniak, "KABA Subject Heading System: An Example of the Collective Effort of Academic Libraries in Poland," in Jadwiga Woźniak and Robert C. Miller (eds.), *Research Libraries: Cooperation in Automation*, Warsaw: Center for Formats and Authority Control, Warsaw University, 1999, pp. 75–80.

72. Kruszyńska, Krystyna, "Experience of Libraries Creating the KABA Subject Headings System," in Jadwiga Woźniak and Robert C. Miller (eds.), *Research Libraries: Cooperation in Automation*, Warsaw: Center for Formats and Authority Control, Warsaw University, 1999, pp. 81–83.

73. Hayes, Robert M., *Narrative Report on the Series of Visits to Institutions Participating in the Study of Staffing Patterns in Academic Libraries of Central and Eastern Europe*, Los Angeles: UCLA, December 20, 1998.

74. See "Final Report (Penultimate Draft)," Association for Library Collections and Technical Services Commission on Cataloging: Description and Access Task Force on the Harmonization of ISBD(ER) and AACR2, Jan. 23, 1999, rev. June 14, 1999, http://www.library.yale.edu/cataloging/aacrer/tf-harm21.htm.

75. Czermiński, Jurand, "VTLS in Polish Academic and Research Libraries," *Task Quarterly*, 2(1), 1998, 121–28.

76. Czermiński, Jurand, "A Coordinator's View of the Impact of Policy on the Financing and Managing of Library Automation," in Lass, A. and Quandt, R. E. (eds), *Library Automation in Transitional Societies: Lessons from Eastern Europe*, New York: Oxford University Press, 1999, p. 50.

77. Hollender, Henryk, "The Sequencing of Automation in Central and East European Libraries," in Lass, Andrew and Richard E. Quandt (eds.), *Library Automation in Transitional Societies: Lessons from Eastern Europe*, New York: Oxford University Press, 2000, pp. 335–44.

78. E-mail from Błażej Feret to Richard E. Quandt, dated November 28, 2000.

79. For an analysis of a sample on non-East European consortia see Wade, Rona, "The Very Model of a Modern Library Consortium," *Library Consortium Management*, 1 (1/2), 1999, 5–18.

80. Faculty of Information Studies, University of Toronto. See e-mail from Nadia Caidi to Richard E. Quandt dated January 18, 2001.

81. Caidi, Nadia, "The Information Infrastructure in Central and Eastern Europe: A Case Study of the Library Community," unpublished paper based on a doctoral dissertation at the Graduate School of Education and Information Studies, Los Angeles: UCLA, October 31, 2000.

Chapter 10

1. The Daugavpils population that is not ethnic Latvian consists of Russians, Ukrainians and Belorussians.

2. Johnson, Eric A., "Baltic Research Libraries," unpublished paper, Washington, D.C.: Library of Congress, 1994. Latvian and Lithuanian are Indo-European languages, whereas Estonian is Finno-Ugric. Its only relations in Europe are Finnish and Hungarian; even Finnish and Estonian have evolved along separate paths for the past 1000 years and the connection with Hungarian is even more distant. As a private hobby, I have been trying to find Estonian and Hungarian words that have the same roots, and have so far only found four: (1) "hand" in English is käsi in Estonian and kéz in Hungarian; (2) "water" is vesi in Estonian and víz in Hungarian; (3) "push" is lükka in Estonian and lök in Hungarian; and (perhaps) (4) "fire" is tuli in Estonian and tűz in Hungarian. Finns and Estonians "sort of" understand one another, but similar-sounding phrases in the two languages can mean very different things: "Ruumid on alles koristamata" in Estonian means "the rooms are not yet cleaned", while "ruumit ovat koristettumata" in Finnish means "the dead bodies are not yet decorated." Personal communication from Riin Olonen.

3. Smith, Inese A. and Marita V. Grunts, *The Baltic States: Estonia, Latvia, Lithuania*, Oxford: Clio Press, 1993, p. XL.

4. Lieven, Anatole, *The Baltic Revolution: Estonia, Latvia Lithuania and the Path to Independence*, New Haven, CT: Yale University Press, 1993, pp. 64–65.

5. Lieven, Anatole, op. cit., pp. 78–79 and Smith, Graham, "The Resurgence of Nationalism," in Smith, Graham (ed.), *The Baltic States: The National Self-Determination of Estonia, Latvia and Lithuania*, London: MacMillan Press, 1994, pp. 121–43.

6. Plakans, Andrejs, *The Latvians: A Short History*, Stanford, CA: Hoover Institution Press, 1995, pp. 167–83.

7. Iwaskiw, Walter R. (ed.), *Estonia, Latvia, and Lithuania Country Studies*, Washington, D.C.: Library of Congress, 1996, p. xv.

8. Smith, Graham, op. cit., pp. 132–34. Also, personal communication by Valters Nollendorfs.

9. Whitney, Craig R., "Evolution in Europe," *The New York Times*, April 15, 1990.

10. See also "Chronology of Transition," http://www.wdi.bus.umich/transiti.htm and Miller, Robert F., "The Baltic States Win Their Independence: Lithuania's Difficult Renaissance," http://english-server.nss.emu.edu/history/baltic-states-independence.txt, February 17, 1992.

11. *PlanEcon*, 7/47–8, December 23, 1991.

12. Personal communication from Valters Nollendorfs and also "Baltics. Them and Us," *The Economist*, August 17, 1996, http://www.economist.com/archive/view.cgi.

13. Nedyalkova, Yordanka, "Latvia Gets New Language Regulations," *Transitions Online*, August 28, 2000; http://www.tol.cz/week.html.

14. Kukk, Kalek, "Five Years in the Monetary Development of the Baltic States: Differences and Similarities," Tallinn: Bank of Estonia, 1997, http://www.www.ee/epbe/bulletin97/article/article1.

15. Stepan, Alfred, "When Democracy and the Nation State are Competing Logics: Reflections on Estonia," *European Journal of Sociology*, 35, 1994, 127–41. Stepan was the first rector of the Central European University.

16. *Estonia: Implementing the* EU *Accession Agenda*, Washington, D.C.: The World Bank, 1999.

17. "U.S. SEED Act Assistance Phase-Out Estonia," USAID Fact Sheet, Washington, D.C.: USAID, May 11, 1994. The total assistance provided to Estonia under the SEED Act through March 31, 1994 amounted to $14.7 million.

18. In Latvia: Ilona Dobelniece (library director at the University of Agriculure), Alija Janbicka (library director at Riga Technical University), Edvīns Karnītis (director of the Academic Library), Ināra Lejniece (library director at the Medical Academy, now Stradiņš University), Gunārs Mangulis (director of the University of Latvia Library), Andris Vilks (national librarian), Olafs Brinkmanis (director of the Riga Business School), Pēteris Bušmanis (prorector of the University of Agriculture), Raita Karnīte (Economics Institute of the Latvian Academy of Science), Egons Lavendelis (rector of the Riga Technical University), Tālis Millers (Latvian Academy of Science), Voldemārs Strīķis (rector of the University of Agriculture), Vita Matiss (executive director of the Soros Foundation-Latvia), R. Kondratovichs (prorector of the University of Latvia), Juris Krumiņš (prorector of the University of Latvia), Juris Zaķis (rector of the University of Latvia), Elizabeth Bruksle (office of the Prime Minister), Gunta Veismane (director of the Latvian School of Economics); in Estonia: Ivi Eenmaa (national librarian), Mait Klaassen (rector of the Agricultural University), Ingrid Saare (Tartu University Library), Rein Taagepera (dean of the School of Social Sciences, Tartu University), Asko Tamme (Academy of Sciences library), Peeter Tulviste

(rector of Tartu University), Madis Habakuk (president of the Estonian Business School), Avo Merits (vice president of the Estonian Business School), Olav Arna (rector of the Tallinn Technical University), Mall Hellam (director of the Open Estonia Fund), Henn Käärik, School of Social Sciences, Tartu University); in Lithuania: Juozas Marcinkievicius (director, Lithuanian Academy of Sciences library), Povilas Saudargas (deputy director, Lithuanian Academy of Sciences library), Vladas Bulavas (national librarian), Vytas Gruodis (director, Open Society Foundation of Lithuania), Vladas Domarkas (minister of education), Rimantas Urbonas (rector, Agricultural Academy), Kestutis Krisciunas (rector, Kaunas Technical University), Rolandas Pavilionis (rector, Vilnius University), Bendiktas Juodka (prorector, Vilnius University).

19. Letter from Gundar J. King to Jānis Gaigulis, dated August 28, 1994.

20. Nollendorfs, Valters, "Restoring the Core: Democratization through Education in Post-Communist Europe," *Baltic Studies Newsletter*, 16/4, December 1992, 4–8.

21. Nollendorfs, Valters, "Overhaul through Shortfall: One Summer's Observations and Ruminations about the Future of Educational Reform in Latvia," *Baltic Studies Newsletter*, 17/4, December 1994, 2–6.

22. Nollendorfs, Valters, " 'A Dreadful Thing': Latvian Education Policy Problems Revisited and Revived," *Baltic Studies Newsletter*, 24/4, No. 96, December 2000, 8–10.

23. Simple as this itinerary sounds, it was not without its minor drama. I had specifically requested a visa from the Latvian Embassy in Washington that would permit me multiple entries and exits from Latvia; but when prior to my departure my passport was returned to me by the embassy, it turned out that my visa was valid specifically for a single entry and exit. I called the embassy about this and was assured that I should not worry about it, because it was perfectly all right to have such a visa and still be permitted multiple exits and entries. Later in Riga, I was having dinner with U.S. Ambassador Ints Silins and I asked him whether he thought that this was something that I needed to worry about, and he emphatically said yes, that when I tried to reenter a few days later from Estonia, I might well be turned away at the border. The next morning I had an appointment with Prime Minister Gailis, and I mentioned this little conundrum to his adviser, Ms. Bruksle, who promised to "fix" the problem. Unfortunately, after trying for half an hour to get through on the telephone to the consular office of the Foreign Ministry, she was unable to do so, the reason being that this office had moved two months earlier, but the prime minister's office did not yet know about it. Thanks to the legwork by AABS, my problem was fixed in the end, but the episode left a Kafkaesque flavor in my mouth.

24. Quandt, Richard E., *Report on the Baltics*, New York: The Andrew W. Mellon Foundation, April 1, 1995.

25. *Directory of Higher Educational Institutions in the Baltics: Latvia*, Riga: Baltic Academic Center, 1995.

26. Private communication from U.S. Ambassador Ints Silins, March 14, 1995.

27. Johnson, Eric A., "Baltic Research Libraries," Internet Report, Washington, D.C.: Library of Congress, December 14, 1994.

28. See *Annual Report 1992–1993, The Soros Foundation Latvia*, Riga: 1993; *The Open Estonia Foundation in 1992*, Tallinn: 1993; *The Open Estonia*

Foundation in 1993, Tallinn: 1994; *The Open Estonia Foundation in 1994*, Tallinn: 1995; *Current Projects and Programs 1994*, Tallinn: Open Estonia Foundation, 1994; *Annual Report 1994 Open Society Fund–Lithuania*, Vilnius: 1995.

29. A second more or less general university exists in Lithuania: Vytautas Magnus University, a relatively small private institution concentrating on business and management, environmental studies, fine arts, humanities, social sciences, computer science, and Catholic theology.

30. Personal communication from Valters Nollendorfs.

31. See, for example, Karnītis, Edvīns, *Latvian Information Infrastructure: The Present Stage and Trends of Development*, Riga, Latvijas Akadēmiskā Bibliotēka, 1996.

32. Kurm, Margit, "Using Internet in the Baltic States," unpublished IREX Report, 1994.

33. E-mail from Harijs Bondars to Richard E. Quandt, dated July 11, 2000.

34. See, for example, *Baltic States Co-operation for Quality Assurance in Higher Education*, Report of the Regional Advisory Mission, Riga, October 24–26, 1994, Strasbourg: Council of Europe, DECS LRP 94/32, February 14, 1995.

35. E-mail from Valters Nollendorfs to Richard E. Quandt, dated July 15, 2000.

36. Personal communication from Valters Nollendorfs.

37. E-mail from Valters Nollendorfs to Richard E. Quandt, dated July 17, 2000. With regard to business, King and Zulauf have said "The commonly shared values in Latvia today include the acceptance of autocratic management and fatalistic tolerance of graft . . . and a belief in education and training for professional and vocational purposes," but this observation is also applicable to higher education. See King, Gundar and Dwight Zulauf, "Business in Latvia," *European Business Journal*, 8/2, 1996, 19.

38. Bollag, Burton, "Baltic Universities Struggle to Modernize Their Programs," *The Chronicle of Higher Education*, June 6, 1997, p. A39.

39. *Latvian Research: An International Evaluation*, Copenhagen: The Danish Research Councils, 1992.

40. Evaluations assigned the grades of excellent, very good, good, fair, and poor. Excellent and very good were generally understood to mean research that is significant today on an international level and is on the cutting edge of the discipline. Fair and poor were not generally intended to imply that the work was riddled with elementary errors, but rather that it was not original, fundamental, or promising by international standards.

41. Miljan, Toivo, "EuroFaculty: Tartu-Riga-Vilnius Report," *Baltic Studies Newsletter*, 18/3, 1994, 5.

42. E-mail from Gundar King to Richard E. Quandt, dated July 12, 2000.

43. http://www.rbi.lv/english/MBA/framet.htm.

44. Bruce, Sam, "The Riga Business School at Riga Technical University—Excellence in Business Education," *Baltic Studies Newsletter*, 20/2, 1996, 13.

45. E-mail from Gundar King to Richard E. Quandt, dated July 15, 2000; also see Quandt, Richard E., op. cit.

46. http://www.sseriga.edu.lv/

47. Personal communication from Valters Nollendorfs.

48. http://www.bentley.edu/resource/international/achieve.htm.

49. http://www.ebs.ee/university_facts/.

50. "Euro-faculty in the Baltic States," *Baltic Studies Newsletter*, 17/3, 1993, p. 12.

51. Miljan, Toivo, "Excerpts of the 1995–96 Annual Report on the Euro-Faculty Programme," *Baltic Studies Newsletter*, 21/1, 1997, 8–15.

52. Geoffrey Alderman (Middlesex University), Sven Caspersen (Aalborg University), Bernhard Kempen (University of Würzburg), Jacques Pertek (University of Valenciennes), "Excerpts of the Report of the EuroFaculty Evaluation Committee, *Baltic Studies Newsletter*, 21/2, 1997, 5–11.

53. Macevicience, Vida and Lee G. Burchinal, "Central European Conference and Exhibition for Academic Libraries and Informatics," Conference Report, Vilnius: September 27–29, 1993.

54. For a description of the library situation in the Baltic countries see Johnson, Eric A., "Baltic Research Libraries," unpublished paper, Washington, D.C.: Library of Congress, 1994 and Quandt, Richard E., *Report on the Baltics*, New York: The Andrew W. Mellon Foundation, 1995, pp. 44–58.

55. Karnītis, Edvīns, *Electronic Information Services in Libraries in Latvia: The Concept and Its Implementation*, Riga: Latvijas Akadēmiskā Bibliotēka, 1997.

56. For a position paper by two librarians using ALISE and LIBER respectively, see Janbicka, Aija and Edvīns Karnītis, "Concept of Provision of Scientific and Technical Information for Latvia," http://educate.lib.chalmers.se/IATUL/proceed contents/fullpaper/karnitis.html; undated, but probably 1997.

57. *The Proposal for Grant Addressed to the Andrew W. Mellon Foundation by the Consortium of Estonian Libraries Network*, Tallinn: ELNET, June 1996.

58. This is an unusual library which has the distinction that its holdings of Estonian materials are 98% complete for the pre-1917 period and 100% complete thereafter. Among others, it has such rarities as Stahl's 1637 *Estonian Grammar* and Balthasar Russow's 1578 and 1584 chronicles of Estonia and Livonia.

59. Instead of formal authority files, they used lists of equivalencies such as a list that would indicate that Tchaikovsky, Chaikovsky, Czajkowski, Csajkovsky were all the same composer.

60. Letter from Mihkel Reial to Richard E. Quandt, dated April 2, 1998.

61. One minor complaint was that the Russian character set was not yet 100% operational—but then I had never encountered a system about which there were no character set complaints.

62. A few examples will illustrate this point. (1) In searching for entries, a relevant issue is the "scoping" of the search algorithm; i.e., the levels of categories you could search. Ideally, one would want to search in arbitrary order; for example, first on "piano sonatas," and then on "Mozart." But in INNOPAC, you can first search only on the author's name, not in the reverse order. The Music Academy seemed entirely willing to wait and see how troublesome this would turn out to be. (2) Some of the libraries could not use the circulation algorithm, because they are closed stack libraries and the user's request would need to be printed at several staff workstations at once, something that INNOPAC did not do. (3) The circulation of periodicals presented some problems, because unlike U.S. practice, one year's issues of a journal are often not bound together; hence, you would either have to make a separate bibliographic entry for each issue, or you could not circulate the journal. (4) Some university accounting

systems required that in the case of fines for overdue books, a special form had to be filled in, in which the Value Added Tax on the fine needed to be calculated; something that INNOPAC was not equipped to do.

63. E-mail from Edvīns Karnītis to Richard E. Quandt, dated December 28, 1994.

64. "Conception of Organization of the United Library Information Network of Latvia," Riga: Latvian Development Agency, September 25, 1994.

65. E-mail from Andris Vilks to Richard E. Quandt, dated June 15, 1996.

66. An excellent description of the libraries and the LATNET network is contained in Dobelniece, I., Edvīns Karnītis, Gunārs Mangulis, I. Lejniece, Andris Vilks, A. Buholte, V. Poznaka, A. Janbicka, "Integrated Informational Network of Latvian Libraries," Riga: March 14, 1995.

67. It had to (1) be based on client-server architecture and provide service for the following clients: Windows95, X-terminals, Z39.50, www; (2) be based on generally accepted database standards such as Oracle or Informix and run in both Unix and Windows environments; (3) catalog according to ISBD and AACR2; (4) handle Latin, Latvian, and Russian characters; (5) handle cataloging, authority control, circulation, interlibrary loan, acquisitions, data exchange in full MARC format, union catalog, multilevel bibliography, serials control; (6) have a report generator; (7) have automatic conversion between USMARC, UKMARC and UNIMARC, and satisfy data export and import and security requirements.

68. Fax from A. Virtmanis to William G. Bowen, dated February 21, 1997.

69. E-mail from Richard E. Quandt to Andris Vilks, dated April 28, 1997. Capitalization in the original.

70. "Single Information Network of Latvia's Libraries," Riga: Association of Latvia's Academic Libraries, 1996. I did not, in fact, see a printed copy of this project plan until June, when I was sent a copy for my evaluation by A. Priekulis, the deputy minister of education.

71. E-mail from Jānis Gaigulis to Richard E. Quandt, dated April 28, 1997. Capitalization in the original.

72. E-mail from Gundar King to Jānis Gaigulis, dated May 1, 1997.

73. Letter from Richard E. Quandt to Andris Šķēle, dated May 5, 1997.

74. He was not alone in misunderstanding the objectives and methods of American foundations and the extent to which U.S. government regulations concerning grant details were irrelevant for us. I was asked numerous times in Eastern Europe whether travel funded by the Mellon Foundation had to employ U.S. flag carriers or whether the hardware and software purchased with grant moneys had to be of U.S. make. But if Karnītis had checked, he would have realized that Mellon funds had already been used to purchase Aleph as well as TINLIB.

75. Quandt, Richard E., *Report on the Latvian Library Consortium*, New York: The Andrew W. Mellon Foundation, May 1997.

76. By this time, my opinion of ALISE had sunk to new depths: the day before, at the National Library, I used ALISE to search for a name in a database of 20,000 items and the search took 56 seconds. See Quandt, Richard E., op. cit.

77. Letter from T. Dennis Sullivan to Andris Vilks, dated May 22, 1997.

78. E-mail from Valters Nollendorfs to Richard E. Quandt, dated June 18, 1997.

79. Letter from Richard E. Quandt to A. Priekulis, dated June 27, 1997. Emphasis in the original.

80. Letter from Aleksandrs Kiršteins to Richard E. Quandt, dated July 18, 1997.

81. She was succeded in 1999 by Margarita Marcinkeviča, another extremely able and conscientious person, who has been acting director of LINC since Ābele's resignation in 2000.

82. An intermediate plan of making AABS the actual grantee and have it administer the grant was ultimately rejected.

83. Neuman, Tanya R., "Who Will Step Down First? Latvia's Government in Crisis—Again," *The Baltic Times*, 2/69, July 24, 1997.

84. Letter from Atis Sausnītis to Richard E. Quandt, dated September 17, 1997.

85. Letter from Ramona Umblija to Richard E. Quandt, dated September 26, 1997.

86. Preliminary questionnaires had already been sent to ALISE, III, Ex Libris, VTLS, Dynix/Horizon, and LIBER.

87. Ryan, Alan, "My Way," *The New York Review of Books*, 47/13, August 10, 2000, 47–50.

88. Quandt, Richard E., *Report on Trip to Estonia, Latvia, and Poland*, New York: The Andrew W. Mellon Foundation, October 1997.

89. Letter from Richard E. Quandt to Ramona Umblija, dated October 21, 1997. Emphasis in the original.

90. Letter from Aldis Ābele to Richard E. Quandt, dated December 21, 1997. Ābele reports that the meeting took place on November 16; however, the minutes reveal that this is a typo and it took place on December 16.

91. Freedman, Maurice J. and Ellen Starkman, *Report: The Library Information Network of Latvia*, Riga: USIS, May 22, 1998.

92. E-mail from Valters Nollendorfs to Richard E. Quandt, dated April 20, 1998.

93. Letter from Richard E. Quandt to Aldis Ābele, dated April 18, 1998.

94. Letter from Aldis Ābele to Richard E. Quandt, dated June 3, 1998.

95. Quandt, Richard E., *Report on Trip to Latvia, Slovakia, Hungary, May 20–June 1, 1999*, New York: The Andrew W. Mellon Foundation, June, 1999.

96. E-mail from Gundar King to Richard E. Quandt, dated August 1, 2000.

97. Taagepera, Rein, "Is the USA a Model for Europe?" *Baltic Studies Newsletter*, 33/1, March 1999, 1–8.

98. And there is some reason to believe that much progress has taken place in the last six months.

99. Although CALICO had a surfeit of other problems, as is evident from Table 10-2.

Chapter 11

1. http://www.sggw.waw.pl/plusnew/ae/ae_sgn_iso.html/.

2. The modern roots of teaching economics and business in Hungary go back to the creation in 1891 of the Eastern Commercial Course (Keleti Kereskedelmi Tanfolyam), a two-year post-secondary course of study. For an excellent history of business and economics teaching in Hungary, see Szögi, László and Vilmos Zsidi (eds.), *Tanulmányok a magyarországi közgazdasági felsőoktatás történe-*

téből (Studies on the history of Hungarian higher education in economics), Budapest: Budapest University of Economics, 1995.

3. http://www.bkae.hu.

4. http://www.bkae.hu/~phd/phdk.html.

5. http://www.kvif.hu/old.

6. http://www.kkf.hu.

7. Dietl, Jerzy, "Education for Business Management," in Cichocki, Krzysztof and Paul Marer, *Education for Transition to Market Economy in Countries of Central and Eastern Europe*, Warsaw: Polish Fulbright Alumni Association; Polish-U.S. Fulbright Commission, 1966, pp. 12–26.

8. See http://akson.sgh.waw.pl/mba-ca/greetings.htm and http://www.csom.umn.edu/wwwpages/IPS2/offshore/wem/wemprs396.html. Another example is provided by the Warsaw University of Technology. It established a business school in 1991 with the assistance of the London Business School, the École des Hautes Études Commerciales and the Norwegian School of Economics and Business Administration. In October 1992, it started its international graduate program in business administration; in that year it also launched a one-year full-time MSC program in business; in 1995 it began a part-time Master of Hautes Études Commerciales en Sciences de Gestion Approfondies program for French-speaking students; in 1996 it launched an Executive MBA program. Its various programs have enjoyed broad support from governments, the business community, some additional academic institutions (such as Harvard and MIT), the Soros Foundations, and the European Foundation for Management Development. http://www.sbus.pw.edu.pl/szkola.html.

9. Grayson, Leslie E., *Paving the Road to Prosperity*, Vienna: International Institute for Applied Systems Analysis, 1993.

10. http://www.imc.hu.

11. http://www.cmc.cz.

12. Portwood, Jim, "International Management Center MBA Program Evaluation," unpublished paper, Philadelphia: Temple University, 1996.

13. http://www.iemba.cz/About%20IEMBAP.htm.

14. "Report of the Dean and CEO to the Board of Trustees," Budapest: International Management Center, March 8, 1996.

15. "Inside Track Business Education: Weatherhead Severs Links with Budapest School," *Financial Times*, September 18, 2000. See also http:www.ft.com.

16. "Inside Track /260 Business Education: A School in Transition," *Financial Times*, GATEust 21, 2000.

17. However, the not-for-profit Czech Association of MBA Schools (CAMBAS), devoted to maintaining the highest possible standards, lists only the Brno Business School at the Technical University of Brno, the Masaryk Institute of Advanced Studies at the Czech Technical University (Prague), and the Prague International Business School associated with the University of Economics. *Informational Brochure 1998*, Prague: CAMBAS, 1998.

18. E-mail from Irina Rybacek to Richard E. Quandt, dated November 15, 2000.

19. For example, at the Warsaw School of Economics, the Universities of Łódż, Gdańsk, and Szczecin, at Jagiellonian University, the Warsaw University of Agriculture, Warsaw University, Warsaw Technical University, the Technical

University of Gdańsk, Marie Curie Skłodowska University, Silesian University, the University of Agriculture of Szczecin, the Academy of Economics in Poznań, the Malopolska School of Economics, and undoubtedly others. See http://www.ceebd.co.uk/ceeed/poland/bus/man.htm.

20. http://www.cofund.org.pl/bkkk/plwin/edmanag/buig_old/ilist.html.

21. http://www.efmd.be.

22. "Mid-Career Development Inter-University, Post-Graduate Program," Kraków, Progress and Business Foundation, GATEust 5, 1993, Joint Statement in Support of KISM by Andrzej Pelczar, Jan Janowski, Jerzy Mikułowski-Pomorski (rectors of Jagiellonian University, the Academy of Mining and Metallurgy, and the Academy of Economics, respectively).

23. To wit: "The demand for management education is undeniably accelerating in Poland and other East Central European countries. The long-term goal of KISM is to attract students throughout the region, in order to build trade and investment links through shared education. In addition, as an accredited program, KISM will also attract Western students thereby promoting additional multi-national interaction which is crucial to the demands of today's global business environment. Furthermore, the KISM project addresses the need to develop faculty in business principles and teaching methods, including developing curriculum and case studies that apply to transition economies." Op. cit.

24. Letter from Václav Klaus to Richard Rosett, dated October 20, 1990. Underlining in the original.

25. "Response to RFA Number SPO/EE-91–002, Part II, Cost," Rochester, NY: U.S. Business School in Prague, October 20, 1990. These were not courses that required several months of prior preparation by the professor; they were rather bread-and-butter exercises for an experienced academic. As an extreme calculation, that works out to a remuneration $625/hour; even if we assume that a teacher would have spent every working hour of the 15-day teaching period on his or her course, it works out to a per diem remuneration of $1,667, which I considered a grossly excessive figure. I obviously had unrealistic expectations about the extent to which American, university-inspired projects would be willing to forego the normal commercial per diems that top-level academics can command.

26. http://www.usbsp.cz/about/shortly.htm.

27. In his valuable assessment of management education in the region, Leslie Grayson, op. cit., tends to favor one-year programs and executive MBA programs on the sensible grounds that high-level executives cannot afford the time for longer programs.

28. Grayson, Leslie E. (ed), *Cases in Management in Eastern and Central Europe and Russia*, Laxenburg, Austria and Charlottesville, VA: International Institute for Applied Systems Analysis and University of Virginia, 1993.

29. Grayson, Leslie E, *Paving the Road to Prosperity*, p. 3.

30. E-mail from Krzysztof Obłój to Richard E. Quandt, dated October 7, 2000.

31. *A U.S. University Consortium for Management Education at Jagiellonian University and the Academy of Economics at Kraków, Poland—Fourth Grant Request*, Hartford, CT: University of Hartford, February 26, 1993.

32. *A Consortium for Management Education at Jagiellonian University, Final Progress Report—First Grant*, Hartford, CT: University of Hartford, July 15, 1992.

33. *Interim Report—Phases V through* VII, Hartford, CT: Consortium for Management Education, February 26, 1993.

34. *A Consortium for Management Education at Jagiellonian University and the Academy of Economics—Interim Report, Third Grant*, Hartford, CT: University of Hartford, February 26, 1993.

35. Letter from William J. Horne to Robert Hager, dated May 27, 1992.

36. *U.S. Consortium for Management Education in Central and Eastern Europe, Funding Proposal for the Center for Faculty Development in Business Economics Studies*, Hartford, CT: University of Hartford, March 29, 1994.

37. Letter from Marek Szymoński to Demetrios Giannaros, dated June 20, 1994. (Szymoński was vice rector of Jagiellonian University.)

38. *U.S. Consortium for Management Education in Central and Eastern Europe—Final Report*, Hartford, CT: University of Hartford, April 30, 1998.

39. E-mail from Tadeusz Borkowski to Richard E. Quandt, dated September 14, 2000.

40. Paciorek, Anna, "Górą warszawska SGH i wrocławska AE" (Warsaw School of Economics tops Wrocław Academy of Economics), *Rzeczpospolita*, June 2, 2000.

41. Bochniarz, Zbigniew, "International Team Teaching for Central and Eastern European Executives: Assessing Experiences of Warsaw and Other Executive Programs," in Sulejewicz, Aleksander and Mahmood Zaidi (eds.), *Polish-American MBA at the Warsaw School of Economics*, Warsaw: Warsaw School of Economics Press, 2000, pp. 52–73.

42. Sulejewicz, Aleksander, Grażyna Leśniak-Łebkowska, Mahmood Zaidi, "Development and Delivery of an American Style MBA Degree at the Warsaw School of Economics," in Sulejewicz, Aleksander and Mahmood Zaidi (eds.), *Polish-American MBA at the Warsaw School of Economics*, Warsaw: Warsaw School of Economics Press, 2000, pp. 23–51.

43. Quandt, Richard E., *Report on Trip to Poland, October 21–31*, 1993, p. 4.

44. The grant had allocated only $10,000 for this purpose.

45. See http://www.csom.umn.edu/wwwpages/IPS2/offshore/wem/wemdes.html.

46. USAID's Management Training and Economic Education Program in Central and Eastern Europe had funded 572 courses, but only 16 of these were in human resource management and none in industrial relations. See *The Human Side of Enterprise: Human Resource Management and Industrial Relations Education in the Czech and Slovak Republics*, Proposal submitted to The Andrew W. Mellon Foundation, Ithaca, NY: Cornell University, January 1993.

47. Letter from Richard E. Quandt to Linda Gasser, dated February 24, 1994.

48. Quandt, Richard E., *Trip Report on Projects in Eastern Europe: June 1994*, New York: The Andrew W. Mellon Foundation, 1994.

49. Personal communication from Ĺubica Blajziková, October 23, 2000.

50. Gasser, Linda, "CEHREI Project Reflections: Perspectives on the Success of an Academic Partnership for Transformation," unpublished paper, Ithaca: ILR School, Cornell University, October 2000.

51. http://www.soce.org/.

52. *Proposal for the Loyola Marymount University Center for Technology*

Management in Poland, Los Angeles: Loyola Marymount University, March 10, 1995.

53. Oppenheim, Bohdan W. and Zbigniew H. Przasnyski, *Final Report on the LMU Center for Technology Management in Poland*, Los Angeles: Loyola Marymount University, February 2, 1998.

54. Oppenheim, Bohdan W. and Zbigniew H. Przasnyski, "Transfer of Technology Management and Total Quality Knowledge to Poland," unpublished paper, Los Angeles: Loyola Marymount University, November 1998. This appears to be reminiscent of the disagreement that arose between the University of Minnesota and the Warsaw School of Economics, but was not resolved as amicably.

55. The integration of older faculty and courses into new programs was frequently difficult, including at WSE. However, at that institution, students "voted with their feet" and abandoned the older courses; moreover, faculty at WSE—the premier institution of its kind in Poland—probably felt less threatened by the new knowledge than those at less prestigious, provincial institutions.

56. Oppenheim, Bohdan W. and Zbigniew H. Przasnyski, "Reflections and Lessons Learned," unpublished paper, Loyola Marymount University, February 2, 1998.

57. Letter from Bohdan Oppenheim and Zbigniew Przasnyski to Stefan Berczyński, dated February 18, 1998.

58. Oppenheim, Bohdan W. and Zbigniew H. Przasnyski, "Transfer of Technology Management and Total Quality Knowledge to Poland," unpublished paper, Loyola Marymount University, November 1998.

59. Sontheimer, Kevin, "A Brief Report with Exhibits Concerning Financial Management Problems at the University of Economics Bratislava," unpublished paper, Pittsburgh: University of Pittsburgh, USAID MTEE-Slovakia Grant project, October 21, 1996.

60. Letter from Richard E. Quandt to William G. Bowen, dated October 29, 1996.

61. http://portal.research.bell-labs.com/cgi-wald/dbaccess/411?terms=&country=poland&org=&maxhits=.

62. *New Initiatives for the 1990s and Beyond, 1991 Annual Report*, Stamford, CT: International Executive Service Corps, 1991, p. 3. See also http://www.iesc.org.

63. http://www.heifer.org/about_hpi/index.htm.

64. Morris, Adele and Vijaya Ramachandran, "HPI and Rabbit Farming in Poland: A Status Report for the Mellon Foundation," unpublished paper, New York: The Andrew W. Mellon Foundation, November 15, 1994.

65. *Final Report to the Andrew W. Mellon Foundation on the Accomplishments of the Center for Promotion and Support of Agricultural Enterprise, Sandomierz, Poland*, Norwalk, CT: TechnoServe, February 1999.

66. Gelzleichter, Richard E., "Purdue/Poland Project," unpublished paper, West Lafayette, IN: Purdue University, May 2000.

67. An area under active development in Hungary by the Friends of Hungarian Higher Education Foundation; see chapter 5.

68. See *Grant Proposal to The Andrew W. Mellon Foundation, Exchange Development and Education Exchange Program in Poland for January 1, 1995–December 31, 1995*, West Lafayette, IN: Purdue University, January 1995 and *Grant Proposal to The Andrew W. Mellon Foundation, Purdue/Poland Extension, Research, and Academic Education Exchange Program for March 1, 1996–December 31*, 1996, West Lafayette, IN: Purdue University, March 1996.

69. Limerick was also partnering with Case Western Reserve University in a project to create an indigenous capability to teach small business management at the Universities of Pécs, Veszprém and Miskolc, also funded by Mellon.

70. O'Connor, Laurence, "Overall Review of Activities and Accomplishments, 1994–1998," in *Andrew W. Mellon Foundation Programme Hungarian Agricultural Project 1994–1998, Final Report*, Limerick: University of Limerick, 1998.

71. Village tourism was generally considered to be a potential growth area by providing low-cost vacations for Hungarians as well as foreigners. Mellon had funded the International Management Center in 1993 to prepare a television series on village tourism. Both extension activities and rural development in Hungary needed to take into account the substantial regional income inequalities. See Brown, David L. and László Kulcsár, "Household Pluriactivity in Rural Hungary," paper presented at the meeting of the American Sociological Association, San Francisco, 1998.

72. Hogan, John D., "The Master of Business Administration Learning Experience—Emerging Role, Mission, Goals and Expectations," paper presented at the International Conference on Master of Business Administration Programs—Polish and Foreign Experience and Plans for Future Development, Łódź, February 23–24, 1996.

73. See, for example, the dated, but still relevant work by Gordon, Robert Aaron and James Edwin Howell, *Higher Education for Business*, New York: Columbia University Press, 1959.

74. The Ford Foundation has moved its activities further east, the Pew Charitable Trusts have terminated its East European program, the Rockefeller Brothers Fund is reportedly pulling out of the region, The Andrew W. Mellon Foundation has shifted resources to major commitments in South Africa and to the development of digital libraries, USAID has largely moved out of the western edge of Eastern Europe, and even the TEMPUS program is slated to end at the end of 2000.

Chapter 12

1. As was done by SUNY–Albany in an early project; see chapter 4.

2. Behrman, Jack, "A Decade of Assistance to Private Enterprises in Transitioning Economies 1990–2000," Chapel Hill, NC: Consortium for the MBA Enterprise Corps, 2000.

3. Sergesketter, Terri M., "The River Company," unpublished report to the MBA Enterprise Corps, Kraków: 1993.

4. Letter from Kent Peterson to Richard E. Quandt, dated February 25, 1993.

5. "Students for European Development, A Proposal Submitted to The Andrew W. Mellon Foundation," Washington, D.C.: Georgetown University, April 14, 1992.

6. "Students for East/Central European Development, Proposal Prepared for The Andrew W. Mellon Foundation," Washington, D.C.: Georgetown University, April 1993.

7. "A Program of Management Development for Czechoslovak Executives," Program draft proposal, Washington, D.C.: Georgetown University, July 1990.

8. I had no objections to activities that would benefit Americans as such, but

the mandate I received for the program I was directing was that the bulk of the benefit should accrue to East European countries.

9. The concluding report of MBAEC incorrectly reports the sum of Mellon grants as $625,000. See Behrman, Jack, op. cit., p. 67.

10. Based on data in Behrman, Jack, op. cit., pp. 52–66.

11. Memorandum from Zack A. Kollias to Lynne Gerber, dated December 8, 1993.

12. This is an average over all three projects discussed in this section; the per person cost of MBAEC by itself was higher, which is not surprising in the light of the much greater organizational effort required by that project.

13. Siegel, Daniel and Jenny Yancey, *The Rebirth of Civil Society: The Development of the Nonprofit Sector in East Central Europe and the Role of Western Assistance*, New York: Rockefeller Brothers Fund, 1992.

14. Peterson, George E., "Decentralization in Poland, Hungary, and the Czech Republic: What Have We Learned?" in Peterson, George E., *Decentralization in Central and Eastern Europe*, Washington, D.C.: The Urban Institute, 2000, p. 21.

15. But all taxes in the Czech Republic are national taxes. See Kamenicková, Vera and George E. Peterson, "Fiscal Decentralization in the Czech Republic," in Peterson, George E., *Decentralization in Central and Eastern Europe*, Washington, D.C.: The Urban Institute, 2000, p. 1.

16. Created as recently as 1991 by merging the former College of Education and Institute of Technology into a single institution with university status.

17. The connection between Ewert and Holenda rested on a coincidence: at a dinner party in Roanoke, VA, in November 1991, Ewert met a person of Czech origin who asked him whether he would be interested in helping with the development of West Bohemia. Letter from Bern Ewert to Richard E. Quandt, dated March 9, 1992.

18. Letter from Jiří Holenda and Zdeněk Mraček to Petr Pithart, dated January 27, 1992.

19. Quandt, Richard E., *Report on Trip, June 1993*, New York: The Andrew W. Mellon Foundation, June 1993.

20. Ewert, H. Bernhard, M. Kunz and Jiří Vacek, *Grant of the Mellon Foundation to* BOGMU *to Implement Budgeting and Restructuring in the City of Plzeň*, Final Report 1994–95, Plzeň: George Mason University, December 1995, pp. 3–4.

21. Letter from Richard E. Quandt to William G. Bowen, dated February 14, 1995.

22. "Mellon Foundation Grant Proposal from George Mason University, Economic and Management Reforms in the Czech Republic," Fairfax, VA: George Mason University, October 8, 1993. Emphasis in the original.

23. "Virginia Community Certification Program," Richmond, VA: Virginia Department of Economic Development, October 1, 1992.

24. Letter from Richard E. Quandt to William G. Bowen, dated October 13, 1993.

25. Letter from Robert Posner to Richard E. Quandt, dated March 2, 1994.

26. Czechinvest was an organization responsible for promoting foreign investment in the Czech Republic and was attached to the Ministry of Trade and Industry; the two most important participants from CzechInvest were Director Jan Havelka and Hana Chlebna, who directed CzechInvest's Regions Department. See http://www.czechinvest.org/ci/ci_an.nsf/?Open.

27. See Brown, Brack, *Final Report to The Andrew W. Mellon Foundation on The Czech Economic Development Certification Program 1993–1997*, Fairfax, VA: 1998, p. 5. Much of the material in this section is drawn from that outstanding report.

28. Memorandum from Bern Ewert and Michael Kunz to Richard E. Quandt, dated October 26, 1994.

29. Letter from Brack Brown to Richard E. Quandt, dated July 14, 1994.

30. E-mail from Richard E. Quandt to Brack Brown, dated November 21, 1994. Capitalization in the original.

31. Letter from Brack Brown to Richard E. Quandt, dated December 9, 1994. For the Financial Services Volunteer Corps, see later in this chapter.

32. Memorandum from Bern Ewert to Brack Brown and Louise White, dated December 8, 1994.

33. Letter from Richard E. Quandt to Karel Dyba, dated December 16, 1994.

34. Letter from Karel Dyba to Richard E. Quandt, dated January 4, 1995.

35. Fax report from Back Brown to Richard E. Quandt, dated November 17, 1994.

36. Letter from Brack Brown to Richard E. Quandt, dated February 6, 1995.

37. Letter from Zdeněk Prosek to Bern Ewert, dated March 21, 1995. The presentation was was favorably received, but the sale did not go through, because Mercedes decided to build the new factory in Germany.

38. Letter from Brack Brown to Zdeněk Degen, dated September 21, 1995.

39. E-mail from Zdeněk Degen to Richard E. Quandt, dated October 12, 1995.

40. Letter from Louise White to Bern Ewert and Michael Kunz, dated December 5, 1995.

41. Letter from Richard E. Quandt to Václav Klaus, dated November 22, 1996.

42. E-mail from Brack Brown to Richard E. Quandt, dated January 3, 2001.

43. Brown, Brack, op. cit., p. 17. This comment describes attitudes that are curiously analogous to those of some of the KOLIN librarians discussed in chapter 8.

44. Morris, Anne, "Management Training Program in Market Economy Logistics," New York: City University of New York, Center for Logistics and Transportation, July 23, 1992.

45. Such as the Central Institute of Transport in Prague, the Prague School of Economics, the Institute of Logistics and Transport Praha, Ltd., the University of Transport and Communications in Žilina, the Bratislava School of Economics, the Ministry of Transport of Slovakia, the Ministry of Economy, Czechoslovak State Railways, and others.

46. Morris, Anne, "Final Report: Executive Management Program in Strategic Logistics," New York: City University of New York, Center for Logistics and Transportation, February 28, 1996.

47. Letter from Richard E. Quandt to Rudolf Andorka, dated April 6, 1996.

48. Letter from Rudolf Andorka to Richard E. Quandt, dated May 21, 1996: "sajnos a csehek kultúrája 'nyugatibb' mint a mienk, nota bene évszázadok óta nyugatibb."

49. Stypułkowski, Cezary, "Building Human Resources in Banking," in Cichocki, Krzysztof and Paul Marer, *Education for Transition to Market Economy in Countries of Central and Eastern Europe*, Warsaw: Polish Fulbright Alumni Association; Polish-U.S. Fulbright Commission, 1966, pp. 28–31.

50. http://www.nira.go.jp/ice/tt-info/nwdtt99/c1236.html.

51. Spiro, Gene, "The Banking Finance Assistance Centre, Summary Report and Evaluation," unpublished paper, January 2000.

52. *Annual Report 1988–89*, New York: IEWS, 1989, p. 2.

53. *Annual Report 1989–90*, New York: IEWS, 1990, p. 1.

54. The year 1991–92 was of prime importance for the work of IEWS: in 1991 it organized a conference on privatization at the request of the Polish Prime Minister, Jan Krzysztof Bielecki, and in May 1992 it released a major report entitled *Report by the Task Force on Western Assistance to Hungary, Poland and the* CSFR.

55. *Annual Report 1991–92*, New York: IEWS, 1992.

56. Former undersecretary of the Treasury and former president of the Federal Reserve Bank of New York.

57. Letter from Richard E. Quandt to William G. Bowen, dated April 15, 1993.

58. Külkereskedelmi Bank.

59. Országos Takarékpénztár.

60. Magyar Hitelbank.

61. Kereskedelmi és Hitelbank.

62. One small Mellon grant was for holding a BFAC conference on privatization in Slovakia.

63. "Application of Financial Services Volunteer Corps for Support from Andrew W. Mellon Foundation," New York: FSVC, February 25, 1991.

64. "Quarterly Report 4th Quarter, 1998," New York: FSVC, 1999.

65. "Evaluation of the Financial Services Volunteer Corps (FSVC) Project in Russia," unpublished paper, Arlington, VA: Carana Corpporation, February 29, 2000.

66. Letter from John C. Whitehead to William G. Bowen, dated March 2, 1999.

67. Regulski, Jerzy, "Building Democracy in Poland: The State Reform of 1998," Discussion Paper. No. 9, Open Society Institute, Budapest: 1998.

68. Pigey, Juliana H. (with assistance from Judith Kálmán), "Fiscal Decentralization and Local Government Finance in Hungary 1990–1999," in Peterson, George E. (ed.), *Decentralization in Central and Eastern Europe*, Washington, D.C.: The Urban Institute, 2000.

69. Kopányi, Mihály, Samir El Daher, Deborah Wetzel, Michel Noel, Anita Papp, "Hungary: Modernizing the Subnational Government Level," World Bank Discussion Paper No. 417, Washington, D.C.: May 2000.

70. Levitas, Anthony, "Possible Futures for the Municipal Development Program," unpublished paper, RDF, September 13, 1993.

71. "Draft Outline for Follow-up Roundtable on Local Infrastructure Debt Finance," Budapest: Ministry of Interior, June 1994.

72. Letter from Richard E. Quandt to William G. Bowen, dated November 23, 1994.

73. Jókay, Charles, Gábor Szepesi, György Szmetana, "Municipal Debt Management and Bankruptcy Intervention in Hungary: Case Studies and Lessons Learned (1996–2000), unpublished paper prepared for the World Bank, June 2000, rev. October 2000.

74. Cochran, Thomas H., "Reflections on the International Public Finance Program, 1993–1996," unpublished paper, New York: October 2000.

75. Quigley, Kevin F. F., "Lofty Goals, Modest Results: Assisting Civil Society in Eastern Europe," in *Funding Virtue: Civil Society Aid and Democracy Promotion*, ed. by Marina Ottaway and Thomas Carothers, Washington, D.C.: Carnegie Endowment for International Peace, 2000, p. 192.

76. Johnson, Erik C., "Central Europe's Think Tanks: A Voice for Reform," paper presented at CIPE Symposium, Budapest, 1996; see http://www.cipe.org/ert/e21/ceeE21.html.

77. For a compendium of public policy institutes and "think tanks," see *Think Tanks in Central and Eastern Europe: A Comprehensive Directory*, Budapest: Freedom House, 1997.

78. Regulska, Joanna, "Building Democracy at the Local Level: FSLD-LDP Partnership," unpublished paper, New Brunswick, NJ: Rutgers University, January 2001.

79. *Ten Years (1989–1999) Local Democracy Poland*, New Brunswick, NJ: Rutgers University, 2000, p. 5.

80. *Annual Report 1990*, Warsaw: Foundation in Support of Local Democracy, May 1991.

81. FSLD has also been supported by the German Marshall Fund of the United States, the Rockefeller Brothers Fund, the Pew Charitable Trusts, the Charles Stuart Mott Foundation, the Ford Foundation, USAID, USIA, and PHARE, as well as the European business community. See Regulska, Joanna, op. cit.

82. http://www.fdrl.org.pl/info1missang.htm.

83. *Annual Report 1998*, Warsaw: Foundation in Support of Local Democracy, 1999.

84. Mellon supported FCS with several minor grants.

85. *Annual Report 1990*, New York, The Charter 77 Foundation, 1991.

86. *1993–1994 Report*, New York, The Foundation for a Civil Society, 1995.

87. *Annual Report 1991*, New York: The Charter 77 Foundation, 1992.

88. *Annual Report 1992*, New York: The Charter 77 Foundation, 1993. The first Executive Committee of this program consisted of Madeleine Albright, Zdeněk Drabek, Robert Hormats, Rita Klimová, Bevis Longstreth, Wendy Luers, John Petty, and myself. The program engaged and assigned experts for terms of up to two years to agencies such as the Czech Ministry of the Economy, the Czech Ministry of Agriculture, the Slovak National Bank, the Czech Environment Management Center, the Slovak Academic Information Agency, the Office of the Chief Architect of Prague, the Slovak Ministry of Privatization, the office of the mayor of Český Krumlov, the Regional Development Agency at Horna Nitra, the Slovak Foreign Policy Association, and many others.

89. Williams, John, "Learning from Their Neighbors: A Survey of Economic Policy Transfer Projects in Central and Eastern Europe," New York: The Foundation for a Civil Society, 1995.

90. "Business in the Community Program," Washington, DC: Central Europe Institute, September 13, 1994.

91. "Business in the Community Program," Washington, DC: Central Europe Institute, September 13, 1994, p. 6.

92. "Business in the Community/Entrepreneurship Center Program," Washington, D.C.: The Central Europe Institute, December 28, 1995.

Appendix: Links to Institutions

Academia Istropolitana Nova	http://www.ainova.sk/
Academy of Mining and Metallury	http://www.agh.edu.pl
Adam Mickiewicz University	http://www.amu.edu.pl/welcome.html .en
ALISE	http://www.alise.lv/ALISE/pages.nsf
Ameritech Library Services	http://www.ameritech.com
Budapest University of Economic Sciences	http://www.bkae.hu
CASLIN	http://www.caslin.cz:7777/
Center for Economic Research and Graduate Education (CERGE-EI)	http://www.cerge.cuni.cz/
Central European University	http://www.ceu.hu/
Charles University	http://www.cuni.cz
Comenius University	http://www.uniba.sk/webuk/www/
Copernicus University Library	http://katalog.bu.uni.torun.pl
CYFRONET	http://www.cyfronet.krakow.pl
Czech Academy of Sciences	http://www.lib.cas.cz/
CzechInvest	http://www.czechinvest.org/ci/ci_an. nsf/?Open
Czechoslovak Management Center	http://www.cmc.cz
Debrecen Universitas	http://www.cic.klte.hu/universitas/u_ home.htm
Dynix	http://www.epixtech.com/product/ dynix.asp
Eötvös Lóránd University	http://www.elte.hu
Estonian Business School	http://www.ebs.ee/university_facts/
Estonian Library Network (ELNET)	http://www.utlib.ee/elnet/index_i.html
Ex Libris	http://www.exlibris.co.il/
Ford Foundation	http://www/fordfound.org
Foundation in Support of Local Democracy	http://www.frdl.org.pl/
German Marshall Fund of the United States	http://www.gmfus.org/
Gödöllő Agricultural University	http://www.gau.hu/8/Welcome.html
Heifer Project International	http://www.heifer.org/about_hpi/ index.htm
Hungarian Academy of Sciences	http://www.mta.hu/

HUSLONET	http://www.suma.sk/libraries/huslonet.html http://library.upjs.sk/huslonet.htm
IFLA (International Federation of Library Associations)	http://www.ifla.org
Innovative Interfaces, Inc.	http://www.iii.com/
Institute for EastWest Studies	http://www.nira.go.jp/ice/tt-info/nwdtt99/c1236.html
International Executive Service Corps	http://www.iesc.org
International Management Center	http://www.imc.hu
Jagiellonian University	http://www.bj.uj.edu.pl
Janus Pannonius University	http://www.jpte.hu/
József Attila University	http://www.jate.u-szeged.hu
KOLIN (Košice Library Information Network)	http://www.lib.tuke.sk/kolin/KolinOth-e.htm
Kraków Library Group	http://victoria.uci.agh.edu.pl
Latvian Academy Library	http://www.acadlib.lv/
Library Information Network Consortium (LINC)	http://www.linc.lv/main_en.html
Łódź Consortium	http://bg.p.lodz.pl/astr_2.htm
Lublin Consortium	http://priam.umcs.lublin.pl
Marie Curie Skłodowska University	http://www.umcs.lublin.pl/
Masaryk University	http://www.muni.cz
The Andrew W. Mellon Foundation	http://www.mellon.org
MOLIN (Moravian Library Network)	http://www.ics.muni.cz/research/MOLIN.html
National Library of the Czech Republic	http://portico.bl.uk/gabriel/en/countries/czech.html http://www.nkp.cz
National Library of Estonia	http://www.nlib.ee/inglise; shindexi.html
National Library of Latvia	http://portico.bl.uk/gabriel/en/reports/latvia-99-00.html http://vip.latnet.lv.lnb
National Library of Lithuania	http://www.lnb.lt/
National Library of Poland	http://alpha.bn.org.pl/screens/opacmenu.html
National Széchényi Library (Hungary)	http://www.oszk.hu/eng/
Open Society Institute-Budapest	http://www.osi.hu/
Palacký University	http://www.upol.cz/ÚP_En/
Pew Charitable Trusts	http://pewtrusts.com
Polish-American Small Business Advisory Foundation	http://www.paaa.com.pl/historiaeng.htm
Poznań Foundation of Scientific Libraries	http://www.pfsl.poznan.pl
Riga Business School	http://www.rbi.lv/english/framet.htm
Rockefeller Brothers Fund	http://www.rbf.org
Šafárika University	http://www.upjs.sk
SEED Foundation	http://www.hungary.com/seed/

Slovak National Library	http://sab.unisi.it/icau/icau97/urs/sk-snk.htm
Soros Foundations	http://www.bibl.u-szeged.hu/oseas/sorosdir.html
Stockholm School of Economics in Riga	http://www.sseriga.edu.lv/
Szeged Universitas	http://www.theol.u-szeged.hu/~kusi/universitas.html
University of Gdańsk	http://www.univ.gda.pl/indexang.html
University of Limerick	http://www.ul.ie/
University of Nitra	http://www.uniag.sk/
University of Prešov	http://www.unipo.sk/PU/puuk.htm
University of West Bohemia	http://www.zcu.cz
University of Wrocław	http://www.uni.wroc.pl/
US Business School of Prague	http://www.usbsp.cz/about/shortly.htm
VTLS, Inc.	http://www.vtls.com
Warsaw School of Economics	http://www.sgh.waw.pl/
Warsaw University Library	http://www.buw.uw.edu.pl/english/english.htm

References

Academic Computer Center CYFRONET*-Kraków: Guide 1996*, Kraków: CYFRONET, 1996.

Action Program for Hungary's Transformation to Freedom and Prosperity, Blue Ribbon Commission, 1990.

After the Fall: The Human Impact of Ten Years of Transition, Florence, Italy: UNICEF, The MONEE Project, November, 1999.

Agovino, Theresa, "Controversial Rector Divides New University," *The Chronicle of Higher Education*, November 24, 2000.

Alma Mater Jagiellonica, Jagiellonian University, 1999.

Amsterdamski, S., *Perceptions of Dilemmas: Summary of a Qualitative Study*, Vienna: Institut für die Wissenschaften vom Menschen, 1993.

Amsterdamski, S. and Jablecka, J., *Higher Education and Research in Poland: The Inherited Situation and the Reforms*, Vienna: Institut für die Wissenschaften vom Menschen, 1993.

Annual Report 1989, Somerville, MA: Sabre Foundation, Inc., 1990.

Annual Report 1988–89, New York: IEWS, 1989.

Annual Report 1989–90, New York: IEWS, 1990.

Annual Report 1991–92, New York: IEWS, 1992.

Annual Report 1990, New York, The Charter 77 Foundation, 1991.

Annual Report 1991, New York: The Charter 77 Foundation, 1992.

Annual Report 1992, New York, The Charter 77 Foundation, 1993.

Annual Report 1992–1993, Riga: *The Soros Foundation Latvia*, 1993

Annual Report 1990, Warsaw: Foundation in Support of Local Democracy, May 1991.

Annual Report 1998, Warsaw: Foundation in Support of Local Democracy, 1999.

Annual Report, July 1, 1998–June 30, 1999, New Haven, CT: Civic Education Project, 2000.

Annual Report 1994 Open Society Fund—Lithuania, Vilnius: 1995.

1993–1994 Report, New York: The Foundation for a Civil Society, 1995.

1999 Annual Report to the Conference of European National Librarians, Warsaw: Biblioteka Narodowa w Warszawie, 2000.

Application to The Andrew W. Mellon Foundation for a Support to Building of the Comenius University Bratislava's Computer Network [*sic*], Bratislava: Comenius University, July 26, 1991.

Ash, Timothy Garton, "Refolution: The Springtime of Two Nations," *The New York Review of Books*, June 15, 1989.

———, *The Magic Lantern*, New York: Random House, 1990.

———, "The Revolution of the Magic Lantern," *The New York Review of Books*, January 18, 1990.

———, *The Uses of Adversity*, New York: Random House, 1989.

Ashenfelter, Orley, Erik Berglof, Konrad Stahl, *Evaluation of CERGE-EI*, Prague: September 3, 1999.

"Az Országos Atheneum Bizottság beszámolója az MTA 1991. májusi közgyűlése számára az egyetemekkel történt eddigi megállapodásokról" (Report of the National Atheneum Commission to the May, 1991 meeting concerning the agreement with the universities), Budapest, April 25, 1991.

Babouf, Charif and Jitka Charvátová, *Catalogue of Arabic Manuscripts of the National Library of the Czech Republic*, Prague: Albertina icome Praha, 1998.

Bakonyi, Géza, "VOCAL—a Corvina könyvtárak osztott katalogizálási rendszere,"(VOCAL—the distributed cataloging system of the Corvina libraries) *Könyvtári Figyelő*, 45, 1999/2; http://www.oszk.hu/kiadvany/kf/1999/2/bakonyi_h.html.

Bakonyi, Géza and Károly Kokas, "Könyvtári integrált rendszerek és hazai alkalmazásuk" (Integrated library systems and their domestic use), October 1996, http://dis.gau.hu/tinlib/tinbgkk.html.

Balczerowicz, Leszek, "Thoughts on Changes in Education in Post-socialist Countries," in *Education for Transition to Market Economy in Countries of Central and Eastern Europe*, ed. by Cichocki, Krzysztof and Paul Marer, Warsaw: Polish—U.S. Fulbright Commission, 1996.

Balcerowicz, Leszek, "Clamping Down on Corruption," *The Wall Street Journal Europe*, October 20–21, 2000, p. 11.

Balík, Vojtěch, "CASLIN, Priorities of the Czech National Library, and Government Policy on Information and Libraries,"," in Lass, Andrew and Richard E. Quandt (eds.), *Library Automation in Transitional Societies: Lessons from Eastern Europe*, New York: Oxford University Press, 2000, pp. 120–33.

Baltic States Co-operation for Quality Assurance in Higher Education, Report of the Regional Advisory Mission, Riga, October 24–26, 1994, Strasbourg: Council of Europe, DECS LRP 94/32, February 14, 1995.

"Baltics. Them and Us," *The Economist*, August 17, 1996, http://www.economist.com/archive/view.cgi.

Barnes, Harry G., Memorandum of March 7, 1990.

Becker, Jonathan A. (ed.), *Assessing the Impact of Book & Journal Donations to Central & Eastern Europe*, Prague and New Haven, CT: Civic Education Project, 1994.

Behrman, Jack, "A Decade of Assistance to Private Enterprises in Transitioning Economies 1990–2000," Chapel Hill, NC: Consortium for the MBA Enterprise Corps, 2000.

Bejger, Peter, "American Books for Eastern Europe," *Publishers Weekly*, 237(36), September 7, 1990.

Belkindas, Misha, Dinc Mustafa and Olga Ivanova, "Statistical Systems Need Overhaul in Transition Economies," *Transition*, 10(4), August 1999.

Bennett, Scott, "Information Based Productivity," in Ekman, Richard and Richard E. Quandt (eds.), *Technology and Scholarly Communication*, Berkeley, CA: The University of California Press, 1999, pp. 73–94.

Bernasek, Anna, "Hungarian Rhapsody," *Fortune*, Nov. 8, 1999, pp. 46–48.

Beschel, Robert P., "Western Proposals for Assistance to Poland and Hungary: An Evolving Catalog," Carnegie Corporation of New York, December 1989.

Blazyca, G., "Industrial Structure and the Economic Problems of Industry in a Centrally Planned Economy: The Polish Case," *The Journal of Industrial Economics*, 28 (1980), 313–26.

Blomström, Magnus and Ari Kokko, "Foreign Investment as a Vehicle for International Technology Transfer," in *Creation and Transfer of Knowledge*, ed. by G. Barba Navaretti, P. Dasgupta, K.-G. Mäler, D. Siniscalco; Berlin: Springer, 1998.

Bok, Derek, "Report to the Parliamentary Subcommittee Concerning Draft Law on Higher Education," unpublished paper resulting from a visit to Hungary, October 4–9, 1992.

———, *Universities in Hungary*, Washington, DC: Citizens Democracy Corps, 1992.

Bollag, Burton, "World Bank Promoting a Western Approach to Higher Education in Eastern Europe," *The Chronicle of Higher Education*, July 31, 2001.

———, "Science Academies in Eastern Europe Start to Lose Grip on Research With the Emergence of Competitive Grants," *The Chronicle of Higher Education*, August 14, 1991.

———, "Baltic Universities Struggle to Modernize Their Programs," *The Chronicle of Higher Education*, June 6, 1997, p. A39.

———, "Reforms in Higher Education Disappoint Eastern Europeans," *The Chronicle of Higher Education*, December 3, 1999; http://chronicle.com/weekly/v46/i15/15a05501.htm.

———, "Private Colleges Reshape Higher Education in Eastern Europe and Former Soviet States," *The Chronicle of Higher Education*, June 11, 1999; http://chronicle.com/weekly/v45/i40/40a04301.htm.

———, "Enrollments Rise 50% at Russian Universities," *The Chronicle of Higher Education*, February 4, 2000; http://chronicle.com/daily/2000/02/2000020402n.htm.

———, "Romania Suspends University Rectors Over Allegations of Selling Diplomas," *The Chronicle of Higher Education*, October 2, 2000, http://chronicle.com/daily/2000/10/2000100205n.htm.

———, "Euopean Ministers Affirm their Commitment to a Unified Higher Education System," *The Chronicle of Higher Education*, May 23, 2001, http://chronicle.com/cgi2-bin/printable.cgi.

Borgman, Christine L., "Könyvtári automatizálás és hálózatok, Részlet a felsőoktatási világbanki 1,3 moduljával kapcsolatos jelentésből" (Library automation and networks, fragment from the report on the World Bank higher education module No. 1.3), *Tudományos és Műszaki Tájékoztatás*, 43(1996), 95–104.

———, "Library Automation in Central and Eastern Europe: Progress and Prospects," in *International Conference on Library Automation, Conference Proceedings, Budapest, April 10–13, 1996*, ed. by Monika Segbert, Katarina Steinwachs, Peter Burnett; Luxembourg: European Commission, DG XIII/E-4, 1997, pp. 13–22.

Bowen, William G., *Inside the Boardroom*, New York: John Wiley and Sons, 1994.

Brada, Joseph C. and Arthur E. King, "Sequencing Measures for the Transformation of Socialist Economies to Capitalism: Is There a J-Curve for Economic Reforms?", Conference on the East European Transformation, Princeton University, May 3–4, 1991.

Brickell, Anthony, "Unicode/ISO 10646 and the Chase Project," in *International Conference on Library Automation, Conference Proceedings, Budapest, April 10–13, 1996*, ed. by Monika Segbert, Katarina Steinwachs, Peter Burnett; Luxembourg: European Commission, DG XIII/E-4, 1997, pp. 119–25.

Bridge, Frank R., "Automated System Marketplace 1992," *Library Journal*, April 1, 1992.

Brown, Brack, "Unbalanced Sectoral Development in the Czech Republic: Democracy and the Civic, Market, & State Sectors," Fairfax, VA: George Mason University, 1994.

Brown, Brack, *Final Report to The Andrew W. Mellon Foundation on the Czech Economic Development Certification Program 1993–1997*, Fairfax, VA: 1998.

Brown, David L. and László Kulcsár, "Household Pluriactivity in Rural Hungary," paper presented at the meeting of the American Sociological Association, San Francisco, 1998.

Bruce, Sam, "The Riga Business School at Riga Technical University—Excellence in Business Education," *Baltic Studies Newsletter*, 20/2, 1996, 13.

Brunovska, Alena, "Standpoint Regarding the Inspection Record Carried out at Academia Istropolitana," Bratislava: Academia Istropolitana, 1995.

Brynjolfsson, Erik and Lorin M. Hitt, "Beyond Computation: Information Technology, Organizational Transformation and Business Performance," *The Journal of Economic Perspectives*, 14 (Fall, 2000), 23–48.

Building Open Societies: Soros Foundations 1994 Report, New York: Open Society Institute, 1995

Building Open Societies: Soros Foundations 1995 Report, New York: Open Society Institute, 1996

Building Open Societies: Soros Foundations Network 1996 Report, New York: Open Society Institute, 1997

Building Open Societies: Soros Foundations Network 1997 Report, New York: Open Society Institute, 1998.

Building Open Societies: Soros Foundations Network 1998 Report, New York: Open Society Institute, 1999.

Burchard, Maria, "Union Authority File: 1993–1998," in Jadwiga Woźniak and Robert C. Miller (eds.), *Research Libraries: Cooperation in Automation*, Warsaw: Center for Formats and Authority Control, Warsaw University, 1999, pp. 37–43.

Burmeister, Erzsébet, "Számitástechnika a miskolci egyetem könyvtárában az 1974. és 1994 között eltelt 20 évben" (Computer technology in the library of the University of Miskolc during the 20 years between 1974 and 1994), Miskolc: University of Miskolc, January 1995; http://www.lib.uni-miskolc.hu/publ/erzsi/szamtech.htm.

Burnett, Peter, "Retrospective Conversion: Prerequisites and Problems," in *International Conference on Library Automation, Conference Proceedings, Budapest, April 10–13, 1996*, ed. by Monika Segbert, Katarina Steinwachs, Peter Burnett; Luxembourg: European Commission, DG XIII/E-4, 1997, pp. 177–86.

"Business in the Community Program," Washington, DC: Central Europe Institute, September 13, 1994.

Caidi, Nadia, "The Information Infrastructure in Central and Eastern Europe: A Case Study of the Library Community," unpublished paper based on a doctoral dissertation at the Graduate School of Education and Information Studies, Los Angeles: UCLA, October 31, 2000.

Camp, Robert C., *Benchmarking*, Milwaukee, WI: ASQC Quality Press, 1989.

Campbell, Colin G. and Ralf Dahrendorf (eds.), *Changes in Central Europe: Challenges and Perspectives for Higher Education and Research*, Vienna: Institut für die Wissenschaften vom Menschen, 1994.

CAROLINA, e-mail news service (carolina@mbox.fsv.cuni.cz), Prague, Dec. 10, 1999.

Carter, Anne P., Michael Grossman, Michael S. McPherson, Sherwin Rosen, Michael Rothschild, "Site Visit Report," Report to the New York State Education Department, Bureau of Doctoral Program Review, January 27–28, 2000.

Central European Small Business Enterprise Development Commission, Annual Report 1992, Washington, DC, 1993.

Chapman, Bruce, "A Recipe for a Free-Market Hungary," *The Wall Street Journal*, April 6, 1990.

Chapman, David W., "Hungarian Higher Education Management Seminar Series: A Report to the Mellon Foundation," Albany, NY: SUNY-Albany, July 1992.

Charemza, W., Gronicki, M., Quandt, R. E., "Modelling Parallel Markets in Centrally Planned Economics: The Case of the Automobile Market in Poland," *European Economic Review*, 32 (1988), 861–83.

Charemza, W. and Quandt, R. E., "Models and Estimation of Disequilibrium in Centrally Planned Economies," *Review of Economic Studies*, 64 (1982), 109–116.

Charnes, Abraham, William W. Cooper, Arie Y. Lewin, Lawrence M. Seiford, *Data Envelopment Analysis: Theory, Methodology, and Application*, Boston: Kluwer, 1994.

Chressanthis, G. A. and J. D. Chressanthis, "The Determinants of Library Subscription Prices of the Top-ranked Economics Journals: An Econometric Analysis," *Journal of Economic Education*, 25 (1994), 367–82.

The Chronicle of Higher Education, December 15, 2000; http://chronicle.com/free/v47/i16/16a04702.htm.

"Chronology of Transition," http://www.wdi.bus.umich/transiti.htm

Chwirot, Stanisław, "Accreditation at Polish Universities," Paper presented at the Quality Assurance at Polish Universities Conference, Kraków, March 27, 1999; http://www.amu.edu.pl/~ects/uka/referat2.html.

Civic Education Project, New Haven, CT: CEP Pamphlet, 1993.

Concept for Higher Education Development in Hungary (ed. István Bakos), Budapest: Co-ordination Office for Higher Education, Ministry of Education and Culture, 1991.

Conference of Rectors of Academic Schools in Poland Directory, Kraków: Nowa Galicja, 1998.

Cochran, Thomas H., "Reflections on the International Public Finance Program, 1993–1996," unpublished paper, New York: October 2000.

"Conception of Organization of the United Library Information Network of Latvia," Riga: Latvian Development Agency, September 25, 1994.

A Consortium for Management Education at Jagiellonian University, Final Progress Report—First Grant, Hartford, CT: University of Hartford, July 15, 1992.

A Consortium for Management Education at Jagiellonian University and the Academy of Economics—Interim Report, Third Grant, Hartford, CT: University of Hartford, February 26, 1993.

Cooke, Ian and Paul Mayes, *Introduction to Innovation and Technology Transfer*, Boston and London: Artech House, Inc., 1996.

Coricelli, F., *Macroeconomic Policies and the Development of Markets in Transition Economies*, Budapest: CEU Press, 1998.

Crane, Keith, "Institutional Legacies and the Economic, Social, and Political Environment for Transition in Hungary and Poland," *American Economic Review, Papers and Proceedings*, 81 (May, 1991), 318–22.

Cummings, M. Anthony, Marcia L. Witte, William G. Bowen, Laura O. Lazarus, Richard H. Ekman, *University Libraries and Scholarly Communication*, Association of Research Libraries, 1992.

Current Projects and Programs 1994, Tallinn: Open Estonia Foundation, 1994.

Czermiński, Jurand, "VTLS in Polish Academic and Research Libraries," *Task Quarterly*, 2(1), 1998, 121–28.

———, "A Coordinator's View of the Impact of Policy on the Financing and Managing of Library Automation," in Lass, A. and Quandt, R. E. (eds), *Library Automation in Transitional Societies: Lessons from Eastern Europe*, New York: Oxford University Press, 2000.

Dahl, Robert, "Transitions to Democracy," Paper presented at the US—Hungarian Political Scientists First Roundtable, The International Research & Exchanges Board, San Francisco, CA, August 28–29, 1990.

Dahrendorf, Ralf, *Reflections on the Revolution in Europe*, London: Chatto & Windus Ltd, 1990.

———, "Main Problems of the Reform of Higher Education and Research in Former Eastern Europe," in Campbell, Colin G. and Ralf Dahrendorf (eds.) *Changes in Central Europe: Challenges and Perspectives for Higher Education and Research*, Vienna: Institut für die Wissenschaften vom Menschen, 1994.

Darvas, Peter, "Governmental Reforms in Hungarian Higher Education: Historical Traditions and New Actors," in *Reform and Change in Higher Education*, ed. by Mauch, James E. and Paula L. W. Sabloff, New York: Garland Publishing, 1995.

———, "Institutional Innovations in Higher Education in Central Europe. Agenda for Research and Cooperation," Vienna: Institute for Human Sciences, 1996.

Deák, Nóra, "Current Library Management Issues in Hungarian Libraries or How Do Hungarian Libraries Cope with the Changes?," in Ian R. M. Mowat and Maria Śliwińska (eds.), *Library Management: East-West Relation*, Proceedings of the seminar held in Toruń, November 6–12, 1994, Toruń: Nicholas Copernicus University, 1995, pp. 137–41.

A Debreceni Református Kollégium Könyvtáráról, Debrecen; undated.

Deehy, Suzanne M., *Interim Grant Report to The Andrew W. Mellon Foundation*, Cambridge, MA: Sabre Foundation, Inc., June 27, 1994.

Demszky, Gábor, "Building a Market Economy in Hungary, *Uncaptive Minds*, IV(2), 1991, pp. 44–52.

Desruisseaux, Paul, "Scholar's Remark Leads to Journal Donations for Eastern Europe," *The Chronicle of Higher Education*, 39(42), June 23, 1993.

Dialogue of East European Partners (DEEP), Budapest: National Széchényi Library, 1995.

Dietl, Jerzy, "Education for Business Management," in Cichocki, Krzysztof and Paul Marer, *Education for Transition to Market Economy in Countries of Central and Eastern Europe*, Warsaw: Polish Fulbright Alumni Association; Polish-U.S. Fulbright Commission, 1966, pp. 12–26.

Directory of Higher Educational Institutions in the Baltics: Latvia, Riga: Baltic Academic Center, 1995.

Dobelniece, I., Edvīns Karnītis, Gunārs Mangulis, I. Lejniece, Andris Vilks, A. Buholte, V. Poznaka, A. Janbicka, "Integrated Informational Network of Latvian Libraries," Riga: March 14, 1995.

Dobrzyńska-Lankosz, Ewa, "The Kraków Library Group as an Example of a Local Consortium,"," in *International Conference on Library Automation, Conference Proceedings, Budapest, April 10–13, 1996*, ed. by Monika Segbert, Katarina Steinwachs, Peter Burnett; Luxembourg: European Commission, DG XIII/E-4, 1997, pp. 289–95.

———, "The Polish VTLS Consortium: Six Years of Cooperation," in Jadwiga Woźniak and Robert C. Miller (eds.), *Research Libraries: Cooperation in Automation*, Warsaw: Center for Formats and Authority Control, Warsaw University, 1999, pp. 23–29.

———, "The Kraków Project: Problems of Management," in Lass, Andrew and Richard E. Quandt (eds.), *Library Automation in Transitional Societies: Lessons from Eastern Europe*, New York: Oxford University Press, 2000, pp. 360–73.

"Draft Outline for Follow-up Roundtable on Local Infrastructure Debt Finance," Budapest: Ministry of Interior, June 1994.

Dudzińska, Elżbieta, "Import of Serials at a Time of Economic Change," *Polish Libraries Today*, 3 (1995), 49–71.

The East & Central Europe Journal Donation Project, Proposal to the Carnegie Corporation of New York, New York: New School for Social Research, March 2000.

East European Initiative, Summary of USIA Programs, Washington, DC: United States Information Agency, October 1991.

Education for Transition, Civic Education Project and Institut für die Wissenschaften vom Menschen, 1997; Part I. *International Cooperation in Social Science Higher Education in Central and Eastern Europe;* Part II. *Social Science Teaching at Central and East European Universities* (prepared by Daniel Calingaert); Part III. *Higher Education Policy in Central and Eastern Europe.*

Edwards, Heather, "GAELIC, the Gauteng and Environs Library Consortium, South Africa," http://sunsite.wits.ac.za/gaelic/ital_article.htm, April 18, 1999.

ELP Newsletter, 1, No. 1, Autumn 1999.

ERASMUS and LINGUA (Action II) Guidelines for Applicants, Brussels: Commission of the European Communities, 1991.

Establishing an Environmental Assistance Program for Central and Eastern Eu-

rope, Washington, DC: The German Marshall Fund of the United States, August 15, 1990.

Estonia: Implementing the EU Accession Agenda, Washington, DC: The World Bank, 1999.

"Euro-faculty in the Baltic States," *Baltic Studies Newsletter*, 17/3, 1993.

European Bank for Reconstruction and Development, *Annual Report*, London: The Bank, 1991, 1992, 1993, 1994, 1995, 1996, 1997, 1998.

"Evaluation of the Financial Services Volunteer Corps (FSVC) Project in Russia," unpublished paper, Arlington, VA: Carana Corporation, February 29, 2000.

Ewert, H. Bernhard, M. Kunz and Jiří Vacek, *Grant of the Mellon Foundation to GMU to Implement Budgeting and Restructuring in the City of Plzeň*, Final Report 1994–95, Plzeň: George Mason University, December 1995.

"Excerpts of the Report of the EuroFaculty Evaluation Committee, *Baltic Studies Newsletter*, 21/2, 1997, 5–11.

Fábry, G., *Higher Education and Research in Hungary during the Period of Social Transformation: 1990–1992*, Vienna: Institut für die Wissenschaften vom Menschen, 1993.

Feret, Błazej and Marzena Marcinek, "The Future of the Academic Library and the Academic Librarian: A Delphi Study," *Library Career Development*, 7/10 (1999), 91–107.

Final Report Academic Libraries Automation Project, Łódż: Technical University of Łódż, December 31, 2000.

"Final Report (Penultimate Draft)," Association for Library Collections and Technical Services Commission on Cataloging: Description and Access Task Force on the Harmonization of ISBD(ER) and AACR2, Jan. 23, 1999, rev. June 14, 1999, http://www.library.yale.edu/cataloging/aacrer/tf-harm21.htm.

Final Report to The Andrew W. Mellon Foundation from CERGE-EI, Ann-Arbor, MI: CERGE-EI Foundation, November 1999.

Final Report to The Andrew W. Mellon Foundation, Poznań: Poznań Foundation of Scientific Libraries, March 2000.

Final Report to the Andrew W. Mellon Foundation on the Accomplishments of the Center for Promotion and Support of Agricultural Enterprise, Sandomierz, Poland, Norwalk, CT: TechnoServe, February 1999.

Flemming, John and John Micklewright, *Income Distribution, Economic Systems and Transition*, Florence, Italy: UNICEF, Innocenti Occasional Papers, Economic and Social Policy Series No. 70, May, 1999.

Fogarassy, Miklós, "Special Role of the Hungarian National Library in the Formation and Management of Consortia," in Lass, A. and Quandt, R. E. (eds), *Library Automation in Transitional Societies: Lessons from Eastern Europe*, New York: Oxford University Press, 2000, pp. 420–26.

Fogel, Daniel S. and James E. Mauch, "Restructuring Higher Education in the Czech and Slovak Republics: *An Institutional Perspective*," in *Reform and Change in Higher Education*, ed. by Mauch, James E. and Paula L. W. Sabloff, New York: Garland Publishing, 1995.

"Forint Votes" for Civil Society Organizations Studies, Series ed. László Harsányi; Budapest: Research Project on Nonprofit Organizations, 2000.

Freedman, Maurice J. and Ellen Starkman, *Report: The Library Information Network of Latvia*, Riga: USIS, May 22, 1998.

Freeman, Kassie, "Equality in Higher Education in Post-Communist Hungary and Poland: *Challenges and Prospects*," in *Reform and Change in Higher Education*, ed. by Mauch, James E. and Paula L. W. Sabloff, New York: Garland Publishing, 1995.

Friedrich, Alex, "Bad Market Ends Well, Kozeny Laughs," *The Prague Post Online*, January 7, 1998; http://www.praguepost.cz/archive/busi10798a.html

Frydman, R., Rapaczynski, A., and Earle, J. S., *The Privatization Process in Central Europe*, Budapest: CEU Press, 1993.

Füredi, Mihály, " 'Szubjektiv' gondolatok a hazai könyvtárgépesitésről," ('Subjective' thoughts about domestic library automation), *Könyvtári Figyelő*, 38 (1992/4), 636–644.

Galbally, James F., Margaret A. Miller, Anthony M. Morgan, Timothy R. Warner, "Report on Hungarian Higher Education and the Technical University of Budapest," Washington, DC: Citizens Democracy Corps, July 1993.

Gasser, Linda, "CFHREI Project Reflections: Perspectives on the Success of an Academic Partnership for Transformation," unpublished paper, Ithaca, NY: ILR School, Cornell University, October 2000.

Gelzleichter, Richard E., "Purdue/Poland Project," unpublished paper, West Lafayette, IN: Purdue University, May 2000.

The German Marshall Fund of the United States 1995 Annual Report, Washington, DC: 1995, 1996, 1997, 1998.

Giannaros, Demetrios, "Did the 'Shock Therapy' Approach Work in the Economic Restructuring of Eastern Europe? Some Evidence from Poland and Russia: A Brief Review," *Global Business & Economics Review*, 2/1(2000), 53–66.

Glenday, Christine, "An Overview of Science and Technology Reform in Central and Eastern Europe 1990–1993," NSF/Europe, http://www.nsf.gov/home/int/europe/reports/67.htm.

Głowacka, Teresa and Jadwiga Woźniak, "KABA Subject Heading System: An Example of the Collective Effort of Academic Libraries in Poland," in Jadwiga Woźniak and Robert C. Miller (eds.), *Research Libraries: Cooperation in Automation*, Warsaw: Center for Formats and Authority Control, Warsaw University, 1999, pp. 75–80.

Goldfeld, S. M. and Quandt, R. E., "Budget Constraints, Bailouts, and the Firm Under Central Planning," *Journal of Comparative Economics*, 12 (1988), 502–20.

———, "Output Targets, the Soft Budget Constraint, and the Firm Under Central Planning, *Journal of Economic Behavior and Optimization*, 14 (1990), 205–22.

———, "Rationing, Defective Inputs and Bayesian Updates Under Central Planning," *Economics of Planning*, 23 (1990), 161–74.

———, "Effects of Bailouts, Taxes and Risk Aversion on the Enterprise," *Journal of Comparative Economics*, 16 (1992), 150–67.

———, "Uncertainty, Bailouts, and the Kornai Effect," *Economics Letters*, 41 (1993), 113–119.

Gomba, Olga, "A Retrospective Conversion: The Case of Debrecen," Lass, Andrew and Richard E. Quandt (eds.), *Library Automation in Transitional Societies: Lessons from Eastern Europe*, New York: Oxford University Press, 2000, pp. 183–95.

Goodman, S. E., L. I. Press, S. R. Ruth and A. M. Rutkowski, "The Global Diffusion of the Internet: Patterns and Problems, *Communications of the ACM*, 37/8, (1994), 27–31.

Gordon, Robert Aaron and James Edwin Howell, *Higher Education for Business*, New York: Columbia University Press, 1959.

Górny, Mirosław, Bogdan Maruszewski, Andrzej Nikisch, "In Search of an Optimum Model of Development," in Lass, Andrew and Richard E. Quandt (eds.), *Library Automation in Transitional Societies: Lessons from Eastern Europe*, New York: Oxford University Press, 2000, pp. 83–90.

Grant Proposal to The Andrew W. Mellon Foundation, Exchange Development and Education Exchange Program in Poland for January 1, 1995–December 31, 1995, West Lafayette, IN: Purdue University, January 1995.

Grant Proposal to The Andrew W. Mellon Foundation, Purdue/Poland Extension, Research, and Academic Education Exchange Program for March 1, 1996–December 31, 1996, West Lafayette, IN: Purdue University, March 1996.

Grayson, Leslie E. (ed.), *Cases in Management in Eastern and Central Europe and Russia*, Laxenburg, Austria and Charlottesville, VA: International Institute for Applied Systems Analysis and University of Virginia, 1993.

Grayson, Leslie E., *Paving the Road to Prosperity*, Vienna: International Institute for Applied Systems Analysis, 1993.

Greenberg, Janet, ACLS *Manual for International Book and Journal Donations*, New York: American Council of Learned Societies, 1993.

Gruntorád, Jan, "Research and Academic Networking in the Czech and Slovak Federal Republic," Prague: Czech Technical University, 1992.

Gyüre, Péter, "Osztott katalogizálás a gyakorlatban—a "VOCAL" rendszer," (Shared cataloging in practice—the VOCAL system), http://www.iif.hu/rendezvenyek/networkshop/98/eloadas/html/d/pgyure/pgyure.htm.

Halm, Johan van, "Sabre Foundation, Inc. Scientific Assistance Project—Outside Evaluation," Amersfoort, Netherland: Johan van Halm Information Consultancy, February, 1994.

Handke, Mirosław, "Financing Higher Education: Its Limits and Possibilities in Poland,"" *Zarządzanie Współczesną Szkołą Wyższą* (Promoting Effectiveness of Higher Education: Issues of Institutional Cooperation and Development), Polish-American Seminar, Kraków, March 10–11, 1994.

Hankiss, Elemér, "European Paradigms: East and West, 1945–1944," *Dœadalus*, 123, No. 3, Summer 1994, pp. 155–26.

Hare, Paul G., "Hungary: In Transition to a Market Economy," *The Journal of Economic Perspectives*, 5 (1991), 195–201.

Harrison, Lawrence E. and Samuel P. Huntington (eds.), *Culture Matters: How Values Shape Human Progress*, New York: Basic Books, 2000.

Hayes, Robert M., *Narrative Report on the Series of Visits to Institutions Participating in the Study of Staffing Patterns in Academic Libraries of Central and Eastern Europe*, Los Angeles: UCLA, December 20, 1998.

———, "Staffing Patterns for Academic Libraries of Central and Eastern Europe, Russia, and CIS Countries," in *Library Automation in Transitional Societies: Lessons from Eastern Europe*, ed. by Andrew Lass and Richard E. Quandt; New York: Oxford University Press, 2000, pp. 374–98.

Healey, Robin, "Teaching Morality," *Central Europe Review*, 1, No. 20, November 8, 1999; http://www.ce-review.org/99/20/healey20.html.

Hladky, Rostislav, "Towards the 21st Century: A University in Transition," Olomouc: Palacký University paper, February 2, 2000.

Hogan, John D., "The Master of Business Administration Learning Experience—Emerging Role, Mission, Goals and Expectations," paper presented at the International Conference on Master of Business Administration Programs—Polish and Foreign Experience and Plans for Future Development, Łódż, February 23–24, 1996.

Hollender, Henryk, "The Sequencing of Automation in Central and East European Libraries," in Lass, Andrew and Richard E. Quandt (eds.), *Library Automation in Transitional Societies: Lessons from Eastern Europe*, New York: Oxford University Press, 2000), pp. 335–44.

———, "National Union Catalogue: Report #1 for the Mellon Foundation," Warsaw: Warsaw University, November 14, 2000.

Hoy, Frank and Przemysław Kulawczuk, "Outcomes Assessment of Small Business Assistance in Poland," http://www.sbaer.uca.edu/docs/proceedings/96ICSS101.txt.

Hranaiova, Jana, *Delivery Options in Futures Contracts and Basis Behavior*, Dissertation in Partial Fulfillment of the Requirements for the Degree of Doctor of Philosphy, Ithaca, NY: Cornell University, August 2000.

Hrubos, Ildikó, "Transformation of Hungarian Higher Education," *International Higher Education*, No. 20, Summer 2000, pp. 8–10.

Hryniewicz, Janusz, Bohdan Jałowiecki, Agnieszka Mync, *The Brain Drain in Poland*, Warsaw: University of Warsaw, European Institute for Regional and Local Development, September 1992.

Huber, Robert, "The Reorganization of Research in Central and Eastern Europe," NCEEER *Quarterly Newsletter*, Issue Five, April 2000.

The Human Side of Enterprise: Human Resource Management and Industrial Relations Education in the Czech and Slovak Republics, Proposal submitted to The Andrew W. Mellon Foundation, Ithaca, NY: Cornell University, January 1993.

Hunyady, Zsolt, "Archival and Manuscript Resources in Hungary," Paper delivered at the 33rd International Congress of Medieval Studies, Western Michigan University, Kalamazoo, MI; http://www.asu.edu/clas/acmrs/archivalresources.htm.

Információ Technológia, Vol. 1, No. 3, 1992.

Informational Brochure 1998, Prague: CAMBAS, 1998.

"Inside Track Business Education: Weatherhead Severs Links with Budapest School," *Financial Times*, September 18, 2000.

"Inside Track \260 Business Education: A School in Transition," *Financial Times*, August 21, 2000.

Interim Report—Phases V through VII, Hartford, CT: Consortium for Management Education, February 26, 1993.

IREX *News in Brief*, 1, No. 1, Febr. 1991, p. 1.

Iwaskiw, Walter R. (ed), *Estonia, Latvia, and Lithuania Country Studies*, Washington, DC: Library of Congress, 1996.

Janbicka, Aija and Edvīns Karnītis, "Concept of Provision of Scientific and Technical Information for Latvia," http://educate.lib.chalmers.se/IATUL/proceedcontents/fullpaper/karnitis.html; undated, but probably 1997.

Jařab, Joseph, "Higher Education and Research in the Czech Republic," *Higher Education Management*, 5, November 1993, 309–16.

Johnson, Eric A., "Baltic Research Libraries," unpublished paper, Washington, DC: Library of Congress, 1994.
Johnson, Erik C., "Central Europe's Think Tanks: A Voice for Reform," paper presented at CIPE Symposium, Budapest, 1996; see http://www.cipe.org/ert/e21/ceeE21.html.
Johnson, Simon and Gary W. Loveman, *Starting Over in Eastern Europe*, Cambridge, MA: Harvard Business School Press, 1995, pp. 28–32.
Jókay, Charles, Gábor Szepesi, György Szmetana, "Municipal Debt Management and Bankruptcy Intervention in Hungary: Case Studies and Lessons Learned (1996–2000), unpublished paper prepared for the World Bank, June 2000, rev. October 2000.
Jordan, P. and Przbylska, B., "BORIS," *TransAtlantic Perspectives*, Autumn 1994, No. 30, pp. 13–15.
Kabat, Ladislav, "The Czech and Slovak Higher Education System in the Period of Transition," unpublished paper, Nitra: University of Nitra, May 2000.
Kadlecová, Ivana, "Current State of Automation," unpublished paper, Prague: Czech Academy of Sciences, 2000.
Kadlecová, Ivana and Antonín Vítek, "Publikační Aktivita v Akademii Věd České Republiky," (Publication activities at the Czech Academy of Science), Prague: Czech Academy of Science, 1993.
Kamiński, Stanisław, "Contemporary Problems of Acquisitions in Polish Libraries (Based on the Example of Warsaw University Library BUW)," *Polish Libraries Today*, 3 (1995), 35–48.
Kantůrková, Eva, "The Archives of One Foundation," in Paton, Derek and Jitka Uhdeová (eds.), *The Charta 77 Foundation [Twenty Years]*, Brno: Atlantis, 1998.
Kaplan, R. D., *Balkan Ghosts*, New York: St. Martin's Press, 1993.
Karnītis, Edvīns, *Latvian Information Infrastructure: The Present Stage and Trends of Development*, Riga, Latvijas Akadēmiskā Bibliotēka, 1996.
———, *Electronic Information Services in Libraries in Latvia: The Concept and Its Implementation*, Riga: Latvijas Akadēmiskā Bibliotēka, 1997.
Karwacki, Andrzej, "New Tasks and the Need for Professional Administration in Polish Higher Education Institutions, *Zarządzanie Współczesną Szkołą Wyższą* (Promoting Effectiveness of Higher Education: Issues of Institutional Cooperation and Development) Polish-American Seminar II, Kraków, May 26–27, 1994.
Kemme, David M. "Economic Transition in Eastern Europe and the Soviet Union: Issues and Strategies," Institute for East-West Security Studies, New York, November 1990.
King, Gundar and Dwight Zulauf, "Business in Latvia," *European Business Journal*, 8/2, 1996.
Kirk, Mary E., *Where Walls Once Stood: U.S. Responses to New Opportunities for Academic Cooperation with East Central Europe*, New York: Institute of International Education, 1992.
Klimová, Milena, "Main Problems of the Slavonic Library at the National Library in Prague," unpublished paper, Prague: Slovenská knihovna, 1991.
Klvačová, Eva, "Nečekejme na Godota, nepřijde" (Don't wait for Godot, he is not coming), Interview with Václav Klaus, *Ekonom*, 24/1999, 12–14.
Knoll, Alfred and Stanislav Psohlavec (Project mgrs.), *Chronicon Concilii Con-*

stantiniensis, Memoria Mundi Series Bohemica 2, Prague: Národni knihovna and Albertina icome Praha, 1995.

———, *Antiphonarium Sedlecense*, Memoria Mundi Series Bohemica 1, Prague: Národni knihovna and Albertina icome Praha, 1995.

———, *Digitization of Rare Library Materials, Storage of and Access to Data: The Solution for Compound Documents, Manuscripts and Old Books*, Memoria Mundi Series Bohemica, Prague: Národni knihovna and Albertina icome Praha, 1997.

———, *Digitization of Rare Library Materials, Storage and Access to Data*, Memoria Mundi Series Bohemica, Prague: Národni knihovna and Albertina icome Praha, 1999.

Kocsis, Károly, "Main Problems in Restructuring of Hungarian Higher Education System," Paper presented at the Round Table Discussion on Selected Problems of Restructuring of Hungarian Higher Education, organized by the Hungarian Rectors' Conference and the Citizens Democracy Corps, Budapest, October 14, 1991.

———, "Business Education and Restructuring the Higher Education System in Hungary," Paper given at the Business Education's Response to Eastern Europe Conference, Columbia University, May 1, 1991.

Koj, Aleksander, "Jagiellonian University in the Period of Transition from State-owned to State-supported Institution of Higher Education," *Zarządzanie Współczesną Szkołą Wyższą* (Promoting Effectiveness of Higher Education: Issues of Institutional Cooperation and Development), Polish-American Seminar, Kraków, March 10–11, 1994.

Kopányi, Mihály, Samir El Daher, Deborah Wetzel, Michel Noel, Anita Papp, "Hungary: Modernizing the Subnational Government Level," World Bank Discussion Paper No. 417, Washington, DC: May 2000.

Kornai, J., "Resource Constrained vs. Demand Constrained," *Econometrica*, 47 (1979), 801–820.

———, *The Economics of Shortage*, Amsterdam: North Holland, 1980.

———, *Growth, Shortage and Efficiency*, Oxford: Blackwell, 1982.

———, "The Soft Budget Constraint," *Kyklos*, 39 (1986), 3–30.

———, "The Hungarian Reform Process: Visions, Hopes, and Reality," *Journal of Economic Literature*, 24 (1986), 1687–1737.

———, *Struggle and Hope: Essays on Stabilization and Reform in a Post-socialist Economy*, Cheltenham, UK: Edward Elgar, 1997.

———, *From Socialism to Capitalism: What is Meant by the 'Change of the System'*, London: The Social Market Foundation, 1998.

Kosztolányi, Gusztáv, "Acquatic Chernobyl—Requiem for the Tisza and Szamos: Part I," *Central Europe Review*, Vol. 2, No. 7, February 21, 2000, http://www.ce-review.org/00/7/csardas7.html.

Kotrba, Josef, "Privatization Process in the Czech Republic: Players and Winners," in *The Czech Republic and Economic Transition in Eastern Europe*, ed. By Jan Svejnar, New York: Academic Press, 1995.

Koucký, J. and Hendrichová, J., *Higher Education and Research in the Czech Republic: Major Changes since 1989*, Vienna: Institut für die Wissenschaften vom Menschen, 1993.

Kozma, Tamas, "Higher Education in Hungary: Facing the Political Transition," *European Journal of Education*, 25, No. 4, 1990, 379–90.

Kruszyńska, Krystyna, "Experience of Libraries Creting the KABA Subject Headings System," in Jadwiga Woźniak and Robert C. Miller (eds.), *Research Libraries: Cooperation in Automation*, Warsaw: Center for Formats and Authority Control, Warsaw University, 1999, pp. 81–83.

Kučera, Vladimír, "Project EARN/Bitnet Connection & Computer Services in UTIA, Final Report," Prague: Institute for Information Theory and Automation, June, 1992.

Kukk, Kalek, "Five Years in the Monetary Development of the Baltic States: Differences and Similarities," Tallinn: Bank of Estonia, 1997, http://www.www.ee/epbe/bulletin97/article/article1.

Kurm, Margit, "Using Internet in the Baltic States," unpublished IREX Report, 1994.

Laber, Jeri, "Fighting Back in Prague," *The New York Review of Books*, April 27, 1989.

Lackó, Zsuzsa and Anikó Soltész, "A SEED Alapitvány múltja, jelene, jövője," (The SEED Foundation: Its Past, Present, Future), Working paper, Budapest: The SEED Foundation, January, 2000.

Lajos, Tamás, "Perspectives, Hopes and Disappointments: Higher Education Reform in Hungary," *European Journal of Education*, 28, No. 4, 1993, pp. 403–11.

Landes, David, "Culture Makes Almost All the Difference," in Harrison, Lawrence E. and Samuel P. Huntington (eds.), *Culture Matters: How Values Shape Human Progress*, New York: Basic Books, 2000, pp. 2–13.

Lass, Andrew, "CSFR/US National Library Project: Priorities and Strategic Planning," Mount Holyoke College Working paper, October 15, 1991.

———, "Portable Worlds: On the Limits of Replication in the Czech and Slovak Republics," in Buraway, Michael and Verdery, Katherine, *Uncertain Transition*, Lanham: Rowman & Littlefield Publishers, 1999, pp. 273–300.

———, "Managing Delays: The Micropolitics of Time in the Czech and Slovak Automation Project," in Lass, Andrew and Richard E. Quandt (eds.), *Library Automation in Transitional Societies: Lessons from Eastern Europe*, New York: Oxford University Press, 2000, pp. 345–59.

Lass, Andrew and Richard E. Quandt, (eds.), *Library Automation in Transitional Societies: Lessons from Eastern Europe*, New York: Oxford University Press, 2000.

Latvian Research: An International Evaluation, Copenhagen: The Danish Research Councils, 1992.

Lévay, Irén, "Library Integration of Debrecen University," unpublished paper, Debrecen: Kossuth Lajos University, 2000.

Levitas, Anthony, "Possible Futures for the Municipal Development Program," unpublished paper, RDF, September 13, 1993.

Lewartowski, Kazimierz, *Late Helladic Simple Graves: A Study of Mycenaean Burial Customs*, as reported in *Newsletter*, American School of Classical Studies at Athens, No. 43, Summer 1999, p. 3

Lieven, Anatole, *The Baltic Revolution: Estonia, Latvia Lithuania and the Path to Independence*, New Haven, CT: Yale University Press, 1993.

Lipton, David and Jeffrey Sachs, Creating a Market Economy in Eastern Europe: The Case of Poland," *Brookings Papers on Economic Activity*, 1, 1990, pp. 75–147.

Lovatt, Catherine, "Gold and Cyanide," *Central Europe Review*, Vol. 2, No. 7, February 21, 2000, http://www.ce-review.org/00/7/lovatt7.html.

Lyman, Randall, "Czech Libraries to Enter 20th Century," *Prognosis*, June 11–24, 1993.

Macevicience, Vida and Lee. G. Burchinal, "Central European Conference and Exhibition for Academic Libraries and Informatics," Conference Report, Vilnius: September 27–29, 1993.

MacWilliams, Bryon, "Conflicts over Language Impede Ukraine's Higher-Education System," *The Chronicle of Higher Education*, February 2, 2001, pp. A42–A44.

Mader, Béla, "Automation and Academic Libraries in Hungary: Theory and Practice in the Period of New Challenges,", in Lass, Andrew and Richard E. Quandt (eds.), *Library Automation in Transitional Societies: Lessons from Eastern Europe*, New York: Oxford University Press, 2000, pp. 93–103.

"A Magyar Felsőoktatás Fejlesztése 2000-ig" (Development of Hungarian Higher Education until 2000), Budapest: Secretariat of the Hungarian Rectors' Conference, January, 1992.

A Magyar Tudományos Akadémia Könyvtára, Budapest: Hungarian Academy of Sciences, 1988.

Making Change Last: Lessons Learned in Providing Management Training and Economics Education to Central and Eastern Europe, Highlights of a Conference, Washington, DC: Business-Higher Education Forum, December 1–2, 1992.

Malova, Darina and Erik Lastic, "Higher Education in Slovakia: A Complicated Restoration of Liberal Rules," *East European Constitutional Review*, 9/2, Summer 2000, 100–104.

Management and Economics Education Project 180–0029 (State University of New York) Final Evaluation, Budapest University of Economic Sciences, November 1998.

Management Training and Economics Education Report, AID Grant # EUR-G-00-91-00061-00; Second Quarterly Report (Seventh Year), Period January 1, 1998 through March 31, 1998 and Third Quarterly Report (Seventh Year) Period April 1, 1998 through June 30, 1998, State University of New York, Albany.

Management and Economics Education Project 180–0029 (State University of New York) Final Evaluation, Budapest University of Economic Sciences, November 1998.

Manikowski, Adam, "Library Automation—Fortunes and Miseries in Poland," Chapter 1 in Lass, Andrew and Richard E. Quandt (eds.), *Library Automation in Transitional Societies: Lessons from Eastern Europe*, New York: Oxford University Press, 2000.

Marer, Paul, "Foreign Economic Liberalization in Hungary and Poland," *American Economic Review, Papers and Proceedings*, 81 (May, 1991), 329–33.

McCrudden, Christopher, Jackson Newell, Karen Schilling, Karen Spear, "CDC Consultants' Report on Hungarian Higher Education," Washington, DC: Citizens Democracy Corps, July 1993.

McCrudden, Christopher, Anthony Morgan, Timothy Warner, "Report on Follow-up Conferences to the Site Visits by American Volunteers," Budapest: Hungarian Higher Education Project, August, 1994.

"Mellon Foundation Grant Proposal from George Mason University, Economic and Management Reforms in the Czech Republic," Fairfax, VA: George Mason University, October 8, 1993.

Micgiel, John and Stanisław Wellisz, *Final Report on the Warsaw University-Columbia University Cooperative Program in Economics*, New York: Columbia University, March 29, 1999.

"Mid-Career Development Inter-University, Post-Graduate Program," Kraków, Progress and Business Foundation, August 5, 1993.

Miljan, Toivo, "EuroFaculty: Tartu-Riga-Vilnius Report," *Baltic Studies Newsletter*, 18/3, 1994, 3–10.

Miljan, Toivo, "Excerpts of the 1995–96 Annual Report on the EuroFaculty Programme," *Baltic Studies Newsletter*, 21/1, 1997, 8–15.

Miller, Robert F., "The Baltic States Win Their Independence: Lithuania's Difficult Renaissance," http://english-server.nss.emu.edu/history/baltic-states-independence.txt, February 17, 1992.

Mirsky, Arlene Elgart, "The New Global Community," *The Metropolitan Corporate Counsel*, 8/7–8 (December 2000/January 2001).

Mokhtari, M. and Caner, S., "The Problem of Arrears in the Russian Federation," United States Government Technical Assistance, Working paper, June 1999.

Mooney, Paul, "Students at Chinese University Reportedly Pay Large Bribes to Switch Majors," *The Chronicle of Higher Education*, November 22, 2000; see http://chronicle.com/cgi2-bin/printable.cgi.

Morgan, Anthony, "Reform in Hungarian Higher Education," *International Higher Education*, No. 19, Spring 2000, pp. 23–24.

Morgan, Anthony and Amy Aldous Bergerson, "Importing Organizational Reform: The Case of Lay Boards in Hungary," Paper presented at The Association for the Study of Higher Education International Forum, San Antonio, TX, November 18, 1999; final version published in *Higher Education*, 40(2000), 423–48.

Morris, Adele and Vijaya Ramachandran, "HPI and Rabbit Farming in Poland: A Status Report for the Mellon Foundation," unpublished paper, New York: The Andrew W. Mellon Foundation, November 15, 1994.

Morita, Tsuneo, "The Hidden Growth Potential of EU Candidates," *Transition*, 10, No. 5, Oct. 1999, 5–7.

Morris, Anne, "Management Training Program in Market Economy Logistics," New York: City University of New York, Center for Logistics and Transportation, July 23, 1992.

———, "Final Report: Executive Management Program in Strategic Logistics," New York: City University of New York, Center for Logistics and Transportation, February 28, 1996.

Mott Foundation Civil Society Grants in Central Eastern Europe/Russia and Central Eastern Europe/FSU, Flint, MI: The Charles H. Mott Foundation, 1999.

Muraközy, László, "Szubjektiv emlékek a debreceni egyetemi integrációról," (Subjective memories of the Debrecen integration), unpublished working paper, Debrecen: Debrecen University, October 2000.

Murányi, Lajos, "Das ungarische Informations-und Dokumentationswesen nach 1990" (The state of Hungarian information and documentation technology after 1990), *Newsletter*, Sozialwissenschaften in Osteuropa, November 1994, http://www.berlin.iz-soz.de/publications/newsletter/socsci-eastern-europe/nl 94-95/nl94-4-5.htm.

Navaretti, Giorgio Barba and Andrea Bigano, "R&D Inter-firm Agreements in Developing Countries. Where? Why? How?," in *Creation and Transfer of Knowledge*, ed. by G. Barba Navaretti, P. Dasgupta, K.-G. Mäler, D. Siniscalco; Berlin: Springer, 1998.

Nedyalkova, Yordanka, "Latvia Gets New Language Regulations," *Transitions Online*, August 28, 2000; http://www.tol.cz/week.html.

Neuman, Tanya R., "Who Will Step Down First? Latvia's Government in Crisis—Again," *The Baltic Times*, 2/69, July 24, 1997.

New Initiatives for the 1990s and Beyond, 1991 Annual Report, Stamford, CT: International Executive Service Corps, 1991

Newberg, Paula R. and Thomas Carothers, "Aiding—and Defining—Democracy," *World Policy Journal*, Spring 1996.

Noll, Roger and W. E. Steinmueller, "An Economic Analysis of Scientific Journal Prices: A Preliminary Analysis," *Serials Review*, 18 (1992), 32–37.

Nollendorfs, Valters, "Restoring the Core: Democratization through Education in Post-Communist Europe," *Baltic Studies Newsletter*, 16/4, December 1992, 4–8.

———, "Overhaul through Shortfall: One Summer's Observations and Ruminations About the Future of Educational Reform in Latvia," *Baltic Studies Newsletter*, 17/4, December 1994, 2–6.

———, " 'A Dreadful Thing': Latvian Education Policy Problems Revisited and Revived," *Baltic Studies Newsletter*, 24/4, No. 96, December 2000, 8–10.

Nowak, Grzegorz, "The History of Library Automation Project for Lublin Consortium Libraries," unpublished paper, Lublin: Marie Curie Skłodowska University, 1999.

O'Connor, Laurence, "Overall Review of Activities and Accomplishments, 1994–1998," in *Andrew W. Mellon Foundation Programme Hungarian Agricultural Project 1994–1998, Final Report*, Limerick: University of Limerick, 1998.

Olenick, Peter, "A Primer for Wide Area Computer Networks," Princeton University, CIT, August, 1992.

On libraries, on the supplements of Act 27:1987 (Digest) of the Slovak National Council on the state care of historical monuments and on the amendments and supplements of Act No. 68:1997 (Coll) on Matica slovenská, Act No, 183, Bratislava, May 12, 2000.

The Open Estonia Foundation in 1992, Tallinn: 1993.

The Open Estonia Foundation in 1993, Tallinn: 1994.

The Open Estonia Foundation in 1994, Tallinn: 1995.

Oppenheim, Bohdan W. and Zbigniew H. Przasnyski, *Final Report on the LMU Center for Technology Management in Poland*, Los Angeles: Loyola Marymount University, February 2, 1998.

———, "Reflections and Lessons Learned," unpublished paper, Loyola Marymount University, February 2, 1998.

———, "Transfer of Technology Management and Total Quality Knowledge to Poland," unpublished paper, Loyola Marymount University, November 1998.

Orbán, Viktor, "The Case Against Restitution," *Uncaptive Minds*, IV(2), 1991, 33–38.

Oriental Collection of the Library of the Hungarian Academy of Sciences, Budapest: Hungarian Academy of Sciences, 1991.

Paciorek, Anna, "Górą warszawska SGH i wrocławska AE" (Warsaw School of Economics tops Wrocław Academy of Economics), *Rzeczpospolita*, June 2, 2000.

Padziński, Andrzej, "The Special Collections of the Hieronim Łopaciński Voivodship Public Library in Lublin," Lublin: Łopaciński Library, 2000.

Palmer, Tom G., *Philanthropy in Central and Eastern Europe: A Resource Book for Foundations, Corporations, and Individuals*, Fairfax, VA: The Institute for Humane Studies, George Mason University, 1991.

Paluszkiewicz, Anna, "Evolution of Authority File Work in Poland," in Lass, Andrew and Richard E. Quandt (eds.), *Library Automation in Transitional Societies: Lessons from Eastern Europe*, New York: Oxford University Press, 2000, pp. 224–34.

Paluszkiewicz, Ewa, "The Retrospective Conversion of the Catalog in the Library of the American Studies Center, Warsaw University,"," in Lass, Andrew and Richard E. Quandt (eds.), *Library Automation in Transitional Societies: Lessons from Eastern Europe*, New York: Oxford University Press, 2000, pp. 113–19.

PASBAF Proposals to The Andrew W. Mellon Foundation, dated March 3, 1994, May 30, 1995, and July 3, 1996.

Pasztaleniec-Jarzyńska, Joanna and Halina Tchórzewska-Kabata, *The National Library in Warsaw*, Warsaw, The National Library Publishing House, 1998.

Pelczar, Andrzej, "Challenges Facing Universities in Central and Eastern Europe: A Polish Perspective," Paper delivered at the 41st Biennial CRE Conference, Maynooth, Dublin, May 5–8, 1993; also *in* CRE-*Action*, No. 102, 1993, 41–53.

———, "Autonomy, Competition and Cooperation in Higher Education," *Zarządzanie Współczesną Szkołą Wyższą* (Promoting Effectiveness of Higher Education: Issues of Institutional Cooperation and Development), Polish-American Seminar, Kraków, March 10–11, 1994.

Peterson, George E., "Decentralization in Poland, Hungary, and the Czech Republic: What Have We Learned?" in Peterson, George E. (ed.), *Decentralization in Central and Eastern Europe*, Washington, DC: The Urban Institute, 2000.

Pew Charitable Trusts Grants List, Philadelphia: The Pew Charitable Trusts, 1999.

Phare 1994 Annual Report.

Pigey, Juliana H. (with assistance from Judith Kálmán), "Fiscal Decentralization and Local Government Finance in Hungary 1990–1999," in Peterson, George E. (ed.), *Decentralization in Central and Eastern Europe*, Washington, DC: The Urban Institute, 2000.

Plakans, Andrejs, *The Latvians: A Short History*, Stanford, CA: Hoover Institution Press, 1995, pp. 167–83

PlanEcon, 7/47–48, December 23, 1991.

Porter, Michael E., "Attitudes, Values, Beliefs, and the Microeconomics of Prosperity," in Harrison, Lawrence E. and Samuel P. Huntington, *Culture Matters: How Values Affect Human Progress*, New York: Basic Books, 2000, pp. 14–28.

Portes, Richard, "Economic Reforms in Hungary," *American Economic Review, Papers and Proceedings*, 60 (May, 1970), 307–13.
Portes, Richard, Richard E. Quandt, David Winter, Stephen Yeo, "Macroeconomic Planning and Disequilibrium: Estimates for Poland, 1955–1980," *Econometrica*, 55 (1987), 19–41.
——— "Estimation de la taille des erreurs de planification," *Annales de l'Insee*, No. 55/56, (1984), 245–55.
Portes, Richard, Richard E. Quandt and Stephen Yeo, "Tests of the Chronic Shortage Hypothesis: The Case of Poland," *Review of Economics and Statistics*, LXX/2 (1988), 288–95.
Portes, Richard and Winter, David, "Disequilibrium Estimates for Consumption Goods Markets in Centrally Planned Economies," *The Review of Economic Studies*, 47 (1980), 137–59.
Portwood, Jim, "International Management Center MBA Program Evaluation," unpublished paper, Philadelphia: Temple University, 1996.
Postma, Renée, "Slovakia not out of Trouble yet after Defeat Meciar," October 1, 1998, http://www.rnw.nl/budapest/html/slovakia0110.html.
Program Guidelines, Philadelphia: The Pew Charitable Trusts, 1994.
"A Program of Management Development for Czechoslovak Executives," Program draft proposal, Washington, DC: Georgetown University, July 1990.
Programs Supporting Reform in Central and Eastern Europe, The German Marshall Fund of the United States, Washington, DC: 1995.
Prokop, Igor, Milan Rákus, and Daniela Slížová, "Realization of the CASLIN Project in the Slovak National Library and Its Impact on the Automation of Libraries in Slovakia," in Lass, Andrew and Richard E. Quandt (eds.), *Library Automation in Transitional Societies: Lessons from Eastern Europe*, New York: Oxford University Press, 2000, pp. 153–79.
The Proposal for Grant Addressed to the Andrew W. Mellon Foundation by the Consortium of Estonian Libraries Network, Tallinn: ELNET, June 1996.
Proposal for the Loyola Marymount University Center for Technology Management in Poland, Los Angeles: Loyola Marymout University, March 10, 1995.
Quandt, Richard E., *The Econometrics of Disequilibrium*, Oxford: Blackwell, 1988.
——— "Recent Economic Developments in Eastern Europe," Princeton University, Financial Research Center Memorandum No. 125, May 1991.
——— *Report on Grants in Eastern Europe*, The Andrew W. Mellon Foundation, October 8, 1991.
——— *Report on Trip, June 1993*, New York: The Andrew W. Mellon Foundation, June 1993.
——— *Report on Trip to Poland, October 21–31, 1993*: New York: The Andrew W. Mellon Foundation, 1993.
——— *Report on Projects in Eastern Europe, June 1994*, New York: The Andrew W. Mellon Foundation, 1994.
——— *Trip Report on Projects in Eastern Europe: June 1994*, New York: The Andrew W. Mellon Foundation, 1994.
——— *Report on Trip to the Czech Republic, Slovakia and Hungary, May 20-June 2*, New York: The Andrew W. Mellon Foundation, 1995.
——— *Report on the Baltics*, New York: The Andrew W. Mellon Foundation, 1995.

——— *Report on Trip to the Czech Republic, Slovakia and Hungary, May 18, 1996 through June 1, 1996*, New York: The Andrew W. Mellon Foundation, 1996.

——— *Report on the Latvian Library Consortium*, New York: The Andrew W. Mellon Foundation, May 1997.

——— *Report on Trip to Estonia, Latvia, and Poland*, New York: The Andrew W. Mellon Foundation, October 1997.

——— *Report on Trip to Latvia, Slovakia, Hungary, May 20–June 1, 1999*, New York: The Andrew W. Mellon Foundation, June, 1999.

——— "A Simulation Model for Journal Subscription by Libraries," *Journal of the American Society for Information Science*, 47/8 (1996), 610–617.

——— "On the Optimum Number of Library Software Licenses," *Journal of Economic Behavior and Organization*, 38(1999), 349–56.

——— "The Kornai Effect Revisited," in Eric Maskin and András Simonovits (eds.), *Planning, Shortage and Transformation: Essays in Honor of János Kornai*, Cambridge, MA: The MIT Press, 2000, pp. 209–229.

"The Question of Imagination; An Interview with Professor Andrzej Kajetan Wróblewski, Rector of Warsaw University," *Polityka*, Oct. 21, 1992.

Quigley, Kevin F. F., *Conversations on Democracy Assistance*, Washington, DC: East European Studies, The Woodrow Wilson International Center for Scholars, 1996.

———, *For Democracy's Sake: Foundations and Democracy Assistance in Central Europe*, Washington, DC, The Woodrow Wilson Center Press, 1997.

———, "For Democracy's Sake: Background Paper," Paper delivered at the Bratislava Workshop, November 2–3, 1995.

———, "For Democracy's Sake: How Funders Fail—and Succeed," *World Policy Journal*, Spring 1996, pp. 109–18.

———, "Lofty Goals, Modest Results: Assisting Civil Society in Eastern Europe," in *Funding Virtue: Civil Society Aid and Democracy Promotion*, ed. by Marina Ottaway and Thomas Carothers, Washington, DC: Carnegie Endowment for International Peace, 2000, pp. 191–215.

Reddy, Allan C., *A Macro Perspective of Technology Transfer*, Westport, CT: Quorum Books, 1996.

Regulska, Joanna, "Building Democracy at the Local Level: FSLD-LDP Partnership," unpublished paper, New Brunswick, NJ: Rutgers University, January 2001.

Regulski, Jerzy, "Building Democracy in Poland: The State Reform of 1998," Discussion Paper. No. 9, Open Society Institute, Budapest: 1998.

Request for Grant Addressed to The Andrew W. Mellon Foundation, Warsaw, National Library of Poland, January 20, 1994.

Report 1991–1995, Academia Istropolitana, Bratislava, January, 1996.

"Report of the Dean and CEO to the Board of Trustees," Budapest: International Management Center, March 8, 1996.

"Response to RFA Number SPO/EE-91-002, Part II, Cost," Rochester, NY: US Business School in Prague, October 20, 1990.

Roberts, Rachel, *Library Development in Central and Eastern Europe from Assistance to Cooperation: An Investment in the Future*, Proceedings of a Workshop held in Strasbourg, February 3–4, 1994, Luxembourg: European Commission, 1994.

Rockefeller Brothers Fund Annual Report 1990, New York: 1991.

Rockefeller Brothers Fund Annual Report 1991, New York: 1992.

Rockefeller Brothers Fund Annual Report 1992, New York: 1993.

Rocki, Marek, "Basic Issues Concerning the Costs of Higher Education," *Zarządzanie Współczesną Szkołą Wyższą* (Promoting Effectiveness of Higher Education: Issues of Institutional Cooperation and Development), Polish-American Seminar, Kraków, March 10–11, 1994.

Róna-Tas, András, "Accreditation in Hungary and the Hungarian Accreditation Committee," *Accreditation and Background Information*, Budapest, Hungarian Accreditation Committee, 1993.

Rondinelli, Dennis A. and Martin R. Fellenz, "Privatization and Private Enterprise Development in Hungary: An Assessment of Market-Reform Policies," *Business & The Contemporary World*, 5 (Autumn, 1993), 75–88.

Rondinelli, Dennis A. and Jay Yurkiewicz, "Privatization and Economic Restructuring in Poland: An Assessment of Transition Policies," *American Journal of Economics and Sociology*, 55 (April 1996), 145–60.

Rupnik, Jacques, "Higher Education and the Reform Process in Central and Eastern Europe," *European Journal of Education*, 27, No. 1–2, 1992, 45–51.

Ruth, Stephen, "Internet-related Grants in Central Europe: Insights and Lessons Learned," Fairfax, VA: George Mason University, December, 1999.

———, "Using Academic Networks in Developing Nations: A Value-Added Services Perspective," in *International Information Systems*, San Diego, CA: HBJ Professional Publishing, July 1992, pp. 87–99.

———, "Measuring Long Term Effects of Technology Transfer in Developing Nations: The Case of Internet Training at the Romanian Academy of Sciences," *Information Technology for Development*, 9 (2000), 105–21.

Ruth, Stephen R. and Raul Gouet, "Must Invisible Colleges be Invisible? An Approach to Examining Large Communities of Network Users," *Internet Research*, 3/1 (1993), 36–53.

Ruth, Stephen and Robert Schware, "Can this Connectivity Project Succeed? Predicting and Assessing Success in Internet Implementations in Developing Countries," http://www.isoc.org/inet96/proceedings/g5/g5_4.thm.

Ruys, Theodore and Olga Ruys, "Belarus—Book Donation Project in Belarus," *Dialogue of East European Partners (DEEP)*, Budapest: National Széchényi Library, 1995, pp. 9–14.

Ryan, Alan, "My Way," *The New York Review of Books*, 47/13, August 10, 2000, 47–50.

Řepišová, Zuzana, "The Implementation of an Automated System at the University Library in Bratislava, Slovakia," Lass, Andrew and Richard E. Quandt (eds.), *Library Automation in Transitional Societies: Lessons from Eastern Europe*, New York: Oxford University Press, 2000, pp. 204–18.

Sarmir, E. and Zajac, S., *Higher Education in the Slovak Republic: Major Changes since 1989*, Vienna: Institut für die Wissenschaften vom Menschen, 1993.

Schöpflin, George, "Postcommunism: The Problems of Democratic Construction," *Dœdalus*, 123, No. 3, Summer 1994, pp. 127–41.

SEED *Act Implementation Report Fiscal Year 1998*, Washington, DC: Department of State, March, 1999.

Sergesketter, Terri M., "The River Company," unpublished report to the MBA Enterprise Corps, Kraków: 1993.

Shim, Wonsik and Paul B. Kantor, "A Novel Approach to the Evaluation of Research Libraries," Final Report, Alexandria Project Laboratory, SCILS, Rutgers University.

Siegel, Daniel and Jenny Yancey, *The Rebirth of Civil Society: The Development of the Nonprofit Sector in East Central Europe and the Role of Western Assistance*, New York: Rockefeller Brothers Fund, 1992.

"Single Information Network of Latvia's Libraries," Riga: Association of Latvia's Academic Libraries, 1996.

Śliwińska, Maria, "Staff Preparation for Automation," in *Library Automation in Transitional Societies: Lessons from Eastern Europe*, ed. by Andrew Lass and Richard E. Quandt; New York: Oxford University Press, 2000, pp. 399–406.

———, "Nicholas Copernicus University Library Computerization," unpublished paper, Toruń: Copernicus University, 2000.

Slovak News Update, The Foundation for a Civil Soviety, New York, December 1999.

Smith, Inese A. and Marita V. Grunts, *The Baltic States: Estonia, Latvia, Lithuania*, Oxford: Clio Press, 1993.

Smith, Graham, "The Resurgence of Nationalism," in Smith, Graham (ed.), *The Baltic States: The National Self-Determination of Estonia, Latvia and Lithuania*, London: MacMillan Press, 1994, pp. 121–43.

Smith, Graham (ed.), *The Baltic States: The National Self-Determination of Estonia, Latvia and Lithuania*, London: MacMillan Press, 1994

Software License Agreement between Ex Libris LTD *and Bibliotēku Informacijās Tīklu Konsorcijs* (Library information Network Consortium of Latvia), October 24, 1998.

Sontheimer, Kevin, "A Brief Report With Exhibits Concerning Financial Management Problems at the University of Economics Bratislava," unpublished paper, Pittsburgh: University of Pittsburgh, USAID MTEE-Slovakia Grant project, October 21, 1996.

———, "A Proposal Submitted to Katerina Vajdova Executive Director and the Board of Trustees of Academia Istropolitana Nova," University of Pittsburgh, Economic Policy Institute, December 1, 1999.

Soros, G., *Opening the Soviet System*, London: Weidenfeld and Nicolson, 1990.

———, *Soros on Soros—Staying Ahead of the Curve*, New York: Wiley, 1995.

———, *Underwriting Democracy*, New York: The Free Press, 1990, 1991.

Souborné katalogy: organizace a služby (Union catalogues: their organization and services), Prague: Národni knihovna ČR, 2000.

Spiro, Gene, "The Banking Finance Assistance Centre, Summary Report and Evaluation," unpublished paper, January 2000.

Statistical Yearbooks of Hungary, 1990–98, Budapest: Hungarian Central Statistical Office.

Steinerová, Jela, "Libraries of Central and Eastern Europe: Basic Dilemmas," in *Library Automation in Transitional Societies: Lessons from Eastern Europe*, ed. by Andrew Lass and Richard E. Quandt; New York: Oxford University Press, 2000, pp. 59–73.

Stepan, Alfred, "When Democracy and the Nation State are Competing Logics: Reflections on Estonia," *European Journal of Sociology*, 35, 1994, 127–41.

Stoklasová, Bohdana, "Cataloging Standards in Czech Libraries," in Lass, Andrew and Richard E. Quandt (eds.), *Library Automation in Transitional Societies: Lessons from Eastern Europe*, New York: Oxford University Press, 2000, pp. 219–23.

———, "The Retrospective Conversion in Czech Libraries," Lass, Andrew and Richard E. Quandt (eds.) *Library Automation in Transitional Societies: Lessons from Eastern, Europe*, New York: Oxford University Press, 2000, pp. 196–203.

Stoklasová, Bohdana and Jaroslava Jeřábková, *Czech Books Published in the 20th Century*, Prague: Národni knihovna, 2000.

Stroehlein, Andrew, Jan Culik, Steven Saxonberg, Kazi Stastna, "The Czech Republic 1992 to 1999," *Central Europe Review*, 1, No. 12, September 13, 1999; http://www.ce-review.org/99/12/stroehlein12.html.

Students' Final Research Papers, Nitra: Institute of Economics Studies, University of Agriculture, 1996.

"Students for East/Central European Development, Proposal Prepared for The Andrew W. Mellon Foundation," Washington, DC: Georgetown University, April 1993.

"Students for European Development, A Proposal Submitted to The Andrew W. Mellon Foundation," Washington, DC: Georgetown University, April 14, 1992.

Stypułkowski, Cezary, "Building Human Resources in Banking," in Cichocki, Krzysztof and Paul Marer, *Education for Transition to Market Economy in Countries of Central and Eastern Europe*, Warsaw: Polish Fulbright Alumni Association; Polish-U.S. Fulbright Commission, 1966, pp. 28–31.

Sulc, Zdislaw, *Strucne dejiny ekonomickych reforem*, Brno: Doplnek, 1998, pp. 86–7.

Sulejewicz, Aleksander, Grażyna Leśniak-Łebkowska, Mahmood Zaidi, "Development and Delivery of an American Style MBA Degree at the Warsaw School of Economics," in Sulejewicz, Aleksander and Mahmood Zaidi, *Polish-American MBA at the Warsaw School of Economics*, Warsaw: Warsaw School of Economics Press, 2000, pp. 23–51.

Svejnar, Jan (ed.), *The Czech Republic and Economic Transition in Eastern Europe*, New York: Academic Press, 1995.

Svoboda, Martin, "Automation in Czech Libraries—How to Proceed?", Paper presented at LIBER Workshop, Budapest, June, 1992.

———, "Cooperation, Consortia, Compatibility, Connectivity: The Case of CASLIN," in Lass, Andrew and Richard E. Quandt (eds.), *Library Automation in Transitional Societies: Lessons from Eastern Europe*, New York: Oxford University Press, 2000, pp. 134–152.

Szakolczai, G. and Stahl, J., "Increasing or Decreasing Returns to Scale in the Constant Elasticity of Substitution Production Function," *Review of Economics and Statistics*, 51 (1969), 84–90.

Szanton, P. "DIALOG," *TransAtlantic Perspectives*, Autumn 1994, No. 30, pp. 6–9.

Szapiro, Tomasz, "Barriers to Transforming Higher Education and How to Overcome Them," in *Education for Transition to Market Economy in Countries of Central and Eastern Europe*, ed. by Cichocki, Krzysztof and Paul Marer, Warsaw: Polish—U.S. Fulbright Commission, 1996, pp. 37–41.

Szögi, László, *A Short History of Loránd Eötvös University of Budapest, 1635–1985*, Budapest: Loránd Eötvös University, 1985.

Szögi, László and Zsidi Vilmos (eds.), *Tanulmányok a magyarországi közgazdasági felsőoktatás történetéből* (Studies on the history of Hungarian higher education in economics), Budapest: Budapest University of Economics, 1995.

Szörényi, László, *Saecula Hungariae 1701–1796*, Budapest: Széchenyi Arts Center, 1985.

Šmite, Sarma, "Situation in the Latvian Libraries," in Ian R. M. Mowat and Maria Śliwińska (eds.), *Library Management: East-West Relation*, Proceedings of the seminar held in Toruń, November 6–12, 1994, Toruń: Nicholas Copernicus University, 1995.

Taagepera, Rein, "Is the USA a Model for Europe?," *Baltic Studies Newsletter*, 33/1, March 1999, 1–8.

Tapolczai, Ágnes and Péter Popovics, "Novell rendszerben működő integrált könyvtári rendszer (Tinlib) megjelenése www felületen," (The appearance of a Novell-based integrated library system (Tinlib) in the realm of www), November 1997, http://dis.gau.hu/tinlib/tintawww.html.

Tauer, Loren, "Third Annual Report for the Continued Funding of the Institute for Economic Studies," Ithaca, NY: Cornell University, June 1999.

Tauer, Loren W and Lubica Bartova, *Funding for the Institute of Economic Studies, Final Report*, Ithaca, NY: Cornell University, July 2000.

Tauer, Loren and Zdenko Stefanides, "Success in Maximizing Profits and Reasons for Profit Deviation on Dairy Farms," *Applied Economics*, 30 (1998), 151–56.

———, "The Empirical Impact of Bovine Somatotropin on a Group of New York Dairy Farms," *American Journal of Agricultural Economics*, 81 (1999), 95–102.

TEMPUS *Financial Support for Cooperation and Mobility in Higher Education between Central/Eastern Europe and the European Community*, Brussels: Commission of the European Communities, 1991.

Ten Years (1989–1999) Local Democracy Poland, New Brunswick, NJ: Rutgers University, 2000.

Tenner, Arthur R and Irving J. DeToro, *Total Quality Management*, Reading, MA: Addison-Wesley, 1992.

———, *Process Redesign*, Reading, MA: Addison-Wesley, 1996.

Think Tanks in Central and Eastern Europe: A Comprehensive Directory, Budapest: Freedom House, 1997.

Tolnai, György, "A Könyvtárgépesités Helyzete Magyarországon," (Library automation in Hungary), unpublished paper, May, 1992.

———, "A hazai könyvtárgépesités a megváltozott világban," (Hungarian library automation in a changed world), *Tudományos és Műszaki Tájékoztatás*, 39/7–8, 1992, 303–9.

Tomáškov, Petra, "Position and Protection of National Minorities," in *Reports on the State of Human Rights in the Czech Republic 1999*, ed. by Libuše Šilhanová; Prague: Czech Helsinki Committee, 2000, pp. 11–22.

Tříška, D. and Quandt, R. E. (eds.), *Optimal Decisions in Markets and Planned Economies*, Boulder, CO: Westview Press, 1990.

Trojanowska-Bitka, Anna, "Poland," *Dialogue of East European Partners (DEEP)*, Budapest: National Széchényi Library, 1995, pp. 28–34.

Tucker, Aviezer, "Reproducing Incompetence: The Constitution of Czech Higher Education," *East European Constitutional Review*, 9/3, Summer 2000, 94–99.

Twain, Mark, "The £1,000,000 Bank-Note," in *The Complete Short Stories of Mark Twain*, New York: Bantam Books, 1957.

Universitas Carolina Pragensis, Prague: undated publication by Charles University.

U.S. Consortium for Management Education in Central and Eastern Europe—Final Report, Hartford, CT: University of Hartford, April 30, 1998.

U.S. Consortium for Management Education in Central and Eastern Europe, Funding Proposal for the Center for Faculty Development in Business Economics Studies, Hartford, CT: University of Hartford, March 29, 1994.

A U.S. University Consortium for Management Education at Jagiellonian University and the Academy of Economics ay Kraków, Poland—Fourth Grant Request, Hartford, CT: University of Hartford, February 26, 1993.

"U.S SEED Act Assistance Phase-Out Estonia," USAID Fact Sheet, Washington, DC: USAID, May 11, 1994

Vajda, Erik, "Közös (osztott) katalogizálás—közös (központi) katalógus," (Shared (distributed) cataloging—shared (centralized) cataloging), *Könyv, Könyvtár, Könyvtáros*, 2000/3.

Vajda, Ágnes and Éva Kuti, "Citizens' Votes for Nonprofit Activities in Hungary," in *"Forint Votes" for Civil Society Organizations Studies*, Series ed. László Harsányi; Budapest: Research Project on Nonprofit Organizations, 2000.

Varga, Dóra, "Felsőoktatási beruházások világbanki hitel nélkül" (Investment in higher education without World Bank Loans), *Magyar Hirlap*, June 29, 2001.

Veblen, Thorstein, *The Higher Learning in America*, New York: B. W. Huebsch, 1918.

Verrev, Viivi, "Managing the Estonian National Library with Depleting Resources," in Ian R. M. Mowat and Maria Śliwińska (eds.), *Library Management: East-West Relation*, Proceedings of the seminar held in Toruń, November 6–12, 1994, Toruń: Nicholas Copernicus University, 1995.

Vickers, John and George Yarrow, *Privatization: An Economic Analysis*, Cambridge, MA: The MIT Press, 1988.

Virágos, Márta, "The Human Aspects of Library Automation," in *Library Automation in Transitional Societies: Lessons from Eastern Europe*, ed. by Andrew Lass and Richard E. Quandt; New York: Oxford University Press, 2000, pp. 426–42.

"Virginia Community Certification Program," Richmond, VA: Virginia Department of Economic Development, October 1, 1992.

"VOCAL," http://www.lib.klte.hu/rendszer.html.

Vodichev, Evgeny, "The EU TACIS Program as a Tool for Technical Assistance to Russia," Appendix D.3 in *The Humanities and Social Sciences in the Former Soviet Union: An Assessment of Need*, The Kennan Institute, Woodrow Wilson Center, Washington, DC: August 1, 1999.

Wade, Rona, "The Very Model of a Modern Library Consortium," *Library Consortium Management*, 1 (1/2), 1999, 5–18.

Warsaw University Library's Report for The Andrew W. Mellon Foundation, Warsaw: Warsaw University Library, October 19, 1992.

Wheeler, David L., "Falling Behind: U.S. Trails 3 European Nations in Rate of College Graduation," *The Chronicle of Higher Education*, May 26, 2000, p. A63.

Whipple, Tim D. (ed.), *After the Velvet Revolution*, New York, Freedom House, 1991.

Whitney, Craig R., "Evolution in Europe," *The New York Times*, April 15, 1990

Willard, Kristen, "Report of October 18, 1991, Meeting Regarding Czech and Slovak Libraries," New York: The Andrew W. Mellon Foundation, October 18, 1991.

Williams, Frederick and David V. Gibson, *Technology Transfer: A Communication Perspective*, Newbury Park, CA: Sage Publications, 1990.

Williams, John, "Learning from Their Neighbors: A Survey of Economic Policy Transfer Projects in Central and Eastern Europe," New York: The Foundation for a Civil Society, 1995.

Wróblewski, Andrzej K., "The Future of Universities in Poland," unpublished paper, Warsaw University, 1991.

"Wrocław University Library," unpublished paper, Wrocław, 1992.

Zalai, Ernő, "The First Wave of the Recent Reform of Training Management Economics Specialists at the Budapest University of Economic Sciences," Budapest: Budapest University of Economic Sciences, May, 1991.

Żuchowski, Tadeusz J., *Pałac papieski na Watykanie od końca V do początku* XVI *wieku* (The papal palace in the Vatican from the end of the 5th to the beginning of the 16th centuries), Poznań: Adam Mickiewicz University, History of Art Series No. 27, 1999.

Author and Name Index

Subject Index

www.ingramcontent.com/pod-product-compliance
Ingram Content Group UK Ltd.
Pitfield, Milton Keynes, MK11 3LW, UK
UKHW020850150926
475595UK00003B/110

9 780195 146691